Lecture Notes in Computer Scien

T0238410

Commenced Publication in 1973
Founding and Former Series Editors:
Gerhard Goos, Juris Hartmanis, and Jan van Leeuwen

Victor G. Ganzha   Ernst W. Mayr
Evgenii V. Vorozhtsov (Eds.)

Computer Algebra in Scientific Computing

10th International Workshop, CASC 2007
Bonn, Germany, September 16-20, 2007
Proceedings

 Springer

Volume Editors

Victor G. Ganzha
Ernst W. Mayr
Technische Universität München
Institut für Informatik
Garching, Germany
E-mail: {ganzha,mayr}@in.tum.de

Evgenii V. Vorozhtsov
Russian Academy of Sciences
Institute of Theoretical and Applied Mechanics
Novosibirsk, Russia
E-mail: vorozh@itam.nsc.ru

Library of Congress Control Number: 2007934978

CR Subject Classification (1998): I.2, F.2.1-2, G.1, I.3.5, I.2

LNCS Sublibrary: SL 1 – Theoretical Computer Science and General Issues

ISSN   0302-9743
ISBN-10   3-540-75186-6 Springer Berlin Heidelberg New York
ISBN-13   978-3-540-75186-1 Springer Berlin Heidelberg New York

Springer is a part of Springer Science+Business Media

springer.com

© Springer-Verlag Berlin Heidelberg 2007
Printed in Germany

Typesetting: Camera-ready by author, data conversion by Scientific Publishing Services, Chennai, India
Printed on acid-free paper   SPIN: 12163083   06/3180   5 4 3 2 1 0

Dedicated to
Prof. Vladimir P. Gerdt
on the occasion of his 60th birthday

Preface

The outstanding feature of this CASC Workshop is that this is the tenth workshop in the series started in 1998. The general idea of this workshop was to bring together people working in the areas of computer algebra systems(CASs), computer algebra methods and algorithms, and various CA applications in natural sciences and engineering.

The nine earlier CASC conferences, CASC 1998, CASC 1999, CASC 2000, CASC 2001, CASC 2002, CASC 2003, CASC 2004, CASC 2005, and CASC 2006, were held, respectively, in St. Petersburg, Russia, in Munich, Germany, in Samarkand, Uzbekistan, in Konstanz, Germany, in Crimea, Ukraine, in Passau, Germany, in St. Petersburg, Russia, in Kalamata, Greece, and in Chişinău, Moldova, and they proved to be successful.

Since 1998, the topics of papers published in the CASC proceedings accounted both for the development of new excellent computer algebra systems and for expanding the scopes of application of CA methods and techniques. The present volume of the proceedings of CASC 2007 continues this tradition. Among the traditional topics, there are studies in polynomial and matrix algebra, quantifier elimination, and Gröbner bases.

One of the fruitful areas of the application of CA methods and systems is the derivation of new analytic solutions to differential equations, and several papers deal with this topic.

The application of CASs to stability investigation of both differential equations and difference methods for them is also the subject of a number of papers.

Several papers are devoted to the application of computer algebra methods and algorithms to the derivation of new mathematical models in biology and in mathematical physics.

In addition to the accepted submissions, this volume also includes two invited papers. The paper by F. Winkler and E. Shemyakova (RISC, Linz) addresses the theme of extending the range of analytically solvable PDEs with the aid of symbolic and algebraic methods. The key technique used here is the factorization of a differential operator. The authors have introduced the notion of *obstacle* for the factorization of a differential operator, i.e., conditions preventing a given operator from being factorizable.

The other invited lecture, by S. Fritzsche (Max-Planck Institute for Nuclear Physics, Heidelberg), is devoted to the problem of exploring decoherence and entanglement phenomena in quantum information theory. The author presents his Maple-based FEYNMAN program, which was developed recently to support the investigation of the above phenomena. One of the applications presented is the atomic photoionization, where the author shows how the polarization can be transferred from the incoming photons to the emitted photoelectrons, giving rise to a (spin-spin) entanglement between the photoelectron and the remaining (photo-)ion.

All the papers contained in this volume were accepted by the Program Committee after a thorough reviewing process.

The CASC 2007 workshop was supported financially by a generous grant from the Deutsche Forschungsgemeinschaft (DFG). Our particular thanks are due to the members of the CASC 2007 Local Organizing Committee at the University of Bonn: Andreas Weber (Computer Science Department) and Joachim von zur Gathen (B-IT), who ably handled local arrangements in Bonn. We are grateful to W. Meixner for his technical help in the preparation of the camera-ready manuscript for this volume.

July 2007   V.G. Ganzha
                                                                   E.W. Mayr
                                                              E.V. Vorozhtsov

Organization

CASC 2007 was organized jointly by the Department of Informatics at the Technische Universität München, Germany, the Department of Computer Security at the Bonn-Aachen International Center for Information Technology, Bonn, Germany and the Department of Computer Science, Universität Bonn, Germany.

Workshop General Chairs

Vladimir Gerdt (JINR, Dubna) Ernst W. Mayr (TU München)

Program Committee

Alkis Akritas (Volos)
Gerd Baumann (Cairo)
Hans-Joachim Bungartz (Munich)
Andreas Dolzmann (Passau)
Victor Edneral (Moscow)
M'hammed El Kahoui (Marrakech)
Ioannis Emiris (Athens)
Victor Ganzha (Munich, Co-chair)
Evgenii Grebenikov (Moscow)
Jaime Gutierrez (Santander)
Robert Kragler (Weingarten)
Richard Liska (Prague)

Bernard Mourrain (Sophia Antipolis)
Eugenio Roanes-Lozano (Madrid)
Yosuke Sato (Tokyo)
Werner Seiler (Kassel)
Doru Stefanescu (Bucharest)
Stanly Steinberg (Albuquerque)
Serguei Tsarev (Krasnoyarsk)
Evgenii Vorozhtsov (Novosibirsk,
    Co-chair)
Andreas Weber (Bonn)
Song Yan (Coventry)

Local Organizing Committee

Andreas Weber (Bonn) Joachim von zur Gathen (Bonn)

General Organizing Committee

Werner Meixner (Munich, Chair) Annelies Schmidt (Munich, Secretary)

Electronic Address

WWW site: http://wwwmayr.in.tum.de/CASC2007

Table of Contents

Analytic Solutions of Linear Difference Equations, Formal Series, and Bottom Summation

S.A. Abramov[1,*] and M. Petkovšek[2,**]

[1] Russian Academy of Sciences, Dorodnicyn Computing Centre,
Vavilova 40, 119991, Moscow GSP-1, Russia
`sabramov@ccas.ru`
[2] University of Ljubljana, Faculty of Mathematics and Physics,
Jadranska 19, SI-1000 Ljubljana, Slovenia
`marko.petkovsek@uni-lj.si`

Abstract. We consider summation of consecutive values $\varphi(v), \varphi(v+1),$ $\ldots, \varphi(w)$ of a meromorphic function $\varphi(z)$ where $v, w \in \mathbb{Z}$. We assume that $\varphi(z)$ satisfies a linear difference equation $L(y) = 0$ with polynomial coefficients, and that a summing operator for L exists (such an operator can be found – if it exists – by the Accurate Summation algorithm, or alternatively, by Gosper's algorithm when $\operatorname{ord} L = 1$).

The notion of *bottom summation* which covers the case where $\varphi(z)$ has poles in $\mathbb{Z}$ is introduced.

1 Introduction

Similarly to [8,3,5,1], this paper is concerned with the problem of summing the elements of a P-recursive sequence $f(k)$, $k \in \mathbb{Z}$, i.e., a sequence which satisfies a linear difference equation with polynomial coefficients.

Let E_k be the shift operator such that $E_k(f(k)) = f(k+1)$ for sequences $f(k)$ where $k \in \mathbb{Z}$. Let

$$L = a_d(k)E_k^d + \cdots + a_1(k)E_k + a_0(k) \in \mathbb{C}(k)[E_k]. \tag{1}$$

We say that an operator $R \in \mathbb{C}(n)[E_k]$ is a *summing* operator for L if

$$(E_k - 1) \circ R = 1 + M \circ L \tag{2}$$

for some $M \in \mathbb{C}(k)[E_k]$. It is easy to see that if there exists a summing operator for L, then there also exists one of order $< d$ (simply replace R by its remainder when divided by L from the right). Hence we can assume w.l.g. that $\operatorname{ord} R = \operatorname{ord} L - 1 = d - 1$:

$$R = r_{d-1}(k)E_k^{d-1} + \cdots + r_1(k)E_k + r_0(k) \in \mathbb{C}(k)[E_k]. \tag{3}$$

* Partially supported by RFBR under grant 07-01-00482-a.
** Partially supported by ARRS under grant P1-0294.

If a summing operator exists, then it can be constructed by the Accurate Summation algorithm [3] or, when $d = 1$, by Gosper's algorithm [8]. In those cases where $R \in \mathbb{C}[k, E_k]$ exists, equality (2) gives an opportunity to use the discrete Newton-Leibniz formula

$$\sum_{k=v}^{w-1} f(k) = g(w) - g(v) \tag{4}$$

for all integers $v < w$, and for any sequence f such that $L(f) = 0$, taking $g = R(f)$.

However, it was shown in [5] that if R has rational-function coefficients which have poles in $\mathbb{Z}$, then this formula may give an incorrect result (see Example 5 of the present paper). This gives rise to defects in many implementations of summation algorithms. In [5,1] a way was proposed to construct a basis for the space $W_{L,R}$ of all solutions of $L(y) = 0$ for which (4) is valid for all integers $v < w$. It was also proved that $\dim W_{L,R} > 0$ in the case $d = 1$.

In the present paper we give a new sufficient condition for the correctness of definite summation by Gosper's algorithm and by the Accurate Summation algorithm.

In Section 3 below we prove that if a summing operator exists for L with $\operatorname{ord} L = d$, then $\dim W_{L,R} > 0$ regardless of the value of d.

In Section 4 we suppose that L acts on analytic functions:

$$L = a_d(z)E_z^d + \cdots + a_1(z)E_z + a_0(z) \in \mathbb{C}(z)[E_z], \tag{5}$$

where $E_z(\varphi(z)) = \varphi(z+1)$ for analytic functions $\varphi(z)$ where $z \in \mathbb{C}$. We consider the summing operator (if it exists) in the form

$$R = r_{d-1}(z)E_z^{d-1} + \cdots + r_1(z)E_z + r_0(z) \in \mathbb{C}(z)[E_z].$$

Let $\varphi(z)$ be a meromorphic solution of $L(y) = 0$. It turns out that if $\varphi(z)$ has no pole in $\mathbb{Z}$, then $R(\varphi)(z)$ has no pole in $\mathbb{Z}$ as well, and we can use (4) to sum values $\varphi(k)$ for $k = v, v + 1, \ldots, w$. This follows from a stronger statement also proved in Section 4. The fact is that even if $\varphi(z)$ has some poles in $\mathbb{Z}$, the summation task can nevertheless be performed correctly. For any $k \in \mathbb{Z}$ the function $\varphi(z)$ can be represented as

$$\varphi(z) = c_{k,\rho_k}(z - k)^{\rho_k} + c_{k,\rho_k+1}(z - k)^{\rho_k+1} + \ldots$$

with $\rho_k \in \mathbb{Z}$ and $c_{k,\rho_k} \neq 0$. If $L(\varphi) = 0$, then there exists the minimal element ρ in the set of all ρ_k, $k \in \mathbb{Z}$. We associate with $\varphi(z)$ the sequence $f(k)$ such that $f(k) = c_{k,\rho_k}$ if $\rho_k = \rho$, and $f(k) = 0$ otherwise. Then the sequence $f(k)$ satisfies the equation $L(y) = 0$, if we use E_k instead of E_z in L. We associate a sequence $g(k)$ with $R(\varphi)$ in a similar way, and the value of ρ for $R(\varphi)$ will be the same as for φ. Now formula (4) is correct. We call this type of summation *bottom summation*.

Some important auxiliary statements (Section 2) on sequences of power series are based on the idea of the ε-deformation of a difference operator which was first used by M. van Hoeij in [7]; later this idea was used in [4] and in [2] as well.

2   Series-Valued Sequences

We start with some notations and definitions. Let ε be a variable (rather than a "small number"). As usual, $\mathbb{C}[[\varepsilon]]$ is the ring of formal power series in ε and $\mathbb{C}((\varepsilon)) = \mathbb{C}[[\varepsilon]][\varepsilon^{-1}]$ is its quotient field (the field of formal Laurent series in ε).

If $s \in \mathbb{C}((\varepsilon)) \setminus \{0\}$ then we define the *valuation* of s in the following way:

$$\nu(s) = -\min \{m \mid m \in \mathbb{Z}, \ \varepsilon^m s \in \mathbb{C}[[\varepsilon]]\},$$

in addition we set $\nu(0) = \infty$. If $s \in \mathbb{C}((\varepsilon))$, $m \in \mathbb{Z}$ then $[\varepsilon^m]s$ is the coefficient of ε^m in the series s, and $[\varepsilon^\infty]0 = 0$. It follows from the definition of the valuation that if $s, t \in \mathbb{C}((\varepsilon))$ then

$$\nu(st) = \nu(s) + \nu(t), \qquad [\varepsilon^{\nu(st)}](st) = ([\varepsilon^{\nu(s)}]s)([\varepsilon^{\nu(t)}]t), \tag{6}$$

and

$$\nu(s + t) \ \geq \ \min\{\nu(s), \nu(t)\}. \tag{7}$$

If K is a ring, then $K^{\mathbb{Z}}$ denotes the ring of all maps $\mathbb{Z} \to K$, i.e., the ring of all two-sided K-valued sequences. Note that the operator E_k is a ring automorphism of $K^{\mathbb{Z}}$.

If $S \in \mathbb{C}((\varepsilon))^{\mathbb{Z}}$, then $\nu(S)$ denotes the sequence in $\mathbb{Z}^{\mathbb{Z}}$ whose kth element is $\nu(S(k))$. If $m \in \mathbb{Z}$, then $[\varepsilon^m]S$ denotes the sequence in $\mathbb{C}^{\mathbb{Z}}$ whose kth element is $[\varepsilon^m](S(k))$. We say that S is of *bounded depth* if the sequence $\nu(S)$ is bounded from below, i.e., there exists

$$m = \min_k \nu(S(k)). \tag{8}$$

If S is of bounded depth, then m in (8) is the *depth* of S. In this case the *bottom of S*, which is a sequence in $\mathbb{C}^{\mathbb{Z}}$, is defined by

$$\text{bott}(S) = [\varepsilon^m]S.$$

An operator $\Lambda \in \mathbb{C}((\varepsilon))^{\mathbb{Z}}[E_k]$ of the form

$$\Lambda = S_d E_k^d + \cdots + S_1 E_k + S_0, \quad S_0, S_1, \ldots, S_d \in \mathbb{C}((\varepsilon))^{\mathbb{Z}}, \tag{9}$$

defines a map $\mathbb{C}((\varepsilon))^{\mathbb{Z}} \to \mathbb{C}((\varepsilon))^{\mathbb{Z}}$ where $(\Lambda S)(k) = \sum_{j=0}^{d} S_j(k)S(k+j)$. If each sequence S_j has bounded depth m_j for $j = 0, 1, \ldots, d$, then we say that Λ is of bounded depth $m = \min_{0 \leq j \leq d} m_j$. In this case the bottom of Λ is

$$\text{bott}(\Lambda) = \sum_{j=0}^{d} ([\varepsilon^m]S_j)E_k^j \ \in C^{\mathbb{Z}}[E_k].$$

Proposition 1. *Let Λ be an operator of the form (9), of bounded depth. Let $S \in \mathbb{C}((\varepsilon))$ satisfy $\Lambda(S) = 0$. If for all but finitely many $k \in \mathbb{Z}$ we have*

$$\nu(S_0(k)) = \nu(S_d(k)) = \min_{0 \leq j \leq d} \nu(S_j(k)), \tag{10}$$

then S is of bounded depth and $\tilde{\Lambda}(\text{bott}(S)) = 0$, where $\tilde{\Lambda} = \text{bott}(\Lambda)$.

Proof. Fix $k \in \mathbb{Z}$ and $i \in \{0, 1, \ldots, d\}$. From $\Lambda(S) = 0$ it follows that

$$\nu(S_i(k)S(k+i)) = \nu\left(-\sum_{0 \le j \le d, \, j \ne i} S_j(k)S(k+j)\right),$$

so by (6) and (7) we have

$$\nu(S_i(k)) + \nu(S(k+i)) \ge \min_{\substack{0 \le j \le d \\ j \ne i}} \nu(S_j(k)) + \min_{\substack{0 \le j \le d \\ j \ne i}} \nu(S(k+j)). \tag{11}$$

Assume that $\nu(S_i(k)) = \min_{0 \le j \le d} \nu(S_j(k))$. Then it follows from (11) that $\nu(S(k+i)) \ge \min_{\substack{0 \le j \le d \\ j \ne i}} \nu(S(k+j))$. Specializing this to $i = 0$ and $i = d$ and using (10) we obtain that

$$\nu(S(k)) \ge \min_{1 \le j \le d} \nu(S(k+j))$$

and

$$\nu(S(k+d)) \ge \min_{0 \le j \le d-1} \nu(S(k+j))$$

for all but finitely many $k \in \mathbb{Z}$. Therefore, S is of bounded depth. The equality $\tilde{\Lambda}(\text{bott}(S)) = 0$ now follows from (6).   □

Example 1. Let

$$\Lambda = S_1 E_k + S_0, \ S_1(k) = k + 1 + \varepsilon, \ S_0(k) = -k - \varepsilon$$

and

$$S(k) = \begin{cases} -\frac{1}{\varepsilon}, & \text{if } k = 0, \\ \sum_{i=0}^{\infty} \left(-\frac{1}{k}\right)^{i+1} \varepsilon^i, & \text{otherwise.} \end{cases}$$

Then $S_1(k)S(k+1) = -S_0(k)S(k) = -1$ for all k, and $\Lambda(S) = 0$ as a consequence. The depth of S is -1.

We see that

$$\text{bott}(S)(k) = \begin{cases} -1, & \text{if } k = 0, \\ 0, & \text{otherwise,} \end{cases}$$

and $\text{bott}(\Lambda) = (k+1)E_k - k$. It is easy to see that $(k+1)f(k+1) - kf(k) = 0$, where $f(k) = \text{bott}(S)(k)$; so $\tilde{\Lambda}(\text{bott}(S)) = 0$, where $\tilde{\Lambda} = \text{bott}(\Lambda)$.

3   When a Summing Operator Exists

If $\varphi(z) \in \mathbb{C}(z)$, then we write $\hat{\varphi}(k)$ for the sequence $\varphi(k + \varepsilon)$, $k \in \mathbb{Z}$, of rational functions expanded into Laurent series about $\varepsilon = 0$. We associate with every operator

$$N = b_l(z)E_z^l + \cdots + b_1 E_z + b_0(z) \ \in \mathbb{C}(z)[E_z]$$

the operator

$$\hat{N} = \hat{b}_l(k)E_k^l + \cdots + \hat{b}_1(k)E_k + \hat{b}_0(k) \ \in \mathbb{C}((\varepsilon))^{\mathbb{Z}}[E_k]$$

which acts on sequences from $\mathbb{C}((\varepsilon))^{\mathbb{Z}}$.

Proposition 2. *Let $L \in \mathbb{C}[z, E_z]$. Assume that $R \in \mathbb{C}(z)[E_z]$ is a summing operator for L. Let $S \in \mathbb{C}((\varepsilon))^{\mathbb{Z}}$ be such that $\hat{L}(S) = 0$. Then*

$$(E_k - 1)(\hat{R}(S)) = S. \tag{12}$$

Proof. By (2), there is an operator $M \in \mathbb{C}(z)[E_z]$ such that

$$(E_z - 1) \circ R = 1 + M \circ L. \tag{13}$$

The map $N \mapsto \hat{N}$ is a ring homomorphism from $\mathbb{C}(z)[E_z]$ to $\mathbb{C}((\varepsilon))^{\mathbb{Z}}[E_k]$. Therefore, (13) implies

$$(E_k - 1) \circ \hat{R} = 1 + \hat{M} \circ \hat{L}.$$

Applying both sides of this equality to S, we obtain (12). □

Proposition 3. *Let $L \in \mathbb{C}[z, E_z]$, and let $R \in \mathbb{C}(z)[E_z]$ be a summing operator for L. Let $S \in \mathbb{C}((\varepsilon))^{\mathbb{Z}}$ be such that $\hat{L}(S) = 0$. Then $\operatorname{depth}(\hat{R}(S)) = \operatorname{depth}(S)$, and*

$$(E_k - 1)(\operatorname{bott}(\hat{R}(S))) = \operatorname{bott}(S). \tag{14}$$

Proof. It follows from (12) that $\operatorname{depth}(\hat{R}(S)) \leq \operatorname{depth}(S)$. To prove equality, we distinguish two cases.

1. $\operatorname{depth}(\hat{R}(S)) = \nu(\hat{R}(S)(k))$ for all $k \in \mathbb{Z}$.
   Assume that $\operatorname{depth}(\hat{R}(S)) < \operatorname{depth}(S)$. Then $\operatorname{bott}(\hat{R}(S))$ is a non-zero constant sequence. However, since R has rational coefficients, there exists $k_0 \in \mathbb{Z}$ such that for all $k \geq k_0$, the valuation of any coefficient of $\hat{R}$ is non-negative and, as a consequence,

   $$\nu(\hat{R}(S)(k)) \geq \min_{0 \leq i \leq \operatorname{ord} R} \nu(S(k+i)) \geq \operatorname{depth}(S) > \operatorname{depth}(\hat{R}(S))$$

   for all $k \geq k_0$. Then $\operatorname{bott}(\hat{R}(S))(k) = 0$ for all $k \geq k_0$. Hence $\operatorname{bott}(\hat{R}(S))$ is not a non-zero constant sequence. This contradiction implies that $\operatorname{depth}(\hat{R}(S)) = \operatorname{depth}(S)$.
2. $\operatorname{depth}(\hat{R}(S)) = \nu(\hat{R}(S)(k)) < \nu(\hat{R}(S)(k+1))$ or $\operatorname{depth}(\hat{R}(S)) = \nu(\hat{R}(S)(k)) > \nu(\hat{R}(S)(k-1))$, for some $k \in \mathbb{Z}$.
   By (12), also in this case $\operatorname{depth}(\hat{R}(S)) = \operatorname{depth}(S)$.

Now it follows from (12) that (14) is valid. □

Theorem 1. *Let $L \in \mathbb{C}[z, E_z]$, $\operatorname{ord} L = d$, and let*

$$R = r_{d-1}(z)E_z^{d-1} + \cdots + r_1(z)E_z + r_0(z) \in \mathbb{C}(z)[E_z]$$

be a summing operator for L. Denote by V the set of all the poles of $r_0(z), r_1(z), \ldots, r_{d-1}(z)$. Then there exist non-zero $f, g \in \mathbb{C}^{\mathbb{Z}}$ such that

(i) $L(f(k)) = 0$ for all $k \in \mathbb{Z}$,
(ii) $g(k) = r_{d-1}(k)f(k+d-1) + \cdots + r_1(k)f(k+1) + r_0(k)f(k)$ for all $k \in \mathbb{Z} \backslash V$, and

(iii) the discrete Newton–Leibniz formula

$$\sum_{k=v}^{w-1} f(k) = g(w) - g(v)$$

is valid for all integer $v < w$.

Proof. Pick any non-zero $U_1, \ldots, U_d \in \mathbb{C}((\varepsilon))$, and using $\hat{L}$ find a sequence $S \in \mathbb{C}((\varepsilon))^{\mathbb{Z}}$ such that $S(i) = U_i$, $i = 1, 2, \ldots, d$, and $\hat{L}(S) = 0$. So there exists a non-zero sequence S such that $\hat{L}(S) = 0$. Write $f = \text{bott}(S)$, $g = \text{bott}(\hat{R}(S))$. Then (iii) is valid by Proposition 3, and (i) is valid since L has polynomial coefficients. Finally, for all $k \notin V$ we have $g(k) = \text{bott}(\hat{R}(S))(k) = R(\text{bott}(S))(k) = R(f)(k)$, so (ii) is valid.   □

4   The Analytic Case

In the rest of this paper we assume that the sequences under consideration are defined on an infinite interval I of integers, where either $I = \mathbb{Z}$, or

$$I = \mathbb{Z}_{\geq l} = \{k \in \mathbb{Z} \mid k \geq l\}, \ l \in \mathbb{Z}.$$

It is easy to see that Propositions 1 – 3 remain valid if we consider sequences defined on $\mathbb{Z}_{\geq l}$, and define the operators $^{\mathbb{Z}}$ and bott with respect to $\mathbb{Z}_{\geq l}$ instead of with respect to $\mathbb{Z}$.

Let U be an open subset of $\mathbb{C}$ containing I, such that $z \in U \Rightarrow z + 1 \in U$. Denote by $\mathcal{M}(U)$ the set of functions which are meromorphic on U. We associate with $\varphi \in \mathcal{M}(U)$ a sequence $\hat{\varphi} \in \mathbb{C}((\varepsilon))^{\mathbb{Z}}$ whose kth element, $k \in I$, is a (formal) series obtained by expanding $\varphi(\varepsilon + k)$ into Laurent series at $\varepsilon = 0$.

Proposition 4. *Let $L \in \mathbb{C}[z, E_z]$, and let $\varphi \in \mathcal{M}(U)$ satisfy $L(\varphi) = 0$ on U. Then $\hat{L}(\hat{\varphi}) = 0$ everywhere on $\mathbb{Z}$, the sequence $\hat{\varphi} \in \mathbb{C}((\varepsilon))^{\mathbb{Z}}$ is of bounded depth, and $\tilde{L}(\text{bott}(\hat{\varphi})) = 0$ everywhere on $\mathbb{Z}$, where $\tilde{L} = \text{bott}(\hat{L})$.*

Proof. This follows from the trivial fact that the Laurent series of the zero function has only zero coefficients, and from Proposition 1.   □

Corollary 1. *If $a_0, a_1, \ldots, a_d \in \mathbb{C}[z]$ then $\text{bott}(\hat{L}) = a_d(k)E_k^d + \cdots + a_1(k)E_k + a_0(k)$. If in addition $S \in \mathbb{C}((\varepsilon))^{\mathbb{Z}}$ is such that $\hat{L}(S) = 0$, then $L(\text{bott}(S)) = 0$. In particular, if $\varphi \in \mathcal{M}(U)$ is such that $L(\varphi) = 0$ everywhere on U except possibly on a set of isolated points, then $L(\text{bott}(\varphi)) = 0$ everywhere on $\mathbb{Z}$.*

Example 2. In Example 1 we used, in fact, $L = (z+1)E_z - z$, $U = \mathbb{C}$, $\varphi(z) = -\frac{1}{z}$, $\Lambda = \hat{L}$, $S = \hat{\varphi}$.

Theorem 2. (On the bottom summation.) *Let* $L \in \mathbb{C}[z, E_z]$, *and let* $R \in \mathbb{C}(z)[E_z]$ *be a summing operator for* L. *Let* $\varphi \in \mathcal{M}(U)$ *satisfy* $L(\varphi) = 0$ *on* U, *and let* $\psi = R(\varphi)$. *Then the* bottom summation *formula*

$$\sum_{k=v}^{w-1} \mathrm{bott}(\hat{\varphi})(k) = \mathrm{bott}(\hat{\psi})(w) - \mathrm{bott}(\hat{\psi})(v) \tag{15}$$

is valid for any $v < w$, $v, w \in I$. *In particular, if* φ *has no pole in* $\mathbb{Z}$ *(i.e.,* $\mathrm{depth}(\hat{\varphi}) = 0$*), then the function* $\psi = R(\varphi) \in \mathcal{M}(U)$ *has no pole in* $\mathbb{Z}$, *and the discrete Newton–Leibniz formula*

$$\sum_{k=v}^{w-1} \varphi(k) = \psi(w) - \psi(v) \tag{16}$$

is valid for any $v < w$, $v, w \in I$.

Proof. The statement follows from Propositions 4 and 3. □

Consider some known examples in the context of Theorem 2.

Example 3. The function $\varphi(z) = z\Gamma(z+1)$ satisfies the equation $L(y) = 0$ where $L = zE_z - (z+1)^2$. We have $R = \frac{1}{z}$, $\mathrm{ord}\,R = 0$, and $\psi(z) = R(\varphi)(z) = \Gamma(z+1)$. Evidently $\varphi(z)$ has finite values when $z = 0, 1, \dots$, and has simple poles when $z = -1, -2, \dots$. If we consider $I = \mathbb{Z}$ then $\mathrm{depth}(\hat{\varphi}) = \mathrm{depth}(\hat{\psi}) = -1$ and

$$\mathrm{bott}(\hat{\varphi})(k) = \begin{cases} \frac{(-1)^{k+1}k}{(-k-1)!}, & \text{if } k < 0, \\ 0, & \text{if } k \geq 0, \end{cases}$$

$$\mathrm{bott}(\hat{\psi})(k) = \begin{cases} \frac{(-1)^{k+1}}{(-k-1)!}, & \text{if } k < 0, \\ 0, & \text{if } k \geq 0. \end{cases}$$

As a consequence of (15) we have

$$\sum_{k=v}^{w-1} \frac{(-1)^k k}{(-k-1)!} = \frac{(-1)^w}{(-w-1)!} - \frac{(-1)^v}{(-v-1)!}$$

for any $v < w \leq 0$, or equivalently

$$\sum_{k=v}^{w-1} \frac{(-1)^k k}{(k-1)!} = \frac{(-1)^{w+1}}{(w-2)!} - \frac{(-1)^{v+1}}{(v-2)!}$$

for any $1 \leq v < w$.

If $I = \mathbb{Z}_{\geq 0}$ then $\mathrm{depth}(\hat{\varphi}) = \mathrm{depth}(\hat{\psi}) = 0$, and by (16) we have $\sum_{k=v}^{w-1} k\Gamma(k+1) = \Gamma(w+1) - \Gamma(v+1)$ for any $0 \leq v < w$ or, equivalently, $\sum_{k=v}^{w-1} k \cdot k! = w! - v!$.

Example 4. The rational function $\varphi(z) = \frac{1}{z(z+1)}$ satisfies the equation $L(y) = 0$ where $L = (z+2)E_z - z$. We have $R = -z - 1$, and $\psi(z) = R(\varphi)(z) = -\frac{1}{z}$. If we consider $I = \mathbb{Z}$ then $\mathrm{depth}(\hat{\varphi}) = \mathrm{depth}(\hat{\psi}) = -1$ and

$$\mathrm{bott}(\hat{\varphi})(k) = \delta_{0,k} - \delta_{-1,k},$$

$$\mathrm{bott}(\hat{\psi})(k) = -\delta_{0,k},$$

where δ is the Kronecker delta. A simple direct check shows that (15) is valid.

If $I = \mathbb{Z}_{\geq 1}$ then $\mathrm{depth}(\hat{\varphi}) = \mathrm{depth}(\hat{\psi}) = 0$, and by (16) we have $\sum_{k=v}^{w-1} \frac{1}{k(k+1)} = -\frac{1}{w} + \frac{1}{v}$ for any $0 \leq v < w$.

The following example demonstrates a conflict between combinatorial and analytic definitions of the symbol $\binom{p}{q}$.

Example 5. Consider the hypergeometric sequence

$$t(k) = \frac{\binom{2k-3}{k}}{4^k} \tag{17}$$

which satisfies the equation $2(k+1)(k-2)t(k+1) - (2k-1)(k-1) = 0$. It has been noticed in [5] that even though Gosper's algorithm succeeds on this sequence, producing $R(k) = \frac{2k(k+1)}{k-2}$, and $t(k)$ is defined for all $k \in \mathbb{Z}$, the discrete Newton–Leibniz formula

$$\sum_{k=0}^{w-1} t(k) = R(w)t(w) - R(0)t(0) = \frac{2w(w+1)\binom{2w-3}{w}}{(w-2)4^w} \tag{18}$$

is not correct. If we assume that the value of $\binom{2k-3}{k}$ is 1 when $k = 0$ and -1 when $k = 1$ (as is common practice in combinatorics) then the expression on the right gives the true value of the sum only at $w = 1$. However, assume that the value of $\binom{2k-3}{k}$ is defined as

$$\lim_{z \to k} \frac{\Gamma(2z-2)}{\Gamma(z+1)\Gamma(z-2)}. \tag{19}$$

This limit exists for all $k \in \mathbb{Z}$, but

$$\lim_{z \to 0} \frac{\Gamma(2z-2)}{\Gamma(z+1)\Gamma(z-2)} = \frac{1}{2} \neq 1$$

and

$$\lim_{z \to 1} \frac{\Gamma(2z-2)}{\Gamma(z+1)\Gamma(z-2)} = -\frac{1}{2} \neq -1.$$

Set

$$\varphi(z) = \frac{\Gamma(2z-2)}{\Gamma(z+1)\Gamma(z-2)4^z}$$

and

$$\psi(z) = \frac{2z(z+1)}{z-2} \varphi(z).$$

Then formula (16) gives the correct result

$$\sum_{k=0}^{w-1} \frac{\Gamma(2k-2)}{\Gamma(k+1)\Gamma(k-2)4^k} = \frac{2w(w+1)\Gamma(2w-2)}{(w-2)\Gamma(w+1)\Gamma(w-2)4^w} \tag{20}$$

for all $w \geq 1$, provided that the values of the summand and of the right-hand side are defined by taking appropriate limits.

Note that if $\alpha k_0 + \beta$ is a non-positive integer, then we can often avoid a direct computation of limits using the asymptotic equality

$$\Gamma(\alpha z + \beta) \sim \frac{(-1)^{\alpha k_0 + \beta}}{(-\alpha k_0 - \beta)! \cdot \alpha \cdot (z - k_0)}, \quad z \to k_0,$$

instead. If $\alpha \neq 0$ and $-\frac{\beta}{\alpha}$ is an integer γ, then $\Gamma(\alpha z + \beta)$ has integer poles at $\gamma, \gamma - 1, \ldots$ if $\alpha > 0$ and $\gamma, \gamma + 1, \ldots$ if $\alpha < 0$.

The following example is related to the case ord $L > 1$.

Example 6. For the operator $L = (z - 3)(z - 2)(z + 1)E_z^2 - (z - 3)(z^2 - 2z - 1)E_z - (z - 2)^2$ there exists the summing operator

$$R = zE_z + \frac{1}{z - 3}$$

([5]). By [6] the equation $L(y) = 0$ has solutions holomorphic in the half-plane $\operatorname{Re} z > 2$. Denote by $\varphi(z)$ an arbitrary solution of this kind. By Theorem 2, formula (16) must be correct for the case $I = \mathbb{Z}_{\geq 3}$ in spite of the fact that one of the coefficients of R has a pole at $z = 3$. This implies that $\varphi(z)$ vanishes at $z = 3$. This can be easily confirmed by the substitution of $z = 3$ into $L(\varphi) = 0$, which results in $-\varphi(3) = 0$. The algorithm from [4] yields $\varphi(z) = (\varphi(4) + 4\varphi(5))(z - 3) + O((z - 3)^2)$, and formula (16) gives the correct result for $3 \leq v < w$.

5   Conclusion

Indiscriminate application of the discrete Newton–Leibniz formula to the output of Gosper's algorithm or of the Accurate Summation algorithm in order to compute a definite sum can lead to incorrect results. This can be observed in many implementations of these algorithms in computer algebra systems.

In the present paper it is shown, in particular, that such undesirable phenomena cannot occur if the elements of the sequence under summation are the values $\varphi(k)$, $k \in \mathbb{Z}$, of an analytic function $\varphi(z)$, which satisfies (in the complex plane $\mathbb{C}$) the same difference equation with polynomial coefficients as does the original sequence (at integer points).

A practical consequence of this result is as follows. If the conditions formulated above are satisfied, then a computer-algebra-system user can be sure that the obtained sum was computed correctly.

On the more theoretical side, if $\varphi(z)$ mentioned above has some poles at integer points, then one can nevertheless find the sum of a sequence which, however, is not the sequence of values of $\varphi(k)$, $k \in \mathbb{Z}$, but is associated with $\varphi(z)$ in a natural way. This can yield an interesting (and, probably, unexpected) identity. We call this sequence associated with $\varphi(z)$, the *bottom* of $\varphi(z)$. If $\varphi(z)$ is defined for all $z \in \mathbb{Z}$ then its bottom coincides with the sequence $\varphi(k)$, $k \in \mathbb{Z}$.

References

1. Abramov, S.A.: On the summation of P-recursive sequences. In: Proc. ISSAC'06 (Int. Symp. on Symbolic and Algebraic Computation). Genova, Italy; 9–12 July, 2006, pp. 17–22 (2006)
2. Abramov, S.A., Barkatou, M.A., van Hoeij, M.: Apparent singularities of linear difference equations with polynomial coefficients. AAECC (Applicable Algebra in Engineering, Communication and Computing) 17, 117–133 (2006)
3. Abramov, S.A., van Hoeij, M.: Integration of solutions of linear functional equations. Integral Transforms and Special Functions 8, 3–12 (1999)
4. Abramov, S.A., van Hoeij, M.: Set of poles of solutions of linear difference equations with polynomial coefficients. J. Comput. Math. Math. Phys 43, 57–62 (2003) (Translated from Zhurnal Vychislitel'noi Matematiki i Matematicheskoi Fiziki 43, 60–65 (2003))
5. Abramov, S.A.: Petkovšek, M.: Gosper's Algorithm, Accurate Summation, and the discrete Newton–Leibniz formula. In: ISSAC'05 (Int. Symp. on Symbolic and Algebraic Computation). Beijing, China; 24–27 July, 2005, pp. 5–12 (2005)
6. Barkatou, M.A.: Contribution à l'étude des équations différentielles et aux différences dans le champ complexe. PhD Thesis, INPG Grenoble, France (1989)
7. van Hoeij, M.: Finite singularities and hypergeometric solutions of linear recurrence equations. J. Pure Appl. Algebra 139, 109–131 (1999)
8. Gosper Jr., R.W.: Decision procedure for indefinite hypergeometric summation. Proc. Natl. Acad. Sci.ᵡ75, 40–42 (1978)

Computations in Modules over Commutative Domains

Alkiviadis G. Akritas[1] and Gennadi I. Malaschonok[2]

[1] University of Thessaly, Department of Computer and Communication Engineering,
GR-38221 Volos, Greece
`akritas@uth.gr`
[2] Tambov State University, Internatsionalnaya 33, Tambov 392000, Russia
`malaschonok@ya.ru`

Abstract. This paper is a review of results on computational methods of linear algebra over commutative domains. Methods for the following problems are examined: solution of systems of linear equations, computation of determinants, computation of adjoint and inverse matrices, computation of the characteristic polynomial of a matrix.

1 Introduction

Let $\mathbf{R}$ be a commutative domain with identity, $\mathbf{K}$ the field of quotients of $\mathbf{R}$. This paper is devoted to the review of effective matrix methods in the domain $\mathbf{R}$ for a solution of standard linear algebra problems. The problems are: solving linear systems in $\mathbf{K}$, computing the adjoint and inverse matrix, computing the matrix determinant and computing the characteristic polynomial of a matrix.

The standard used to tackle these problems in commutative domain $\mathbf{R}$ consists of the using the field of fractions $\mathbf{K}$ of this domain. The ring $\mathbf{R}$ may be canonically immersed in the field $\mathbf{K}$. To solve a problem in the commutative domain any algorithm that is applicable over the field of fractions of this domain you can be applied.

Unfortunately this way results in algorithms with suitable complexity *only* in the case where the cost of operations in the field does not depend on the value of the operands. As an example consider the finite fields. But in the general case the cost of operations in the field depends on the value of the operands. More over this cost, in general, grows very quickly. For example, Gauss' method in the ring of integer numbers results in an algorithm that has exponential growth of complexity — instead of cubic.

So the main aim in commutative domains is to construct algorithms with controlled intermediate results.

The algorithms presented here have two main features:

- The intermediate elements in the algorithms are minors of the initial matrix. So the growth of these elements is bounded by the maximal value of the minors of the initial matrix.

V.G. Ganzha, E.W. Mayr, and E.V. Vorozhtsov (Eds.): CASC 2007, LNCS 4770, pp. 11–23, 2007.

- With the exception of the last algorithm, which requires $O(n^3)$ operations, the number of operations in all other algorithms is the same as that of the algorithm for matrix multiplication.

We denote by $O(n^\beta)$ or by $\gamma n^\beta + o(n^\beta)$ the number of multiplication operations, necessary for the multiplication of square matrices of order n. For the standard matrix multiplication algorithm we have $\beta = 3$ and $\gamma = 1$, whereas for Strassen's algorithm [21] the values are $\beta = \log_2 7$ and $\gamma = 1$, when the order of the matrix is some power of 2. For the best algorithm today we have $\beta < 2.376$ and γ unknown [8].

In the second section we present methods for solving systems of linear equations and performing determinant computations. Included are: Dodgson's method [9], the method of Forward and Backward Direction [12], the One-pass method [14], [15] and the Recursive Method [17],[18]. Corresponding methods for determinant computations with some generalization are discussed in [16].

Methods for computing the adjoint and inverse matrices are presented in the third section [20].

In the forth section a method for computing the characteristic polynomial of a matrix is presented [19]. This method was developed in [5].

Finally, in the conclusion we present the best complexity bounds available today (in commutative domains) for the methods presented above.

2   System of Linear Equations

Let R be a commutative domain, F be the field of fractions of R, $A \in R^{n \times m}, c \in R^n, n \leq m, A^* = (A, c) = (a_{ij})$ and,

$$Ax = c$$

be a system of linear equations over R.

Solving the above system with Cramer's rule we obtain

$$x_i = (\delta^n_{i,m+1} - \textstyle\sum_{j=n+1}^m x_j \delta^n_{ij})(\delta^n)^{-1}, \ i = 1 \ldots n,$$

where $x_j, j = n + 1, \ldots, m$, are the free variables, and the determinant $\delta^n \neq 0$.

We denote by $\delta^k, k = 1, \ldots, n$ the left upper corner minor of matrix A of order k, and by δ^k_{ij} the corner minor of matrix A where columns i and j have been interchanged. We assume that all corner minors $\delta^k, k = 1, \ldots, n$ are different from 0.

2.1   Dodgson's Algorithm

The determinant identity

$$\begin{vmatrix} a & b & c \\ d & e & f \\ g & h & k \end{vmatrix} = \frac{\begin{vmatrix} \begin{vmatrix} a & b \\ d & e \end{vmatrix} & \begin{vmatrix} b & c \\ e & f \end{vmatrix} \\ \begin{vmatrix} d & e \\ g & h \end{vmatrix} & \begin{vmatrix} e & f \\ h & k \end{vmatrix} \end{vmatrix}} \cdot e^{-1}$$

or in the more general form

$$\hat{a}_{ij}^{k+1} = \begin{vmatrix} \hat{a}_{i-1,j-1}^{k} & \hat{a}_{i-1,j}^{k} \\ \hat{a}_{i,j-1}^{k} & \hat{a}_{ij}^{k} \end{vmatrix} \cdot (\hat{a}_{i-1,j-1}^{k-1})^{-1}$$

where

$$\hat{a}_{ij}^{k+1} = |A_{j-k,\ldots,j:(columns)}^{i-k,\ldots,i:(rows)}|$$

is an instance of Sylvester's identity [2]. Dodgson [9] used it for the computation (condensation) of determinants and the solution of systems of linear equations computing a sequence of minors for

$$k = 2,\ldots, n-1, \ i = k+1,\ldots,n, \ j = k+1,\ldots,m.$$

Historical Note: As can be seen from the identity above, Dodgson liked to take the middle element e of the 4 corner minors as the leading minor (element). Later (in their 1945 paper [23]) Waugh and Dwyer took the top-left-corner element, a_{11} as the "middle" element of the 4 minors that are surrounding this a_{11} element.

Subsequent authors [22] and [6] used the same method without references to either Dodgson [9] or Waugh and Dwyer [23]. Other implementations of Dodgson's method can be found in [1] Other implementations of Dodgson method you can see in the book [1].

2.2   Method of Forward and Backward Direction

The forward direction part of this algorithm consists of computing the minors with Dodgson's method; the diagonal is the leading element in every step

$$a_{ij}^{k+1} = (a_{kk}^{k} a_{ij}^{k} - a_{ik}^{k} a_{kj}^{k})(a_{k-1,k-1}^{k-1})^{-1},$$

$$k = 2,\ldots, n-1, \ i = k+1,\ldots,n, \ j = k+1,\ldots,m,$$

where

$$a_{ij}^{k+1} = |A_{1,\ldots,k,j:(columns)}^{1,\ldots,k,i:(rows)}|$$

On one hand, this formula is a determinant identity and on the other hand it is the forward direction algorithm which is reminiscent of Gauss' elimination algorithm. (The only difference is that the leading element is one step behind.) As a result of the forward direction algorithm the matrix of the system becomes

$$\begin{pmatrix} a_{1,1}^{1} & a_{1,2}^{1} & \cdots & a_{1,n-1}^{1} & a_{1,n}^{1} & a_{1,n}^{1} & \cdots & a_{1,m+1}^{1} \\ 0 & a_{2,2}^{2} & \cdots & a_{2,n-1}^{2} & a_{2,n}^{2} & a_{2,n+1}^{2} & \cdots & a_{2,m+1}^{2} \\ \vdots & \vdots & \ddots & \vdots & \vdots & \vdots & \ddots & \vdots \\ 0 & 0 & \cdots & a_{n-1,n-1}^{n-1} & a_{n-1,n}^{n-1} & a_{n-1,n+1}^{n-1} & \cdots & a_{n-1,m+1}^{n-1} \\ 0 & 0 & \cdots & 0 & a_{n,n}^{n} & a_{n,n+1}^{n} & \cdots & a_{n,m+1}^{n} \end{pmatrix}$$

The leading elements $a_{k,k}^{k}$, $k = 1,\ldots,n-1$ cannot be zero.

The backward direction part of the algorithm consists of computing the minors δ_{ij}^k. The minor δ_{ij}^k is the corner minor of order k of the matrix A after column i has been interchanged with column j. The determinant identity of the backward direction algorithm is:

$$\delta_{ij}^n = \left(a_{nn}^n a_{ij}^i - \sum_{k=i+1}^{n} a_{ik}^i \delta_{kj}^n\right)(a_{ii}^i)^{-1}, \quad i = n-1, \ldots 1, \quad j = n+1, \ldots, m.$$

As a result of the backward direction algorithm the matrix of the system becomes:

$$
\begin{pmatrix}
a_{n,n}^n & 0 & \cdots & 0 & 0 & \delta_{1,n}^n & \cdots & \delta_{1,m+1}^n \\
0 & a_{n,n}^n & \cdots & 0 & 0 & \delta_{2,n+1}^n & \cdots & \delta_{2,m+1}^n \\
\vdots & \vdots & \ddots & \vdots & \vdots & \vdots & \ddots & \vdots \\
0 & 0 & \cdots & a_{n,n}^n & 0 & \delta_{n-1,n+1}^n & \cdots & \delta_{n-1,m+1}^n \\
0 & 0 & \cdots & 0 & a_{n,n}^n & \delta_{n,n+1}^n & \cdots & \delta_{n,m+1}^n
\end{pmatrix}
$$

The number of operations, necessary for the procedure of forward and backward direction, is

$$N^m = (9n^2 m - 5n^3 - 3nm - 3n^2 - 6m + 8n)/6,$$
$$N^d = (3n^2 m - n^3 - 3nm - 6n^2 + 13n - 6)/6$$
$$N^a = (6n^2 m - 4n^3 - 6nm + 3n^2 + n)/6.$$

2.3   The One-Pass Method

Another way of computing the minors δ_{ij}^k is given by the following two determinant identities:

$$\delta_{k+1,j}^{k+1} = a_{k+1,k+1}\delta_{kk}^k - \sum_{p=1}^{k} a_{k+1,p}\delta_{pj}^k, \quad j = k+1 \ldots m,$$

$$\delta_{ij}^{k+1} = (\delta_{k+1,k+1}^{k+1}\delta_{i,j}^k - \delta_{k+1,j}^{k+1}\delta_{i,k+1}^k)/\delta_{k,k}^k,$$

$$k = 1, \ldots, n-1, \quad i = 1, \ldots, k, \quad j = k+2, \ldots, m.$$

At the k-th step the coefficient matrix looks like

$$
\begin{pmatrix}
a_{k,k}^k & 0 & \cdots & 0 & \delta_{1,k+1}^k & \cdots & \delta_{1,m+1}^k \\
0 & a_{k,k}^k & \cdots & 0 & \delta_{2,k+1}^k & \cdots & \delta_{2,m+1}^k \\
\vdots & \vdots & \ddots & \vdots & \vdots & \ddots & \vdots \\
0 & 0 & \cdots & a_{k,k}^k & \delta_{k,k+1}^k & \cdots & \delta_{k,m+1}^k \\
a_{k+1,1} & a_{k+1,2} & \cdots & a_{k+1,k} & a_{k+1,k+1} & \cdots & a_{k+1,m+1} \\
\vdots & \vdots & \ddots & \vdots & \vdots & \ddots & \vdots \\
a_{n,1} & a_{n,2} & \cdots & a_{n,k} & a_{n,k+1} & \cdots & a_{n,m+1}
\end{pmatrix}
$$

The number of operations, necessary for the one-pass algorithm, is

$$N^m = (9n^2m - 6n^3 - 3nm - 6m + 6n)/6,$$
$$N^d = (3n^2m - 2n^3 - 3nm - 6m + 2n + 12)/6$$
$$N^a = (6n^2m - 4n^3 - 6nm + 3n^2 + n)/6.$$

When the number of equations and unknowns in the system is the same and equal to n, the last two algorithms can be compared

Number of Operations			
Method	Multiplications	Divisions	Add./Substr.
FB	$\frac{(4n^3+3n^2-n-6)}{6}$	$\frac{(2n^3-6n^2+10n-6)}{6}$	$\frac{(2n^3+3n^2-5n)}{6}$
OP	$\frac{(n^3+2n^2-n-2)}{2}$	$\frac{(n^3-7n+6)}{6}$	$\frac{(2n^3+3n^2-5n)}{6}$

2.4   The Recursive Method

The minors δ_{ij}^k and a_{ij}^k are elements of the following matrices

$$A_{k,c}^{r,l,(p)} = \begin{pmatrix} a_{r+1,k+1}^p & a_{r+1,k+2}^p & \cdots & a_{r+1,c}^p \\ a_{r+2,k+1}^p & a_{r+2,k+2}^p & \cdots & a_{r+2,c}^p \\ \vdots & \vdots & \ddots & \vdots \\ a_{l,k+1}^p & a_{l,k+2}^p & \cdots & a_{l,c}^p \end{pmatrix},$$

$$G_{k,c}^{r,l,(p)} = \begin{pmatrix} \delta_{r+1,k+1}^p & \delta_{r+1,k+2}^p & \cdots & \delta_{r+1,c}^p \\ \delta_{r+2,k+1}^p & \delta_{r+2,k+2}^p & \cdots & \delta_{r+2,c}^p \\ \vdots & \vdots & \ddots & \vdots \\ \delta_{l,k+1}^p & \delta_{l,k+2}^p & \cdots & \delta_{l,c}^p \end{pmatrix},$$

$G_{k,c}^{r,l,(p)}, A_{k,c}^{r,l,(p)} \in \mathbf{R}^{(l-r)\times(c-k)}$, $0 \le k < n$, $k < c \le n$, $0 \le r < m$, $r < l \le m$, $1 \le p \le n$.

We describe one recursive step reducing the matrix $\tilde{A} = A_{k,c}^{k,l,(k+1)}$ to the diagonal form

$$\tilde{A} \to (\delta^l I_{l-k}, \hat{G})$$

where

$$\tilde{A} = A_{k,c}^{k,l,(k+1)}, \quad \hat{G} = G_{l,c}^{k,l,(l)}$$

$0 \le k < c \le m$, $k < l \le n$, $l < c$. Note that if $k = 0$, $l = n$ and $c = m$, then we obtain the solution of the original system.

Description of One Step of the Recursive Method

$$\tilde{A} = \begin{pmatrix} A^1 \\ A^2 \end{pmatrix} \to_1 \begin{pmatrix} \delta^s I_{s-k} & G_2^1 \\ A_1^2 & A_2^2 \end{pmatrix} \to_2 \begin{pmatrix} \delta^s I_{s-k} & G_2^1 \\ 0 & \hat{A}_2^2 \end{pmatrix} \to_3$$

$$\rightarrow_3 \begin{pmatrix} \delta^s I_{s-k} & G^1_{2'} & G^1_{2''} \\ 0 & \delta^l I_{l-s} & \hat{G}^2_{2''} \end{pmatrix} \rightarrow_4 \begin{pmatrix} \delta^l I_{s-k} & 0 & \hat{G}^1_{2''} \\ 0 & \delta^l I_{l-s} & \hat{G}^2_{2''} \end{pmatrix} = \begin{pmatrix} \delta^l I_{l-k} & \hat{G} \end{pmatrix}$$

We may choose arbitrary numbers s: $k < s < l$ and write the matrix $\tilde{A}$ as follows:

$$\tilde{A} = \begin{pmatrix} A^1 \\ A^2 \end{pmatrix},$$

where $A^1 = A^{k,s,(k+1)}_{k,c}$ is the upper part of the matrix $\tilde{A}$ consisting of $s - k$ rows and $A^2 = A^{s,l,(k+1)}_{k,c}$ is the lower part of the matrix $\tilde{A}$.

$$A^1 \rightarrow (\delta^s I_{s-k}, G^1_2), \qquad (I)$$

where $A^1 \in \mathbf{R}^{(s-k)\times(c-k)}$, $G^1_2 = G^{k,s,(s)}_{s,c}$.

Let $A^2 = (A^2_1, A^2_2)$ where $A^2_1 = A^{s,l,(k+1)}_{k,s}$ and $A^2_2 = A^{s,l,(k+1)}_{s,c}$ consisting of $s - k$ and $c - s$ columns respectively, $\delta^k \neq 0$. The matrix $\hat{A}^2_2 = A^{s,l,(s+1)}_{s,c}$ is computed with the help of the matrix identity

$$\hat{A}^2_2 = (\delta^s \cdot A^2_2 - A^2_1 \cdot G^1_2)(\delta^k)^{-1}. \qquad (II)$$

$$\hat{A}^2_2 \rightarrow (\delta^l I_{l-s}, \hat{G}^2_{2''}), \qquad (III)$$

where $\hat{A}^2_2 \in \mathbf{R}^{(l-s)\times(c-s)}$ and $\hat{G}^2_{2''} = G^{s,l,(l)}_{l,c}$.

Let $G^1_2 = (G^1_{2'}, G^1_{2''})$, where the blocks $G^1_{2'} = G^{k,s,(s)}_{s,l}$ and $G^1_{2''} = G^{k,s,(s)}_{l,c}$ contain $l - s$ and $c - l$ columns respectively, and $\delta^s \neq 0$.

The matrix $\hat{G}^1_{2''} = G^{k,s,(l)}_{l,c}$ is computed with the help of the matrix identity

$$\hat{G}^1_{2''} = (\delta^l \cdot G^1_{2''} - G^1_{2'} \cdot \hat{G}^2_{2''})(\delta^s)^{-1}. \qquad (IV)$$

In the result we obtain δ^l and

$$\hat{G} = \begin{pmatrix} \hat{G}^1_{2''} \\ \hat{G}^2_{2''} \end{pmatrix}$$

Complexity of the Recursive Method is $O(mn^{\beta-1})$

We can obtain an exact estimate. For $n = 2^N, m = n + 1$ and $\beta = \log_2 7$ the number of multiplication operations is

$$\frac{7}{15} n^{\log_2 7} + n^2 (\log_2 n - \frac{2}{3}) + n(2 \log_2 n + \frac{1}{5}).$$

For $n = 2^N$, $\beta = 3$ the number of multiplications and divisions is

$$N^m = (6n^2 m - 4n^3 + (6nm - 3n^2) \log_2 n - 6nm + 4n)/6,$$
$$N^d = ((6nm - 3n^2) \log_2 n - 6nm - n^2 + 6m + 3n - 2)/6.$$

The number of multiplication operations for $m = n + 1$ is $(1/3)n^3 + O(n^2)$. The estimations for the previous two methods are, respectively, $n^3 + O(n^2)$ and $(2/3)n^3 + O(n^2)$.

3   Adjoint Matrix

The best method for computing the matrix determinant and adjoint matrix in the arbitrary commutative ring was suggested in the papers by Kaltofen [10] and Kaltofen and Villard [11]. Its complexity is $O(n^{\beta+1/3} \log n \log \log n)$; see also [3].

Here we describe the best method for computing adjoint matrices in commutative domains. Let $\mathcal{A} = \begin{pmatrix} A & C \\ B & D \end{pmatrix}$ be an invertible matrix and A an invertible block. Then

$$\mathcal{A}^{-1} = \begin{pmatrix} I & -A^{-1}C \\ 0 & I \end{pmatrix} \begin{pmatrix} I & 0 \\ 0 & (D - BA^{-1}C)^{-1} \end{pmatrix} \begin{pmatrix} I & 0 \\ -B & I \end{pmatrix} \begin{pmatrix} A^{-1} & 0 \\ 0 & I \end{pmatrix}$$

is the factorization of the inverse matrix. This requires two multiplication operations and two inversions of blocks. In case $n = 2^p$ it will take 2^{p-1} inversions of 2×2 blocks and 2^{p-k} multiplications of $2^k \times 2^k$ blocks.

Overall, $n^{\log 7} - n/2$ multiplication operations will be needed, if we use Strassen's multiplication algorithm. In general, if the complexity of matrix multiplication is $O(n^\beta)$, then the computation of the factors of the inverse matrix can be done in time $O(n^\beta)$.

Let R be a commutative ring, and let $\mathcal{A} = (a_{i,j})$ be a square matrix of order n over the ring R. Let

$$A_t^{(s)} = (a_{i,j}^s)_{j=s,\dots,t}^{i=s,\dots,t} \text{ and } \mathcal{G}_s^{(t)} = (\delta_{t(i,j)})_{j=t+1,\dots,n}^{i=s,\dots,t}$$

Theorem 1. *Let $\mathcal{A}$ be a square block matrix of order n over the ring R; that is,*

$$\mathcal{A} = \begin{pmatrix} A & C \\ B & D \end{pmatrix},$$

where A is a square block of order s, $(1 < s < n)$, the determinant of which, δ_s, is neither zero nor a zero divisor in R. Then, the adjoint matrix $\mathcal{A}^*$ can be written as the product

$$\mathcal{A}^* = \begin{pmatrix} \delta_s^{-1}\delta_n I & -\delta_s^{-1}FC \\ 0 & I \end{pmatrix} \begin{pmatrix} I & 0 \\ 0 & G \end{pmatrix} \begin{pmatrix} I & 0 \\ -B & \delta_s I \end{pmatrix} \begin{pmatrix} F & 0 \\ 0 & I \end{pmatrix}, \qquad (*)$$

where $F = A^*$, $G = \delta_s^{-n+s+1} A_n^{(s+1)*}$, I is the identity matrix and we have the identity

$$A_n^{(s+1)} = \delta_s D - BFC.$$

Theorem 2. *Let $\mathcal{A}_n^{(s+1)}$ be a square block matrix of order $n - s$, $(s > 0, n - s > 2)$, over the ring R; that is,*

$$\mathcal{A}_n^{(s+1)} = \begin{pmatrix} \mathbf{A} & \mathbf{C} \\ \mathbf{B} & \mathbf{D} \end{pmatrix},$$

where $\mathbf{A}$ is a square block of order $t - s$, $(1 < s < t < n)$, and δ_s and δ_t are neither zero nor zero dividers in R. Then, the matrix $\delta_s^{-n+s+1} \mathcal{A}_n^{(s+1)*}$ can be written as the product

$$\begin{pmatrix} \delta_t^{-1}\delta_n I & -\delta_t^{-1}\mathbf{FC} \\ 0 & I \end{pmatrix} \begin{pmatrix} I & 0 \\ 0 & \delta_s^{-1}\mathbf{G} \end{pmatrix} \begin{pmatrix} I & 0 \\ -\mathbf{B} & \delta_t I \end{pmatrix} \begin{pmatrix} \mathbf{F} & 0 \\ 0 & I \end{pmatrix}, \qquad (**)$$

where $\mathbf{F} = \delta_s^{-t+s+1} \mathcal{A}_t^{(s+1)*}$, $\mathbf{G} = \delta_t^{-n+t+1} \mathcal{A}_n^{(t+1)*}$, I is the identity matrix and we have the identity

$$\mathcal{A}_n^{(t+1)} = \delta_s^{-1}(\delta_t \mathbf{D} - \mathbf{BFC}).$$

Remark 1. If $n = s+2$, then, $\mathcal{A}_n^{(s+1)*} = \begin{pmatrix} a_{n,n}^{s+1} & -a_{n-1,n}^{s+1} \\ -a_{n,n-1}^{s+1} & a_{n-1,n-1}^{s+1} \end{pmatrix}$. And if $n = s+1$, then $\mathcal{A}_n^{(s+1)*} = 1$.

3.1   Dichotomic Process

The dimensions of the upper left block A (of the initial square block matrix $\mathcal{A}$) may be chosen arbitrarily. The case will be examined when the dimensions of block A are powers of two.

Let n be the order of the matrix $\mathcal{A}$, $2^h < n \leq 2^{h+1}$ and assume that all minors δ_{2i}, $i = 1, 2, \ldots$ are not zero or zero dividers of the ring R. According to Theorems 1 and 2 we are going to sequentially compute adjoint matrices for the upper left blocks of order $2, 4, 8, 16, \ldots$ of matrix $\mathcal{A}$.

1. For the block of order 2 we have:

$$A_{2,2}^2 = (a_{i,j})_{i,j=1,2}, \quad \delta_2 = \det A_{2,2}^1,$$

$$A_{2,2}^{2*} = \begin{pmatrix} a_{2,2} & -a_{1,2} \\ -a_{2,1} & a_{1,1} \end{pmatrix}.$$

2. For the block of order 4 we have:

$$A_{4,4}^{4*} = \begin{pmatrix} \delta_2^{-1}\delta_4 I & -\delta_2^{-1}FC \\ 0 & I \end{pmatrix} \begin{pmatrix} I & 0 \\ 0 & G \end{pmatrix} \begin{pmatrix} I & 0 \\ -B & \delta_2 I \end{pmatrix} \begin{pmatrix} F & 0 \\ 0 & I \end{pmatrix},$$

$F = A_{2,2}^{2*}$, $B = (a_{i,j})_{j=1,2}^{i=3,4}$,
$C = (a_{i,j})_{j=3,3}^{i=1,2}$, $D = (a_{i,j})_{i,j=3,4}$, $\mathcal{A}_4^{(3)} = \delta_2 D - BFC = (a_{i,j}^3)_{i,j=3,4}$, $G = \delta_2^{-1}\mathcal{A}_4^{(3)*}$, $\delta_4 = \det G$.

3. For the block of order 8 we have:

$$A_{8,8}^{8*} = \begin{pmatrix} \delta_4^{-1}\delta_8 I & -\delta_4^{-1}FC \\ 0 & I \end{pmatrix} \begin{pmatrix} I & 0 \\ 0 & G \end{pmatrix} \begin{pmatrix} I & 0 \\ -B & \delta_4 I \end{pmatrix} \begin{pmatrix} F & 0 \\ 0 & I \end{pmatrix},$$

$$F = A_{4,4}^{4*},\ B = (a_{i,j})_{j=1,\ldots,4}^{i=5,\ldots,8},\ C = (a_{i,j})_{j=5,\ldots,8}^{i=1,\ldots,4},\ D = (a_{i,j})_{i,j=5,\ldots,8},$$

$$G = \begin{pmatrix} \delta_6^{-1}\delta_8 I & -\delta_6^{-1}\mathbf{FC} \\ 0 & I \end{pmatrix} \begin{pmatrix} I & 0 \\ 0 & \delta_4^{-1}\mathbf{G} \end{pmatrix} \begin{pmatrix} I & 0 \\ -\mathbf{B} & \delta_6 I \end{pmatrix} \begin{pmatrix} \mathbf{F} & 0 \\ 0 & I \end{pmatrix},$$

$$\mathcal{A}_8^{(5)} = \delta_4 D - BFC = (a_{i,j}^5)_{i,j=5,\ldots,8},\ \mathbf{F} = \delta_4^{-3}\mathcal{A}_6^{(5)*},\ \delta_6 = \det \mathbf{F},\ \mathbf{B} = (a_{i,j}^5)_{j=5,6}^{i=7,8},$$

$$\mathbf{C} = (a_{i,j}^5)_{j=7,8}^{i=5,6},\ \mathbf{D} = (a_{i,j}^5)_{i,j=7,8},\ \mathcal{A}_8^{(7)} = \delta_4^{-1}(\delta_6 \mathbf{D} - \mathbf{BFC}) = (a_{i,j}^7)_{i,j=7,8},$$

$$\mathbf{G} = \delta_6^{-1}\mathcal{A}_8^{(7)*},\ \delta_8 = \det \mathbf{G}.$$

Complexity Estimation

Let $\gamma n^\beta + o(n^\beta)$ be an asymptotic estimation of the number of operations for multiplying two matrices of order n. Then the complexity of computing the adjoint matrix of order $n = 2^p$ is

$$F(n) = 6\gamma n^\beta \frac{1 - (n/2)^{1-\beta}}{2^\beta - 2} + o(n^\beta)$$

4   Characteristic Polynomial

In the case of an arbitrary commutative ring, the best algorithms for computing the characteristic polynomial are Chistov's algorithm [7] and the improved Berkowitz algorithm [4]. The complexity of these methods is $O(n^{\beta+1}\log n)$. We present the best method to date — for computations in commutative domains — which has complexity $O(n^3)$.

Let $A = (a_{ij})$ be an $n \times n$ matrix over the ring R. If all the diagonal minors δ_k $(k = 1, \ldots, n-1)$ of matrix A are *not* zero, then the following identity holds

$$A_u = \tilde{L}A,$$

where A_u is an upper triangular matrix and $\tilde{L}$ is a lower triangular matrix with determinant different from zero, such that

$$\tilde{L} = \mathbf{D}_{n-2}^{-1}\tilde{\mathbf{L}}_{n-1}\cdots\mathbf{D}_1^{-1}\tilde{\mathbf{L}}_2\tilde{\mathbf{L}}_1$$

$\tilde{\mathbf{L}}_k = \mathrm{diag}(I_{k-1}, \tilde{L}_k)$, $\mathbf{D}_k = \mathrm{diag}(I_k, D_k)$, where I_k is the identity matrix of order k, $D_k = \delta_k I_{n-k}$,

$$L_k = \begin{pmatrix} \delta_k & 0 \\ v_k & I_{n-k} \end{pmatrix},\ \tilde{L}_k = \begin{pmatrix} 1 & 0 \\ -v_k & \delta_k I_{n-k} \end{pmatrix}$$

$v_k = (a_{k+1,k}^k, \ldots, a_{n,k}^k)^T$, $A_u = (a_{i,j}^{(n)})$ is an $n \times n$ matrix, and $a_{i,j}^{(n)} = a_{i,j}^i$, for $i \leq j$, $a_{i,j}^{(n)} = 0$, for $i > j$.

The proof is based on Sylvester's identity

$$a^{k-1}_{k-1,k-1}a^{k+1}_{i,j} = a^k_{k,k}a^k_{i,j} - a^k_{i,k}a^k_{k,j}.$$

The factorization of matrix A into upper and lower triangular matrices is the result of the forward direction part — of the forward and backward direction algorithm.

Let $A^{(k)}_u = (a^{(k)}_{i,j})$ be an $n \times n$ matrix, $k = 1, \ldots, n$, with $a^{(k)}_{i,j} = a^i_{i,j}$ for $i \le j < k$, $a^{(k)}_{i,j} = a^k_{i,j}$, $i \ge k$, $j \ge k$, and the remaining elements zero. Then $A_u = \tilde{L}A$ reduces to the identities

$$A^{(2)}_u = \tilde{\mathbf{L}}_1 A; \quad A^{(k+1)}_u = \mathbf{D}^{-1}_{k-1}\tilde{\mathbf{L}}_k A^{(k)}_u, \quad k = 2, \ldots, n - 1,$$

which subsequently enable the computation of matrices $A^{(k)}_u$, $k = 2, 3, \ldots, n$, such that all the elements of the matrices $\mathbf{D}_k$ and $\tilde{\mathbf{L}}_k$ are elements of the matrix $A^{(k)}_u$.

The requirement that the diagonal minors δ_k $(k = 1, 2, \ldots, n - 1)$ be different from zero may be weakened. If a diagonal minor δ_k of order k is equal to zero, and in column v_k there is a nonzero element $a^k_{i,k}$, then rows i and k must be interchanged; that is, multiply on the left the matrix of interchanges $P_k = P_{(i,k)} = I_n + E_{ik} + E_{ki} - E_{kk} - E_{ii}$, where E_{ik} denotes a matrix in which all elements are zero except element (i, k), which is equal to one.

And if $\delta_k = 0$ and $v_k = 0$, then necessarily $P_k = \tilde{\mathbf{L}}_k = \mathbf{D}_{k-1} = I_n$, $\mathbf{D}_k = \mathbf{D}_{k-1}$.

The factorization formula remains as before, only now

$$\tilde{\mathbf{L}}_k = \operatorname{diag}(I_{k-1}, \tilde{L}_k)P_k.$$

Note the following identities, which will be subsequently needed:

$$\tilde{L}_k L_k = D_k, \quad \tilde{L}L = T,$$

where

$$L = \mathbf{L}_1 \mathbf{L}_2 \cdots \mathbf{L}_{n-1},$$

$\mathbf{L}_k = P^{-1}_k \operatorname{diag}(I_{k-1}, L_k)$, T is a diagonal matrix defined by, $T = S_1 S_2$, where $S_1 = \operatorname{diag}(1, S)$, and $S_2 = \operatorname{diag}(S, 1)$, with $S = \operatorname{diag}(\delta_1, \delta_2, \ldots, \delta_{n-1})$.

To indicate the matrix A from which a given triangular or diagonal matrix was computed, we write $L = L(A), T = T(A)$.

4.1   Computation of Similar p-Trianular Matrix

Let $A = \begin{pmatrix} \mathbf{a} \ \mathbf{b} \\ \mathbf{c} \ \mathbf{d} \end{pmatrix}$ be a matrix over R with blocks $\mathbf{a}$ of order $p \times p$ and $\mathbf{d}$ of order $n \times n$. We will call matrix A upper p-triangular, if the block $(\mathbf{c}, \mathbf{d})$ looks like an upper triangular matrix.

We will denote with calligraphic letters block-diagonal matrices of order $(n + p) \times (n + p)$ of the type $\operatorname{diag}(I_p, G) = \mathcal{G}$, where G is a $p \times p$ matrix.

Let G be some $p \times p$ matrix and let $\tilde{L} = \tilde{L}((\mathbf{c},\mathbf{d})\mathcal{G})$, and $T = T(((\mathbf{c},\mathbf{d})\mathcal{G}))$. If we take now $G = L$, $\tilde{\mathcal{L}} = \mathrm{diag}(I_p, \tilde{L})$, $\mathcal{L} = \mathrm{diag}(I_p, L)$, then the matrix

$$A_u = \tilde{\mathcal{L}} A \mathcal{L}$$

will become an upper p-triangular matrix, and matrix $T^{-1}A_u$ will be similar to A.

The cofactors L and $\tilde{L}$ of the matrix can be computed sequentially. Since $((\mathbf{c},\mathbf{d})\mathcal{G}) = (\mathbf{c},\mathbf{d}L)$ and the first p of the columns of the matrix $(\mathbf{c},\mathbf{d}L)$ constitute block $\mathbf{c}$ and are independent from L, then using them we can compute sequentially the first p cofactors of the matrix $\tilde{L} : \tilde{\mathbf{L}}_1, \mathbf{D}_1, \tilde{\mathbf{L}}_2, \ldots, \mathbf{D}_{p-1}, \tilde{\mathbf{L}}_p$. From these we can write the first p cofactors of matrix L , can compute p columns of the matrix $\mathbf{d}L$ and after that the following p cofactors of matrix $\tilde{L}$, etc. For $p = 1$ we obtain a quasi-triangular matrix, that is a matrix with zero elements under the second diagonal, which is obtained by the elements $a_{2,1}, a_{3,2}, \ldots, a_{n,n-1}$.

Let us denote by A_k $(1 \le k \le n)$ the corner minors of order k of the quasi-triangular matrix $A = (a_{i,j}), a_{i,j} = 0$ for $i \ge 2, j \le i - 1$, and assume $A_0 = 1$. Then its determinant can be computed as shown

$$\det(A_n) = a_{nn} \det(A_{n-1}) + \sum_{i=1}^{n-1} a_{i,n} \det(A_{i-1}) \prod_{j=i+1}^{n-1} (-a_{j,j-1}).$$

The complexity of this method is $\frac{5}{3}n^3 + O(n^2)$ — multiplicative operations.

5   Conclusion

For computations over commutative domains we have the following results:

- The complexity of the $O(n^3)$ methods (FB) and (OP) for solving systems of linear equations of size $n \times m$ is

$$M_{(FB)} = (1/2)(4n^2m - 2n^3 - 2nm - 3n^2 - 2m + 7n - 2),$$

$$M_{(OP)} = (1/6)(12n^2m - 8n^3 - 6nm - 9n^2 - 12m + 8n + 12).$$

Suppose that the complexity of the given method for matrix multiplications is $\gamma n^\beta + o(n^\beta)$, where γ and β are constants, and n is the order of the matrix. Then, the complexity of the recursive methods for solving systems of size $n \times m$ is

$$S(n,m) = \gamma \frac{n^\beta}{2^\beta} \left[(4\frac{m}{n} - 2)\frac{1 - n^{2-\beta}}{1 - 2^{2-\beta}} - \frac{1 - n^{1-\beta}}{1 - 2^{1-\beta}} \right] + o(n^{\beta-1}m).$$

- The complexity of the method for the computation of the determinant of a matrix of order n is $S(n,n)$. The complexity of the method for the computation of the kernel of a linear operator is $S(n,m)$.

- The complexity of the method for the computation and the factorization of the adjoint matrix is

$$F(n) = 6\gamma n^\beta \frac{1 - (n/2)^{1-\beta}}{2^\beta - 2} + o(n^\beta)$$

- Finally, the complexity of the best method we know today for the computation of the characteristic polynomial of a matrix of order n is $\frac{5}{3}n^3 + O(n^2)$.

References

1. Akritas, A.: Elements of Computer Algebra with Applications. John Wiley Interscience, New York (1989)
2. Akritas, A.G., Akritas, E.K., Malaschonok, G.I: Various proofs of Sylvester's (determinant) identity. Mathematics and Computations in Simulation 42, 585–593 (1996)
3. Akritas, A.G., Malaschonok, G.I.: Computation of the Adjoint Matrix. In: Alexandrov, V.N., van Albada, G.D., Sloot, P.M.A., Dongarra, J.J. (eds.) ICCS 2006. LNCS, vol. 3992, pp. 486–489. Springer, Heidelberg (2006)
4. Abdeljaoued, J.: Berkowitz Algorithm, Maple and computing the characteristic polynomial in an arbitrary commutative ring. Computer Algebra MapleTech 4(3), Birkhauser Boston (1997)
5. Abdeljaoued, J., Malaschonok, G.I.: Efficient Algorithms for Computing the Characteristic Polynomial in a Domain. Journal of Pure and Applied Algebra 156(2-3), 127–145 (2001)
6. Bareiss, E.H.: Sylvester's Identity and Multistep Integer-Preserving Gaussian Elimination. Math. Comp. 22(103), 565–578 (1968)
7. Chistov, A.L.: Fast parallel calculation of the rank of matrices over a field of arbitrary characteristic. In: Budach, L. (ed.) FCT 1985. LNCS, vol. 199, pp. 147–150. Springer, Heidelberg (1985)
8. Coppersmith, D., Winograd, S.: Matrix multiplication via arithmetic progressions. Journal of Symbolic Computation 9, 251–280 (1990)
9. Dodgson, C.L.: Condensation of determinants, being a new and brief method for computing their arithmetic values. Proc. Royal Soc. Lond. A15, 150–155 (1866)
10. Kaltofen, E.: On Computing Determinants of Matrices Without Divisions. In: ISSAC'92. Proc. Internat. Symp. Symbolic Algebraic Comput, pp. 342–349. ACM Press, New York (1992)
11. Kaltofen, E., Villard, G.: On the complexity of computing determinants. In: ASCM 2001. Proc. Fifth Asian Symposium on Computer Mathematics, pp. 13–27 (2001) (extended abstract)
12. Malashonok, G.I.: Solution of a system of linear equations in an integral domain. USSR Journal of computational Mathematics and Mathematical Physics 23(6), 1497–1500 (1983)
13. Malashonok, G.I.: On the solution of a linear equation system over commutative rung. Math. Notes of the Acad. Sci. USSR 42(4), 543–548 (1987)
14. Malashonok, G.I.: A new solution method for linear equation systems over commutative rung. In: International Algebraic Conference, Theses on the ring theory, algebras and modules, Novosibirsk, p. 82 (August 21-26, 1989)

15. Malashonok, G.I.: Algorithms for the solution of systems of linear equations in commutative rings; Effective methods in algebraic geometry (Castiglioncello, 1990), Progr. Math., Birkhauser Boston, Boston, 94, 289–298 (1991)

16. Malashonok, G.I.: Argorithms for Computing Determinants in commutative rings; Diskretnaya Matematika 7(4), 68–76 (1995) transl. in: Discrete Math. Appl. 5(6), 557–566 (1996)

17. Malaschonok, G.I.: On the solution of systems of linear equations; Computational Commutative Algebra, COCOA-IV, Abstracts, Genova, 32 (May 29-June 2, 1995)

18. Malaschonok, G.I.: Recursive Method for the Solution of Systems of Linear Equations; Computational Mathematics. In: Proc. of the 15th IMACS World Congress, I, Berlin, August 1997, Wissenschaft & Technik Verlag, Berlin, pp. 475–480 (1997)

19. Malaschonok, G.I.: A Computation of the Characteristic Polynomial of an Endomorphism of a Free Module; Zap. Nauchnyh Sem. S.-Peterburg. Otdel. Mat. Inst. Steklov (POMI), 258 (1999). Teor. Predst. Din. Sist. Komb. i Algoritm. Metody 4, 101–114 (1999)

20. Malaschonok, G.I.: Effective Matrix Methods in Commutative Domains; Formal Power Series and Algebraic Combinatorics, pp. 506–517. Springer, Heidelberg (2000)

21. Strassen, V.: Gaussian Elimination is not optimal; Numerische Mathematik 13, 354–356 (1969)

22. Sasaki, T., Murao, H.: Efficient Gaussian elimination method for symbolic determinants and linear systems. A.C.M. Trans. Math. Software 8(4), 277–289 (1968)

23. Waugh, F.V., Dwyer, P.S.: Compact computation of the inverse of a matrix. Annals of Mathemaical Statistic 16, 259–271 (1945)

Advances on the Continued Fractions Method Using Better Estimations of Positive Root Bounds

Alkiviadis G. Akritas[1], Adam W. Strzeboński[2], and Panagiotis S. Vigklas[1]

[1] University of Thessaly, Department of Computer and Communication Engineering,
GR-38221 Volos, Greece
{akritas, pviglas}@uth.gr
[2] Wolfram Research, Inc., 100 Trade Center Drive, Champaign, IL 61820, USA
adams@wolfram.com

Abstract. We present an implementation of the Continued Fractions (CF) real root isolation method using a recently developed upper bound on the positive values of the roots of polynomials. Empirical results presented in this paper verify that this implementation makes the CF method *always* faster than the Vincent-Collins-Akritas bisection method[1], or any of its variants.

1 Introduction

We begin by first reviewing some basic facts about the continued fractions method for isolating the positive roots of polynomials. This method is based on Vincent's theorem of 1836, [Vincent 1836], which states:

Theorem 1. *If in a polynomial, $p(x)$, of degree n, with rational coefficients and without multiple roots we perform sequentially replacements of the form*

$$x \leftarrow \alpha_1 + \frac{1}{x}, x \leftarrow \alpha_2 + \frac{1}{x}, x \leftarrow \alpha_3 + \frac{1}{x}, \ldots$$

where $\alpha_1 \geq 0$ is a random non negative integer and $\alpha_2, \alpha_3, \ldots$ are random positive integers, $\alpha_i > 0$, $i > 1$, then the resulting polynomial either has no sign variations or it has one sign variation. In the last case the equation has exactly one positive root, which is represented by the continued fraction

$$\alpha_1 + \cfrac{1}{\alpha_2 + \cfrac{1}{\alpha_3 + \cfrac{1}{\ddots}}}$$

whereas in the first case there are no positive roots.

---

[1] Misleadingly referred to (by several authors) initially as "modified Uspensky's method" and recently as "Descartes' method".

V.G. Ganzha, E.W. Mayr, and E.V. Vorozhtsov (Eds.): CASC 2007, LNCS 4770, pp. 24–30, 2007.

Note that if we represent by $\frac{ax+b}{cx+d}$ the continued fraction that leads to a transformed polynomial $f(x) = (cx+d)^n p(\frac{ax+b}{cx+d})$, with one sign variation, then the single positive root of $f(x)$—in the interval $(0,\infty)$—corresponds to *that* positive root of $p(x)$ which is located in the open interval with endpoints $\frac{b}{d}$ and $\frac{a}{c}$. These endpoints are *not* ordered and are obtained from $\frac{ax+b}{cx+d}$ by replacing x with 0 and ∞, respectively. See the papers by Alesina & Galuzzi, [Alesina and Galuzzi 1998] and Chapter 7 in [Akritas 1989] for a complete historical survey of the subject and implementation details respectively[2].

Cauchy's method, for computing bounds on the positive roots of a polynomial, was mainly used until now in the Continued Fraction (CF) real root isolation method, [Akritas and Strzeboński 2005]. In the SYNAPS implementation of the CF method, [Tsigaridas and Emiris 2006], Emiris and Tsigaridas used Kioustelidis method, [Kioustelidis 1986] for computing such bounds and independently verified the results obtained in [Akritas and Strzeboński 2005].

Both implementations of the CF method showed that its "Achilles heel" was the case of a big number of very large rational roots. In this case the CF method was up to 4 times slower than REL—the fastest implementation of the Vincent-Collins-Akritas bisection method, [Collins and Akritas 1976], developed by Rouillier and Zimmermann, [Rouillier and Zimmermann 2004]. Table 1 presented below, is an exact copy of the last table (Table 4), found in [Akritas and Strzeboński 2005].

Table 1. Products of factors (x-randomly generated integer root). All computations were done on a 850 MHz Athlon PC with 256 MB RAM; (s) stands for time in seconds and (MB) for the amount of memory used, in MBytes.

Roots (bit length)	Degree	No. of roots	CF T (s)/M (MB)	REL T (s)/M (MB)
10	100	100	0.8/1.82	0.61/1.92
10	200	200	2.45/2.07	10.1/2.64
10	500	500	33.9/3.34	878/8.4
1000	20	20	0.12/1.88	0.044/1.83
1000	50	50	16.7/3.18	4.27/2.86
1000	100	100	550/8.9	133/6.49

The last three lines of Table 1 demonstrate the weaker performance of CF in the case of a big number of very large rational roots. However, we recently generalized and extended a theorem by Ştefănescu, [Ştefănescu 2005], and developed a new method for computing upper bounds on the positive roots of polynomials, [Akritas, Strzeboński & Vigklas 2006]. As was verified, this method provides even sharper upper bounds on the positive roots of polynomials. In this paper,

---

[2] Alesina and Galuzzi point out in their work that Vincent's theorem can be implemented in various ways; the Vincent-Collins-Akritas bisection method is also one such implementation.

we incorporated into CF this new method for computing upper bounds for positive roots[3]. It turns out that with this modification, the CF algorithm is now *always* faster than that of Vincent-Collins-Akritas, or any of its variants.

2   Algorithmic Background

In this section we present the CF algorithm (where we correct a misprint in Step 5 that appeared in [Akritas and Strzeboński 2005] and explain where the new bound on the positive roots is used.

2.1   Description of the Continued Fractions Algorithm CF

Using the notation of the paper [Akritas and Strzeboński 2005], let $f \in Z[x] \setminus \{0\}$. By $sgc(f)$ we denote the number of sign changes in the sequence of nonzero coefficients of f. For nonnegative integers a, b, c, and d, such that $ad - bc \neq 0$, we put

$$intrv(a, b, c, d) := \Phi_{a,b,c,d}((0, \infty))$$

where

$$\Phi_{a,b,c,d} : (0, \infty) \ni x \longrightarrow \frac{ax + b}{cx + d} \in (min(\frac{a}{c}, \frac{b}{d}), max(\frac{a}{c}, \frac{b}{d}))$$

and by *interval data* we denote a list

$$\{a, b, c, d, p, s\}$$

where p is a polynomial such that the roots of f in $intrv(a, b, c, d)$ are images of positive roots of p through $\Phi_{a,b,c,d}$, and $s = sgc(p)$.

The value of parameter α_0 used in step 4 below needs to be chosen empirically. In our implementation $\alpha_0 = 16$.

Algortihm Continued Fractions (CF)
Input: A squarefree polynomial $f \in Z[x] \setminus \{0\}$
Output: The list *rootlist* of positive roots of f.

1. Set *rootlist* to an empty list. Compute $s \leftarrow sgc(f)$. If $s = 0$ return an empty list. If $s = 1$ return $\{(0, \infty)\}$. Put interval data $\{1, 0, 0, 1, f, s\}$ on *intervalstack*.
2. If *intervalstack* is empty, return *rootlist*, else take interval data $\{a, b, c, d, p, s\}$ off *intervalstack*.
3. Compute a lower bound α on the positive roots of p.
4. If $\alpha > \alpha_0$ set $p(x) \leftarrow p(\alpha x)$, $a \leftarrow \alpha a$, $c \leftarrow \alpha c$, and $\alpha \leftarrow 1$.
5. If $\alpha \geq 1$, set $p(x) \leftarrow p(x + \alpha)$, $b \leftarrow \alpha a + b$, and $d \leftarrow \alpha c + d$. If $p(0) = 0$, add $[b/d, b/d]$ to *rootlist*, and set $p(x) \leftarrow p(x)/x$. Compute $s \leftarrow sgc(p)$. If $s = 0$ go to step 2. If $s = 1$ add $intrv(a, b, c, d)$ to *rootlist* and go to step 2.

---

[3] Note that the computed bounds are integers rather than powers of two.

6. Compute $p_1(x) \leftarrow p(x+1)$, and set $a_1 \leftarrow a$, $b_1 \leftarrow a+b$, $c_1 \leftarrow c$, $d_1 \leftarrow c+d$, and $r \leftarrow 0$. If $p_1(0) = 0$, add $[b_1/d_1, b_1/d_1]$ to $rootlist$, and set $p_1(x) \leftarrow p_1(x)/x$, and $r \leftarrow 1$. Compute $s_1 \leftarrow sgc(p_1)$, and set $s_2 \leftarrow s - s_1 - r$, $a_2 \leftarrow b$, $b_2 \leftarrow a + b$, $c_2 \leftarrow d$, and $d_2 \leftarrow c + d$.

7. If $s_2 > 1$, compute $p_2(x) \leftarrow (x+1)^m p(\frac{1}{x+1})$, where m is the degree of p. If $p_2(0) = 0$, set $p_2(x) \leftarrow p_2(x)/x$. Compute $s_2 \leftarrow sgc(p_2)$.

8. If $s_1 < s_2$, swap $\{a_1, b_1, c_1, d_1, p_1, s_1\}$ with $\{a_2, b_2, c_2, d_2, p_2, s_2\}$.

9. If $s_1 = 0$ goto step 2. If $s_1 = 1$ add $intrv(a_1, b_1, c_1, d_1)$ to $rootlist$, else put interval data $\{a_1, b_1, c_1, d_1, p_1, s_1\}$ on $intervalstack$.

10. If $s_2 = 0$ goto step 2. If $s_2 = 1$ add $intrv(a_2, b_2, c_2, d_2)$ to $rootlist$, else put interval data $\{a_2, b_2, c_2, d_2, p_2, s_2\}$ on $intervalstack$. Go to step 2.

Please note that the lower bound, α, on the positive roots of $p(x)$ is computed in Step 3, and used in Step 5.

To compute this bound we generalized Ştefănescu's theorem, [Ştefănescu 2005], in the sense that Theorem 2 (see below) applies to polynomials with any number of sign variations; moreover we have introduced the concept of *breaking up* a positive coefficient into several parts to be paired with negative coefficients (of lower order terms).

Theorem 2. *Let* $p(x)$

$$p(x) = \alpha_n x^n + \alpha_{n-1} x^{n-1} + \ldots + \alpha_0, \quad (\alpha_n > 0) \tag{1}$$

be a polynomial with real coefficients and let $d(p)$ *and* $t(p)$ *denote the degree and the number of its terms, respectively.*

Moreover, assume that $p(x)$ *can be written as*

$$p(x) = q_1(x) - q_2(x) + q_3(x) - q_4(x) + \ldots + q_{2m-1}(x) - q_{2m}(x) + g(x), \tag{2}$$

where all the polynomials $q_i(x)$, $i = 1, 2, \ldots, 2m$ *and* $g(x)$ *have only positive coefficients. In addition, assume that for* $i = 1, 2, \ldots, m$ *we have*

$$q_{2i-1}(x) = c_{2i-1,1} x^{e_{2i-1,1}} + \ldots + c_{2i-1,t(q_{2i-1})} x^{e_{2i-1,t(q_{2i-1})}}$$

and

$$q_{2i}(x) = b_{2i,1} x^{e_{2i,1}} + \ldots + b_{2i,t(q_{2i})} x^{e_{2i,t(q_{2i})}},$$

where $e_{2i-1,1} = d(q_{2i-1})$ *and* $e_{2i,1} = d(q_{2i})$ *and the exponent of each term in* $q_{2i-1}(x)$ *is greater than the exponent of each term in* $q_{2i}(x)$. *If for all indices* $i = 1, 2, \ldots, m$, *we have*

$$t(q_{2i-1}) \geq t(q_{2i}),$$

then an upper bound of the values of the positive roots of $p(x)$ *is given by*

$$ub = \max_{\{i=1,2,\ldots,m\}} \left\{ \left(\frac{b_{2i,1}}{c_{2i-1,1}} \right)^{\frac{1}{e_{2i-1,1}-e_{2i,1}}}, \ldots, \left(\frac{b_{2i,t(q_{2i})}}{c_{2i-1,t(q_{2i})}} \right)^{\frac{1}{e_{2i-1,t(q_{2i})}-e_{2i,t(q_{2i})}}} \right\},$$

for any permutation of the positive coefficients $c_{2i-1,j}$, $j = 1, 2, \ldots, t(q_{2i-1})$. Otherwise, for each of the indices i for which we have

$$t(q_{2i-1}) < t(q_{2i}),$$

*we **break up** one of the coefficients of $q_{2i-1}(x)$ into $t(q_{2i}) - t(q_{2i-1}) + 1$ parts, so that now $t(q_{2i}) = t(q_{2i-1})$ and apply the same formula (3) given above.*

For a proof of this theorem and examples comparing its various implementations, see [Akritas, Strzeboński & Vigklas 2006]. It turns out that all existing methods (i.e. Cauchy's, Lagrange-McLaurent, Kioustelidis's, etc) for computing upper bounds on the positive roots of a polynomial, are special cases of Theorem 2.

In this recent paper of ours, we also presented two new implementations of Theorem 2, the combination of which yields the best upper bound on the positive roots of a polynomial. These implementation are:

(a) **"first–λ" implementation of Theorem 2.** For a polynomial $p(x)$, as in (2), with λ negative coefficients we first take care of all cases for which $t(q_{2i}) > t(q_{2i-1})$, by breaking up the last coefficient $c_{2i-1,t(q_{2i})}$, of $q_{2i-1}(x)$, into $t(q_{2i}) - t(q_{2i-1}) + 1$ *equal* parts. We then pair each of the first λ positive coefficients of $p(x)$, encountered as we move in non-increasing order of exponents, with the first unmatched negative coefficient.

(b) **"local-max" implementation of Theorem 2.** For a polynomial $p(x)$, as in (1), the coefficient $-\alpha_k$ of the term $-\alpha_k x^k$ in $p(x)$ —as given in Eq. (1)— is paired with the coefficient $\frac{\alpha_m}{2^t}$, of the term $\alpha_m x^m$, where α_m is the largest positive coefficient with $n \geq m > k$ and t indicates the number of times the coefficient α_m has been used.

As an upper bound on the positive roots of a polynomial we take the minimum of the two bounds produced by implementations (a) & (b), mentioned above. This minimum of the two bounds is first computed in Step 3 and then used in Step 5 of CF.

3   Empirical Results

Below we recalculate the results of Table 1, comparing the timings in seconds (s) for: (a) the CF using Cauchy's rule (CF_OLD), (b) the CF using the new rule for computing upper bounds (CF_NEW), and (c) REL.

Due to the different computational environment the times differ substantially, but they confirm the fact that now the CF is always faster.

Again, of interest are the last three lines of Table 2, where as in Table 1 the performance of CF_OLD is worst than REL—at worst 3 times slower as the last entry indicates. However, from these same lines of Table 2 we observe that CF_NEW is now always faster than REL—at best twice as fast, as seen in the 5-th line.

Table 2. Products of terms $x - r$ with random integer r. The tests were run on a laptop computer with 1.8 Ghz Pentium M processor, running a Linux virtual machine with 1.78 GB of RAM.

Roots (bit length)	Deg	CF_OLD Time(s) Average (Min/Max)	CF_NEW Time(s) Average (Min/Max)	REL Average (Min/Max)	Memory (MB) CFO/CFN/REL
10	100	0.314 (0.248/0.392)	0.253 (0.228/0.280)	0.346 (0.308/0.384)	4.46/4.48/4.56
10	200	1.74 (1.42/2.33)	1.51 (1.34/1.66)	3.90 (3.72/4.05)	4.73/4.77/5.35
10	500	17.6 (16.9/18/7)	17.4 (16.3/18.1)	129 (122/140)	6.28/6.54/11.8
1000	20	0.066 (0.040/0.084)	0.031 (0.024/0.040)	0.038 (0.028/0.044)	4.57/4.62/4.51
1000	50	1.96 (1.45/2.44)	0.633 (0.512/0.840)	1.03 (0.916/1.27)	5.87/6.50/5.55
1000	100	52.3 (36.7/81.3)	12.7 (11.3/14.6)	17.2 (16.1/18.7)	10.4/11.7/9.17

4   Conclusions

In this paper we have examined the behavior of CF on the special class of polynomials with very many, very large roots—a case where CF exhibited a certain weakness. We have demonstrated that, using our recently developed rule for computing upper bounds on the positive roots of polynomials, CF is speeded up by a considerable factor and is now always faster than any other real root isolation method.

References

[Akritas 1978] Akritas, A.: Vincent's theorem in algebraic manipulation; Ph.D. Thesis, North Carolina State University, Raleigh, NC (1978)

[Akritas 1982] Akritas, A.: Reflections on a pair of theorems by Budan and Fourier. Mathematics Magazine 55(5), 292–298 (1982)

[Akritas 1989] Akritas, A.: Elements of Computer Algebra with Applications. John Wiley Interscience, New York (1989)

[Akritas and Strzeboński 2005] Akritas, A., Strzebonski, A.: A comparative study of two real root isolation methods. Nonlinear Analysis: Modelling and Control 10(4), 297–304 (2005)

[Akritas, Strzeboński & Vigklas 2006] Akritas, A., Strzebonski, A., Vigklas, P.: Implementations of a New Theorem for Computing Bounds for Positive Roots of Polynomials. Computing 78, 355–367 (2006)

[Alesina and Galuzzi 1998] Alesina, A., Galuzzi, M.: A new proof of Vincent's theorem. L'Enseignement Mathematique 44, 219–256 (1998)

[Collins and Akritas 1976] Collins, E.G., Akritas, G.A.: Polynomial real root isolation using Descartes' rule of signs. In: Proceedings of the 1976 ACM Symposium on Symbolic and Algebraic Computations, Yorktown Heights, N.Y., pp. 272–275 (1976)

[Kioustelidis 1986] Kioustelidis, B.: Bounds for positive roots of polynomials. J. Comput. Appl. Math. 16(2), 241–244 (1986)

[Rouillier and Zimmermann 2004] Rouillier, F., Zimmermann, P.: Efficient isolation of polynomial's real roots. Journal of Computational and Applied Mathematics 162, 33–50 (2004)

[Ştefănescu 2005] Ştefănescu, D.: New bounds for positive roots of polynomials. Journal of Universal Computer Science 11(12), 2132–2141 (2005)

[Tsigaridas and Emiris 2006] Tsigaridas, P.E., Emiris, Z.I.: Univariate polynomial real root isolation: Continued fractions revisited. In: Azar, Y., Erlebach, T. (eds.) ESA 2006. LNCS, vol. 4168, pp. 817–828. Springer, Heidelberg (2006)

[Vincent 1836] Vincent, A.J.H.: Sur la resolution des équations numériques; Journal de Mathématiques Pures et Appliquées 1, 341–372 (1836)

An Efficient LLL Gram Using Buffered Transformations

Werner Backes and Susanne Wetzel

Stevens Institute of Technology,
Castle Point on Hudson, Hoboken, 07030 NJ, USA
{wbackes,swetzel}@cs.stevens.edu

Abstract. In this paper we introduce an improved variant of the LLL algorithm. Using the Gram matrix to avoid expensive correction steps necessary in the Schnorr-Euchner algorithm and introducing the use of buffered transformations allows us to obtain a major improvement in reduction time. Unlike previous work, we are able to achieve the improvement while obtaining a strong reduction result and maintaining the stability of the reduction algorithm.

1 Introduction

Lattice theory is of great importance in cryptography. It not only provides effective tools for cryptanalysis, but it is also believed that lattice theory can bring about new cryptographic primitives that exhibit strong security even in the presence of quantum computers. While many aspects of lattice theory are already fairly well-understood, many practical aspects still require further investigation and understanding. With respect to cryptography this is of particular importance as a cryptographic primitive must be secure in both theory and practice.

The goal of lattice basis reduction is to find a basis representing the lattice where the base vectors not only are as small as possible but also are as orthogonal to each other as possible. While the LLL algorithm by Lenstra, Lenstra, and Lovász [11] was the first to allow for the efficient computation of a well-reduced lattice basis in theory, it was not until the introduction of the Schnorr-Euchner variant of the LLL algorithm [19] that lattice basis reduction could efficiently be used in practice (e.g., for cryptanalysis [19,16,17]). Since then, research has focused on improving on the stability and performance of reduction algorithms (e.g., [6,9,10,13,14]).

One can generally identify two main directions of recent work. The first line of research (e.g., [10,14,15,30]) is based on the use of a weaker reduction condition than the original LLL condition. While this allows for an improvement in efficiency it is important to note that it generally results in a less reduced lattice basis. Consequently, this approach cannot be taken in contexts which rely on the strong, proven bounds of the original LLL reduction (e.g., [4,2,5,12]).

In contrast, the second line of research focuses on improving on the stability and performance of lattice basis reduction while maintaining the strong

V.G. Ganzha, E.W. Mayr, and E.V. Vorozhtsov (Eds.): CASC 2007, LNCS 4770, pp. 31–44, 2007.

reduction conditions. It is in this context that this paper focuses on achieving improvements in the reduction time. In particular, this paper introduces an improved variant of the LLL algorithm which uses the Gram matrix to avoid expensive correction steps that are necessary for the Schnorr-Euchner algorithm. While the Gram matrix approach was already used previously [3,22,14,15], the new algorithm provides a major improvement by introducing the use of *buffered transformations*. This new approach allows us to improve the reduction time by up to 40% in comparison to existing methods while obtaining a strong reducti-on result and maintaining the stability of the reduction algorithm at the same time. In contrast, previous work not only relies on a weaker reduction condition [14,15,30] but also suffers from stability problems [30].

Outline: Section 2 provides the definitions and notations used in the remainder of the paper. Then, Section 3 introduces the LLL reduction algorithm using the Gram matrix representation and details ways to improve the running time. Section 4 discusses and analyzes the experiments. The paper closes with some directions for future work.

2   Preliminaries

A *lattice* $L \subset \mathbb{R}^n$ is an additive discrete subgroup of $\mathbb{R}^n$ such that $L = \left\{ \sum_{i=1}^{k} x_i \underline{b}_i | x_i \in \mathbb{Z}, 1 \le i \le k \right\}$ with linear independent vectors $\underline{b}_1, \ldots, \underline{b}_k \in \mathbb{R}^n$ ($k \le n$). $B = (\underline{b}_1, \ldots, \underline{b}_k) \in \mathbb{R}^{n \times k}$ is the *lattice basis* of L with dimension k. The basis of a lattice is not unique. However, different bases B and B' for the same lattice L can be transformed into each other by means of a *unimodular trans-formation*, i.e., $B' = BU$ with $U \in \mathbb{Z}^{n \times k}$ and $|\det U| = 1$. Typical unimodular transformations are the exchange of two base vectors—referred to as swap—or the adding of an integral multiple of one base vector to another one—generally referred to as translation.

Unlike the lattice basis, the *determinant* of a lattice is an invariant, i.e., it is independent of a particular basis: For a lattice $L \in \mathbb{R}^n$ with basis $B \in \mathbb{R}^{n \times k}$ the determinant $\det(L)$ is defined as $\det(L) = |\det(B^T B)|^{\frac{1}{2}}$. The Hadamard inequality $\det(L) \le \prod_{i=1}^{k} \|\underline{b}_i\|$ (where $\|.\|$ denotes the Euclidean length of a vector) gives an upper bound for the determinant of the lattice. Equality holds if B is an orthogonal basis.

The *orthogonalization* $B^* = (\underline{b}_1^*, \ldots, \underline{b}_k^*)$ of a lattice basis $B = (\underline{b}_1, \ldots, \underline{b}_k) \in \mathbb{R}^{n \times k}$ can be computed by the Gram-Schmidt method: $\underline{b}_1^* = \underline{b}_1$, $\underline{b}_i^* = \underline{b}_i - \sum_{j=1}^{i-1} \mu_{i,j} \underline{b}_j^*$ for $2 \le i \le k$ where $\mu_{i,j} = \frac{\langle \underline{b}_i, \underline{b}_j^* \rangle}{\|\underline{b}_j^*\|}$ for $1 \le j < i \le k$ where $\langle ., . \rangle$ defines the scalar product of two vectors. It is important to note that for a lattice $L \subset \mathbb{R}^n$ with basis $B = (\underline{b}_1, \ldots, \underline{b}_k) \in \mathbb{R}^{n \times k}$ a vector $\underline{b}_i^*$ of the orthogonalization $B^* = (\underline{b}_1^*, \ldots, \underline{b}_k^*) \in \mathbb{R}^{n \times k}$ is not necessarily in L. Furthermore, computing the orthogonalization B^* of a lattice basis using the Gram-Schmidt method strongly depends on the order of the basis vector of the lattice basis B.

The *defect of a lattice basis* $B = (\underline{b}_1, \ldots, \underline{b}_k) \in \mathbb{R}^{n \times k}$ defined as $\mathrm{dft}(B) = \frac{\prod_{i=1}^{n} \|\underline{b}_i\|}{\det(L)}$ allows one to compare the quality of different bases. Obviously,

$\mathrm{dft}(B) \geq 1$ and $\mathrm{dft}(B) = 1$ for an orthogonal basis. The goal of lattice basis reduction is to determine a basis with smaller defect. That is, for a lattice $L \subset \mathbb{R}^n$ with bases B and $B' \in \mathbb{R}^{n \times k}$, B' is better reduced than B if $\mathrm{dft}(B') < \mathrm{dft}(B)$. The most well-known and most-widely used lattice basis reduction method is the LLL reduction method [11]:

Definition 1. *For a lattice $L \subseteq \mathbb{Z}^n$ with basis $B = (\underline{b}_1, \ldots, \underline{b}_k) \in \mathbb{Z}^{n \times k}$, corresponding Gram-Schmidt orthogonalization $B^* = (\underline{b}_1^*, \ldots, \underline{b}_k^*) \in \mathbb{Z}^{n \times k}$ and coefficients $\mu_{i,j}$ with $1 \leq j < i \leq k$, the basis B is LLL-reduced if*

$$|\mu_{i,j}| \leq \frac{1}{2} \qquad\qquad \text{for } 1 \leq j < i \leq k \text{ and} \qquad (1)$$

$$\|\underline{b}_i^* + \mu_{i,i-1}\underline{b}_{i-1}^*\|^2 \geq y\|\underline{b}_{i-1}^*\|^2 \qquad\qquad \text{for } 1 < i \leq k. \qquad (2)$$

The reduction parameter y may arbitrarily be chosen in $\left(\frac{1}{4}, 1\right)$. Condition (1) is generally referred to as size-reduction [3,18]. The Schnorr-Euchner algorithm [19,1] allows for an efficient computation of an LLL-reduced lattice basis in practice.

Algorithm 1: SchnorrEuchnerLLL(B,y)

INPUT: Lattice basis $B = (\underline{b}_1, \ldots, \underline{b}_k) \in \mathbb{Z}^{n \times k}$, $y \in [\frac{1}{2}, 1)$
OUTPUT: LLL-reduced lattice basis

```
(1)   APPROX_BASIS(B', B)
(2)   B₁ = ‖b'₁‖², i = 2, Fc = false, Fr = false
(3)   while (i ≤ k) do
(4)     μii = 1, Bi = ‖b'ᵢ‖²
(5)     for (2 ≤ j < i) do
(6)       if (|⟨b'ᵢ, b'ⱼ⟩| < 2^(r/2) ‖b'ᵢ‖‖b'ⱼ‖) then        /* correction step 1 */
(7)         s = APPROX_VALUE(⟨bᵢ, bⱼ⟩)
(8)       else
(9)         s = ⟨b'ᵢ, b'ⱼ⟩
(10)      μij = (s − Σ_{m=1}^{j−1} μjm μim Bm)/Bj
(11)      Bi = Bi − μij² Bj
(12)    for (i > j ≥ 1) do                                 /* size-reduction */
(13)      if (|μij| > ½) then
(14)        Fr = true
(15)        if (|⌈μij⌋| > 2^(r/2)) then                    /* correction step 2 */
(16)          Fc = true
(17)        bᵢ = bᵢ − ⌈μij⌋bⱼ
(18)        for (1 ≤ m ≤ j) do                             /* update μ matrix */
(19)          μim = μim − ⌈μij⌋μjm
(20)    if (Fr = true) then
(21)      APPROX_VECTOR(b'ᵢ, bᵢ)
(22)      Fr = false
(23)    if (Fc = true) then
(24)      i = max{i − 1, 2}
(25)      Fc = false
(26)    else
(27)      if (Bi < (y − μ²ᵢᵢ₋₁)Bᵢ₋₁) then                  /* check LLL condition */
(28)        SWAP(bᵢ₋₁, bᵢ)
(29)        if (i = 2) then
(30)          B₁ = ‖b'₁‖²
(31)        i = max{i − 1, 2}
(32)      else
(33)        i = i + 1
```

In order to make LLL reduction practical, the Schnorr-Euchner algorithm uses floating-point approximations of vectors and the basis (APPROX_BASIS and APPROX_VECTOR). For stability reasons, this requires employing suitable correction steps (see [19] for details). These corrections include either the computation of exact scalar products (see Line (7)) as part of the Gram-Schmidt orthogonalization or a step-back (see Line (25)) due to a large μ_{ij} used as part of the the size-reduction (see Line (17)). In order to prevent the corruption of the lattice, an exact data type is used to modify the actual lattice basis (see Line (19)). (In the algorithm, r denotes the bit precision of the data type used to approximate the lattice basis.)

3   LLL Reduction Using the Gram Matrix

The performance of the Schnorr-Euchner algorithm for a given approximation data type strongly depends on the number of correction steps (computation of exact scalar products and step-backs) needed in the reduction process. Experiments show [28,22] that it is the sheer number of exact scalar products along with their high computational costs that have a main impact on the reduction time. In turn, the number of step-backs is negligible compared to the number of exact scalar products and the total number of reduction steps. In order to speed up the reduction process, the goal is to minimize the number of correction steps, in particular the computation of exact scalar products.

In this context, in the NTL implementation [30] of the Schnorr-Euchner LLL algorithm (LLL_FP), the original measures for when to compute exact scalar products or perform step-backs have been modified. LLL_FP uses a lower bound for the computation of exact scalar products and the step-backs have been replaced by a heuristic that, if necessary, recomputes the Gram-Schmidt coefficients using an approximation data type with extended precision. In addition, the first condition for LLL-reduced bases (see Equation (1) in Definition 1) may be relaxed in order to avoid infinite loops. While these changes result in a major speedup of the reduction, they also have a negative effect on the stability of the reduction algorithm itself. For details see Figure 5 in Section 4.1.

Another approach to avoid the computation of exact scalar products is to perform the LLL reduction based on the Gram matrix instead of the original lattice basis [3,22]:

Definition 2. *For a lattice L with basis $B = (\underline{b}_1, \ldots, \underline{b}_k) \in \mathbb{R}^{n \times k}$, the corresponding Gram matrix G is defined as $G = B^T B$.*

Obviously, the Gram matrix inherently provides the necessary scalar products for the reduction process. Recently, Nguyen and Stehlé used the same approach for their L^2 algorithm [14,23]. In addition to using the Gram matrix, Nguyen and Stehlé also used ideas introduced in the NTL code for their size-reduction [14,15]. While they can prove that their algorithms yields an (δ, η)-LLL-reduced basis with $\eta > 0.5$, it lacks the stronger size-reduction criterion thus yielding a lesser reduced basis than the original LLL algorithm (see Definition 1) which, in contrast, yields an $(\delta, 0.5)$-LLL-reduced basis with $0.5 \leq \delta < 1$.

Our new algorithm—designed to address the challenges associated with exact scalar products—is also based on the LLL for Gram matrices [3,22] and adapts the computation of the Gram matrix and the LLL condition check of the L^2 algorithm introduced in [14] (see Line (28) of Algorithm 2). In contrast to the L^2 algorithm, we keep the stronger LLL condition and the second type of correction step of the original Schnorr-Euchner algorithm (see Line (17) of Algorithm 1). The challenge with using the Gram matrix instead of reducing the original basis lies in the fact that most applications of lattice basis reduction require a reduced lattice basis and not just a reduced Gram matrix. It therefore is necessary to either apply all transformation to both the Gram matrix and the exact basis (while basing all necessary decisions solely on the Gram matrix) or alternatively collect all transformations in a transformation matrix which is then applied to the original basis at the end of the reduction process. Both approaches have drawbacks. In the first approach all transformations are performed twice (once on the Gram matrix and once on the original basis). In the second method the bit length of the entries of a transformation matrix increases and often surpasses the size of the entries of the lattice basis. Our algorithm therefore introduces a solution that achieves a major improvement by buffering transformations, thus allowing the use of a transformation matrix with machine-type integers only (see Section 3.1).

We now first introduce the basic outline of our new variant of the Schnorr-Euchner LLL using the Gram matrix representation. In particular, we detail the Gram matrix updates which are crucial for the algorithm. In Section 3.1 we will then introduce the optimizations that in practice allow for a vast improvement of the running time.

Algorithm 2: LLL_GRAM(B)

INPUT: Lattice basis $B = (\underline{b}_1, \ldots, \underline{b}_k) \in \mathbb{Z}^{n \times k}$
OUTPUT: LLL-reduced lattice basis B

```
(1)   COMPUTE_GRAM(A, B)
(2)   APPROX_BASIS_GRAM(A', A)
(3)   R₁₁ = A'₁₁, i = 2, Fc = false, Fr = false
(4)   while (i ≤ k) do
(5)      μii = 1, S₁ = Rii                        /* orthogonalization */
(6)      for (2 ≤ j < i) do
(7)         Rij = A'ji − Σ(j−1, m=1) Rim μim
(8)         μij = Rij / Rjj
(9)         Rii = Rii − Rij μij
(10)        Sj+1 = Rii
(11)     for (i > j ≥ 1) do                       /* size-reduction */
(12)        if (|μi,j| > 1/2) then
(13)           Fr = true
(14)           bi = bi − ⌈μij⌋ bj
(15)           REDUCE_GRAM(A, i, ⌈μij⌋, j)
(16)           if (|μij| > 2^(r/2)) then          /* correction step 2 */
(17)              Fc = true
(18)              for (1 ≤ m ≤ j) do             /* update μ matrix */
(19)                 μim = μim − ⌈μij⌋ μim
(20)        if (Fr = true) then
```

```
(21)        APPROX_VECTOR_GRAM(A', A, i)
(22)        F_r = false
(23)     if (F_c = true) then
(24)        i = max(i − 1, 2)
(25)        F_c = false
(26)     else
(27)        i' = i
(28)        while ((i > 1) ∧ (y · R_{(i−1)(i−1)} > S_{i−1})) do      /* check LLL condition */
(29)           b_i ↔ b_{i−1}
(30)           SWAP_GRAM(A, i − 1, i)
(31)           SWAP_GRAM(A', i − 1, i)
(32)           i = i − 1
(33)        if (i ≠ i') then
(34)           if (i = 1) then
(35)              R_11 = A'_11
(36)              i = 2
(37)        else
(38)           i = i + 1
```

Unlike the Gram version of the LLL algorithm introduced in [3,22] we only use the upper triangular (including the diagonal) of the Gram matrix. This allows us to take advantage of the symmetric properties of the Gram matrix in order to improve the running time of the reduction algorithm. Consequently, we define the subroutines APPROX_BASIS_GRAM and APPROX_VECTOR_GRAM as follows:

Algorithm 3:
APPROX_BASIS_GRAM(A',A)

INPUT: Gram matrix A
OUTPUT: Approximate Gram matrix A'

(1) **for** $(1 \leq i \leq k)$ **do**
(2)   **for** $(i \leq j \leq k)$ **do**
(3) $A'_{i,j} = $ APPROX_VALUE$(A_{i,j})$

Algorithm 4:
APPROX_VECTOR_GRAM(A',A, l)

INPUT: Gram matrix A, vector index l
OUTPUT: Approximate Gram matrix A'

(1) **for** $(1 \leq i < l)$ **do**
(2) $A'_{i,l} = $ APPROX_VALUE$(A_{i,l})$
(3) **for** $(l \leq i \leq k)$ **do**
(4) $A'_{l,i} = $ APPROX_VALUE$(A_{l,i})$

The size-reduction of the LLL_GRAM described in [22] is slightly modified to work with the upper triangular Gram matrix. The new size-reduction for LLL_GRAM (see Algorithm 5) is only slightly more expensive than the equivalent step in the original Schnorr-Euchner algorithm.

Algorithm 5:
REDUCE_GRAM$(A, l, \lceil \mu_{ij} \rceil, j)$

INPUT: Gram matrix A, indices $l, j, \lceil \mu_{ij} \rceil$
OUTPUT: Gram matrix A

(1) $T = A_{l,l} - 2 \cdot \lceil \mu_{ij} \rceil \cdot A_{j,l} - \lceil \mu_{ij} \rceil^2 \cdot A_{j,j}$
(2) **for** $(1 \leq m < j)$ **do**
(3) $A_{m,l} = A_{m,l} - \lceil \mu_{ij} \rceil \cdot A_{m,j}$
(4) **for** $(j \leq m < l)$ **do**
(5) $A_{m,l} = A_{m,l} - \lceil \mu_{ij} \rceil \cdot A_{j,m}$
(6) **for** $(l + 1 \leq m < k)$ **do**
(7) $A_{l,m} = A_{l,m} - \lceil \mu_{ij} \rceil \cdot A_{j,m}$
(8) $A_{l,l} = T$

Algorithm 6:
SWAP_GRAM(A,i,j)

INPUT: Gram matrix A, indices $i, j, i < j$
OUTPUT: Gram matrix A

(1) **for** $(1 \leq m < j)$ **do**
(2) $A_{m,i} \leftrightarrow A_{m,j}$
(3) **for** $(j \leq m < j)$ **do**
(4) $A_{m,i} \leftrightarrow A_{j,m}$
(5) **for** $(j \leq m < i)$ **do**
(6) $A_{i,m} \leftrightarrow A_{j,m}$
(7) $A_{i,i} \leftrightarrow A_{i,j}$

Swapping basis vectors in the Gram matrix representation is, in practice, more expensive than in the original Schnorr-Euchner algorithm (see Algorithm 6). This is due to the fact that we now have to swap n elements (dimension of the lattice basis vectors) for the Gram matrix representation, while for the original Schnorr-Euchner algorithm we only have to swap the two pointers to the respective basis vectors.

3.1   Optimizations

In this section we introduce techniques to optimize Algorithm 2. We concentrate on the operations with the exact data type (usually long integer arithmetic like GMP [7,26]) and on the overhead for updating both the Gram matrix and the lattice basis. The goal is to reduce the number of expensive operations, such as multiplications or operations involving the long integer arithmetic. To accomplish this goal, we either use the far more efficient machine-type integer operations (assuming the respective values fit within the limits of machine-type integers) or we make use of the specialized and more efficient functions for combined operations like `mpz_addmul` instead of $a = a + b \cdot c$ [7].

Buffered Matrix Transformations. The basic idea of this new technique is to reduce the overhead due to the amount of long integer operations by using machine-type integers to buffer the lattice basis transformations until the limit of the machine-type integer (typically 32 or 64 bit) is reached. The buffered transformations are then applied to the lattice basis at once and the buffer is flushed. This allows us to considerably reduce the number of long integer operations and instead replace them by far more efficient machine-type integer operations.

Implementing buffered matrix transformations requires replacing the update of the lattice basis (see Line (16) in Algorithm 2) with a new subroutine called BUFFERED_TRANSFORM as well as adding a number of initializations and update steps. In the following let m be the bit size of the machine-type integers. $T = (\underline{t}_1, \ldots, \underline{t}_n)$ is used to buffer the matrix transformations, $Tmax_i$ for $1 \leq i \leq n$ contains an estimate for the maximum value in $\underline{t}_i$ and is used to check for possible overflows. pos_{min} and pos_{max} are used to indicate for which vectors the transformations have occurred and consequently allow us to limit the matrix multiplication to these vectors when flushing the transformation buffer.

In addition, the following modifications to Algorithm 2 have to be made. Before the main while-loop in Line (5) of Algorithm 2 we have to initialize $T = I_n$, $Tmax = (1, \ldots, 1)^T$, $pos_{min} = k$ and $pos_{max} = 1$. In Lines (30) - (34) we have to add the swap operations $Tmax_i \leftrightarrow Tmax_{i-1}$ and $\underline{t}_i \leftrightarrow \underline{t}_{i-1}$.

Algorithm 7: BUFFERED_TRANSFORM($B, i, \lceil \mu_{ij} \rfloor, j$)

INPUT: Lattice Basis $B = (\underline{b}_1, \ldots, \underline{b}_k) \in \mathbb{Z}^{n \times k}$, indices $i, j, \lceil \mu_{ij} \rfloor$
OUTPUT: Lattice Basis B

(1)   **if** $((Tmax_i + |\lceil \mu_{ij} \rfloor| \cdot Tmax_j) > 2^{m-1} - 1)$ **then** /* check for possible overflow */
(2)   **for** $(pos_{min} \leq x \leq pos_{max})$ **do**   /* perform $B' = B \cdot T$ */
(3)   **for** $(1 \leq z \leq n)$ **do**
(4) $B'_{xz} = 0$

$$
\begin{aligned}
&(5) \qquad \textbf{for } (pos_{min} \leq y \leq pos_{max}) \textbf{ do} \\
&(6) \qquad\quad \textbf{for } (1 \leq z \leq n) \textbf{ do} \\
&(7) \qquad\qquad B'_{xz} = B'_{xz} + T_{xy} \cdot B_{yz} \\
&(8) \quad B \leftrightarrow B' \\
&(9) \quad T = I_n \\
&(10) \quad Tmax = (1, \ldots, 1)^T \\
&(11) \quad pos_{max} = i \\
&(12) \quad pos_{min} = j \\
&(13) \quad \textbf{if } (|\lceil \mu_{ij} \rfloor| > 2^{m-1} - 1) \textbf{ then} \\
&(14) \qquad \underline{b}_i = \underline{b}_i - \lceil \mu_{ij} \rfloor \cdot \underline{b}_j \\
&(15) \qquad \textbf{return} \\
&(16) \quad \underline{t}_i = \underline{t}_i - \lceil \mu_{ij} \rfloor \cdot \underline{t}_j \\
&(17) \quad Tmax_i = Tmax_i + |\lceil \mu_{ij} \rfloor| \cdot Tmax_j \\
&(18) \quad \textbf{if } (pos_{max} < i) \textbf{ then} \\
&(19) \qquad pos_{max} = i \\
&(20) \quad \textbf{if } (pos_{min} > j) \textbf{ then} \\
&(21) \qquad pos_{min} = j
\end{aligned}
$$

Comments:
- (9) /* reset transformation buffer */
- (14) /* long integer computation */
- (16) /* machine integer computation */
- (18) /* update pos_{max} */
- (20) /* update pos_{min} */

The advantage of writing the partial matrix multiplication as shown above is that for the loop in Lines (6) - (7) the factor T_{xy} is constant in each iteration of the inner loop. This allows us to use additional optimizations which we present in the next section.

Further Optimizations. We can split the additional optimizations into two categories. The first is to avoid unnecessary operations like a multiplication with 1 or addition with 0 within a loop. This kind of optimization has also been used in Victor Shoup's NTL code [30]. The second kind is to take advantage of features of modern CPUs which include the support of certain multimedia streaming extensions [8,25]. These can efficiently be used to speed up some of the vector operations, like Line (19) in Algorithm 1.

As an example for the first category we show how to avoid unnecessary multiplications in Algorithm 7, Lines (6) - (7). (Algorithm 5 can be modified accordingly.) We can rewrite the loop as follows:

$$
\begin{aligned}
&(7) \quad \textbf{if } (T_{xy} \neq 0) \textbf{ then} \\
&(8) \qquad \textbf{if } (T_{xy} = 1) \textbf{ then} \\
&(9) \qquad\quad \textbf{for } (1 \leq z \leq n) \textbf{ do} \\
&(10) \qquad\qquad B'_{xz} = B'_{xz} + B_{yz} \\
&(11) \qquad \textbf{else} \\
&(12) \qquad\quad \textbf{if } (T_{xy} = -1) \textbf{ then} \\
&(13) \qquad\qquad \textbf{for } (1 \leq z \leq n) \textbf{ do} \\
&(14) \qquad\qquad\quad B'_{xz} = B'_{xz} - B_{yz} \\
&(15) \qquad\quad \textbf{else} \\
&(16) \qquad\qquad \textbf{for } (1 \leq z \leq n) \textbf{ do} \\
&(17) \qquad\qquad\quad B'_{xz} = B'_{xz} + T_{xy} \cdot B_{yz}
\end{aligned}
$$

This technique is efficient only if T_{xy} stays constant throughout the loop and $T_{xy} = 0$, $T_{xy} = 1$ or $T_{xj} = -1$ for a sufficient number of cases. Both conditions are dependent on the context in which they are used. In case of the buffered transformations, the majority of matrix entries is expected to be 0. For a sufficiently large n (dimension of the lattice basis vector) this optimization has the potential to reduce the running time even for the machine-type integers, e.g., Line (16) in Algorithm 7. In case the mantissa of the data type used for the approximation of the lattice basis fits into a machine-type integer then one can also avoid an expensive arbitrary long integer multiplication by splitting the values of the data type used for the approximation into sign, mantissa, and exponent. In case of a large μ_{ij} in Line (16) of Algorithm 2 we can then replace the expensive multiplication of two long integer values with a cheaper multiplication of a machine-type integer and a long integer value followed by a bit shift.

In order to allow for the second type of additional optimizations one can either use a compiler like those from Intel or Sun which already provide built-in support for auto-vectorization or, like in the case of the current version of GCC [20,24], one needs to assist the compiler in order for it to be able to take advantage of multimedia extensions. Vector operations on machine data types with limited dependencies, for example Line (16) in Algorithm 7, are ideal candidates for the use of multimedia streaming extensions. Using these streaming extensions for loops where values of the current loop iteration are dependent on previous iterations is far more difficult. For example, Line (16) in Algorithm 7 can be rewritten as follows:

(16)   **for** $(1 \leq l \leq n; l + = 4)$ **do**
(17) $T_{i,l} = T_{i,l} - \lceil \mu_{ij} \rfloor \cdot T_{j,l}$
(18) $T_{i,l+1} = T_{i,l+1} - \lceil \mu_{ij} \rfloor \cdot T_{j,l+1}$
(19) $T_{i,l+2} = T_{i,l+2} - \lceil \mu_{ij} \rfloor \cdot T_{j,l+2}$
(20) $T_{i,l+3} = T_{i,l+3} - \lceil \mu_{ij} \rfloor \cdot T_{j,l+3}$

The number of statements within the loop (here four statements) is dependent on the processor used and the available multimedia extension and has to be derived experimentally. The loop has to be adjusted accordingly in case the vector dimension n is not a multiple of the number of statements within the loop.

4   Experiments

The experiments in this paper focus on unimodular lattices. For one, these lattices are more difficult to reduce than knapsack or random lattices with the same dimension and length of base vectors [1]. Furthermore the result of the reduction can be easily verified since the reduced bases have a defect of 1. Unimodular lattice bases can be easily generated by multiplying together lower and upper triangular matrices with determinant 1. That is, entries in the diagonal are set to 1 while the lower (respectively upper part) of the matrix is selected at random. Using lower triangular matrices U_j, upper triangular matrices V_j with $1 \leq j \leq 2$ and permutation matrices P_i for $1 \leq i \leq 4$, we considered the following three variants of $n \times n$ dimensional unimodular lattice bases:

$$M_1 : B = (U_1 \cdot V_1)$$
$$M_2 : B = (U_1 P_1 \cdot V_1 P_2)$$
$$M_3 : B = (U_1 P_1 \cdot V_1 P_2) \cdot (V_2 P_3 \cdot U_2 P_4)$$

We generated 1000 unimodular bases for each type and dimension with $n = 5, 10, 15, \ldots, 100$. In the following, we compare our new Gram variant of the Schnorr-Euchner algorithm, called xLiDIA, with LLL_FP from NTL 5.4 [30] and the so-called **proved variant** in fpLLL 1.3 (with the default $\eta = 0.51$) [14,23]. Computer algebra systems like Magma [29] often use one or a combination of the aforementioned LLL algorithms [21][1]. We did not consider the LLL reduction algorithms introduced in [9,10,6] (implementation provided by [6]) which use Householder reflections for the orthogonalization due to stability problems when reducing unimodular lattice bases. The instability is caused by the fact that the first correction step in the Schnorr-Euchner algorithm (see Algorithm 1) cannot be adapted to Householder reflections or Givens rotations.

All experiments were performed on a Sun X4100 server with two dual core AMD Opteron processors (2.2GHz) and 4GB of main memory using Sun Solaris 10 OS. We compiled all programs with GCC 4.1.1 [24] using the same optimization flags. In the xLiDIA, NTL, and fpLLL implementation of the LLL algorithm we used GMP 4.2.1 [26] with the AMD64 patch [27] as long integer arithmetic and machine-type doubles for the approximation of the lattice basis. The following figures show the average reduction time (with reduction parameter $\delta = 0.99$) of the 1000 unimodular bases per dimension.

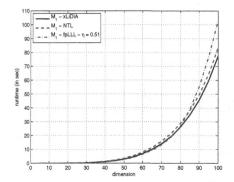
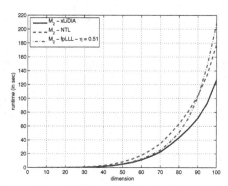

Fig. 1. Reduction times for basis type M_1 for NTL, fpLLL, and xLiDIA

Fig. 2. Reduction times for basis type M_2 for NTL, fpLLL, and xLiDIA

Figures 1 – 3 show the reduction times for unimodular bases of types M_1, M_2, and M_3. One can easily see that the reduction times for M_3-type bases are higher than those for M_2-type bases which are higher than those for M_1-type bases (for the same dimension n). That is M_3-type bases are more difficult to reduce than those bases of types M_1 or M_2. The relative improvement in reduction time of the xLiDIA implementation in comparison to NTL and fpLLL not only increases with the dimension of the lattice bases but also depends on the difficulty in reducing a lattice basis. For example, for bases of dimension 100, the reduction time with xLiDIA is 25% lower than that of fpLLL for M_1-type bases, 34% lower for M_2-type bases and roughly 45% lower for M_3-type bases.

While for smaller dimensions the reduction time of fpLLL is comparable to xLiDIA, for higher dimensions fpLLL catches up with the slower NTL. In fact,

---

[1] Magma is using $\eta = 0.501$ for the provable LLL variant (`Proof:=true`) which is based upon fpLLL.

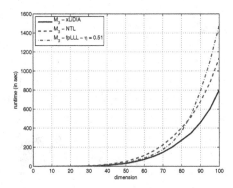

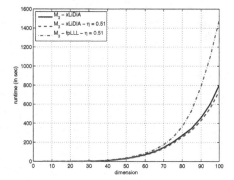

Fig. 3. Reduction times for basis type M_3 for NTL, fpLLL, and xLiDIA

Fig. 4. Reduction times for fpLLL and xLiDIA using different η

at dimension 85 for M_1, dimension 95 for M_2, and dimension 85 for M_3, the reduction time of fpLLL even exceeds that of NTL. This behavior of fpLLL is due to the overhead caused by updating two matrices (Gram matrix and lattice basis) for each transformation in the reduction process. The newly-introduced concept of buffered transformations as part of xLiDIA prevents this kind of behavior.

It is important to recall that fpLLL uses a relaxed reduction condition with $\eta = 0.51$ instead of $\eta = 0.5$ as in the original LLL algorithm and the xLiDIA implementation used for Figures 1 – 3. In order to demonstrate the impact of the relaxed reduction condition on the reduction time, we compare a modified implementation of xLiDIA with $\eta = 0.51$ to fpLLL with $\eta = 0.51$ and the original xLiDIA with $\eta = 0.5$. Figure 4 clearly shows that a relaxed reduction condition, i.e., a larger η results in an additional decrease of the reduction time. Furthermore, Figure 4 demonstrates that under the same reduction conditions (i.e., when the relaxed reduction condition is used for both fpLLL and xLiDIA) our newly-introduced variant xLiDIA outperforms fpLLL even further.

4.1   Stability

Aside from allowing for the analysis of the different algorithms based on their reduction times, our experiments also show the effectiveness of the various heuristics. In particular, it can be seen that the heuristics used in the NTL implementation [30] of the LLL reduction algorithm do not work for all types of bases. One can generally identify two serious and one minor problem. The serious problems are the reduction process running into an infinite loop or not providing a correctly reduced lattice basis. The minor problem identified is that of using a relaxed reduction condition without providing any feedback of such upon completion of the reduction.

Figure 5 shows the failure rates of the NTL implementation for the unimodular bases of types M_2 and M_3. (For lattice bases of type M1 the NTL implementation did not exhibit any failures.) **NTL-infinite loop** accounts for those

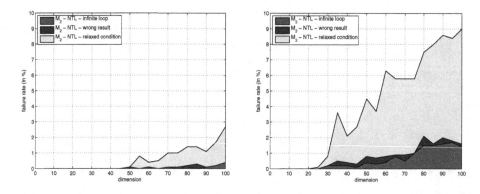

Fig. 5. NTL failure rates

cases in which the reduction process did not yield a reduction result within one hour. In fact, in these cases NTL even issued a warning indicating that the algorithm might have run into an infinite loop. NTL-wrong results accounts for those cases in which the reduction algorithm did not compute a permutation of the unit vectors $\pm e_i$ for $1 \leq i \leq n$ as the reduced lattice basis. (The reduction times for bases resulting in serious failure were not included in the timings for Figures 1 – 3. In order to avoid infinite loops, NTL employs the heuristic of relaxing the reduction condition 1 in Definition 1 for LLL reduced bases and NTL-relaxed condition accounts for those cases where this heuristic was used. Figure 5 clearly shows that the failure rates are increasing both with the dimension and the difficulty to reduce a lattice basis. Furthermore, it is obvious that the infinite loop prevention heuristic does not work effectively.

In contrast to NTL, our xLiDIA implementation and the proved variant of fpLLL did not exhibit any stability problems. However, testing the fast and heuristic variants (also included in the fpLLL package) led to an infinite loop on both algorithms even when reducing small unimodular lattice bases of dimension 10 with entries of maximum bit length of 100 bits.

5   Conclusion and Future Work

In this paper we introduced a new LLL variant using the Gram matrix representation which significantly outperforms the implementations of NTL and fpLLL. In particular, with our new variant we have shown that it is possible to considerably decrease the running time of LLL reduction without weakening its reduction conditions nor sacrificing the stability of the reduction process. It is important to note that the optimizations introduced in this paper could also be applied to the proved variant of fpLLL without affecting its respective correctness proof [14].

Future work includes further optimizing the reduction algorithms to take advantage of newly-introduced features in today's computers such as dual or quad core CPUs. We are also striving to find ways to extend the use of machine-type doubles for lattice bases with larger entries in higher dimensions.

Acknowledgments

We would like to thank Jared Cordasco for his valuable comments that helped us improve this paper. This work was partially supported by the Sun Microsystems Academic Excellence Grant Program.

References

1. Backes, W., Wetzel, S.: Heuristics on Lattice Basis Reduction in Practice. ACM Journal on Experimental Algorithms 7 (2002)
2. Bleichenbacher, D., May, A.: New Attacks on RSA with Small Secret CRT-Exponents. In: Yung, M., Dodis, Y., Kiayias, A., Malkin, T.G. (eds.) PKC 2006. LNCS, vol. 3958, pp. 1–13. Springer, Heidelberg (2006)
3. Cohen, H.: A Course in Computational Algebraic Number Theory. In: Undergraduate Texts in Mathematics, Springer, Heidelberg (1993)
4. Coster, M., LaMacchia, B., Odlyzko, A., Schnorr, C.: An Improved Low-Density Subset Sum Algorithm. In: Davies, D.W. (ed.) EUROCRYPT 1991. LNCS, vol. 547, pp. 54–67. Springer, Heidelberg (1991)
5. Ernst, M., Jochemsz, E., May, A., de Weger, B.: Partial Key Exposure Attacks on RSA up to Full Size Exponents. In: Cramer, R.J.F. (ed.) EUROCRYPT 2005. LNCS, vol. 3494, pp. 371–384. Springer, Heidelberg (2005)
6. Filipovic, B.: Implementierung der Gitterbasenreduktion in Segmenten. Master's thesis, University of Frankfurt am Main (2002)
7. Granlund, T.: GNU MP: The GNU Multiple Precision Arithmetic Library. SWOX AB, 4.2.1 edition (2006)
8. Klimovitski, A.: Using SSE and SSE2: Misconceptions and Reality. Intel Developer UPDATE Magazine (March 2001),
   http://www.intel.com/technology/magazine/computing/sw03011.pdf
9. Koy, H., Schnorr, C.: Segment LLL-Reduction of Lattice Bases. In: Silverman, J.H. (ed.) CaLC 2001. LNCS, vol. 2146, pp. 67–80. Springer, Heidelberg (2001)
10. Koy, H., Schnorr, C.: Segment LLL-Reduction with Floating Point Orthogonalization. In: Silverman, J.H. (ed.) CaLC 2001. LNCS, vol. 2146, pp. 81–96. Springer, Heidelberg (2001)
11. Lenstra, A., Lenstra, H., Lovàsz, L.: Factoring Polynomials with Rational Coefficients. Math. Ann. 261, 515–534 (1982)
12. May, A.: Cryptanalysis of NTRU (preprint) (1999)
13. Nguyen, P., Stehlè, D.: Low-Dimensional Lattice Basis Reduction Revisited. In: Buell, D.A. (ed.) Algorithmic Number Theory. LNCS, vol. 3076, pp. 338–357. Springer, Heidelberg (2004)
14. Nguyen, P., Stehlè, D.: Floating-Point LLL Revisited. In: Cramer, R.J.F. (ed.) EUROCRYPT 2005. LNCS, vol. 3494, pp. 215–233. Springer, Heidelberg (2005)
15. Nguyen, P., Stehlè, D.: LLL on the Average. In: Hess, F., Pauli, S., Pohst, M. (eds.) Algorithmic Number Theory. LNCS, vol. 4076, pp. 238–256. Springer, Heidelberg (2006)

16. Nguyen, P.Q.: Cryptanalysis of the Goldreich-Goldwasser-Halevi Cryptosystem from Crypto '97. In: Wiener, M.J. (ed.) CRYPTO 1999. LNCS, vol. 1666, pp. 288–304. Springer, Heidelberg (1999)

17. Nguyen, P.Q., Stern, J.: Lattice Reduction in Cryptology: An Update. In: Bosma, W. (ed.) Algorithmic Number Theory. LNCS, vol. 1838, pp. 85–112. Springer, Heidelberg (2000)

18. Pohst, M.E., Zassenhaus, H.: Algorithmic Algebraic Number Theory. Cambridge University Press, Cambridge (1989)

19. Schnorr, C., Euchner, M.: Lattice Basis Reduction: Improved Practical Algorithms and Solving Subset Sum Problems. In: Budach, L. (ed.) FCT 1991. LNCS, vol. 529, pp. 68–85. Springer, Heidelberg (1991)

20. Stallman, R.M., Community, G.D.: GNU Compiler Collection. Free Software Foundation, Inc. (2005)

21. Stehlé, D.: The New LLL Routine in the Magma Computational Algebra System. Magma 2006 Conference (2006), http://magma.maths.usyd.edu.au/Magma2006/

22. Wetzel, S.: Lattice Basis Reduction Algorithms and their Applications. PhD thesis, Universität des Saarlandes (1998)

23. fpLLL - Homepage (Damien Stehlé) (July 2007), http://www.loria.fr/~stehle/

24. GCC - Homepage (July 2007), http://gcc.gnu.org

25. Auto-Vectorization in GCC (July 2007), http://gcc.gnu.org/projects/tree-ssa/vectorization.html

26. GMP - Homepage, http://gmplib.org/

27. AMD64 patch for GMP 4.2 (July 2007), http://www.loria.fr/~gaudry/mpn_AMD64/index.html

28. Lattice Basis Reduction Experiments (July 2007), http://www.cs.stevens.edu/~wbackes/lattice/

29. Magma - Homepage (July 2007), http://magma.maths.usyd.edu.au/

30. NTL - Homepage (July (2007), http://www.shoup.net/ntl/

On the Computation of A_∞-Maps[*]

Ainhoa Berciano[1], María José Jiménez[2], and Pedro Real[2]

[1] Dpto. Matemática Aplicada, Estadística e Investigación Operativa,
Universidad del País Vasco, Barrio Sarriena s/n, 48940 Leioa (Vizcaya), Spain
ainhoa.berciano@ehu.es
[2] Dpto. de Matemática Aplicada I, Universidad de Sevilla,
Avda. Reina Mercedes s/n, 41012 Sevilla, Spain
{majiro, real}@us.es

Abstract. Starting from a chain contraction (a special chain homotopy equivalence) connecting a differential graded algebra A with a differential graded module M, the so-called homological perturbation technique "tensor trick" [8] provides a family of maps, $\{m_i\}_{i \geq 1}$, describing an A_∞-algebra structure on M derived from the one of algebra on A. In this paper, taking advantage of some annihilation properties of the component morphisms of the chain contraction, we obtain a simplified version of the existing formulas of the mentioned A_∞-maps, reducing the computational cost of computing m_n from $O(n!^2)$ to $O(n!)$.

Keywords: A_∞-algebra, contraction, Basic Perturbation Lemma, transference, computation.

1 Introduction

At present, A_∞-structures (or strong homotopy structures) find natural applications not only in Algebra, Topology and Geometry but also in Mathematical Physics, related to topics such as string theory, homological mirror symmetry or superpotentials [14,17,18]. Nevertheless, there are few methods for computing explicit A_∞-structures, being the better known technique the *tensor trick* [8]. This tool is used in the context of Homological Perturbation Theory. Starting from a chain contraction c (a special chain homotopy equivalence, also called strong deformation retract) from a differential graded algebra A onto a differential graded module M, the tensor trick technique gives explicit formulas for computing a family of higher maps $\{m_i\}_{i \geq 1}$ that provides an A_∞-algebra structure on M (derived from the algebra structure on A). However, the associated computational costs are extremely high (see [11,12,1]). In this paper, we are concerned about finding a more cost-effective formulation of the family of maps transferred to M. As it is shown in section 3, the use of annihilation properties of the component morphisms of the chain contraction allows to reformulate the A_∞–maps on M (which depend on the mentioned component morphisms). Afterwards, in

---

[*] Partially supported by the PAICYT research project FQM-296 and by a project of University of the Basque Country "EHU06/05".

section 4 we carry out a theoretical study of the time and space invested in computing m_n, presenting the computational savings obtained, in comparison with the original formulas defined by the Basic Perturbation Lemma.

The results can be extended to the case of A being an A_∞-algebra (then, another A_∞-algebra structure is also induced on M). We remark that such a transference can also be performed in the case of c being a general explicit chain homotopy equivalence.

Of course, all the results given in this paper can be easily translated into the context of coalgebras and A_∞–coalgebras.

2   Notations and Preliminaries

We briefly recall here some basic definitions in Homological Algebra as well as the notations used throughout the paper. See [3] or [16] for further explanations.

Take a commutative unital ring Λ. Let (M, d) be a DG-module, that is, a Λ– module graded on the non-negative integers ($M = \bigoplus_{n \geq 0} M_n$) and endowed with a differential d (of degree -1). An element $x \in M_n$ has degree n, what will be expressed by $|x| = n$. In the case that $M_0 = \Lambda$, M is called *connected* and if, besides, $M_1 = 0$, then it is called *simply connected*. Given a connected DG–module, M, the *reduced* module $\overline{M}$ is the one with $\overline{M}_n = M_n$ for $n > 1$ and $\overline{M}_0 = 0$.

We will denote the module $M \otimes \overset{n}{\cdots} \otimes M$ by $M^{\otimes n}$, with $M^{\otimes 0} = \Lambda$ and the morphism $f \otimes \overset{n}{\cdots} \otimes f : M^{\otimes n} \to N^{\otimes n}$ by $f^{\otimes n}$. We adhere to Koszul convention for signs. More concretely, given $f : M \to M'$, $h : M' \to M''$, $g : N \to N'$ and $k : N' \to N''$ DG–module morphisms, then

$$(h \otimes k)(f \otimes g) = (-1)^{|k||f|}(hf \otimes kg).$$

On the other hand, if $f : M^{\otimes i} \to M$ is a DG–module morphism and n is a non–negative integer, we will denote by $f^{[n]} : M^{\otimes n} \to M^{\otimes n-i+1}$ the morphism

$$f^{[n]} = \sum_{j=0}^{n-i} 1^{\otimes j} \otimes f \otimes 1^{\otimes n-i-j}$$

and the morphism $f^{[]} : \bigoplus_{j \geq i} M^{\otimes j} \to \bigoplus_{k \geq 1} M^{\otimes k}$ will be the one such that $f^{[]}|_{M^{\otimes n}} = f^{[n]}$.

We will denote by $\uparrow$ and $\downarrow$ the *suspension* and *desuspension* operators, which shift the degree by $+1$ and -1, respectively. A given morphism of graded modules of degree k, $f : M \to N$, induces another one between the suspended modules $sf : sM \to sN$, given by $sf = (-1)^k \uparrow f \downarrow$.

Given a DG-module (M, d), the *tensor module* of M, $T(M)$, is the DG–module

$$T(M) = \bigoplus_{n \geq 0} T^n(M) = \bigoplus_{n \geq 0} M^{\otimes n}$$

whose differential structure is provided by $d_M^{[]}$. Every morphism of DG-modules $f : M \to N$ induces another one $T(f) : T(M) \to T(N)$, such that $T(f)|_{M^{\otimes n}} = f^{\otimes n}$.

A *DG–algebra*, (A, d_A, μ_A), is a DG–module endowed with an associative product, μ_A, compatible with the differential d_A and which has a unit $\eta_A : \Lambda \to A$, that is, $\mu_A(\eta_A \otimes 1) = \mu_A(1 \otimes \eta_A) = 1$. If there is no confusion, subscripts will be omitted. A *DG–coalgebra* (C, d_C, Δ_C) is a DG–module provided with a compatible coproduct and counit $\xi_C : C \to \Lambda$ (so, $(\xi_C \otimes 1)\Delta_C = (1 \otimes \xi_C)\Delta_C = 1$).

In the case of the tensor module $T(M)$, a product, μ, and a coproduct, Δ, can be naturally defined on an element $a_1 \otimes \cdots \otimes a_n \in T^n(M)$, as follows:

$$\mu((a_1 \otimes \cdots \otimes a_n) \otimes (a_{n+1} \otimes \cdots \otimes a_{n+p})) = a_1 \otimes \cdots \otimes a_{n+p};$$
$$\Delta(a_1 \otimes \cdots \otimes a_n) = \sum_{i=0}^{n}(a_1 \otimes \cdots \otimes a_i) \otimes (a_{i+1} \otimes \cdots \otimes a_n).$$

Therefore, $T(M)$ acquires both structures of DG–algebra (denoted by $T^a(M)$) and DG–coalgebra $(T^c(M))$, though they are not compatible to each other (that is, $(T(M), \mu, \Delta)$ is not a Hopf algebra).

We recall here two equivalent definitions of A_∞–algebra (resp. A_∞–coalgebra) [13,19].

– An A_∞-algebra (respectively, A_∞-coalgebra), is a DG-module (M, m_1) (resp. (M, Δ_1)) endowed with a family of maps

$$m_i : M^{\otimes i} \to M \text{ (resp., } \Delta_i : M \to M^{\otimes i})$$

of degree $i - 2$ such that, for $n \geq 1$,

$$\sum_{n=1}^{i}\sum_{k=0}^{i-n}(-1)^{n+k+nk}m_{i-n+1}(1^{\otimes k} \otimes m_n \otimes 1^{\otimes i-n-k}) = 0, \qquad (1)$$

$$\text{(resp., } \sum_{n=1}^{i}\sum_{k=0}^{i-n}(-1)^{n+k+nk}(1^{\otimes i-n-k} \otimes \Delta_n \otimes 1^{\otimes k})\Delta_{i-n+1} = 0). \qquad (2)$$

– An A_∞-algebra (resp., A_∞–coalgebra) is a graded module M endowed with a morphism of modules $m : T(sM) \to M$ (resp., $\Delta : M \to T(s^{-1}M)$) such that the morphism $d = -(\uparrow mT(\downarrow))^{[]}$ (resp., $d = -(T(\downarrow)\Delta \uparrow)^{[]}$) makes $T^c(sM)$ (resp., $T^a(s^{-1}M)$) to be a DGA–coalgebra (resp., DGA-algebra).

The *reduced bar construction* of a connected DG–algebra A, $\bar{B}(A)$, is a DG–coalgebra whose module structure is given by

$$T(s\bar{A}) = \bigoplus_{n \geq 0}(s\bar{A}\otimes \overset{n\ times}{\cdots} \otimes s\bar{A}).$$

The total differential d_B is given by the sum of the tensor differential, d_t (which is the natural one on the tensor product) and the *simplicial differential*, d_s (that depends on the product on A):

$$d_t = -\sum_{i=0}^{n-1} 1^{\otimes i}\otimes \uparrow d_A \downarrow \otimes 1^{\otimes n-i-1}; \qquad d_s = \sum_{i=0}^{n-2} 1^{\otimes i}\otimes \uparrow \mu_A \downarrow^{\otimes 2} \otimes 1^{\otimes n-i-2}.$$

The coproduct $\Delta_B : \bar{B}(A) \to \bar{B}(A) \otimes \bar{B}(A)$ is the natural one on the tensor module.

In the context of homological perturbation theory, the main input data are *contractions* [4,9,15,7,10]: a contraction $c : \{N, M, f, g, \phi\}$ from a DG-module N to a DG-module M, consists in a particular homotopy equivalence determined by two DG-module morphisms, $f : N_\star \to M_\star$ and $g : M_\star \to N_\star$ and a homotopy operator $\phi : N_\star \to N_{\star+1}$ such that $fg = 1_M$, and $\phi d_N + d_N \phi + gf = 1_N$. Moreover, these data are also required to satisfy the anihilation properties:

$$f\phi = 0, \qquad \phi g = 0, \qquad \phi\phi = 0.$$

Given a DG–module contraction $c : \{N, M, f, g, \phi\}$, one can establish the following ones [7,8]:

- The *suspension contraction* of c, sc, which consists of the suspended DG–modules and the induced morphisms:

$$sc : \{sN, sM, sf, sg, s\phi\},$$

  being $sf = \uparrow f \downarrow$, $sg = \uparrow g_1 \downarrow$ and $s\phi = - \uparrow \phi \downarrow$, which are briefly expressed by f, g and $-\phi$.
- The *tensor module contraction*, $T(c)$, between the tensor modules of M and N:

$$T(c) : \{T(N), T(M), T(f), T(g), T(\phi)\},$$

  where

$$T(\phi)|_{T^n(N)} = \phi^{[\otimes n]} = \sum_{i=0}^{n-1} 1^{\otimes i} \otimes \phi \otimes (g\,f)^{\otimes n-i-1}.$$

A morphism of graded modules $f : N \to N$ is called *pointwise nilpotent* whenever for all $x \in N$, $x \neq 0$, there exists a positive integer n such that $f^n(x) = 0$. A *perturbation of a DG-module* N consists in a morphism of graded modules $\delta : N \to N$ of degree -1, such that $(d_N + \delta)^2 = 0$. A *perturbation datum* of the contraction $c : \{N, M, f, g, \phi\}$ is a perturbation δ of the DG-module N satisfying that the composition $\phi\delta$ is pointwise nilpotent.

The main tool when dealing with contractions is the **Basic Perturbation Lemma** [2,5,15], which is an algorithm whose input is a contraction of DG–modules $c : \{N, M, f, g, \phi\}$ and a perturbation datum δ of c and whose output is a new contraction $c_\delta : \{(N, d_N + \delta), (M, d_M + d_\delta), f_\delta, g_\delta, \phi_\delta\}$ defined by the formulas

$$d_\delta = f\,\delta\,\Sigma_c^\delta\,g; \qquad f_\delta = f\,(1 - \delta\,\Sigma_c^\delta\,\phi); \qquad g_\delta = \Sigma_c^\delta\,g; \qquad \phi_\delta = \Sigma_c^\delta\,\phi;$$

where $\Sigma_c^\delta = \sum_{i\geq 0}(-1)^i\,(\phi\delta)^i$.

The pointwise nilpotency of the composition $\phi\delta$ guarantees that the sums are finite for each particular element.

3   Transferring A_∞–Algebras Via Homological Perturbation Theory

A_∞–algebras were first introduced by Stasheff in [20]. They are, roughly speaking, algebras which are associative "up to homotopy" (also called strongly homotopy associative algebras).

In the papers of Gugenheim, Stasheff and Lambe [6,9,8], they describe a technique called *tensor trick* by which, starting from a contraction between a DG–algebra A and a DG–module M, an A_∞–algebra structure is induced on M. This transference also exists in the case that A is an A_∞–algebra. Moreover, in the case that a general homotopy equivalence is established between A and M, it is also possible to derive a formulation for an A_∞–algebra structure on M. We will mainly focus our efforts on obtaining computational improvements in the first case.

3.1   Transference Via Contractions

Let us consider the contraction

$$c : \{A, M, f, g, \phi\},$$

where A is a connected DG–algebra and M a DG–module. The first step consists in tensoring, in order to obtain the underlying graded module of the bar construction of A,

$$T(sc) : \{T^c(s\bar{A}), T^c(s\bar{M}), Tf, Tg, T(-\phi)\};$$

and then, considering the simplicial differential, d_s, which is a perturbation datum for this contraction, and using the Basic Perturbation Lemma, a new contraction is obtained,

$$\{\bar{B}(A), (T^c(s\bar{M}), \tilde{d}), \tilde{f}, \tilde{g}, \tilde{\phi}\},$$

where $(T^c(s\bar{M}), \tilde{d})$ is called the *tilde bar construction* of M [20], denoted by $\widetilde{B}(M)$. Then, the perturbed differential $\tilde{d}$ induces a family of maps $m_n : M^{\otimes n} \to M$ of degree $n - 2$ that provides an A_∞–algebra structure on M.

The transference of an A_∞–algebra structure was also studied by Kadeishvili in [13] for the case $M = H(A)$. Using this technique, in the following theorem, an expression of a family of A_∞–operations is given with regard to the component morphisms of the initial contraction. Although this formulation is implicitly derived from the mentioned papers [13] and [8], an explicit proof is given in [12].

Theorem 1. *[13,8] Let (A, d_A, μ) and (M, d_M) be a connected DG–algebra and a DG–module, respectively and $c : \{A, M, f, g, \phi\}$ a contraction between them. Then the DG–module M is provided with an A_∞–algebra structure given by the operations*

$$m_1 = -d_M$$

$$m_n = (-1)^{n+1} f \, \mu^{(1)} \, \phi^{[\otimes 2]} \, \mu^{(2)} \cdots \phi^{[\otimes n-1]} \, \mu^{(n-1)} \, g^{\otimes n}, \qquad n \geq 2 \qquad (3)$$

where

$$\mu^{(k)} = \sum_{i=0}^{k-1} (-1)^{i+1} 1^{\otimes i} \otimes \mu_A \otimes 1^{\otimes k-i-1}.$$

As far as the computation of these formulas is concerned, we can take advantage of the annihilation properties of f, g and ϕ to deduce a more economical formulation for m_n.

Theorem 2. *Any composition of the kind $\phi^{[\otimes s]} \mu^{(s)}$ ($s = 2, \ldots, n-1$) in the formula (3), which is given by*

$$\left(\sum_{j=0}^{s-1} 1^{\otimes j} \otimes \phi \otimes (g\,f)^{\otimes s-j-1} \right) \circ \left(\sum_{i=0}^{s-1} (-1)^{i+1} 1^{\otimes i} \otimes \mu_A \otimes 1^{\otimes s-i-1} \right),$$

can be reduced to the following sum

$$\sum_{i=0}^{s-1} (-1)^{i+1} 1^{\otimes i} \otimes \phi \mu_A \otimes 1^{\otimes s-i-1}. \qquad (4)$$

Moreover, given a composition of the kind

$$\left(\phi^{[\otimes s-1]} \mu^{(s-1)} \right) \circ \left(\phi^{[\otimes s]} \mu^{(s)} \right) \qquad s = 3, \ldots, n-2,$$

for every index i in the sum (4) of $\phi^{[\otimes s]} \mu^{(s)}$, the formula of $\phi^{[\otimes s-1]} \mu^{(s-1)}$ in such a composition can be reduced to

$$\sum_{j=i-1,\, j \geq 0}^{s-2} (-1)^{j+1} 1^{\otimes j} \otimes \phi \mu_A \otimes 1^{\otimes s-j-2}. \qquad (5)$$

In other words, the whole composition $(\phi^{[\otimes 2]} \mu^{(2)}) \circ \cdots \circ (\phi^{[\otimes n-1]} \mu^{(n-1)})$ in the formula of m_n can be expressed by

$$\sum_{i_{n-1}=0}^{n-2} \left(\sum_{i_{n-2}=i_{n-1}-1}^{n-3} \left(\cdots \left(\sum_{i_2=i_3-1}^{1} (\phi\mu)^{(2,i_2)} \right) \cdots \right) (\phi\mu)^{(n-2,i_{n-2})} \right) (\phi\mu)^{(n-1,i_{n-1})},$$

where $(\phi\mu)^{(k,j)} = (-1)^{j+1} 1^{\otimes j} \otimes \phi \mu_A \otimes 1^{\otimes k-j-1}$ and each addend exists whenever the corresponding index $i_k \geq 0$.

Proof. Let us prove the formula 4 of $\phi^{[\otimes s]} \mu^{(s)}$ for any $s = n-1, n-2, \ldots, 2$, by induction over the number $k = n - s$ of factors of the type $\phi^{[\otimes *]} \mu^{(*)}$ that are composed, following the scheme

$$m_n = (-1)^{n+1} f\, \mu^{(1)}\, \phi^{[\otimes 2]}\, \mu^{(2)} \cdots \underbrace{\phi^{[\otimes n-2]}\, \mu^{(n-2)}\, \underbrace{\underbrace{\phi^{[\otimes n-1]}\, \mu^{(n-1)}\, g^{\otimes n}}_{k=1}}_{k=2}}_{k=n-2} \quad (6)$$

At the same time, we will prove the major reduction of terms given by (5) for $s = n-2, \ldots, 2$.

- $\boxed{k=1}$ The composition of morphisms $\phi^{[\otimes n-1]}\, \mu^{(n-1)}\, g^{\otimes n}$ can be written as

$$\left(\sum_{j=0}^{n-2} 1^{\otimes j} \otimes \phi \otimes (gf)^{\otimes n-j-2} \right) \circ \left(\sum_{i=0}^{n-2} (-1)^{i+1} g^{\otimes i} \otimes \mu_A\, g^{\otimes 2} \otimes g^{\otimes n-i-2} \right).$$

Now, using the facts that $fg = 1$ and $\phi g = 0$, it is simple to see that the only non null elements are those where ϕ is applied over μ_A, so the original formula of $\phi^{[\otimes n-1]}\, \mu^{(n-1)}$ is simplified to

$$\sum_{i=0}^{n-2} (-1)^{i+1} 1^{\otimes i} \otimes \phi\mu_A \otimes 1^{\otimes n-i-2}.$$

- $\boxed{k=2}$ In this case, taking into account the formula obtained for $k=1$,

$$\phi^{[\otimes n-1]}\, \mu^{(n-1)}\, g^{\otimes n} = \sum_{i=0}^{n-2} (-1)^{i+1} g^{\otimes i} \otimes \phi\mu_A g^{\otimes 2} \otimes g^{\otimes n-i-2} \quad (7)$$

and that $\phi^{[\otimes n-2]}\mu^{(n-2)}$ is the composition

$$\left(\sum_{j=0}^{n-3} 1^{\otimes j} \otimes \phi \otimes (gf)^{\otimes n-j-3} \right) \circ \left(\sum_{i=0}^{n-3} (-1)^{i+1} 1^{\otimes i} \otimes \mu_A \otimes 1^{\otimes n-i-3} \right),$$

we can use the anihilation properties $\phi g = 0$ and $\phi^2 = 0$, to conclude that the factor ϕ in $\phi^{[\otimes n-2]}$ has to be applied over μ_A and hence,

$$\phi^{[\otimes n-2]}\, \mu^{(n-2)} = \sum_{j=0}^{n-3} (-1)^{j+1} 1^{\otimes j} \otimes \phi\mu_A \otimes (gf)^{\otimes n-j-3}. \quad (8)$$

Now, considering the composition of the sum (7) with (8), one can observe that, since $f\phi = 0$, for each index i in the sum (7), the only addends of (8) that have to be considered for the composition are those $j \geq i-1$. On the other hand, $fg = 1$ is also satisfied, so

$$\phi^{[\otimes n-2]}\, \mu^{(n-2)} = \sum_{j=i-1}^{n-3} (-1)^{j+1} 1^{\otimes j} \otimes \phi\mu_A \otimes 1^{\otimes n-j-3}.$$

$-$ $\boxed{k = m}$ Finally, let us assume that the proposition is true for $\phi^{[\otimes n-k]}\mu^{(n-k)}$ for all $k = 1, \ldots, m - 1$. Now, considering, on one hand, $\phi^{[\otimes n-m]}\mu^{(n-m)}$,

$$\left(\sum_{j=0}^{n-m-1} 1^{\otimes j} \otimes \phi \otimes (gf)^{\otimes n-j-m-1} \right) \left(\sum_{i=0}^{n-m-1} (-1)^{i+1} 1^{\otimes i} \otimes \mu_A \otimes 1^{\otimes n-i-m-1} \right)$$

and that, on the other hand, the composition of morphisms

$$\phi^{[\otimes n-m+1]}\mu^{(n-m+1)} \cdots \phi^{[\otimes n-1]}\mu^{(n-1)} g^{\otimes n}$$

by induction hypothesis, is a sum of elements that are tensor product of factors of the type ϕ(something) or g, using again the annihilation properties, it follows that

$$\phi^{[\otimes n-m]}\mu^{(n-m)} = \sum_{j=0}^{n-m-1} (-1)^{j+1} 1^{\otimes j} \otimes \phi\mu_A \otimes (gf)^{\otimes n-j-m-1}.$$

Since, by induction hypothesis,

$$\phi^{[\otimes n-m+1]}\mu^{(n-m+1)} = \sum_{i=0}^{n-m} (-1)^{i+1} 1^{\otimes i} \otimes \phi\mu_A \otimes 1^{\otimes n-m-i},$$

taking into account that $fg = 1$ and the fact that $f\phi = 0$, again we can reduce the number of terms of $\phi^{[\otimes n-m]}\mu^{(n-m)}$ to

$$\sum_{j=i-1}^{n-m-1} (-1)^{i+1} 1^{\otimes i} \otimes \phi\mu_A \otimes 1^{\otimes n-m-i-1},$$

where i is the index corresponding to the term of the preceding sum that is being composed with $\phi^{[\otimes n-m]}\mu^{(n-m)}$.

We can generalize the results showed above to the case that the "big" DG-module of a given contraction is an A_∞-algebra. The stability of the A_∞-structures with respect to the contractions follows from the paper [8]. In fact, it is possible to extract the next theorem as an implicit consequence of the results there.

Theorem 3. *Given* $c : \{A, M, f, g, \phi\}$ *a contraction, where* $(A, m_1, m_2, \ldots)$ *is a connected* A_∞-*algebra and* M *is a DG-module, then* M *inherits an* A_∞-*algebra structure.*

Proof. The proof follows the same scheme as in theorem 1 (and for that reason, we will only sketch it slightly) , making use of the tensor trick and the Basic Perturbation Lemma, with the difference that, now, the perturbation datum for the contraction

$$T(sc) : \{T^c(s\bar{A}),\, T^c(s\bar{M}),\, T(f),\, T(g)\,T(-\phi)\}$$

is the one induced by the A_∞–maps

$$d_m|_{(s\bar{A})^{\otimes n}} = -\sum_{k=2}^{n}\sum_{i=0}^{n-k} 1^{\otimes i} \otimes \uparrow m_k \downarrow^{\otimes k} \otimes 1^{\otimes n-k-i}.$$

Since the family of maps $\{m_i\}_{i\geq 1}$ defines an A_∞-algebra structure on A, $d_{\tilde{B}} = d_t + d_m$ is a differential on $T^c(s\bar{A})$ (in fact, $(T^c(s\bar{A}), d_{\tilde{B}})$ is the tilde bar construction of A). On the other hand, the pointwise nilpotency of $T(-\phi)\,d_m$ follows because d_m reduces the simplicial dimension, while $T(-\phi)$ keeps it the same.

Thanks to the Basic Perturbation Lemma, a new differential is obtained on $T^c(s\bar{M})$, $\tilde{d}$, given by the formula:

$$\tilde{d} = d_t + T(f)\,d_m \sum_{i\geq 0}(-1)^i (T(-\phi)\,d_m)^i\,T(g).$$

This way, $\tilde{d}$ induces a family of maps $\{m_i^M\}_{i\geq 1}$ on M, where m_n^M, up to sign, can be expressed by

$$f\,m_n\,g^{\otimes n} + \sum_{l=1}^{n-2}\ \sum_{2\leq k_1 < \ldots < k_l \leq n-1} \pm f\,m_{k_1}\,(\phi^{[\otimes k_1]}\,m^{(k_1)}_{k_2-k_1+1})\cdots(\phi^{[\otimes k_l]}\,m^{(k_l)}_{n-k_l+1})\,g^{\otimes n}$$

where $m^{(k)}_{n-k+1} : A^{\otimes n} \to A^{\otimes k}$ is given by

$$m^{(k)}_{n-k+1} = \sum_{i=0}^{n-k+1} 1^{\otimes i} \otimes m_{n-k+1} \otimes 1^{\otimes k-i-1}.$$

Notice that, since m_i is a map of degree $i-2$, m_n^M has degree $n-2$.

If we examine the formula above in low dimensions, we obtain, up to sign:

$$m_2^M = \pm f\,m_2\,g^{\otimes 2};$$
$$m_3^M = \pm f\,m_3\,g^{\otimes 3} \pm f\,m_2\,\phi^{[\otimes 2]}\,m_2^{(2)}\,g^{\otimes 3};$$
$$m_4^M = \pm f\,m_4\,g^{\otimes 4} \pm f\,m_2\,\phi^{[\otimes 2]}\,m_3^{(2)}\,g^{\otimes 4} \pm f\,m_3\,\phi^{[\otimes 3]}\,m_2^{(3)}\,g^{\otimes 4}$$
$$\pm f\,m_2\,\phi^{[\otimes 2]}\,m_2^{(2)}\,\phi^{[\otimes 3]}\,m_2^{(3)}\,g^{\otimes 4}.$$

Notice that only the last addend of each map is the one induced in the case of A being an algebra, instead of the 2^{n-2} addends generated in these cases (the number of subsets of a set of $n-2$ elements). At each addend of each A_∞–map, one can obtain a reduction in number of terms, of the same nature than the one showed in theorem 2.

Theorem 4. *Any composition of the kind $\phi^{[\otimes s]} m_r^{(s)}$ in the formula of m_n^M, which is given by*

$$\left(\sum_{j=0}^{s-1} 1^{\otimes j} \otimes \phi \otimes (g\,f)^{\otimes s-j-1} \right) \circ \left(\sum_{i=0}^{r} 1^{\otimes i} \otimes m_r \otimes 1^{\otimes s-i-1} \right),$$

can be reduced to the following sum

$$\sum_{i=0}^{r} 1^{\otimes i} \otimes \phi\, m_r \otimes 1^{\otimes s-i-1}.$$

Proof. This proof is completely dual to the one of theorem 2, so it is left to the reader.

3.2 Transference Via Homotopy Equivalences

In [10], a general chain homotopy equivalence e between two DG-modules M and M' is considered as a pair of chain contractions $\{\hat{M}, M, f, g, \phi\}$ and $\{\hat{M}, M', f', g', \phi'\}$, where $\hat{M}$ is a "big" DG-module obtained from e. Our interest here is to compute the A_∞-algebra structure on M' derived from that of M. Having at hand the mentioned characterization of chain homotopy equivalence and the results of the previous subsection, our task is then reduced to determine the transferring of A_∞-structures via chain contractions in the sense from-small-to-big. In a more formal way, our main problem here is the transference of the A_∞-algebra structure from a "small" DG-module N to a "big" DG-module M via the chain contraction $c : \{M, N, f, g, \phi\}$. The following propositions are straightforward and, in particular, allow to design an algorithmic method for transferring A_∞-structures via chain homotopy equivalences:

Proposition 1. *Let $c : \{M, N, f, g, \phi\}$ be a chain contraction and let (N, μ) be a DG-algebra with product μ. Then, M has a structure of DG-algebra, provided by the product $\mu_M = g\,\mu\,(f \otimes f)$.*

Proposition 2. *Let $c : \{M, N, f, g, \phi\}$ be a chain contraction and let (N, μ) be an A_∞-algebra with higher maps $(n_1, n_2, n_3, \ldots)$. Then, the DG-module M inherits a structure of A_∞-algebra, given by the maps $(g\,n_1\,f, g\,n_2\,f^{\otimes 2}, g\,n_3\,f^{\otimes 3}, \ldots)$.*

4 Computational Advantages: Theoretical Study

In this section we are concerned about the theoretical study of the time and space invested in computing the maps of an A_∞-algebra structure induced by a contraction $c : \{A, M, f, g, \phi\}$. We will focus on the case of A being an algebra. In particular, we will make a comparison between the original formulas defined by the Basic Perturbation Lemma and the reduced formulas obtained in the previous section.

Regarding the original formulas of the A_∞–algebra maps, we must say that experimental results can be obtained with [1], a software developed in order to perform low dimension computations. This software is based on the initial formulation for the map $m_n : M^{\otimes n} \to M$ given in theorem 1:

$$m_n = (-1)^{n+1} f\, \mu^{(1)}\, \phi^{[\otimes 2]}\, \mu^{(2)} \cdots \phi^{[\otimes n-1]}\, \mu^{(n-1)}\, g^{\otimes n}\,, \quad n \geq 2\,.$$

We will take $n \geq 3$, since no improvement is obtained in the case $n = 2$.

As for complexity in space, let us consider the number of addends generated in the sum above. Taking into account that

$$\phi^{[\otimes k]} = \sum_{i=0}^{k-1} 1^{\otimes i} \otimes \phi \otimes (g\,f)^{\otimes k-i-1} \text{ and } \mu^{(k)} = \sum_{i=0}^{k-1} (-1)^{i+1} 1^{\otimes i} \otimes \mu_A \otimes 1^{\otimes k-i-1}\,,$$

the result of applying m_n to an element $x_1 \otimes x_2 \otimes \cdots x_n$ has $(n-1)!^2$ addends.

Concerning complexity in time, let us assume that each component morphism of the initial contraction, f, g and ϕ, consumes a unit of time when applied (that is, each one of these morphisms is considered a basic operation); we will also make this assumption for the composition $g\,f$ which is applied in different terms of the morphisms $\phi^{[\otimes k]}$.

Notice that applying $g^{\otimes n}$ is $O(n)$ in time.

On the other hand, the number of operations of each addend of the form $1^{\otimes i} \otimes \phi \otimes (g\,f)^{\otimes k-i-1}$ is $k-i$ and the one of each addend $1^{\otimes i} \otimes \mu_A \otimes 1^{\otimes k-i-1}$ is 1. That is, the number of basic operations can be expressed by

$$n + 2\,(n-1)!^2 + (n-1)! \sum_{k_i \in \{1,2,\ldots,i\}} (k_2 + 1 + k_3 + 1 + \cdots + k_{n-1} + 1)\,,$$

where n comes from $g^{\otimes n}$, $2\,(n-1)!^2$ from the two operations $f\,\mu$ at the end of each addend and the big sum corresponds to the operations on the composition

$$\phi^{[\otimes 2]}\, \mu^{(2)} \cdots \phi^{[\otimes n-1]}\, \mu^{(n-1)}\,.$$

Notice that the sum is multiplied by $(n-1)!$ because of all the possibilities for taking an addend $1^{\otimes i} \otimes \mu_A \otimes 1^{\otimes k-i-1}$ of each $\mu^{(k)}$. The sum above can be expressed by

$$n + n\,(n-1)!^2 + \frac{(n+3)(n-2)}{4}\,(n-1)!^2.$$

Therefore, the complexity of the algorithm becomes $O(n!^2)$ in time.

Now, taking into account the first reduction of terms in the sums involved in m_n (theorem 2), any composition of morphisms of the form $\phi^{[\otimes s]} \mu^{(s)}$, which had s^2 addends, is reduced to a sum with s terms. So, the total number of addends is now $(n-1)!$.

As for the number of operations, now it is $O(n)$ for each addend. Moreover, the number of operations is, exactly,

$$n + (n-1)!\,(2n-2)\,,$$

and hence $O(n!)$ in time.

Finally, considering that the upla of indexes $(i_2, i_3, \ldots i_{n-1})$ for the sums must be taken so that $i_k \geq i_{k+1} - 1$, we eliminate (for $n \geq 4$)

$$S_n = \sum_{k=1}^{n-3} \sum_{i=1}^{k} i \cdot k! = \sum_{k=1}^{n-3} \frac{k(k+1)!}{2}$$

addends, so the number of addends becomes $(n-1)! - S_n$. Now, taking into account that $(n-1)!$ can be expressed by

$$(n-1)! = 2 + \sum_{k=1}^{n-3}(k+1)! + \sum_{k=1}^{n-3} k(k+1)!,$$

it is easy to see that

$$\frac{(n-1)!}{2} < (n-1)! - S_n < (n-1)!,$$

so the algorithm is still $O((n-1)!)$ in space. However, the final number of addends, $(n-1)! - S_n$, is much "closer" to $\frac{(n-1)!}{2}$ than to $(n-1)!$, as it is shown in the following comparative table.

n	5	10	50	100	1000
$((n-1)! - S_n)/(n-1)!$	$0,708333$	$0,563704$	$0,510421$	$0,505103$	$0,500050$

Summing up, the order of complexity in time and space of the original formula versus the new one is presented in the following table.

	original formula		new formula	
	time	space	time	space
m_n	$O(n!^2)$	$O((n-1)!^2)$	$O(n!)$	$O((n-1)!)$

References

1. Berciano, A., Sergeraert, F.: Software to compute A_∞-(co)algebras: Araia Craic, http://www.ehu.es/aba/araia-craic.htm
2. Brown, R.: The twisted Eilenberg–Zilber theorem, Celebrazioni Archimedee del secolo XX, Simposio di topologia, 34–37 (1967)
3. Cartan, H., Eilenberg, S.: Homological Algebra. Princeton University Press, Princeton (1956)
4. Eilenberg, S., Mac Lane, S.: On the groups $H(\pi, n)$–I. Annals of Math. 58, 55–106 (1953)
5. Gugenheim, V.K.A.M.: On the chain complex of a fibration. Illinois J. Math. 3, 398–414 (1972)
6. Gugenheim, V.K.A.M.: On Chen's iterated integrals, Illinois J. Math, 703–715 (1977)

7. Gugenheim, V.K.A.M., Lambe, L.A.: Perturbation theory in Differential Homological Algebra I. Illinois J. Math. 33(4), 566–582 (1989)
8. Gugenheim, V.K.A.M., Lambe, L.A., Stasheff, J.D.: Perturbation Theory in Differential Homological Algebra II. Illinois J. Math. 35(3), 357–373 (1991)
9. Gugenheim, V.K.A.M., Stasheff, J.: On Perturbations and A_∞–structures. Bull. Soc. Math. Belg. 38, 237–246 (1986)
10. Huebschmann, J., Kadeishvili, T.: Small models for chain algebras. Math. Z. 207, 245–280 (1991)
11. Jiménez, M.J.: A_∞–estructuras y perturbación homológica, Tesis Doctoral de la Universidad de Sevilla, Spain (2003)
12. Jiménez, M.J., Real, P.: Rectifications of A_∞–algebras. In Communications in Algebra (to appear)
13. Kadeishvili, T.: On the Homology Theory of Fibrations. Russian Math. Surveys 35(3), 231–238 (1980)
14. Kontsevich, M.: Homological algebra of mirror symmetry. In: Proceedings of ICM (Zurich, 1994), Birkhäuser, Basel, pp. 120–139 (1995) [alg-geom/9411018]
15. Lambe, L.A., Stasheff, J.D.: Applications of perturbation theory to iterated fibrations. Manuscripta Math. 58, 367–376 (1987)
16. Mac Lane, S.: Homology, Classics in Mathematics. Springer, Berlin (1995) (Reprint of the 1975 edition)
17. Nakatsu, T.: Classical open-string field theory: A_∞-algebra, renormalization group and boundary states. Nuclear Physics B 642, 13–90 (2002)
18. Polishchuk, A.: Homological mirror symmetry with higher products, math. AG/9901025. Kontsevich, M., Soibelman, Y. Homological mirror symmetry and torus fibrations, math.SG/0011041
19. Prouté, A.: Algebrès diffeérentielles fortement homotopiquement associatives (A_∞-algèbre), Ph. D. Thesis, Université Paris VII (1984)
20. Stasheff, J.D.: Homotopy Associativity of H-spaces I, II. Trans. A.M.S 108, 275–312 (1963)

Algebraic Visualization of Relations Using RelView

Rudolf Berghammer[1] and Gunther Schmidt[2]

[1] Institut für Informatik, Christian-Albrechts-Universität Kiel
Olshausenstraße 40, 24098 Kiel, Germany
rub@informatik.uni-kiel.de
[2] Fakultät für Informatik, Universität der Bundeswehr München
85577 Neubiberg, Germany
gunther.schmidt@unibw.de

Abstract. For graphs there exist highly elaborated drawing algorithms. We concentrate here in an analogous way on visualizing relations represented as Boolean matrices as, e.g., in RelView. This means rearranging the matrix appropriately, permuting rows and columns simultaneously or independently as required. In this way, many complex situations may successfully be handled in various application fields. We show how relation algebra and RelView can be combined to solve such tasks.

1   Introduction

Although graphs as well as relations are frequently used as modeling tools, the theory of graphs is far more broadly known than the theory of relations. While complexity considerations in graph theory, e.g., aim at asymptotic behaviour of algorithms, graph drawing has its main impact for graphs of small or moderate size. There exist specific application areas where people work with such graphs to model practical situations using graph drawing as a supporting technique. A lot of highly elaborated graph drawing algorithms and implemented tools help getting an impression on how the graph in question is structured; cf. e.g., [9,16].

Since many years relation algebra is used as a means for problem solving in mathematics, computer science, engineering, and some other disciplines. A lot of practical applications can be found in [7,8]. As demonstrated in [18] for example, graph theory and relation theory interact in many ways. But relations are geared towards an algebraic treatment, ultimately leading to the structure of a relation algebra in the sense of Tarski (see [21]) and, if required, its mechanization via appropriate Computer Algebra systems. While one may draw relations in a multitude of versions when interpreting them as graphs, we here aim at depicting them as Boolean matrices. This is often very useful for visual editing and for discovering structural properties that are not evident from a graph representation. While this approach does not bring much additional visualization in simple cases (i.e., for most of those presented here due to the page limit), it helps when several algebraic conditions are supposed to hold. As an example,

V.G. Ganzha, E.W. Mayr, and E.V. Vorozhtsov (Eds.): CASC 2007, LNCS 4770, pp. 58–72, 2007.

we mention that for games a decomposition into loss, draw, and win positions may be obtained, which means a block decomposition as follows:

$$\begin{pmatrix} 0 & 0 & * \\ 0 & \text{total} & * \\ \text{total} & * & * \end{pmatrix}$$

The Boolean matrix can be rearranged appropriately by permuting rows and columns simultaneously or independently as required, so as to have a more or less immediate impression of what it stands for. Easily visible rectangular zones of zeros or ones are helpful as well as arrangements along the diagonal. Here, submatrices will be obtained that do not admit empty rows. But there remains the question whether this is possible, whether it is justified algebraically, and how the desired form can be obtained algorithmically.

The observation concerning relations occurring in practice is that they are "not too big"; row or column numbers do often not exceed 40 or 50. Even if an algorithm turns out to be rather inefficient when considered asymptotically, it seems possible to handle that size with our actual computer equipment. A competent overview on Multi-Criteria Decision Aid of this type is given in [6] for example. It contains a considerable number of practical examples of tables that are limited in size but lend themselves to be investigated with algebraic methods with respect to several criteria. They resemble evaluations with patient material or the multi-purpose transnational water system of, e.g., Lago Maggiore.

This is where our investigation started. Using some basic algorithms, we have developed a tool set of relation- algebraic expressions describing rearrangements of Boolean matrices / relations into specific forms. These could immediately be translated into the programming language of the specific purpose Computer Algebra system RELVIEW [1,3,4]. While we have handled, among others, symmetric idempotent relations, matching decompositions, independent and/or covering pairs of sets, implication structure decompositions, equivalences, we restrict to presenting here all the types of orderings traditionally used by decision makers. Several pictures show how the system then can help in visualization.

2   A Motivating Example

As a rather trivial initial example consider the following relation on elements $1, 2, \ldots 7$, represented with the RELVIEW tool as a Boolean matrix.

A black square stands for the matrix entry 1 (or *true*) and a white square stands for the entry 0 (or *false*). It is easy to check that the relation R represented by the matrix is a *pre-order relation*, i.e., is *reflexive* ($I \subseteq R$) and *transitive*

($RR \subseteq R$). With a graph drawing algorithm one would obtain something as shown in the following pictures. Again they are produced by RELVIEW; the picture on the left uses the hierarchical polyline drawing algorithm of [13] and that on the right is the result of the spring-embedder algorithm of [15].

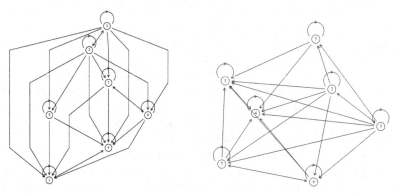

But these drawings do not give an appropriate impression of what the relation R really expresses. More intuitive is the left one of the following two RELVIEW-matrices. It is given as an upper right triangle of rectangles of either 1's or 0's, where the four rectangle-forming parts correspond to the four sets $\{2,3\}$, $\{5\}$, $\{6,7\}$ and $\{1,4\}$ of indices of the original matrix.

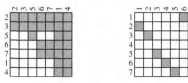

Later we will show how such a rearrangement can be obtained from the original matrix. The key of our procedure will be the relation-algebraic specification of a permutation relation P; for the above example it is represented by the REL-VIEW-matrix on the right. Then the new version is obtained by multiplying R with the transpose of P from the left and with P from the right. The former rearranges the rows of R accordingly and the latter does so for the columns.

3   Relation Algebra

We write $R : X \leftrightarrow Y$ if R is a relation with domain X and range Y, i.e., a subset of the direct product $X \times Y$. If the base sets X and Y of R's type $X \leftrightarrow Y$ are finite and of size m and n, respectively, we may consider R as a Boolean $m \times n$ matrix. This interpretation is well suited for many purposes. As it is also used by REL-VIEW to depict relations, we will often use matrix terminology and notation, i.e., speak about rows and columns and write $R_{x,y}$ instead of $\langle x, y \rangle \in R$ or $x\,R\,y$. We assume the reader to be familiar with the basic operations on relations, viz. R^{T} (*transposition*), $\overline{R}$ (*complement*), $R \cup S$ (*union*), $R \cap S$ (*intersection*), and RS (*multiplication*), the predicate $R \subseteq S$ (*inclusion*), and the special relations O (*empty relation*), L (*universal relation*), and I (*identity relation*).

To model sets we will use *vectors*, which are relations v with $v = vL$. Since for the relation modelling a set the range is irrelevant, we consider in the following mostly vectors $v : X \leftrightarrow 1$ with a specific singleton set $1 = \{\perp\}$ as range and omit in such cases the subscript $\perp$, i.e., write v_x instead of $v_{x,\perp}$. Such a vector can be considered as a Boolean column vector, and *represents* the subset $\{x \in X \mid v_x\}$ of X. A non-empty vector v is a *point* if $vv^\mathsf{T} \subseteq I$, i.e., it is *injective*. This means that it represents a singleton subset of its domain or an element from it if we identify a singleton set $\{x\}$ with the element x. In the Boolean matrix model a point $v : X \leftrightarrow 1$ is a Boolean column vector in which exactly one entry is 1.

When dealing with orders, one typically investigates extremal elements. In this paper we only need $least(C, v) = v \cap \overline{C \cup I}\, v$. For a *strict-order relation* $C : X \leftrightarrow X$ i.e., an *asymmetric* $(C \cap C^\mathsf{T} = O)$ and transitive relation, and a vector $v : X \leftrightarrow 1$ this relational function yields either a point that represents the least element of v wrt. C or an empty vector if no least element exists.

With $R \setminus S = \overline{R^\mathsf{T}\, \overline{S}} : Y \leftrightarrow Z$ the *right residual* of $R : X \leftrightarrow Y$ and $S : X \leftrightarrow Z$ is introduced. For $S = R$ this means in particular that $R \setminus R$ has type $Y \leftrightarrow Y$ and for all $x, y \in Y$ the x-column of R is contained in the y-column of R iff $(R \setminus R)_{x,y}$ holds. Hence, $R \setminus R$ coincides with the is-contained relation on the columns of R. The expression $(R^\mathsf{T} \setminus S^\mathsf{T})^\mathsf{T}$ defines the *left residual* $S / R : X \leftrightarrow Y$ of $S : X \leftrightarrow Z$ and $R : Y \leftrightarrow Z$. Here the case $S = R$ yields the transposed is-contained relation, i.e., the contains relation $S / S : X \leftrightarrow X$ on the rows of S.

4 A Short Look at the RELVIEW Tool

RELVIEW is a specific purpose Computer Algebra system for relation algebra. All data it works on are represented as relations which the system visualizes as directed graphs (via several sophisticated graph drawing algorithms) or as Boolean matrices. RELVIEW allows to compute with very large relations, as the system uses a highly efficient implementation of relations via binary decision diagrams (see [3,4]). The user can manipulate and analyse relations by predefined operations and tests. Based on the operations and tests and certain additional control structures relational functions and relational programs may be defined. We exhibit three of them as examples, which we will need later.

The following unary RELVIEW-function `Hasse` computes the Hasse-diagram $H_C = C \cap \overline{CC}$ of a strict-order relation C.

```
Hasse(C) = C & -(C * C).
```

A relational program essentially is a while-program based on the main datatype realized, namely relations. Such a RELVIEW-program has many similarities with a function procedure in programming languages like Pascal or Modula-2. It starts with a head line containing the program's name and the list of formal parameters. Then the local declarations follow. The last part is the body, a sequence of statements which are separated by semicolons and terminated by the RETURN-clause. For an example, we recall E. Szpilrajn who proved in [20] that every partial order relation possesses a linear extension, where R is *linear* if $R \cup R^\mathsf{T} = \mathsf{L}$. This theorem follows from Zorn's lemma and the fact that, given a partial order relation $E : X \leftrightarrow X$ and an incomparable pair $\langle a, b \rangle \in X \times X$, there exists an extension R of E that contains $\langle a, b \rangle$. Using element-wise notation, we can define R for all $x, y \in X$ by $R_{x,y}$ iff $E_{x,y}$ or $E_{x,a}$ and $E_{b,y}$. A relation-algebraic specification of R is $R = E \cup EAE$, where the relation A is a product pq^T of different points $p, q : X \leftrightarrow \mathbf{1}$ representing the elements a and b, respectively. In the finite case the extension may be iterated and this leads to the following RELVIEW-program Szpilrajn for computing a linear extension of E.

```
Szpilrajn(E)
  DECL R, A
  BEG  R = E;
       WHILE -empty(-(R | R^)) DO
         A = atom(-(R | R^));
         R = R | R*A*R OD
       RETURN R
  END.
```

The next example is the RELVIEW-program Classes (formally developed in [2]) that computes for an equivalence relation $R : X \leftrightarrow X$ with set $\mathfrak{C}$ of equivalence classes the canonical epimorphism from X to $\mathfrak{C}$ as a relation $\Phi : X \leftrightarrow \mathfrak{C}$.

```
Classes(R)
  DECL C, x, p
  BEG  C = R*point(Ln1(R));
       x = -C;
       WHILE -empty(x) DO
         p = point(x);
         C = (C^ + (R*p)^)^;
         x = x & -(R*p) OD
       RETURN C
  END.
```

Since Φ is the relational version of the canonical epimorphism, we have for all $x \in X$ and equivalence classes $c \in \mathfrak{C}$ that $\Phi_{x,c}$ iff x belongs to c. Hence, if we consider the columns of the result C of the RELVIEW-program Classes as single vectors, then these vectors are pair-wise disjoint and precisely represent the elements of the set $\mathfrak{C}$. In the literature this is also called a column-wise representation of a set of sets, or the natural projection for the equivalence.

5   Some Simple Rearranging Algorithms

Our rearrangement algorithms are all based on pure relation algebra, supported by one additional fact: Sets X between which the relations hold we work with are necessarily finite; we assume them to be equipped with a linear strict-order relation $\Omega_X : X \leftrightarrow X$, the *base strict-order*. In RELVIEW this order is implicitly given by the internal enumeration of the base set X within the tool. Its Hasse-diagram is $succ : X \leftrightarrow X$. To be more precise, if $x_1, x_2, \ldots, x_n$ is the internal enumeration of X within the tool, then we have $succ_{x_i, x_{i+1}}$ for all i, $1 \leq i \leq n-1$, and the base strict-order Ω_X is obtained as the transitive closure $succ^+$ of $succ$.

5.1   Linear Strict-Order Relations

In what follows, let $C : X \leftrightarrow X$ be any linear strict-order relation on the set X and let $\Omega_X : X \leftrightarrow X$ be the base strict-order on X. Assuming Ω_X to be depicted as a full upper right triangle matrix as the Boolean matrix of the relation $succ^+$ in RELVIEW, how can we permute the set X via a permutation relation $P : X \leftrightarrow X$ so as to see the linear strict-order relation C permuted accordingly as the full upper right triangle? Of course, this is a rather trivial task – but tedious when one has to actually execute it by hand. Subsequently, we show how it can be mechanized and how the algorithm can be formulated in RELVIEW.

In the first step, we compute the two Hasse-diagrams $H_C = C \cap \overline{CC}$ and $H_{\Omega_X} = \Omega_X \cap \overline{\Omega_X \Omega_X}$ of the linear strict-order relations C and Ω_X. Next, we consider the least elements with respect to both these orders, represented by vectors $least(C, \mathsf{L})$ and $least(\Omega_X, \mathsf{L})$, where $\mathsf{L} : X \leftrightarrow \mathbf{1}$. Since L represents the entire base set X, the vectors represent the respective least element of the strict-ordered sets (X, C) and (X, Ω_X). The permutation relation P we are looking for, now is defined iteratively. We start with the relation

$$P_0 = least(C, \mathsf{L}) \, least(\Omega_X, \mathsf{L})^\mathsf{T}$$

that precisely relates the least element of (X, C) with that of (X, Ω_X). It is easy to verify that the extension $P_0 \cup H_C^\mathsf{T} P_0 H_{\Omega_X}$ of the relation P_0 additionally relates the second smallest element of (X, C) with the second smallest element of (X, Ω_X), and no further relationships are introduced. Based on this observation, we successively apply the relational function

$$\tau(R) = R \cup H_C^\mathsf{T} R H_{\Omega_X}$$

to P_0. This leads to a finite chain $P_0 \subset \tau(P_0) \subset \tau^2(P_0) \subset \ldots \subset \tau^{|X|-1}(P_0)$ and the last relation of this chain obviously is the desired permutation relation P. A RELVIEW-implementation of this procedure looks as follows.

```
PermLSO(C)
  DECL L, HC, HO, P, Q
  BEG  L = Ln1(C);
       HC = Hasse(C);
       HO = succ(L);
       P = least(C,L)*least(trans(HO),L)^;
       Q = P | HC^*P*HO;
       WHILE -eq(P,Q) DO
          P = Q;
          Q = P | HC^*P*HO OD
       RETURN P
  END.
```

Having a program at hand for computing this permutation relation P, the following RELVIEW-program for the transformation of C is an immediate consequence.

```
RearrLSO(C)
   DECL P
   BEG  P = PermLSO(C)
        RETURN P^*C*P
   END.
```

The body of `RearrLSO` says that the desired form is obtained via $P^T C P$ since, as already mentioned in Sect. 2, the expression $P^T C$ rearranges the rows of C accordingly and a subsequent multiplication with P from the right does so for the columns.

5.2   Pre-order Relations

As a second kind of orderings, we consider *pre-order relations*. They are defined to be reflexive and transitive and appear quite frequently in optimization problems. The name *quasi-order relation* is also common for pre-order relations.

Each pre-order relation $Q : X \leftrightarrow X$ can be transformed into a form that consists of an upper right triangle of rectangles of 1's and 0's and is shown in the motivating example of Sect. 2. Again the decisive step is the computation of an appropriate permutation relation $P : X \leftrightarrow X$ on the base set X that rearranges the rows and columns accordingly by multiplying Q with P^T from the left and with P from the right. The computation of P is rather straightforward. First, we form the equivalence relation $R = Q \cap Q^T$. In the second step, we remove this relation from Q. It is easy to verify that this yields a strict-order relation $Q \cap \overline{R}$ and, hence, the reflexive closure $(Q \cap \overline{R}) \cup \mathsf{I}$ is a partial order relation on the base set X. In the third step, we compute a linear extension $E : X \leftrightarrow X$ of this partial order relation. Since each permutation relation that transforms the linear strict-order relation $E \cap \overline{\mathsf{I}}$ into a full upper right triangle obviously transforms Q into an upper right triangle of rectangles, the last step consists in the application of the procedure of Sect. 5.1. In the syntax of the RELVIEW tool the computation of the permutation relation P looks as follows.

```
PermPreO(Q)
   DECL I, R, E
   BEG  I = I(Q);
        R = Q & Q^;
        E = Szpilrajn((Q & -R) | I)
        RETURN PermLSO(E & -I)
   END.
```

A formulation of a RELVIEW-program `RearrPreO` for computing the product $P^T Q P$ is completely analogous to that of `RearrLSO` and, therefore, omitted. At this place it should be mentioned that we have used `PermPreO` and `RearrPreO` to generate the last two matrices of the motivating example in Sect. 2.

5.3   Weak-Order Relations

A relation $W : X \leftrightarrow X$ is said to be a *weak-order relation* if it is asymmetric and *negatively transitive*, where the latter property is defined by the inclusion $\overline{W}\,\overline{W} \subseteq \overline{W}$ to hold. Weak-order relations W are precisely those strict-order relations in which the set of indifferent pairs (i.e., the set of pairs $\langle x, y \rangle \in X \times X$ such that neither $W_{x,y}$ nor $W_{y,x}$) forms an equivalence relation $R = \overline{W} \cap \overline{W}^{\mathsf{T}}$ and the equivalence classes of R are linearly ordered by the order of the representatives via W. Informally this means that their Hasse-diagrams are composed by a series of complete bipartite strict-orders, one above another. Due to this property, weak-order relations are often used to model preferences with indifference, for instance in mathematical psychology. See [10,17] for example.

Each weak-order relation $W : X \leftrightarrow X$ can be transformed into an upper right block triangle form. From the single blocks of this form the above mentioned complete bipartite strict-orders and their rearrangement immediately becomes apparent. To obtain a permutation relation on X that transforms W into an upper right block triangle form is possible by performing three steps. First, W is joined with the identity relation $\mathsf{I} : X \leftrightarrow X$. The resulting reflexive closure $E = W \cup \mathsf{I}$ of W is a partial order relation on X. Next, a linear extension E' of E is determined. And, finally, a permutation relation $P : X \leftrightarrow X$ is computed that rearranges the linear strict-order relation $E' \cap \overline{\mathsf{I}}$ into the full upper right triangle $P^{\mathsf{T}}(E' \cap \overline{\mathsf{I}})P$. A little reflection shows that the same permutation relation also transforms the original weak-order relation W into the desired upper right block triangle form $P^{\mathsf{T}}WP$. The following RELVIEW-program `PermWeakO` is a direct translation of the above steps into the programming language of the tool.

```
PermWeakO(W)
    DECL I
    BEG  I = I(W)
         RETURN PermLSO(Szpilrajn(W | I) & -I)
    END.
```

5.4   Semi-order Relations

By definition, a *semi-order relation* $S : X \leftrightarrow X$ is irreflexive ($S \subseteq \overline{\mathsf{I}}$), *semi-transitive* ($SS\overline{S}^{\mathsf{T}} \subseteq S$ or, equivalently, $\overline{S}\,\overline{S} \subseteq \overline{SS}$), and possesses the *Ferrers* property $S\overline{S}^{\mathsf{T}}S \subseteq S$. The interest in this specific kind of orders mainly stems from the following fact, known as Scott-Suppes-Theorem (see [19,11]): Let X be a finite set. Then $S : X \leftrightarrow X$ is a semi-order relation iff there exist a mapping $f : X \to \mathbb{R}$ and a positive $\rho \in \mathbb{R}$ such that the relationship $S_{x,y}$ and the estimation $f(x) + \rho < f(y)$ are equivalent for all $x, y \in X$. The constant ρ can be seen as some sort of threshold. It allows to model preferences with indifference by defining $x, y \in X$ as indifferent iff $|f(x) - f(y)| \leq \rho$.

Irreflexive relations with the Ferrers property are called *interval order relations* since they are strict-orders that have interval representations. We will discuss this later in Sect. 6.2. In the case of semi-order relations the additional assumption of semi-transitivity allows a representation with all intervals of the same positive length, e.g., of length 1.

Each semi-order relation $S : X \leftrightarrow X$ can be rearranged into a threshold in an upper right block triangle form. As in the cases of the transformations of Sects. 5.1 to 5.3, the decisive step here again is the computation of a permutation relation $P : X \leftrightarrow X$ on the base set X that simultaneously transforms rows and columns via $P^\mathsf{T} S P$. To obtain P, we start with $W = \overline{S}^\mathsf{T} S \cup S \overline{S}^\mathsf{T}$. It is not very hard to verify relation-algebraically that W is a weak-order relation on X, which in turn yields the strict-order property for W. A little reflection furthermore shows that we can take as P any permutation relation that transforms W into an upper right block triangle form. In the syntax of RELVIEW the entire procedure looks as follows:

```
PermSemiO(S)
    DECL W
    BEG  W = -S^*S | S*-S^
         RETURN PermWeakO(W)
    END.
```

One may think that in an analogous manner also interval order relations J may be handled by embedding them into some sort of a semi-order closure S, where $S = J \cup J J \overline{J}^\mathsf{T}$, i.e., by adding whatever is missing for semi-transitivity. The relation-algebraic proof that S is indeed a semi-order relation is too long to be included here. But the result of the approach is not as expected. In the following three pictures we see on the left an interval order relation J, in the middle the semiorder relation $S = J \cup J J \overline{J}^\mathsf{T}$, and on the right a permutation relation P that transforms S in an upper right block triangle form $P^\mathsf{T} S P$.

The next two pictures show the two rearranged relations, viz. the Boolean matrix of $P^\mathsf{T} J P$ on the left and that of $P^\mathsf{T} S P$ on the right.

The transformed semi-order relation is in the desired form. However, for the transformed interval order relation we have some objections, as the Ferrers property is not yet visible. In the next section we will demonstrate that a visualization of the latter is also possible, however, at the cost of permuting rows and columns no longer simultaneously.

6   More Complex Rearranging Algorithms

Having developed some simple algorithms which rearrange the matrix representation of specific order relations appropriately by permuting rows and columns simultaneously, we now concentrate on two more complex cases. Here a transformation into an appropriate form requires rows and columns to be permuted independently. We will also show an example that once again exhibits the power of RELVIEW when dealing with the visualization of relations.

6.1   Ferrers Relations

In the theory of partitions of numbers (see e.g., [5]), a partition of a natural number n into a sum $n = a_1 + a_2 + \ldots + a_k$, where $a_1 \geq a_2 \geq \ldots \geq a_k$, is frequently visualized by means of a Ferrers diagram. Such a diagram is drawn as a rectangular array of squares[1] in which the i'th row has the number of squares equal to the number a_i, $1 \leq i \leq k$. All rows are right-justified (or left-justified) and sorted by their lengths (i.e., the number of squares) in decreasing order from the top to the bottom. As a small example, the pictorial representation of the Ferrers diagram for the partition $19 = 7 + 4 + 4 + 2 + 2$ of the number 19 as a RELVIEW-matrix looks as follows.

The black parts of the rows of this 5×7 Boolean matrix exactly correspond to the 5 rows of the Ferrers diagram.

Ferrers diagrams motivated the definition of Ferrers relations by demanding that the latter type of relations can be transformed into upper right staircase block form – the form shown in the above RELVIEW-matrix, but additionally allowing empty columns to occur at the left or empty rows at the bottom – by permuting rows and columns. To clarify that this rearrangement property is equivalent to the simple relation-algebraic definition given in Sect. 5.4, a combination of the predicate logic formulation of the inclusion $R\,\overline{R}^{\mathsf{T}} R \subseteq R$ and a graph interpretation of R is very helpful; see [18] for details. Ferrers relations have a lot of applications in mathematics and computer science. Here we only want to mention their use in the theory of measurement (ranking via Guttmann scaling) and in formal concept analysis; see [12] and [14] for example.

The following two remarks further exhibit that Ferrers relations enjoy important properties: If $R : X \leftrightarrow Y$ is a Ferrers relation, then so are $R\,\overline{R}^{\mathsf{T}} R$, $R\,\overline{R}^{\mathsf{T}}$ as well as $\overline{R}^{\mathsf{T}} R$. For a finite Ferrers relation, there exists a natural number $k \geq 0$ that gives rise to a strictly increasing exhaustion as follows:

$$\mathsf{O} = (R\,\overline{R}^{\mathsf{T}})^k \subset (R\,\overline{R}^{\mathsf{T}})^{k-1} \subset \ldots \subset R\,\overline{R}^{\mathsf{T}} R\,\overline{R}^{\mathsf{T}} \subset R\,\overline{R}^{\mathsf{T}}$$

---

[1] Sometimes also boxes, dots or circles are used instead of squares.

Another characterization says that R has the Ferrers property iff the contains pre-order relation $R\,/\,R$ on the rows (see Sect. 3) is linear. This applies for the is-contained pre-order relation $R\setminus R$ on the columns, too. The relation-algebraic proof of the first fact using the Schröder equivalences (see e.g., [18]) is simple:

$$R\overline{R}^{\mathsf{T}}R \subseteq R \Leftrightarrow \overline{R}R^{\mathsf{T}}\overline{R} \subseteq \overline{R} \Leftrightarrow R\overline{R}^{\mathsf{T}} \subseteq \overline{\overline{R}R^{\mathsf{T}}} \Leftrightarrow \overline{R\,/\,R} \subseteq (R\,/\,R)^{\mathsf{T}}$$

The latter inclusion means that $R\,/\,R$ is indeed linear.

Now, let $R : X \leftrightarrow Y$ be a Ferrers relation for which we intend to develop a relation-algebraic specification of the row permutation relation $P_r : X \leftrightarrow X$ as well as the column permutation relation $P_c : Y \leftrightarrow Y$ such that $P_r^{\mathsf{T}}R$ rearranges the rows of R in decreasing inclusion order from the top to the bottom, and after that $P_r^{\mathsf{T}}RP_c$ rearranges the columns of the intermediate relation $P_r^{\mathsf{T}}R$ in increasing inclusion order from the left to the right. Obviously, one will choose the pre-order relation rearrangement of Sect. 5.2 based on the contains relation $R\,/\,R$ for the rows and the is-contained relation $R\setminus R$ for the columns, respectively. The corresponding RELVIEW-functions look as follows:

```
PermRFerr(R) = PermPreO(R / R).
PermCFerr(R) = PermPreO(R \ R).
```

Finally, the procedure for the upper right staircase block form of R is described by the following RELVIEW-program:

```
RearrFerr(R)
    DECL Pr, Pc
    BEG  Pr = PermRFerr(R);
         Pc = PermCFerr(R)
         RETURN Pr^*R*Pc
    END.
```

6.2   Interval Order Relations

As already mentioned in Sect. 5.4, an interval order relation $J : X \leftrightarrow X$ is an irreflexive relation that possesses the Ferrers property. Hence, the relational programs for transforming Ferrers relations can also be applied to transform the relation J into an upper right staircase block form by permuting rows and columns independently. Since, however, interval order relations are also transitive (and, consequently, specific strict-orders), the 1-blocks of the staircase block form are completely contained in the upper right triangle. Hence, we obtain an upper right block triangle form, as in the case of the relations of Sects. 5.3 and 5.4.

Here is the RELVIEW-function for the permutation of the rows of an interval order relation via a permutation relation $P_r : X \leftrightarrow X$ on the base set X.

```
PermRIntervalO(J) = PermRFerr(J).
```

In exactly the same way we obtain two RELVIEW-functions `PermCIntervalO` and `RearrIntervalO` for the permutation relation $P_c : X \leftrightarrow X$ that rearranges the columns of J and for the desired upper right block triangle form $P_r^{\mathsf{T}}JP_c$.

Each interval order relation can be represented by a set of intervals of a linearly ordered set, ordered by strict left-to-right precedence. Formally we have the following theorem (see [11]): $J : X \leftrightarrow X$ is an interval order relation iff there is a function $f : X \to 2^P$ that assigns to each $x \in X$ a closed interval $f(x) = [a_x, b_x] \subseteq P$ of a linearely ordered set $(P, \leq)$ such that for all $x, y \in X$ we have $J_{x,y}$ iff $b_x < a_y$. In case of $(\mathbb{R}, \leq)$ as $(P, \leq)$ one speaks of a *real interval representation*. Interval representations of interval order relations J via the rows of Boolean matrices may be specified by purely relation-algebraic expressions. The RELVIEW-program resulting from this specification looks as follows; because of lack of space we cannot go into details.

```
IntervalRepr(J)
    DECL fringe(R) = R & -(R*-R^*R);
         coleq(R) = syq(R,R);
         roweq(R) = syq(R^,R^);
         SR, SC, M, C, I, Pc
    BEG  SR = Classes(roweq(J));
         SC = Classes(coleq(J));
         M = SR^*fringe(-J)*SC;
         C = M^*SR^*J*SC;
         I = I(C);
         Pc = PermLSO(C)
         RETURN SR*M*(C | I)^*Pc & SC*(C | I)*Pc
    END.
```

6.3   An Example

Now, let us present a small application of the programs of Sect. 6.2. We consider the following three 13×13 RELVIEW-matrices.

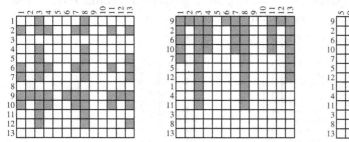

It can easily be checked that the three Boolean matrices represent the same interval order relation R on a 13-element base set. The matrix on the left shows the order's original version, the matrix in the middle the version after sorting the rows in decreasing inclusion order from the top to the bottom, and the matrix on the right the final version, which is an upper right block triangle form. From the latter Boolean matrix we immediately obtain a layered graph-representation of R. The next picture shows the Hasse-diagram of R, drawn by RELVIEW's implementation of the graph drawing algorithm of [13] and slightly prettified by hand to enhance visibility of the minimal elements 5, 9 and 10.

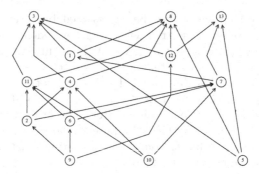

From the attached row and column numbers of the above matrices we directly obtain the following permutation relations P_r for the rows (left matrix) and P_c for the columns (right matrix). Of course, the real way of computation was the other way around. We first computed the permutation relations P_r and P_c and then labeled the rearranged Boolean matrices according to them.

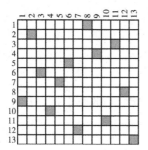 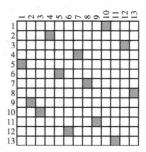

The left one of the following two Boolean matrices is the result of the REL-VIEW-program `IntervalRepr` applied to the interval order relation R, and the matrix right aside is generated from it by sorting the rows according to the first occurrence of the entry 1 via a small RELVIEW-program.

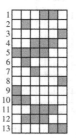 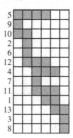

If $\{y_1, y_2, y_3, y_4, y_5, y_6\}$ is the base set of the columns of these Boolean matrices and the linear base strict-order is given by the order of the element's indices, then, e.g., the interval $[y_4, y_5]$ is assigned to the element 1 of the 13-element base set and the singleton interval $[y_2, y_2]$ is assigned to the element 2 of this set. To obtain a real interval representation from the result of `IntervalRepr`, we first have to interpret each black square as a unit interval on the real line, where the left border of the matrix describes 0. This does not yet yield the desired result since it allows comparable intervals to be tangent, whereas the definition demands strict left-to-right precedence. But it is very easy to transform it into a real interval representation. We only have to shorten each interval from the right by a small constant. If, after that, each copy of a multiple interval is accordingly shortened from the left to get different left end points, we even obtain a so-called distinguishing real interval representation. Doing so, e.g., each of the two above matrices yields such a representation $f(i) = \alpha_i$, where the intervals α_i, $1 \leq i \leq 13$, are given as follows:

$$\alpha_1 = [3, 4.9] \quad \alpha_2 = [1, 1.9] \quad \alpha_3 = [5, 5.9] \quad \alpha_4 = [2, 4.9] \quad \alpha_5 = [0, 3.9]$$
$$\alpha_6 = [1.1, 1.9] \quad \alpha_7 = [2, 2.9] \quad \alpha_8 = [5.1, 5.9] \quad \alpha_9 = [0, 0.9] \quad \alpha_{10} = [0, 1.9]$$
$$\alpha_{11} = [2, 4.9] \quad \alpha_{12} = [1, 3.9] \quad \alpha_{13} = [4, 5.9]$$

Here we have shortened each interval from the right by 0.1 and, in addition, two intervals from the left by 0.1 to obtain uniqueness.

7 Conclusion

We have investigated how to visualize relations represented as Boolean matrices using the specific purpose Computer Algebra system RELVIEW. We were able to rearrange Boolean matrices into specific forms which allow to discover structural properties that are not evident in the first place. Such an approach may be useful in various application fields. Starting with basic rearrangement algorithms, we constructed algorithms also for non-trivial cases.

In addition to the types of relations treated in the present paper, we have transformed several other simple types, like injective and univalent relations and (partial) equivalence relations. Relations of the first type can be rearranged into four blocks, where the left upper block looks like an identity relation and the remaining three blocks are empty. Equivalence relations can be transformed into block diagonal form with quadratic blocks of 1's in the diagonal and 0's otherwise. From the latter form one immediately obtains the equivalence classes. In the case of partial equivalence relations we additionally obtain a right lower empty block in the diagonal. We could, due to limited space, not present how to apply our technique to other complex examples. In particular, we have studied the "maximum pair of independent sets rearrangements" of general relations R based on a maximum matching λ contained in R. They are closely related to the term rank of a relation. We have also developed algorithms where quotients are taken according to some congruence relations on the base sets. An example for such a problem is the rearrangement of a relation R according to its so-called difunctional closure $R(R^{\mathsf{T}}R)^+$ into block diagonal form, where the difunctional closure has rectangular blocks of 1's in the diagonal and 0's otherwise.

A future aim is to use the efficiency and visualization power of RELVIEW and the conceptual simplicity of its programming language so as to enable one to scan any given – even real-valued – matrix of moderate size for possibly hidden interesting properties. In the real-valued case, one would use moving so-called cuts at different levels to arrive at Boolean matrices similar to the cuts used in the theory of fuzzy sets. Because of their close relationship to interval order relations we are also interested in the relation-algebraic treatment of interval graphs, e.g., the computation of interval representations or the matrix rearrangement based on perfect elimination orderings.

References

1. Behnke, R., et al.: RELVIEW – A system for calculation with relations and relational programming. In: Astesiano, E. (ed.) ETAPS 1998 and FASE 1998. LNCS, vol. 1382, pp. 318–321. Springer, Heidelberg (1998)
2. Berghammer, R., Hoffmann, T.: Modelling sequences within the RELVIEW system. J. Universal Comput. Sci. 7, 107–123 (2001)
3. Berghammer, R., Leoniuk, B., Milanese, U.: Implementation of relation algebra using binary decision diagrams. In: de Swart, H. (ed.) RelMiCS 2001. LNCS, vol. 2561, pp. 241–257. Springer, Heidelberg (2002)
4. Berghammer, R., Neumann, F.: RELVIEW – An OBDD-based Computer Algebra system for relations. In: Ganzha, V.G., Mayr, E.W., Vorozhtsov, E.V. (eds.) CASC 2005. LNCS, vol. 3718, pp. 40–51. Springer, Heidelberg (2005)
5. Bona, M.: A walk through combinatorics: An introduction to combinatorics and graph theory. World Scientific Publishing, Singapore (2002)
6. Colorni, A., Paruccini, M., Roy, B.: A-MCD-A – Multiple criteria decision aiding. Joint Research Centre of the European Commission (2001)
7. de Swart, H., Orłowska, E., Schmidt, G., Roubens, M. (eds.): Theory and Applications of Relational Structures as Knowledge Instruments. LNCS, vol. 2929. Springer, Heidelberg (2003)
8. de Swart, H., Orłowska, E., Schmidt, G., Roubens, M. (eds.): Theory and Applications of Relational Structures as Knowledge Instruments II. LNCS (LNAI), vol. 4342. Springer, Heidelberg (2006)
9. di Battista, G., Eades, P., Tamassia, R., Tollis, I.G.: Graph drawing – Algorithms for the visualization of graphs. Prentice-Hall, Englewood Cliffs (1999)
10. Fishburn, P.: On the construction of weak orders from fragmentary information. Psychometrika 38, 459–472 (1973)
11. Fishburn, P.: Interval orders and interval graphs. Wiley, Chichester (1985)
12. Guttmann, L.: A basis for scaling qualitative data. American Sociological Review 9, 139–150 (1944)
13. Gansner, E.R., North, S.C., Vo, K.P.: DAG – A program that draws directed graphs. Software – Practice and Experience 17, 1047–1062 (1988)
14. Ganter, B., Wille, R.: Formal concept analysis. Springer, Heidelberg (1999)
15. Kamada, T., Kawai, S.: An algorithm for drawing general undirected graphs. Inf. Proc. Letters 31, 7–15 (1989)
16. Kaufmann, M., Wagner, D. (eds.): Drawing Graphs. LNCS, vol. 2025. Springer, Heidelberg (2001)
17. Öztürk, M., Tsoukias, A., Vincke, P.: Preference modelling. In: Ehrgott, M., Greco, S., Figueira, J. (eds.) Multiple criteria decision analysis: State of the art surveys. Int. Series in Operat. Res. and Manag. Sci, vol. 78, pp. 27–71. Springer, Heidelberg (2005)
18. Schmidt, G., Ströhlein, T.: Relationen und Graphem. Springer. Available also in English: Relations and graphs. In: EATCS Monographs on Theoret. Comput. Sci., Springer, Heidelberg (1993)
19. Scott, D., Suppes, P.: Foundational aspects of the theories of measurement. J. Symbolic Logic 23, 113–128 (1958)
20. Szpilrajn, E.: Sur l'extension de l'ordre partiel. Fund. Math. 16, 386–389 (1930)
21. Tarski, A.: On the calculus of relations. J. Symbolic Logic 6, 73–89 (1941)

Comprehensive Triangular Decomposition

Changbo Chen[1], Oleg Golubitsky[1], François Lemaire[2], Marc Moreno Maza[1], and Wei Pan[1]

[1] University of Western Ontario, London N6A 1M8, Canada
[2] Université de Lille 1, 59655 Villeneuve d'Ascq Cedex, France

Abstract. We introduce the concept of comprehensive triangular decomposition (CTD) for a parametric polynomial system F with coefficients in a field. In broad words, this is a finite partition of the the parameter space into regions, so that within each region the "geometry" (number of irreducible components together with their dimensions and degrees) of the algebraic variety of the specialized system $F(u)$ is the same for all values u of the parameters.

We propose an algorithm for computing the CTD of F. It relies on a procedure for solving the following set theoretical instance of the coprime factorization problem. Given a family of constructible sets $A_1, \ldots, A_s$, compute a family $B_1, \ldots, B_t$ of pairwise disjoint constructible sets, such that for all $1 \leq i \leq s$ the set A_i writes as a union of some of the $B_1, \ldots, B_t$.

We report on an implementation of our algorithm computing CTDs, based on the RegularChains library in MAPLE. We provide comparative benchmarks with MAPLE implementations of related methods for solving parametric polynomial systems. Our results illustrate the good performances of our CTD code.

1 Introduction

Solving polynomial systems with parameters has become an increasing need in several applied areas such as robotics, geometric modeling, stability analysis of dynamical systems and others. For a given parametric polynomial system F, the following problems are of interest.

(P1) Compute the values of the parameters for which F has solutions, or has finitely many solutions.

(P2) Compute the solutions of F as functions of the parameters.

These questions have been approached by various techniques including comprehensive Gröbner bases (CGB) [22,23,14,13,17], cylindrical algebraic decomposition (CAD) [4] and triangular decompositions [24,25,6,7,10,9,20,19,26,5]. Methods based on CGB, or more generally Gröbner bases, are powerful tools for solving problems such as (P1), that is, determining the values u of the parameters such that, the specialized system $F(u)$ satisfies a given property. Methods based on CAD or triangular decompositions are naturally well designed for solving Problem (P2).

In this paper, we introduce the concept of *comprehensive triangular decomposition* for a parametric polynomial system with coefficients in a field. This notion plays the role for triangular decompositions that CGB does for Gröbner bases. With this concept

V.G. Ganzha, E.W. Mayr, and E.V. Vorozhtsov (Eds.): CASC 2007, LNCS 4770, pp. 73–101, 2007.

at hand, we show that Problems (P1) and (P2) can be completely answered by means of triangular decompositions.

Let F be a finite set of polynomials with coefficients in a field $\mathbb{K}$, parameters $U = U_1, \ldots, U_d$, and unknowns $X = X_1, \ldots, X_m$, that is, $F \subset \mathbb{K}[U_1, \ldots, U_d, X_1, \ldots, X_m]$. Let $\overline{\mathbb{K}}$ be the algebraic closure of $\mathbb{K}$, and let $\mathbf{V}(F) \subset \overline{\mathbb{K}}^{d+m}$ be the zero set of F. Let also Π_U be the projection from $\overline{\mathbb{K}}^{d+m}$ on the parameter space $\overline{\mathbb{K}}^d$. For all $u \in \overline{\mathbb{K}}^d$ we define $\mathbf{V}(F(u)) \subseteq \overline{\mathbb{K}}^m$ the zero set defined by F after specializing U at u.

Our first contribution is to show how to compute a finite partition $\mathcal{C}$ of $\Pi_U(\mathbf{V}(F))$ and a family of triangular decompositions $(\mathcal{T}_C, C \in \mathcal{C})$ in $\mathbb{K}[U, X]$ such that for each $C \in \mathcal{C}$ and for each parameter value $u \in C$ the triangular decomposition $\mathcal{T}_C$ specializes at u into a triangular decomposition $\mathcal{T}_C(u)$ of $\mathbf{V}(F(u))$ given by regular chains. Moreover, each "cell" $C \in \mathcal{C}$ is a constructible set given by a family of regular systems in $\mathbb{K}[U]$. We call the pair $(\mathcal{T}_C, C \in \mathcal{C})$ a *comprehensive triangular decomposition* of $\mathbf{V}(F)$, see Section 5.

This is a natural definition inspired by that of a comprehensive Gröbner basis [22] introduced by Weispfenning with the additional requirements proposed by Montes in [14]. From each pair $(C, \mathcal{T}_C)$, we can read geometrical information, such as for which parameter values $u \in C$ the set $\mathbf{V}(F(u))$ is finite; we also obtain a "generic" equidimensional decomposition of $\mathbf{V}(F(u))$, for all $u \in C$. The notion of CTD is also related to the border polynomial of a polynomial system in [26] and the minimal discriminant variety of $\mathbf{V}(F)$ as defined in [12] for the case where $\overline{\mathbb{K}}$ is the field of complex numbers.

Example 1. Let $F = \{vxy + ux^2 + x, uy^2 + x^2\}$ be a parametric polynomial system with parameters $u > v$ and unknowns $x > y$. Then a comprehensive triangular decomposition of $\mathbf{V}(F)$ is:

$$C_1 = \{u(u^3 + v^2) \neq 0\}: \quad \mathcal{T}_{C_1} = \{T_3, T_4\}$$
$$C_2 = \{u = 0\}: \quad \mathcal{T}_{C_2} = \{T_2, T_3\}$$
$$C_3 = \{u^3 + v^2 = 0, v \neq 0\}: \quad \mathcal{T}_{C_3} = \{T_1, T_3\}$$

where

$$T_1 = \{vxy + x - u^2y^2, 2vy + 1, u^3 + v^2\}$$
$$T_2 = \{x, u\}$$
$$T_3 = \{x, y\}$$
$$T_4 = \{vxy + x - u^2y^2, u^3y^2 + v^2y^2 + 2vy + 1\}$$

Here, C_1, C_2, C_3 is a partition of $\Pi_U(\mathbf{V}(F))$ and $\mathcal{T}_{C_i}$ is a triangular decomposition of $\mathbf{V}(F)$ above C_i, for $i = 1, 2, 3$. For different parameter values u, we can directly read geometrical information, such as the dimension of $\mathbf{V}(F(u))$.

By RegSer [19], $\mathbf{V}(F)$ can be decomposed into a set of regular systems:

$$R_1 = \begin{cases} ux + vy + 1 = 0 \\ (u^3 + v^2)y^2 + 2vy + 1 = 0 \;, \\ u(u^3 + v^2) \neq 0 \end{cases} \quad R_2 = \begin{cases} x = 0 \\ y = 0 \;, \\ u \neq 0 \end{cases}$$

$$R_3 = \begin{cases} x = 0 \\ vy + 1 = 0 \\ u = 0 \\ v \neq 0 \end{cases}, \quad R_4 = \begin{cases} 2ux + 1 = 0 \\ 2vy + 1 = 0 \\ u^3 + v^2 = 0 \\ v \neq 0 \end{cases}, \quad R_5 = \begin{cases} x = 0 \\ u = 0 \end{cases}.$$

For each regular system, one can directly read its dimension when parameters take corresponding values. However, the dimension of the input system could not be obtained immediately, since there is not a partition of the parameter space.

By DISPGB [14], one can obtain all the cases over the parameters leading to different reduced Gröbner bases with parameters:

$$u(u^3 + v^2) \neq 0 : \{ux + (u^3v + v^3)y^3 + (-u^3 + v^2)y^2, (u^3 + v^2)y^4 + 2vy^3 + y^2\}$$
$$u(u^3 + v^2) = 0, u \neq 0 : \{ux + 2v^2y^2, 2vy^3 + y^2\}$$
$$u = 0, v \neq 0 : \{x^2, vxy + x\}$$
$$u = 0, v = 0 : \{x\}$$

Here for each parameter value, the input system specializes into a Gröbner basis. Since Gröbner bases do not necessarily have a triangular shape, the dimension may not be read directly either. For example, when $u = 0, v \neq 0$, $\{x^2, vxy + x\}$ is not a triangular set.

In Section 5 we also propose an algorithm for computing the CTD of parametric polynomial system. We rely on an algorithm for computing the difference of the zero sets of two regular systems. Based on the procedures of the TRIADE algorithm [15] and elementary set theoretical considerations, such an algorithm could be developed straightforwardly. We actually tried this and our experimental results (not reported here) shows that this naive approach is very inefficient comparing to the more advanced algorithm presented in Section 3. Indeed, this latter algorithm heavily exploits the structure and properties of regular chains, whereas the former is unable to do so.

This latter procedure, is used to solve the following problem. Given a family of constructible sets, $A_1, \ldots, A_s$ (each of them given by a regular system) compute a family $B_1, \ldots, B_t$ of pairwise disjoint constructible sets, such that for all $1 \leq i \leq s$ the set A_i writes as a union of some the $B_1, \ldots, B_t$. A solution is presented in Section 4. This can be seen as the set theoretical version of the *coprime factorization* problem, see [2,8] for other variants of this problem.

Our second contribution is an implementation report of our algorithm computing CTDs, based on the RegularChains library in MAPLE. We provide comparative benchmarks with MAPLE implementations of related methods for solving parametric polynomial systems, namely: *decompositions into regular systems* by Wang [19] and *discussing parametric Gröbner bases* by Montes [14]. We use a large set of well-known test-problems from the literature. Our implementation of the CTD algorithm can solve all problems which can be solved by the other methods. In addition, our CTD code can solve problems which are out of reach of the other two methods, generally due to memory consumption.

2   Preliminaries

In this section we introduce notations and review fundamental results in the theory of regular chains and regular systems [1,3,11,15,19,21].

We shall use some notions from commutative algebra (such as the dimension of an ideal) and refer for instance to [16] for this subject.

2.1 Basic Notations and Definitions

Let $\mathbb{K}[Y] := \mathbb{K}[Y_1, \ldots, Y_n]$ be the polynomial ring over the field $\mathbb{K}$ in variables $Y_1 < \cdots < Y_n$. Let $p \in \mathbb{K}[Y]$ be a non-constant polynomial. The *leading coefficient* and the *degree* of p regarded as a univariate polynomial in Y_i will be denoted by $\mathrm{lc}(p, Y_i)$ and $\deg(p, Y_i)$ respectively. The greatest variable appearing in p is called the *main variable* denoted by $\mathrm{mvar}(p)$. The degree, the leading coefficient, and the leading monomial of p regarding as a univariate polynomial in $\mathrm{mvar}(p)$ are called the *main degree*, the *initial*, and the *rank* of p; they are denoted by $\mathrm{mdeg}(p)$, $\mathrm{init}(p)$ and $\mathrm{rank}(p)$ respectively.

Let $F \subset \mathbb{K}[Y]$ be a finite polynomial set. Denote by $\langle F \rangle$ the ideal it generates in $\mathbb{K}[Y]$ and by $\sqrt{\langle F \rangle}$ the radical of $\langle F \rangle$. Let h be a polynomial in $\mathbb{K}[Y]$, the *saturated ideal* $\langle F \rangle : h^\infty$ of $\langle F \rangle$ w.r.t h, is the set

$$\{q \in \mathbb{K}[Y] \mid \exists m \in \mathbb{N} \text{ s.t. } h^m q \in \langle F \rangle\},$$

which is an ideal in $\mathbb{K}[Y]$.

A polynomial $p \in \mathbb{K}[Y]$ is a *zerodivisor* modulo $\langle F \rangle$ if there exists a polynomial q such that pq is zero modulo $\langle F \rangle$, and q is not zero modulo $\langle F \rangle$. The polynomial is *regular* modulo $\langle F \rangle$ if it is neither zero, nor a zerodivisor modulo $\langle F \rangle$. Denote by $\mathbf{V}(F)$ the *zero set* (or solution set, or algebraic variety) of F in $\overline{\mathbb{K}}^n$. For a subset $W \subset \overline{\mathbb{K}}^n$, denote by $\overline{W}$ its closure in the Zariski topology, that is the intersection of all algebraic varieties $\mathbf{V}(G)$ containing W for all $G \subset \mathbb{K}[Y]$.

Let $T \subset \mathbb{K}[Y]$ be a *triangular set*, that is a set of non-constant polynomials with pairwise distinct main variables. Denote by $\mathrm{mvar}(T)$ the set of main variables of $t \in T$. A variable in Y is called *algebraic* w.r.t. T if it belongs to $\mathrm{mvar}(T)$, otherwise it is called *free* w.r.t. T. For a variable $v \in Y$ we denote by $T_{<v}$ (resp. $T_{>v}$) the subsets of T consisting of the polynomials t with main variable less than (resp. greater than) v. If $v \in \mathrm{mvar}(T)$, we say T_v is defined. Moreover, we denote by T_v the polynomial in T whose main variable is v, by $T_{\leqslant v}$ the set of polynomials in T with main variables less than or equal to v and by $T_{\geqslant v}$ the set of polynomials in T with main variables greater than or equal to v.

Definition 1. *Let* $p, q \in \mathbb{K}[Y]$ *be two nonconstant polynomials. We say* $\mathrm{rank}(p)$ *is smaller than* $\mathrm{rank}(q)$ *w.r.t Ritt ordering and we write,* $\mathrm{rank}(p) <_r \mathrm{rank}(q)$ *if one of the following assertions holds:*

- $\mathrm{mvar}(p) < \mathrm{mvar}(q)$,
- $\mathrm{mvar}(p) = \mathrm{mvar}(q)$ *and* $\mathrm{mdeg}(p) < \mathrm{mdeg}(q)$.

Note that the partial order $<_r$ is a well ordering. Let $T \subset \mathbb{K}[Y]$ be a triangular set. Denote by $\mathrm{rank}(T)$ the set of $\mathrm{rank}(p)$ for all $p \in T$. Observe that any two ranks in $\mathrm{rank}(T)$ are comparable by $<_r$. Given another triangular set $S \subset \mathbb{K}[Y]$, with $\mathrm{rank}(S) \neq \mathrm{rank}(T)$, we write $\mathrm{rank}(T) <_r \mathrm{rank}(S)$ whenever the minimal element of the symmetric difference $(\mathrm{rank}(T) \setminus \mathrm{rank}(S)) \cup (\mathrm{rank}(S) \setminus \mathrm{rank}(T))$ belongs to $\mathrm{rank}(T)$. By

$\mathrm{rank}(T) \leqslant_r \mathrm{rank}(S)$, we mean either $\mathrm{rank}(T) < \mathrm{rank}(S)$ or $\mathrm{rank}(T) = \mathrm{rank}(S)$. Note that any sequence of triangular sets, of which ranks strictly decrease w.r.t $<_r$, is finite.

Given a triangular set $T \subset \mathbb{K}[Y]$, denote by h_T be the product of the initials of T (throughout the paper we use this convention and when T consists of a single element g we write it in h_g for short). The *quasi-component* $\mathbf{W}(T)$ of T is $\mathbf{V}(T) \setminus \mathbf{V}(h_T)$, in other words, the points of $\mathbf{V}(T)$ which do not cancel any of the initials of T. We denote by $\mathrm{Sat}(T)$ the *saturated ideal of* T: if T is empty then $\mathrm{Sat}(T)$ is defined as the trivial ideal $\langle 0 \rangle$, otherwise it is the ideal $\langle T \rangle : h_T^\infty$.

Let $h \in \mathbb{K}[Y]$ be a polynomial and $F \subset \mathbb{K}[Y]$ a set of polynomials, we write

$$\mathbf{Z}(F, T, h) := (\mathbf{V}(F) \cap \mathbf{W}(T)) \setminus \mathbf{V}(h).$$

When F consists of a single polynomial p, we use $\mathbf{Z}(p, T, h)$ instead of $\mathbf{Z}(\{p\}, T, h)$; when F is empty we just write $\mathbf{Z}(T, h)$. By $\mathbf{Z}(F, T)$, we denote $\mathbf{V}(F) \cap \mathbf{W}(T)$.

Given a family of pairs $\mathbf{S} = \{[T_i, h_i] \mid 1 \leq i \leq e\}$, where $T_i \subset \mathbb{K}[Y]$ is a triangular set and $h_i \in \mathbb{K}[Y]$ is a polynomial. We write

$$\mathbf{Z}(S) := \bigcup_{i=1}^{e} \mathbf{Z}(T_i, h_i).$$

We conclude this section with some well known properties of ideals and triangular sets. For a proper ideal $\mathcal{I}$, we denote by $\dim(\mathbf{V}(\mathcal{I}))$ the dimension of $\mathbf{V}(\mathcal{I})$.

Lemma 1. *Let $\mathcal{I}$ be a proper ideal in $\mathbb{K}[Y]$ and $p \in \mathbb{K}[Y]$ be a polynomial regular w.r.t $\mathcal{I}$. Then, either $\mathbf{V}(\mathcal{I}) \cap \mathbf{V}(p)$ is empty or we have: $\dim(\mathbf{V}(\mathcal{I}) \cap \mathbf{V}(p)) \leq \dim(\mathbf{V}(\mathcal{I})) - 1$.*

Lemma 2. *Let T be a triangular set in $\mathbb{K}[Y]$. Then, we have*

$$\overline{\mathbf{W}(T)} \setminus \mathbf{V}(h_T) = \mathbf{W}(T) \quad \text{and} \quad \overline{\mathbf{W}(T)} \setminus \mathbf{W}(T) = \mathbf{V}(h_T) \cap \overline{\mathbf{W}(T)}.$$

PROOF. Since $\mathbf{W}(T) \subseteq \overline{\mathbf{W}(T)}$, we have

$$\mathbf{W}(T) = \mathbf{W}(T) \setminus \mathbf{V}(h_T) \subseteq \overline{\mathbf{W}(T)} \setminus \mathbf{V}(h_T).$$

On the other hand, $\overline{\mathbf{W}(T)} \subseteq \mathbf{V}(T)$ implies

$$\overline{\mathbf{W}(T)} \setminus \mathbf{V}(h_T) \subseteq \mathbf{V}(T) \setminus \mathbf{V}(h_T) = \mathbf{W}(T).$$

This proves the first claim. Observe that we have:

$$\overline{\mathbf{W}(T)} = \left(\overline{\mathbf{W}(T)} \setminus \mathbf{V}(h_T) \right) \cup \left(\overline{\mathbf{W}(T)} \cap \mathbf{V}(h_T) \right).$$

We deduce the second one.

Lemma 3 ([1,3]). *Let T be a triangular set in $\mathbb{K}[Y]$. Then, we have*

$$\mathbf{V}(\mathrm{Sat}(T)) = \overline{\mathbf{W}(T)}.$$

Assume furthermore that $\mathbf{W}(T) \neq \emptyset$ holds. Then $\mathbf{V}(\mathrm{Sat}(T))$ is a nonempty unmixed algebraic set with dimension $n - |T|$. Moreover, if N is the free variables of T, then for every prime ideal $\mathcal{P}$ associated with $\mathrm{Sat}(T)$ we have

$$\mathcal{P} \cap \mathbb{K}[N] = \langle 0 \rangle.$$

2.2   Regular Chain and Regular System

Definition 2 (Regular Chain). *A triangular set $T \subset \mathbb{K}[Y]$ is a regular chain if one of the following conditions hold:*

- *either T is empty,*
- *or $T \setminus \{T_{\max}\}$ is a regular chain, where $T_{\max}$ is the polynomial in T with maximum rank, and the initial of $T_{\max}$ is regular w.r.t. $\mathrm{Sat}(T \setminus \{T_{\max}\})$.*

It is useful to extend the notion of regular chain as follows.

Definition 3 (Regular System). *A pair $[T, h]$ is a regular system if T is a regular chain, and $h \in \mathbb{K}[Y]$ is regular w.r.t $\mathrm{Sat}(T)$.*

Remark 1. *A regular system in a stronger sense was presented in [19]. For example, consider the polynomial system $[T, h]$ where $T = [Y_1 Y_4 - Y_2]$ and $h = Y_2 Y_3$. Then $[T, h]$ is still a regular system in our sense but not a regular system in Wang's sense. Also we do not restrict the main variables of polynomials in the inequality part. At least our definition is more convenient for our purpose in dealing with zerodivisors and conceptually clear as well. We also note that in the zerodimensional case (no free variables exist) the notion of regular chain and that of a regular set in [19] are the same, see [1,19] for details.*

There are several equivalent characterizations of a regular chain, see [1]. In this paper, we rely on the notion of *iterated resultant* in order to derive a characterization which can be checked by solving a polynomial system.

Definition 4. *Let $p \in \mathbb{K}[Y]$ be a polynomial and $T \subset \mathbb{K}[Y]$ be a triangular set. The iterated resultant of p w.r.t. T, denoted by $\mathrm{res}(p, T)$, is defined as follows:*

- *if $p \in \mathbb{K}$ or all variables in p are free w.r.t. T, then $\mathrm{res}(p, T) = p$,*
- *otherwise, if v is the largest variable of p which is algebraic w.r.t. T, then $\mathrm{res}(p, T) = \mathrm{res}(r, T_{<v})$ where r is the resultant of p and the polynomial T_v.*

Lemma 4. *Let $p \in \mathbb{K}[Y]$ be a polynomial and $T \subset \mathbb{K}[Y]$ be a zerodimensional regular chain. Then the following statements are equivalent:*

- (i) *The iterated resultant $\mathrm{res}(p, T) \neq 0$.*
- (ii) *The polynomial p is regular modulo $\langle T \rangle$.*
- (iii) *The polynomial p is invertible modulo $\langle T \rangle$.*

PROOF. "$(i) \Rightarrow (ii)$" Let $r := \mathrm{res}(p, T)$. Then there exist polynomials $A_i \in \mathbb{K}[Y]$, $0 \leq i \leq n$, such that $r = A_0 p + \sum_{i=1}^{n} A_i T_i$. So $r \neq 0$ implies p is invertible modulo $\langle T \rangle$. Therefore, p is regular modulo $\langle T \rangle$.

"$(ii) \Rightarrow (iii)$" If p is regular modulo $\langle T \rangle$, then p is regular modulo $\sqrt{\langle T \rangle}$. Since T is a zerodimensional regular chain, which implies $\mathrm{Sat}(T) = \langle T \rangle$, we know that $\mathbb{K}[Y]/\sqrt{\langle T \rangle}$ is a direct product of fields. Therefore p is invertible modulo $\sqrt{\langle T \rangle}$, which implies p is invertible modulo $\langle T \rangle$.

"$(iii) \Rightarrow (i)$" Assume $\mathrm{res}(p, T) = 0$, then we claim that p and T have at least one common solution, which is a contradiction to (iii).

We prove our claim by induction on $|T|$.

If $|T| = 1$, we have two cases

(1) If all variables in p are free w.r.t. T, then $\mathrm{res}(p, T) = p = 0$. The claim holds.
(2) Otherwise, we have $\mathrm{res}(p, T) = \mathrm{res}(p, T, \mathrm{mvar}(T)) = 0$. Since $\mathrm{init}(T) \neq 0$, the claim holds.

Now we assume that the claim holds for $|T| = n - 1$. If $|T| = n$, let $v := Y_n$. We have two cases

(1) If v does not appear in p, then $\mathrm{res}(p, T) = \mathrm{res}(p, T_{<v})$. By induction hypothesis, there exist $\xi_1, \xi_2, \cdots, \xi_{n-1} \in \overline{\mathbb{K}}$, such that $\xi' = (\xi_1, \xi_2, \cdots, \xi_{n-1})$ is a common solution of p and $T_{<v}$. Since T is a zerodimensional regular chain, h_{T_v} is invertible modulo $\langle T \rangle$ (by "$(ii) \Rightarrow (iii)$"). So $h_{T_v}(\xi') \neq 0$, which implies that there exists a $\xi_n \in \overline{\mathbb{K}}$, such that $\xi := (\xi_1, \xi_2, \cdots, \xi_{n-1}, \xi_n)$ is a solution of T_v. Therefore ξ is a common solution of p and T.
(2) If v appears in p, then $\mathrm{res}(p, T) = \mathrm{res}(\mathrm{res}(p, T_v, v), T_{<v}) = 0$. Similarly to (1), there exists $\xi' = (\xi_1, \xi_2, \cdots, \xi_{n-1})$, such that $\mathrm{res}(p, T_v, v)(\xi') = T_{<v}(\xi') = 0$ and $h_{T_v}(\xi') \neq 0$. So by the specialization property of resultant, $\mathrm{res}(p(\xi'), T_v(\xi'), v) = 0$, which implies that there exists a $\xi_n \in \overline{\mathbb{K}}$, such that $\xi := (\xi_1, \xi_2, \cdots, \xi_{n-1}, \xi_n)$ is a common solution of p and T_v. Therefore ξ is a common solution of p and T_v.

Theorem 1. *The triangular set T is a regular chain if and only if* $\mathrm{res}(h_T, T) \neq 0$.

PROOF. We start by assuming that T is a zerodimensional regular chain, then the conclusion follows from Lemma 4.

We reduce the general case to the zerodimensional one. First, we introduce a new total ordering $<_T$ on Y defined as follows: if Y_i and Y_j are both in $\mathrm{mvar}(T)$ or both in its complement then $Y_i <_T Y_j$ holds if and only if $Y_i < Y_j$ holds, otherwise $Y_i <_T Y_j$ holds if and only if $Y_j \in \mathrm{mvar}(T)$. Clearly T is also a triangular set w.r.t $<_T$. We observe that h_T, and thus $\mathrm{Sat}(T)$, are unchanged when replacing the variable ordering $<$ by $<_T$. Similarly, it is easy to check that a polynomial $p \in \mathbb{K}[Y]$ reduces to zero by pseudo-division by T w.r.t. $<$ if and only if it reduces to zero by pseudo-division by T w.r.t. $<_T$. Therefore, by applying Theorem 6.1 [1] we deduce that T is a regular chain w.r.t. $<$ if and only if it is a regular chain w.r.t. $<_T$. Similarly, we have $\mathrm{res}(h_T, T) \neq 0$ w.r.t. $<$ if and only if $\mathrm{res}(h_T, T) \neq 0$ w.r.t. $<_T$.

Now we assume that the variables are ordered according to $<_T$. Let N be the set of the variables of Y that do not belong to $\mathrm{mvar}(T)$. The triangular set T is a regular chain in $\mathbb{K}[Y]$ if and only if it is a zerodimensional regular chain when regarded as a triangular set in $\mathbb{K}(N)[Y \setminus N]$ (where $\mathbb{K}(N)$ denotes the field of rational functions with coefficients in $\mathbb{K}$ and variables in N). This is Corollary 3.2 in [3]. Similarly, it is easy to check that $\mathrm{res}(h_T, T) \neq 0$ holds when regarding T in $\mathbb{K}[Y]$ if and only if $\mathrm{res}(h_T, T) \neq 0$ holds when regarding T in $\mathbb{K}(N)[Y \setminus N]$.

Proposition 1. *For every regular system $[T, h]$ we have $\mathbf{Z}(T, h) \neq \emptyset$.*

PROOF. Since T is a regular chain, by Lemma 3 we have $\mathbf{V}(\mathrm{Sat}(T)) \neq \emptyset$. By definition of regular system, the polynomial hh_T is regular w.r.t $\mathrm{Sat}(T)$. Hence, by Lemma 1, the set $\mathbf{V}(hh_T) \cap \mathbf{V}(\mathrm{Sat}(T))$ either is empty, or has lower dimension than $\mathbf{V}(\mathrm{Sat}(T))$. Therefore, the set

$$\mathbf{V}(\mathrm{Sat}(T)) \setminus \mathbf{V}(hh_T) = \mathbf{V}(\mathrm{Sat}(T)) \setminus (\mathbf{V}(hh_T) \cap \mathbf{V}(\mathrm{Sat}(T)))$$

is not empty. Finally, by Lemma 2, the set

$$\mathbf{Z}(T, h) = \mathbf{W}(T) \setminus \mathbf{V}(h) = \overline{\mathbf{W}(T)} \setminus \mathbf{V}(hh_T) = \mathbf{V}(\mathrm{Sat}(T)) \setminus \mathbf{V}(hh_T)$$

is not empty. ∎

Notation 1. *For a regular system $R = [T, h]$, we define* $\mathrm{rank}(R) := \mathrm{rank}(T)$. *For a set $\mathcal{R}$ of regular systems, we define*

$$\mathrm{rank}(\mathcal{R}) := \max\{\mathrm{rank}(T) \mid [T, h] \in \mathcal{R}\}.$$

For a pair of regular systems (L, R), we define $\mathrm{rank}((L, R)) := (\mathrm{rank}(L), \mathrm{rank}(R))$. *For a pair of lists of regular systems, we define*

$$\mathrm{rank}((\mathcal{L}, \mathcal{R})) = (\mathrm{rank}(\mathcal{L}), \mathrm{rank}(\mathcal{R})).$$

For triangular sets $T, T_1, \ldots, T_e$ we write $\mathbf{W}(T) \xrightarrow{D} (\mathbf{W}(T_i), i = 1 \ldots e)$ if one of the following conditions holds:

– *either $e = 1$ and $T = T_1$,*
– *or $e > 1$, $\mathrm{rank}(T_i) < \mathrm{rank}(T)$ for all $i = 1 \ldots e$ and*

$$\mathbf{W}(T) \subseteq \bigcup_{i=1}^{e} \mathbf{W}(T_i) \subseteq \overline{\mathbf{W}(T)}.$$

2.3 Triangular Decompositions

Definition 5. *Given a finite polynomial set $F \subset \mathbb{K}[Y]$, a triangular decomposition of $\mathbf{V}(F)$ is a finite family $\mathcal{T}$ of regular chains of $\mathbb{K}[Y]$ such that*

$$\mathbf{V}(F) = \bigcup_{T \in \mathcal{T}} \mathbf{W}(T).$$

For a finite polynomial set $F \subset \mathbb{K}[Y]$, the TRIADE algorithm [15] computes a triangular decomposition of $\mathbf{V}(F)$. We list below the specifications of the operations from TRIADE that we use in this paper.

Let p, p_1, p_2 be polynomials, and let T, C, E be regular chains such that $C \cup E$ is a triangular set (but not necessarily a regular chain).

– **Regularize**(p, T) returns regular chains $T_1, \ldots, T_e$ such that
  - $\mathbf{W}(T) \xrightarrow{D} (\mathbf{W}(T_i), i = 1 \ldots e)$,
  - for all $1 \le i \le e$ the polynomial p is either 0 or regular modulo $\mathrm{Sat}(T_i)$.

- For a set of polynomials F, **Triangularize**(F, T) returns regular chains $T_1, \ldots, T_e$ such that we have

$$\mathbf{V}(F) \cap \mathbf{W}(T) \subseteq \mathbf{W}(T_1) \cup \cdots \cup \mathbf{W}(T_e) \subseteq \mathbf{V}(F) \cap \overline{\mathbf{W}(T)}.$$

  and for $1 \leq i \leq e$ we have rank$(T_i) <$ rank(T).
- **Extend**$(C \cup E)$ returns a set of regular chains $\{C_i \mid i = 1 \ldots e\}$ such that we have $\mathbf{W}(C \cup E) \xrightarrow{D} (\mathbf{W}(C_i), i = 1 \ldots e)$.
- Assume that p_1 and p_2 are two non-constant polynomials with the same main variable v, which is larger than any variable appearing in T, and assume that the initials of p_1 and p_2 are both regular w.r.t. Sat(T). Then, **GCD**(p_1, p_2, T) returns a sequence

$$([g_1, C_1], \ldots, [g_d, C_d], [\emptyset, D_1], \ldots, [\emptyset, D_e]),$$

where g_i are polynomials and C_i, D_i are regular chains such that the following properties hold:
- $\mathbf{W}(T) \xrightarrow{D} (\mathbf{W}(C_1), \ldots, \mathbf{W}(C_d), \mathbf{W}(D_1), \ldots, \mathbf{W}(D_e))$,
- $\dim \mathbf{V}(\text{Sat}(C_i)) = \dim \mathbf{V}(\text{Sat}(T))$ and $\dim \mathbf{V}(\text{Sat}(D_j)) < \dim \mathbf{V}(\text{Sat}(T))$, for all $1 \leqslant i \leqslant d$ and $1 \leqslant j \leqslant e$,
- the leading coefficient of g_i w.r.t. v is regular w.r.t. Sat(C_i),
- for all $1 \leqslant i \leqslant d$ there exist polynomials u_i and v_i such that we have $g_i = u_i p_1 + v_i p_2 \mod \text{Sat}(C_i)$,
- if g_i is not constant and its main variable is v, then p_1 and p_2 belong to Sat$(C_i \cup \{g_i\})$.

2.4   Constructible Sets

Definition 6 (Constructible set). *A constructible subset of $\overline{\mathbb{K}}^n$ is any finite union*

$$(A_1 \setminus B_1) \cup \cdots \cup (A_e \setminus B_e)$$

where $A_1, \ldots, A_e, B_1, \ldots, B_e$ are algebraic varieties in $\overline{\mathbb{K}}^n$.

Lemma 5. *Every constructible set can write as a union of zero sets of regular systems.*

PROOF. By the definition of constructible set, we only need to prove that the difference of two algebraic varieties can write as a union of zero sets of regular systems. Let $\mathbf{V}(F), \mathbf{V}(G)$, where $F, G \subset \mathbb{K}[Y]$, be two algebraic varieties in $\overline{\mathbb{K}}^n$. With the **Triangularize** operation introduced in last subsection, we write $\mathbf{V}(F)$ as a union of the zero sets of some regular systems

$$\mathbf{V}(F) = \bigcup_{i=1}^{s} \mathbf{W}(T_i) = \bigcup_{i=1}^{s} \mathbf{Z}(T_i, 1).$$

Similarly, we can write $\mathbf{V}(G)$ as

$$\mathbf{V}(G) = \bigcup_{i=1}^{t} \mathbf{Z}(C_i, 1).$$

Then the conclusion follows from the algorithm **DifferenceLR** introduced in next section.

3   The Difference Algorithms

In this section, we present an algorithm to compute the set theoretical difference of two constructible sets given by regular systems. As mentioned in the Introduction, a naive approach appears to be very inefficient in practice. Here we contribute a more sophisticated algorithm, which heavily exploits the structure and properties of regular chains.

Two procedures, **Difference** and **DifferenceLR**, are involved in order to achieve this goal. Their specifications and pseudo-codes can be found below. The rest of this section is dedicated to proving the correctness and termination of these algorithms. For the pseudo-code, we use the MAPLE syntax. However, each of the two functions below returns a sequence of values. Individual value or sub-sequences of the returned sequence are thrown to the flow of output by means of an **output** statement. Hence an **output** statement does not cause the termination of the function execution.

Algorithm 1 Difference($[T, h], [T', h']$)
> **Input** $[T, h]$, $[T', h']$ two regular systems.
> **Output** Regular systems $\{[T_i, h_i] \mid i = 1 \ldots e\}$ such that
>
> $$\mathbf{Z}(T, h) \setminus \mathbf{Z}(T', h') = \bigcup_{i=1}^{e} \mathbf{Z}(T_i, h_i),$$
>
> and $\operatorname{rank}(T_i) \leqslant_r \operatorname{rank}(T)$.

Algorithm 2 DifferenceLR($\mathcal{L}, \mathcal{R}$)
> **Input** $\mathcal{L} := \{[L_i, f_i] \mid i = 1 \ldots r\}$ and $\mathcal{R} := \{[R_j, g_j] \mid j = 1 \ldots s\}$ two lists of regular systems.
> **Output** Regular systems $\mathcal{S} := \{[T_i, h_i] \mid i = 1 \ldots e\}$ such that
>
> $$\left(\bigcup_{i=1}^{r} \mathbf{Z}(L_i, f_i) \right) \setminus \left(\bigcup_{j=1}^{s} \mathbf{Z}(R_j, g_j) \right) = \bigcup_{i=1}^{e} \mathbf{Z}(T_i, h_i),$$
>
> with $\operatorname{rank}(\mathcal{S}) \leqslant_r \operatorname{rank}(\mathcal{L})$.

To prove the termination and correctness of above two algorithms, we present a series of technical lemmas.

Lemma 6. *Let p and h be polynomials and T a regular chain. Assume that $p \notin \operatorname{Sat}(T)$. Then there exists an operation* **Intersect**(p, T, h) *returning a set of regular chains* $\{T_1, \ldots, T_e\}$ *such that*

(i) h *is regular w.r.t* $\operatorname{Sat}(T_i)$ *for all i;*
(ii) $\operatorname{rank}(T_i) <_r \operatorname{rank}(T)$;
(iii) $\mathbf{Z}(p, T, h) \subseteq \cup_{i=1}^{e} \mathbf{Z}(T_i, h) \subseteq (\mathbf{V}(p) \cap \overline{\mathbf{W}(T)}) \setminus \mathbf{V}(h)$;
(iv) *Moreover, if the product of initials h_T of T divides h then*

$$\mathbf{Z}(p, T, h) = \bigcup_{i=1}^{e} \mathbf{Z}(T_i, h).$$

Algorithm 1. Difference($[T, h], [T', h']$)

1: **if** $\text{Sat}(T) = \text{Sat}(T')$ **then**
2:     **output** $\text{Intersect}(h'h_{T'}, T, hh_T)$
3: **else**
4:     Let v be the largest variable s.t. $\text{Sat}(T_{<v}) = \text{Sat}(T'_{<v})$
5:     **if** $v \in \text{mvar}(T')$ and $v \notin \text{mvar}(T)$ **then**
6:         $p' \leftarrow T'_v$
7:         **output** $[T, hp']$
8:         **output** $\text{DifferenceLR}(\text{Intersect}(p', T, hh_T), [T', h'])$
9:     **else if** $v \notin \text{mvar}(T')$ and $v \in \text{mvar}(T)$ **then**
10:         $p \leftarrow T_v$
11:         **output** $\text{DifferenceLR}([T, h], \text{Intersect}(p, T', h'h_{T'}))$
12:     **else**
13:         $p \leftarrow T_v$
14:         $\mathcal{G} \leftarrow \mathbf{GCD}(T_v, T'_v, T_{<v})$
15:         **if** $|\mathcal{G}| = 1$ **then**
16:             Let $(g, C) \in \mathcal{G}$
17:             **if** $g \in \mathbb{K}$ **then**
18:                 **output** $[T, h]$
19:             **else if** $\text{mvar}(g) < v$ **then**
20:                 **output** $[T, gh]$
21:                 **output** $\text{DifferenceLR}(\text{Intersect}(g, T, hh_T), [T', h'])$
22:             **else if** $\text{mvar}(g) = v$ **then**
23:                 **if** $\text{mdeg}(g) = \text{mdeg}(p)$ **then**
24:                     $D'_p \leftarrow T'_{<v} \cup \{p\} \cup T'_{>v}$
25:                     **output** $\text{Difference}([T, h], [D'_p, h'h_{T'}])$
26:                 **else if** $\text{mdeg}(g) < \text{mdeg}(p)$ **then**
27:                     $q \leftarrow \text{pquo}(p, g, C)$
28:                     $D_g \leftarrow C \cup \{g\} \cup T_{>v}$
29:                     $D_q \leftarrow C \cup \{q\} \cup T_{>v}$
30:                     **output** $\text{Difference}([D_g, hh_T], [T', h'])$
31:                     **output** $\text{Difference}([D_q, hh_T], [T', h'])$
32:                     **output** $\text{DifferenceLR}(\text{Intersect}(h_g, T, hh_T), [T', h'])$
33:                 **end if**
34:             **end if**
35:         **else if** $|\mathcal{G}| \geq 2$ **then**
36:             **for** $(g, C) \in \mathcal{G}$ **do**
37:                 **if** $|C| > |T_{<v}|$ **then**
38:                     **for** $E \in \text{Extend}(C, T_{\geqslant v})$ **do**
39:                         **for** $D \in \text{Regularize}(hh_T, E)$ **do**
40:                             **if** $hh_T \notin \text{Sat}(D)$ **then**
41:                               **output** $\text{Difference}([D, hh_T], [T', h'])$
42:                           **end if**
43:                       **end for**
44:                   **end for**
45:                 **else**
46:                   **output** $\text{Difference}([C \cup T_{\geqslant v}, hh_T], [T', h'])$
47:                 **end if**
48:             **end for**
49:         **end if**
50:     **end if**
51: **end if**

Algorithm 2. DifferenceLR(L, R)

1: **if** $L = \emptyset$ **then**
2:     output $\emptyset$
3: **else if** $R = \emptyset$ **then**
4:     output L
5: **else if** $|R| = 1$ **then**
6:     Let $[T', h'] \in R$
7:     **for** $[T, h] \in L$ **do**
8:         output **Difference**$([T, h], [T', h'])$
9:     **end for**
10: **else**
11:     **while** $R \neq \emptyset$ **do**
12:         Let $[T', h'] \in R, R \leftarrow R \setminus \{[T', h']\}$
13:         $S \leftarrow \emptyset$
14:         **for** $[T, h] \in L$ **do**
15:             $S \leftarrow S \cup$ **Difference**$([T, h], [T', h'])$
16:         **end for**
17:         $L \leftarrow S$
18:     **end while**
19:     output L
20: **end if**

PROOF. Let

$$S = \textbf{Triangularize}(p, T),$$
$$\mathcal{R} = \bigcup_{C \in S} \textbf{Regularize}(h, C).$$

We then have

$$\mathbf{V}(p) \cap \mathbf{W}(T) \subseteq \bigcup_{R \in \mathcal{R}} \subseteq \mathbf{V}(p) \cap \overline{\mathbf{W}(T)}.$$

This implies

$$\mathbf{Z}(p, T, h) \subseteq \bigcup_{R \in \mathcal{R},\, h \notin \text{Sat}(R)} \mathbf{Z}(R, h) \subseteq (\mathbf{V}(p) \cap \overline{\mathbf{W}(T)}) \setminus \mathbf{V}(h).$$

Rename the regular chains $\{R \mid R \in \mathcal{R},\ h \notin \text{Sat}(R)\}$ as $\{T_1, \ldots, T_e\}$. By the specification of **Regularize** we immediately conclude that (i), (iii) hold. Since $p \notin \text{Sat}(T)$, by the specification of **Triangularize**, (ii) holds. By Lemma 2, (iv) holds.

Lemma 7. *Let $[T, h]$ and $[T', h']$ be two regular systems. If* $\text{Sat}(T) = \text{Sat}(T')$, *then* $h' h_{T'}$ *is regular w.r.t* $\text{Sat}(T)$ *and*

$$\mathbf{Z}(T, h) \setminus \mathbf{Z}(T', h') = \mathbf{Z}(h' h_{T'}, T, h h_T).$$

PROOF. Since $\text{Sat}(T) = \text{Sat}(T')$ and $h'h_{T'}$ is regular w.r.t $\text{Sat}(T')$, $h'h_{T'}$ is regular w.r.t $\text{Sat}(T)$. By Lemma 2 and Lemma 3, we have

$$
\begin{aligned}
\mathbf{Z}(T, hh'h_{T'}) &= \mathbf{W}(T) \setminus \mathbf{V}(hh'h_{T'}) \\
&= \overline{\mathbf{W}(T)} \setminus \mathbf{V}(hh'h_T h_{T'}) \\
&= \overline{\mathbf{W}(T')} \setminus \mathbf{V}(hh'h_T h_{T'}) \\
&= \mathbf{W}(T') \setminus \mathbf{V}(hh'h_T) \\
&= \mathbf{Z}(T', hh'h_T).
\end{aligned}
$$

Then, we can decompose $\mathbf{Z}(T, h)$ into the disjoint union

$$
\mathbf{Z}(T, h) = \mathbf{Z}(T, hh'h_{T'}) \bigsqcup \mathbf{Z}(h'h_{T'}, T, hh_T).
$$

Similarly, we have:

$$
\mathbf{Z}(T', h') = \mathbf{Z}(T', hh'h_T) \bigsqcup \mathbf{Z}(hh_T, T', h'h_{T'}).
$$

The conclusion follows from the fact that

$$
\mathbf{Z}(T, hh'h_{T'}) \setminus \mathbf{Z}(T', hh'h_T) = \emptyset \quad \text{and} \quad \mathbf{Z}(h'h_{T'}, T, hh_T) \cap \mathbf{Z}(T', h') = \emptyset.
$$

Lemma 8. *Assume that* $\text{Sat}(T_{<v}) = \text{Sat}(T'_{<v})$. *We have*

(i) *If* $p' := T'_v$ *is defined but not* T_v, *then* p' *is regular w.r.t* $\text{Sat}(T)$ *and*

$$
\mathbf{Z}(T, h) \setminus \mathbf{Z}(T', h') = \mathbf{Z}(T, hp') \bigsqcup (\mathbf{Z}(p', T, hh_T) \setminus \mathbf{Z}(T', h')).
$$

(ii) *If* $p := T_v$ *is defined but not* T'_v, *then* p *is regular w.r.t* $\text{Sat}(T')$ *and*

$$
\mathbf{Z}(T, h) \setminus \mathbf{Z}(T', h') = \mathbf{Z}(T, h) \setminus \mathbf{Z}(p, T', h'h_{T'}).
$$

PROOF. (i) As $\text{init}(p')$ is regular w.r.t $\text{Sat}(T'_{<v})$, it is also regular w.r.t $\text{Sat}(T_{<v})$. Since T_v is not defined, we know $v \notin \text{mvar}(T)$. Therefore, p' is also regular w.r.t $\text{Sat}(T)$. On the other hand, we have a disjoint decomposition

$$
\mathbf{Z}(T, h) = \mathbf{Z}(T, hp') \bigsqcup \mathbf{Z}(p', T, hh_T).
$$

By the definition of p', $\mathbf{Z}(T', h') \subseteq \mathbf{V}(p')$ which implies

$$
\mathbf{Z}(T, hp') \bigcap \mathbf{Z}(T', h') = \emptyset.
$$

The conclusion follows.

(ii) Similarly, we know p is regular w.r.t $\text{Sat}(T')$. By the disjoint decomposition

$$
\mathbf{Z}(T', h') = \mathbf{Z}(T', h'p) \bigsqcup \mathbf{Z}(p, T', h'h_{T'}),
$$

and $\mathbf{Z}(T, h) \cap \mathbf{Z}(T', h'p) = \emptyset$, we have

$$
\mathbf{Z}(T, h) \setminus \mathbf{Z}(T', h') = \mathbf{Z}(T, h) \setminus \mathbf{Z}(p, T', h'h_{T'}),
$$

from which the conclusion follows.

Lemma 9. *Assume that* $\text{Sat}(T_{<v}) = \text{Sat}(T'_{<v})$ *but* $\text{Sat}(T_{\leqslant v}) \neq \text{Sat}(T'_{\leqslant v})$ *and that* v *is algebraic w.r.t both* T *and* T'. *Define*

$$\mathcal{G} = \mathbf{GCD}(T_v, T'_v, T_{<v});$$

$$\mathcal{E} = \bigcup_{(g,C)\in\mathcal{G},\ |C|>|T_{<v}|} \mathbf{Extend}(C, T_{\geqslant v});$$

$$\mathcal{R} = \bigcup_{E\in\mathcal{E}} \mathbf{Regularize}(hh_T, E).$$

Then we have

(i)

$$\mathbf{Z}(T, h)$$

$$= \left(\bigcup_{R\in\mathcal{R},\ hh_T\notin\text{Sat}(R)} \mathbf{Z}(R, hh_T) \right) \bigcup \left(\bigcup_{(g,C)\in\mathcal{G},\ |C|=|T_{<v}|} \mathbf{Z}(C\cup T_{\geqslant v}, hh_T) \right).$$

(ii) $\text{rank}(R) <_r \text{rank}(T)$, *for all* $R \in \mathcal{R}$.
(iii) *Assume that* $|C| = |T_{<v}|$. *Then*
(iii.a) $C \cup T_{\geqslant v}$ *is a regular chain and* hh_T *is regular w.r.t it.*
(iii.b) *If* $|\mathcal{G}| > 1$, *then* $\text{rank}(C \cup T_{\geqslant v}) <_r \text{rank}(T)$.

PROOF. By the specification of **GCD** we have

$$\mathbf{W}(T_{<v}) \subseteq \bigcup_{(g,C)\in\mathcal{G}} \mathbf{W}(C) \subseteq \overline{\mathbf{W}(T_{<v})}.$$

That is,

$$\mathbf{W}(T_{<v}) \xrightarrow{D} (\mathbf{W}(C), (g,C) \in \mathcal{G}).$$

From the specification of **Extend** we have: for each $(g,C) \in \mathcal{G}$ such that $|C| > |T_{<v}|$,

$$\mathbf{W}(C\cup T_{\geqslant v}) \xrightarrow{D} (\mathbf{W}(E), E \in \mathbf{Extend}(C \cup T_{\geqslant v})).$$

From the specification of **Regularize**, we have for all $(g,C) \in \mathcal{G}$ such that $|C| > |T_{<v}|$ and all $E \in \mathbf{Extend}(C \cup T_{\geqslant v})$,

$$\mathbf{W}(E) \xrightarrow{D} (\mathbf{W}(R), R \in \mathbf{Regularize}(hh_T, E)).$$

Therefore, by applying the Lifting Theorem [15] we have:

$$\mathbf{W}(T) = \mathbf{W}(T_{<v} \cup T_{\geqslant v})$$

$$\subseteq \left(\bigcup_{R\in\mathcal{R}} \mathbf{W}(R) \right) \bigcup \left(\bigcup_{(g,C)\in\mathcal{G},\ |C|=|T_{<v}|} \mathbf{W}(C\cup T_{\geqslant v}) \right)$$

$$\subseteq \overline{\mathbf{W}(T_{<v} \cup T_{\geqslant v})}$$

$$= \overline{\mathbf{W}(T)},$$

which implies,

$$\mathbf{Z}(T, h) = \mathbf{Z}(T, hh_T)$$

$$\subseteq \left(\bigcup_{R \in \mathcal{R}, \, hh_T \notin \mathrm{Sat}(R)} \mathbf{Z}(R, hh_T) \right) \bigcup \left(\bigcup_{(g,C) \in \mathcal{G}, \, |C| = |T_{<v}|} \mathbf{Z}(C \cup T_{\geqslant v}, hh_T) \right)$$

$$\subseteq \overline{\mathbf{W}(T)} \setminus \mathbf{V}(hh_T) = \mathbf{Z}(T, h).$$

So (i) holds. When $|\mathcal{G}| > 1$, by Notation 1, (ii) and $(iii.b)$ hold.

If $|C| = |T_{<v}|$, by Proposition 5 of [15], we conclude that $(iii.a)$ holds.

Lemma 10. *Assume that* $\mathrm{Sat}(T_{<v}) = \mathrm{Sat}(T'_{<v})$ *but* $\mathrm{Sat}(T_{\leqslant v}) \neq \mathrm{Sat}(T'_{\leqslant v})$ *and that* v *is algebraic w.r.t both* T *and* T'. *Define* $p = T_v$, $p' = T'_v$ *and*

$$\mathcal{G} = \mathbf{GCD}(p, p', T_{<v}).$$

If $|\mathcal{G}| = 1$, *let* $\mathcal{G} = \{(g, C)\}$. *Then the following properties hold*

(i) $C = T_{<v}$.
(ii) *If* $g \in \mathbb{K}$, *then*

$$\mathbf{Z}(T, h) \setminus \mathbf{Z}(T', h') = \mathbf{Z}(T, h).$$

(iii) *If* $g \notin \mathbb{K}$ *and* $\mathrm{mvar}(g) < v$, *then* g *is regular w.r.t* $\mathrm{Sat}(T)$ *and*

$$\mathbf{Z}(T, h) \setminus \mathbf{Z}(T', h')$$

$$= \mathbf{Z}(T, gh) \bigsqcup (\mathbf{Z}(g, T, hh_T) \setminus \mathbf{Z}(T', h')) .$$

(iv) *Assume that* $\mathrm{mvar}(g) = v$.
$(iv.a)$ *If* $\mathrm{mdeg}(g) = \mathrm{mdeg}(p)$, *defining*

$$q' = \mathrm{pquo}(p', p, T'_{<v})$$
$$D'_p = T'_{<v} \cup \{p\} \cup T'_{>v}$$
$$D'_{q'} = T'_{<v} \cup \{q'\} \cup T'_{>v},$$

then we have

$$\mathbf{Z}(T, h) \setminus \mathbf{Z}(T', h') = \mathbf{Z}(T, h) \setminus \mathbf{Z}(D'_p, h'h_{T'}),$$

$\mathrm{rank}(D'_p) < \mathrm{rank}(T')$ *and* $h'h_{T'}$ *is regular w.r.t* $\mathrm{Sat}(D'_p)$.
$(iv.b)$ *If* $\mathrm{mdeg}(g) < \mathrm{mdeg}(p)$, *defining*

$$q = \mathrm{pquo}(p, g, T_{<v})$$
$$D_g = T_{<v} \cup \{g\} \cup T_{>v}$$
$$D_q = T_{<v} \cup \{q\} \cup T_{>v},$$

then we have: D_g *and* D_q *are regular chains such that* $\mathrm{rank}(D_g) < \mathrm{rank}(T)$, $\mathrm{rank}(D_q) < \mathrm{rank}(T)$, hh_T *is regular w.r.t* $\mathrm{Sat}(D_g)$ *and* $\mathrm{Sat}(D_q)$, *and*

$$\mathbf{Z}(T, h) = \mathbf{Z}(D_g, hh_T) \bigcup \mathbf{Z}(D_q, hh_T) \bigcup \mathbf{Z}(h_g, T, hh_T).$$

PROOF. Since $|\mathcal{G}| = 1$, by the specification of the operation **GCD** and Notation 1, (i) holds. Therefore we have

$$\text{Sat}(C) = \text{Sat}(T_{<v}) = \text{Sat}(T'_{<v}) \tag{1}$$

There exist polynomials A and B such that

$$g \equiv Ap + Bp' \quad \text{mod} \quad \text{Sat}(C). \tag{2}$$

From (2), we have

$$\mathbf{V}(\text{Sat}(C)) \subseteq \mathbf{V}(g - Ap - Bp') \tag{3}$$

Therefore, we deduce

$$
\begin{aligned}
& \mathbf{W}(T) \bigcap \mathbf{W}(T') \\
&= \mathbf{W}(T_{<v} \cup p \cup T_{\geqslant v}) \bigcap \mathbf{W}(T'_{<v} \cup p' \cup T'_{\geqslant v}) \\
&\subseteq (\mathbf{W}(T_{<v}) \cap \mathbf{V}(p)) \bigcap (\mathbf{W}(T'_{<v}) \cap \mathbf{V}(p')) \\
&\subseteq \mathbf{V}(\text{Sat}(T_{<v})) \bigcap \mathbf{V}(p) \bigcap \mathbf{V}(p') && \text{by (1)} \\
&\subseteq \mathbf{V}(g - Ap - Bp') \bigcap \mathbf{V}(p) \bigcap \mathbf{V}(p') && \text{by (3)} \\
&\subseteq \mathbf{V}(g).
\end{aligned}
$$

that is

$$\mathbf{W}(T) \bigcap \mathbf{W}(T') \subseteq \mathbf{V}(g). \tag{4}$$

Now we prove (ii). When $g \in \mathbb{K}$, $g \neq 0$, from (4) we deduce

$$\mathbf{W}(T) \bigcap \mathbf{W}(T') = \emptyset. \tag{5}$$

Thus we have

$$
\begin{aligned}
& \mathbf{Z}(T, h) \setminus \mathbf{Z}(T', h') \\
&= (\mathbf{W}(T) \setminus \mathbf{V}(h)) \setminus (\mathbf{W}(T') \setminus \mathbf{V}(h')) \\
&= (\mathbf{W}(T) \setminus \mathbf{V}(h)) && \text{by (5)} \\
&= \mathbf{Z}(T, h).
\end{aligned}
$$

Now we prove (iii). Since $C = T_{<v}$ and $\text{mvar}(g)$ is smaller than or equal to v, by the specification of **GCD**, g is regular w.r.t $\text{Sat}(T)$. We have following decompositions

$$
\begin{aligned}
\mathbf{Z}(T, h) &= \mathbf{Z}(T, gh) \bigsqcup \mathbf{Z}(g, T, hh_T), \\
\mathbf{Z}(T', h') &= \mathbf{Z}(T', gh') \bigsqcup \mathbf{Z}(g, T', h'h_{T'}).
\end{aligned}
$$

On the other hand,

$$\mathbf{Z}(T, gh) \bigcap \mathbf{Z}(T', gh')$$
$$= (\mathbf{W}(T) \cap \mathbf{V}(gh)^c) \bigcap (\mathbf{W}(T') \cap \mathbf{V}(gh')^c)$$
$$\subseteq (\mathbf{W}(T) \cap \mathbf{V}(g)^c) \bigcap (\mathbf{W}(T') \cap \mathbf{V}(g)^c)$$
$$= (\mathbf{W}(T) \cap \mathbf{W}(T')) \bigcap \mathbf{V}(g)^c$$
$$= \emptyset \qquad \text{by (4).}$$

Therefore,

$$\mathbf{Z}(T, h) \setminus \mathbf{Z}(T', h')$$
$$= (\mathbf{Z}(T, gh) \setminus \mathbf{Z}(T', gh')) \bigsqcup (\mathbf{Z}(g, T, hh_T) \setminus \mathbf{Z}(T', h'))$$
$$= \mathbf{Z}(T, gh) \bigsqcup (\mathbf{Z}(g, T, hh_T) \setminus \mathbf{Z}(T', h')).$$

Now we prove $(iv.a)$. First, both h' and h'_T are regular w.r.t $\mathrm{Sat}(C) = \mathrm{Sat}(T_{<v}) = \mathrm{Sat}(T'_{<v})$. From the construction of D'_p, we have $h'h_{T'}$ is regular w.r.t $\mathrm{Sat}(D'_p)$.

Assume that $\mathrm{mvar}(g) = v$ and $\mathrm{mdeg}(g) = \mathrm{mdeg}(p)$. We note that $\mathrm{mdeg}(p') > \mathrm{mdeg}(p)$ holds. Otherwise we would have $\mathrm{mdeg}(g) = \mathrm{mdeg}(p) = \mathrm{mdeg}(p')$ which implies:

$$p \in \mathrm{Sat}(T'_{\geqslant v}) \text{ and } p' \in \mathrm{Sat}(T_{\geqslant v}). \qquad (6)$$

Thus

$$\mathrm{Sat}(T_{\leqslant v}) = \langle T_{\leqslant v} \rangle : h_{T_{\leqslant v}}^\infty = \langle T_{<v} \cup p \rangle : h_{T_{\leqslant v}}^\infty$$
$$\subseteq \mathrm{Sat}(T'_{\leqslant v}) : h_{T_{\leqslant v}}^\infty \qquad \text{by (6)}$$
$$= \mathrm{Sat}(T'_{\leqslant v}),$$

that is $\mathrm{Sat}(T_{\leqslant v}) \subseteq \mathrm{Sat}(T'_{\leqslant v})$. Similarly, $\mathrm{Sat}(T'_{\leqslant v}) \subseteq \mathrm{Sat}(T_{\leqslant v})$ holds. So we have $\mathrm{Sat}(T'_{\leqslant v}) = \mathrm{Sat}(T_{\leqslant v})$, a contradiction.

Hence, $\mathrm{mvar}(q') = v$.

By Lemma 6 [15], we know that D'_p and $D'_{q'}$ are regular chains. Then with Theorem 7 [15] and Lifting Theorem [15], we know

$$\mathbf{Z}(T', h') \subseteq \mathbf{Z}(D'_p, h') \bigcup \mathbf{Z}(D'_{q'}, h') \bigcup \mathbf{Z}(h_p, T', h')$$
$$\subseteq \overline{\mathbf{W}(T')} \setminus \mathbf{V}(h').$$

By Lemma 2, we have

$$\mathbf{Z}(T', h') = \mathbf{Z}(D'_p, h'h_{T'}) \bigcup \mathbf{Z}(D'_{q'}, h'h_{T'}) \bigcup \mathbf{Z}(h_p, T', h'h_{T'}).$$

Since

$$\mathbf{Z}(D'_{q'}, h'h_{T'}) = \mathbf{Z}(D'_{q'}, h_p h'h_{T'}) \bigcup \mathbf{Z}(h_p, D'_{q'}, h'h'_T)$$
$$= \mathbf{Z}(D'_{q'}, ph_p h'h_{T'}) \bigcup \mathbf{Z}(p, D'_{q'}, h_p h'h'_T) \bigcup \mathbf{Z}(h_p, D'_{q'}, h'h'_T)$$

and

$$\mathbf{Z}(p, D'_{q'}, h_p h' h'_T) \subseteq \mathbf{Z}(D'_p, h' h_{T'})$$
$$\mathbf{Z}(h_p, D'_{q'}, h' h'_T) \subseteq \mathbf{Z}(h_p, T', h' h_{T'}),$$

we deduce

$$\mathbf{Z}(T', h') = \mathbf{Z}(D'_p, h' h_{T'}) \bigsqcup \mathbf{Z}(D'_{q'}, p h' h_{T'}) \bigsqcup \mathbf{Z}(h_p, T', h' h_{T'}).$$

Now observe that

$$\mathbf{Z}(T, h) \bigcap \mathbf{Z}(D'_{q'}, p h' h_{T'}) = \emptyset, \text{ and}$$
$$\mathbf{Z}(T, h) \bigcap \mathbf{Z}(h_p, T', h' h_{T'}) = \emptyset.$$

We obtain

$$\mathbf{Z}(T, h) \setminus \mathbf{Z}(T', h') = \mathbf{Z}(T, h) \setminus \mathbf{Z}(D'_p, h' h_{T'}).$$

Finally we prove $(iv.b)$. We assume that $\mathrm{mvar}(g) = v$ and $\mathrm{mdeg}(g) < \mathrm{mdeg}(p)$; this implies $\mathrm{mvar}(q) = v$. Applying Lemma 6 in [15] we know that D_g and D_q are regular chains and satisfy the desired rank condition. Then by Theorem 7 [15] and Lifting Theorem [15] we have

$$\mathbf{Z}(T, h) = \mathbf{Z}(D_g, h h_T) \bigcup \mathbf{Z}(D_q, h h_T) \bigcup \mathbf{Z}(h_g, T, h h_T).$$

This completes the whole proof.

Definition 7. *Given two pairs of ranks* $(\mathrm{rank}(T_1), \mathrm{rank}(T'_1))$ *and* $(\mathrm{rank}(T_2), \mathrm{rank}(T'_2))$, *where* T_1, T_2, T'_1, T'_2 *are triangular sets. We define the product order* $<_p$ *of Ritt order* $<_r$ *on them as follows*

$$(\mathrm{rank}(T_2), \mathrm{rank}(T'_2)) <_p (\mathrm{rank}(T_1), \mathrm{rank}(T'_1))$$
$$\Longleftrightarrow \begin{cases} \mathrm{rank}(T_2) <_r \mathrm{rank}(T_1) & or \\ \mathrm{rank}(T_2) = \mathrm{rank}(T_1), & \mathrm{rank}(T'_2) <_r \mathrm{rank}(T'_1). \end{cases}$$

In the following theorems, we prove the termination and correctness separately. Along with the proof of Theorem 2, we show the rank conditions are satisfied which is part of the correctness. The remained part, say zero set decomposition, will be proved in Theorem 3.

Theorem 2. *Algorithms* **Difference** *and* **DifferenceLR** *terminate and satisfy the rank conditions in their specifications.*

PROOF. The following two statements need to be proved

(i) **Difference** terminates with $\mathrm{rank}(\mathbf{Difference}([T, h], [T', h'])) \leqslant_r \mathrm{rank}([T, h])$,
(ii) **DifferenceLR** terminates with $\mathrm{rank}(\mathbf{DifferenceLR}(\mathcal{L}, \mathcal{R})) \leqslant_r \mathrm{rank}(\mathcal{L})$.

We prove them by induction on the product order $<_p$.

(1) Base case: there are no recursive calls to **Difference** or **DifferenceLR**. The termination of both algorithms is clear. By line 2, 18 of the algorithm **Difference**, rank(**Difference**$([T, h], [T', h'])) \leqslant_r$ rank$([T, h])$. By line 2, 4 of the algorithm **DifferenceLR**, rank(**DifferenceLR**$(\mathcal{L}, \mathcal{R})) \leqslant_r$ rank$(\mathcal{L})$.

(2) Induction hypothesis: assume that both (i) and (ii) hold with inputs whose ranks are smaller than the rank of $([T, h], [T', h'])$ w.r.t. $<_p$.

(3) By (1), if no recursive calls occur in one branch, then (i) and (ii) already hold. When recursive calls occur, by line 8, 11, 21, 25, 30, 31, 32, 41, 46 and Lemma 6, 8, 9, 10, we know the inputs of recursive calls to both **Difference** and **DifferenceLR** have smaller ranks than rank$((([T, h], [T', h']))$ w.r.t $<_p$. By induction hypothesis, (i) holds. Finally, by line 8, 15 of algorithm **DifferenceLR** and (i), (ii) holds.

Theorem 3. *Both* **Difference** *and* **DifferenceLR** *satisfy their specifications.*

PROOF. By Theorem 2, **Difference** and **DifferenceLR** terminate and satisfy their rank conditions. So it suffices to prove the correctness of **Difference** and **DifferenceLR**, that is

(i) $\mathbf{Z}(T, h) \setminus \mathbf{Z}(T', h') = \mathbf{Z}($**Difference**$([T, h], [T', h']))$,
(ii) $\mathbf{Z}(\mathcal{L}) \setminus \mathbf{Z}(\mathcal{R}) = \mathbf{Z}($**DifferenceLR**$(\mathcal{L}, \mathcal{R}))$.

We prove them by induction on the product order $<_p$.

(1) Base case: no recursive calls to **Difference** and **DifferenceLR** occur. First, by line 2, 18 of the algorithm **Difference** and Lemma 6, 7, 10, (i) holds. Second, by line 2, 4 of the algorithm **DifferenceLR**, (ii) holds.

(2) Induction hypothesis: assume that both (i) and (ii) hold with inputs whose ranks are smaller than the rank of $([T, h], [T', h'])$ w.r.t. $<_p$.

(3) By (1), if no recursive calls occur, (i) and (ii) already hold. When there are recursive calls, we first show (i) holds. From the proof of Theorem 2, in **Difference**, the inputs of recursive calls to **Difference** and **DifferenceLR** will have smaller ranks w.r.t. the product order $<_p$. Therefore, by (2), line 7, 8, 11, 20, 21, 25, 30, 31, 32, 41, 46 and Lemma 6, 8, 9, 10, (i) holds.
Finally, by (i) and line $5 - 18$ of algorithm **DifferenceLR**, (ii) holds.

4   Decomposition into Pairwise Disjoint Constructible Sets

We assume that **DifferenceLR**$(\mathcal{L}, \mathcal{R})$ returns a list of regular systems sorted by increasing rank.

Definition 8. *Let $\mathcal{S}$ be a list of regular systems sorted by increasing rank. If $\mathcal{S}$ is empty or consists of a single regular system $[T, h]$, define* **MPD**$(\mathcal{S}) = \mathcal{S}$. *Otherwise, let $\mathcal{S} = \mathcal{L} + \mathcal{R}$, where $|\mathcal{L}| = |\mathcal{R}|$ or $|\mathcal{L}| = |\mathcal{R}| + 1$ (and $+$ denotes concatenation of lists). Define*

$$\mathbf{MPD}(\mathcal{S}) = \mathbf{MPD}(\mathbf{DifferenceLR}(\mathcal{L}, \mathcal{R})) + \mathbf{MPD}(\mathcal{R}).$$

Definition 9. *For a regular system $S = [T, h]$, let $\mathbf{Z}_0(S)$ denote the zero set of S considered as a regular system in $\hat{\mathbb{K}}[\mathrm{mvar}(T)] := \overline{\mathbb{K}(Y \setminus \mathrm{mvar}(T))}[\mathrm{mvar}(T)]$.*

Lemma 11. *For every regular system S, $\mathbf{Z}_0(S)$ is non-empty and finite.*

PROOF. If the regular system $S = [T, h]$ is considered in $\hat{\mathbb{K}}[\mathrm{mvar}(T)]$, it remains to be a regular system and, moreover, T becomes a zero-dimensional regular chain. We have therefore

$$\mathbf{Z}_0(S) = \mathbf{W}_{\hat{\mathbb{K}}}(T) \setminus \mathbf{V}_{\hat{\mathbb{K}}}(h) = \mathbf{V}_{\hat{\mathbb{K}}}(T).$$

Definition 10. *For a finite set of regular systems $S = \{[T_1, h_1], \ldots, [T_k, h_k]\}$ such that $\mathrm{rank}(T_1) = \cdots = \mathrm{rank}(T_k)$, define*

$$\mathbf{Z}_0(S) = \mathbf{Z}_0([T_1, h_1]) \cup \ldots \cup \mathbf{Z}_0([T_k, h_k]).$$

For an arbitrary finite set of regular systems S, let $S_{\mathrm{rank}(S)}$ denote the subset of regular systems of maximal rank. Define $\mathbf{Z}_0(S) = \mathbf{Z}_0(S_{\mathrm{rank}(S)})$.

Lemma 12. *Let S be a list of regular systems sorted by increasing rank represented as a concatenation of two non-empty sublists: $S = \mathcal{L} + \mathcal{R}$. Let $\mathcal{C} = \mathbf{DifferenceLR}(\mathcal{L}, \mathcal{R})$. Then either $\mathrm{rank}(\mathcal{C}) < \mathrm{rank}(S)$, or $|\mathbf{Z}_0(\mathcal{C})| < |\mathbf{Z}_0(S)|$.*

PROOF. If $\mathrm{rank}(\mathcal{L}) < \mathrm{rank}(S)$, then $\mathrm{rank}(\mathcal{C}) < \mathrm{rank}(S)$ by Theorem 2. Otherwise, $\mathrm{rank}(\mathcal{L}) = \mathrm{rank}(S)$ and, since S is sorted by increasing rank, the rank of every system in $\mathcal{R}$ equals $\mathrm{rank}(S)$. By Theorem 2, we have $\mathrm{rank}(\mathcal{C}) \leq \mathrm{rank}(S)$. In case of strict inequality we are done, so assume that $\mathrm{rank}(\mathcal{C}) = \mathrm{rank}(S)$.

Denote $r = \mathrm{rank}(\mathcal{L}) = \mathrm{rank}(\mathcal{C}) = \mathrm{rank}(\mathcal{R}) = \mathrm{rank}(S)$. We have:

$$\bigcup_{C \in \mathcal{C}_r} \mathbf{Z}(C) \subseteq \bigcup_{A \in \mathcal{L}_r} \mathbf{Z}(A) \setminus \bigcup_{B \in \mathcal{R}} \mathbf{Z}(B),$$

which implies

$$\mathbf{Z}_0(\mathcal{C}) \subseteq \mathbf{Z}_0(\mathcal{L}) \setminus \bigcup_{B \in \mathcal{R}} \mathbf{Z}_0(B).$$

Since, by Lemma 11, $\mathbf{Z}_0(S) = \mathbf{Z}_0(\mathcal{L}) \cup \mathbf{Z}_0(\mathcal{R})$ is finite and $\bigcup_{B \in \mathcal{R}} \mathbf{Z}(B) \neq \varnothing$, we obtain the desired $|\mathbf{Z}_0(\mathcal{C})| < |\mathbf{Z}_0(S)|$.

Lemma 13. *For any list S of regular systems, $\mathcal{D} = \mathbf{MPD}(S)$ is well-defined.*

PROOF. We define a well-order on the set of all sorted finite lists of regular systems and prove the statement by induction on this well-order.

For a non-empty list S, let $\phi(S) = (\mathrm{rank}(S), \mathbf{Z}_0(S), |S|)$. Let $\mathcal{L} \prec \mathcal{R}$ iff $\phi(\mathcal{L}) <_{\mathrm{lex}} \phi(\mathcal{R})$. Since $<_{\mathrm{lex}}$ is the lexicographic product of three well-orders, $<_{\mathrm{lex}}$ is a well-order, whence so is $\prec$. Define the empty list to be less than any non-empty list w.r.t. $\prec$.

For empty and singleton lists S, $\mathbf{MPD}(S)$ is well-defined. Let S be a non-singleton and non-empty list. Assume that $\mathbf{MPD}(S')$ is defined for all lists S' such that $S' \prec S$. Let, as in Definition 8, $S = \mathcal{L} + \mathcal{R}$, where $|\mathcal{L}| = |\mathcal{R}|$ or $|\mathcal{L}| = |\mathcal{R}| + 1$. Then by

Lemma 12, **Difference**$(\mathcal{L}, \mathcal{R}) \prec S$. Also, rank$(\mathcal{R}) \leq$ rank$(\mathcal{S})$, $\mathbf{Z}_0(\mathcal{R}) \leq \mathbf{Z}_0(\mathcal{S})$, and $|\mathcal{R}| < |\mathcal{S}|$, whence $\mathcal{R} \prec \mathcal{S}$. This implies that **MPD**$(\mathcal{S})$ is well-defined according to Definition 8.

Note that Definition 8 yields a recursive algorithm for computing **MPD**$(\mathcal{S})$, which terminates according to the previous lemma. The output of this algorithm is a decomposition of the union of zero-sets of regular systems in $\mathcal{S}$ into a disjoint union of zero-sets of regular systems:

Proposition 2. *For all distinct regular systems* $R, S \in \mathcal{D} = $ **MPD**$(\mathcal{S})$, *we have* $\mathbf{Z}(R) \cap \mathbf{Z}(S) = \varnothing$, *and*

$$\bigcup_{R \in \mathcal{S}} \mathbf{Z}(S) = \bigcup_{S \in \mathcal{D}} \mathbf{Z}(D).$$

PROOF. Follows immediately from the definition of **MPD**.

In the following section, to compute comprehensive triangular decompositions, we will see that **SMPD** (strongly make pairwise disjoint) is really required. Given a set of regular systems $A_1, \cdots, A_s$, **SMPD** compute another set of regular systems $B_1, \cdots, B_t$ whose zero sets are pairwise disjoint, such that each $\mathbf{Z}(A_i)$ writes as a union of some of the $\mathbf{Z}(B_1), \cdots, \mathbf{Z}(B_t)$.

Algorithm 3. SMPD$(\mathcal{S})$

1: **if** $|\mathcal{S}| \leq 1$ **then**
2:     output $\mathcal{S}$
3: **end if**
4: Let $[T_0, h_0] \in \mathcal{S}, \mathcal{S} \leftarrow \mathcal{S} \setminus \{[T_0, h_0]\}$
5: $\mathcal{S} \leftarrow$ **SMPD**$(\mathcal{S})$
6: **for** $[T, h] \in \mathcal{S}$ **do**
7:     $\mathcal{A} \leftarrow$ **Difference**$([T, h], [T_0, h_0])$
8:     $\mathcal{B} \leftarrow$ **DifferenceLR**$([T, h], \mathcal{A})$
9:     output **MPD**$(\mathcal{A})$
10:     output **MPD**$(\mathcal{B})$
11: **end for**
12: $\mathcal{C} \leftarrow$ **DifferenceLR**$([T_0, h_0], \mathcal{S})$
13: output **MPD**$(\mathcal{C})$

Proposition 3. *The Algorithm* **SMPD** *terminates and is correct.*

PROOF.   It follows directly from the termination and correctness of algorithms **Difference**, **DifferenceLR** and **MPD**.

5   Comprehensive Triangular Decomposition

In this section we introduce the concept of comprehensive triangular decomposition of an algebraic variety. We propose an algorithm for computing this decomposition and apply it to compute the set of all parameter values at which a given parametric system has an empty or an infinite set of solutions.

Notation 2. *From now on, we assume that $n = m + d$, the variables $Y_1, \ldots, Y_d$ are renamed $U_1, \ldots, U_d$ and viewed as parameters, whereas $Y_{d+1}, \ldots, Y_n$ are renamed $X_1, \ldots, X_m$ and regarded as unknowns.*

If the polynomial set $F \subset \mathbb{K}[Y]$ involves polynomials from $\mathbb{K}[U]$ only, we denote by $\mathbf{V}^U(F)$ its variety in $\overline{\mathbb{K}}^d$. Similarly, if the regular chain $T \subset \mathbb{K}[Y]$ involves polynomials from $\mathbb{K}[U]$ only, we denote by $\mathbf{W}^U(T)$ its quasi-component in $\overline{\mathbb{K}}^d$.

Notation 3. *Let $p \in \mathbb{K}[U][X]$ be a polynomial. We denote by $\mathbf{V}^U(p)$ the variety of $\overline{\mathbb{K}}^d$, consisting of the common roots of the coefficients of p, when p is regarded as a polynomial with variables in X and coefficients in $\mathbb{K}[U]$. Then, we define $\mathbf{V}^U(F)$ as the intersection of all $\mathbf{V}^U(p)$ for $p \in F$.*

For $u \in \overline{\mathbb{K}}^d$, we denote by $p(u)$ the polynomial of $\overline{\mathbb{K}}[X]$ obtained by evaluating p at $U_1 = u_1, \ldots, U_d = u_d$. Clearly, for all $u \in \overline{\mathbb{K}}^d$, the polynomial $p(u)$ is identically null iff $u \in \mathbf{V}^U(p)$. Then, we denote by $F(u)$ the set of all non-zero $p(u)$ for $p \in F$.

Definition 11. *Let $T \subset \mathbb{K}[U, X]$ be a regular chain. The* defining set *of T w.r.t. U, denoted by $\mathbf{D}^U(T)$, is the constructible set of $\overline{\mathbb{K}}^d$ given by*

$$\mathbf{D}^U(T) = \mathbf{W}^U(T \cap \mathbb{K}[U]) \setminus \mathbf{V}^U(\mathrm{res}(h_{T_{>U_d}}, T_{>U_d})).$$

Let $u \in \mathbf{W}^U(T \cap \mathbb{K}[U])$. We say that the regular chain T specializes well at u *if $T(u)$ is a regular chain in $\overline{\mathbb{K}}[X]$ such that $\mathrm{rank}(T(u)) = \mathrm{rank}(T_{>U_d})$.*

Remark 2. *Since $\mathbf{D}^U(T)$ is a constructible set, by Lemma 5, there exists an algorithm to compute a set of regular systems $\mathcal{R}^U(T)$, such that $\mathbf{D}^U(T) = \mathbf{Z}(\mathcal{R}^U(T))$.*

Lemma 14. *Let $T \subset \mathbb{K}[U, X]$ be a regular chain with $\mathrm{mvar}(T) \subseteq X$ and let $u \in \overline{\mathbb{K}}^d$. We have*

$$u \notin \mathbf{V}^U(\mathrm{res}(h_T, T)) \iff \mathrm{res}(h_{T(u)}, T(u)) \neq 0 \text{ and } h_T(u) \neq 0.$$

PROOF. " $\Leftarrow$ " If $h_T(u) \neq 0$ and $\mathrm{res}(h_{T(u)}, T(u)) \neq 0$, then

$$\mathrm{res}(h_{T(u)}, T(u)) = \mathrm{res}(h_T(u), T(u)) \neq 0,$$

which implies $\mathrm{res}(h_T, T)(u) \neq 0$. So $u \notin \mathbf{V}^U(\mathrm{res}(h_T, T))$.

" $\Rightarrow$ " We prove this by induction on $|T|$.

If $|T| = 1$, then $u \notin \mathbf{V}^U(\mathrm{res}(h_T, T))$ implies $h_T(u) \neq 0$ and therefore

$$\mathrm{res}(h_{T(u)}, T(u)) = h_{T(u)} = h_T(u) \neq 0.$$

Now we assume that the conclusion holds for $|T| = n - 1$. If $|T| = n$, let v be the largest variable in $\mathrm{mvar}(T)$. Since $u \notin \mathbf{V}^U(\mathrm{res}(h_T, T))$, we have

$$\mathrm{res}(h_T, T)(u) = \mathrm{res}(h_T, T_{<v})(u) \neq 0.$$

Therefore, $\mathrm{res}(h_{T_{<v}}, T_{<v})(u) \neq 0$. By induction hypothesis, we know $h_{T_{<v}}(u) \neq 0$. By the specialization property of resultant, $\mathrm{res}(h_T(u), T_{<v}(u)) \neq 0$ and therefore $h_T(u) \neq 0$. So $\mathrm{res}(h_T, T)(u) \neq 0$ implies $\mathrm{res}(h_{T(u)}, T(u)) \neq 0$.

Proposition 4. *Let $T \subset \mathbb{K}[U, X]$ be a regular chain and let $u \in \mathbf{W}^U(T \cap \mathbb{K}[U])$. The regular chain T specializes well at $u \in \overline{\mathbb{K}}^d$ if and only if $u \in \mathbf{D}^U(T)$.*

PROOF. Assume that $u \in \mathbf{D}^U(T)$. We prove that T specializes well at u. From Lemma 14 we have

$$\mathrm{res}(h_{T>U_d}(u), T_{>U_d}(u)) \neq 0 \text{ and } h_{T>U_d}(u) \neq 0.$$

With $u \in \mathbf{W}^U(T \cap \mathbb{K}[U])$, which implies $(T \cap \mathbb{K}[U])(u) = \{0\}$, we conclude that $\mathrm{rank}(T(u)) = \mathrm{rank}(T_{>U_d})$. Moreover, by Theorem 1, $T(u)$ is a regular chain. Therefore, the regular chain T specializes well at u. The converse implication is proved similarly.

Definition 12. *Let $T \subset \mathbb{K}[U, X]$ be a regular chain. The comprehensive quasi-component of T w.r.t. U, denoted by $\mathbf{W}_C(T)$, is defined by*

$$\mathbf{W}_C(T) = \mathbf{W}(T) \cap \Pi_U^{-1}(\mathbf{D}^U(T)).$$

Proposition 5. *Let $T \subset \mathbb{K}[U, X]$ be a regular chain. The following properties hold:*

(1) *We have:* $\mathbf{W}_C(T) = \mathbf{W}(T) \setminus \Pi_U^{-1}(\mathbf{V}^U(\mathrm{res}(h_{T>U_d}, T_{>U_d})))$.
(2) *We have:* $\Pi_U(\mathbf{W}_C(T)) = \mathbf{D}^U(T)$.

PROOF. It follows from Definition 11 and Lemma 14.

Definition 13. *Let $F \subset \mathbb{K}[U, X]$ be a finite polynomial set. A comprehensive triangular decomposition of $\mathbf{V}(F)$ is given by :*

1. *a finite partition $\mathcal{C}$ of $\Pi_U(\mathbf{V}(F))$,*
2. *for each $C \in \mathcal{C}$ a set of regular chains $\mathcal{T}_C$ of $\mathbb{K}[U, X]$ such that for $u \in C$ each of the regular chains $T \in \mathcal{T}_C$ specializes well at u and we have for all $u \in C$*

$$\mathbf{V}(F(u)) = \bigcup_{T \in \mathcal{T}_C} \mathbf{W}(T(u)).$$

We will compute the above comprehensive triangular decomposition with the help of the following auxiliary concept:

Definition 14. *Let $F \subset \mathbb{K}[U, X]$ be a finite polynomial set. A pre-comprehensive triangular decomposition (PCTD) of $\mathbf{V}(F)$ is a family of regular chains $\mathcal{T}$ satisfying the following property: for each $u \in \overline{\mathbb{K}}^d$, let $\mathcal{T}_u$ be the subfamily of all regular chains in $\mathcal{T}$ that specialize well at u; then*

$$\mathbf{V}(F(u)) = \bigcup_{T \in \mathcal{T}_u} \mathbf{W}(T(u)).$$

Proposition 6. *Let $F \subset \mathbb{K}[U, X]$ be a finite polynomial set. A triangular decomposition $\mathcal{T}$ of $\mathbf{V}(F)$ is a pre-comprehensive triangular decomposition if and only if*

$$\mathbf{V}(F) = \bigcup_{T \in \mathcal{T}} W_C(T).$$

PROOF. It follows from the definition of $W_C(T)$, Proposition 4 and the definition of pre-comprehensive triangular decomposition.

Algorithm 4. PCTD(F)

Input: A finite set $F \subset \mathbb{K}[U, X]$.
Output: A PCTD of $\mathbf{V}(F)$.
  1: $\mathcal{T} \leftarrow$ **Triangularize**(F)
  2: **while** $\mathcal{T} \neq \emptyset$ **do**
  3: let $T \in \mathcal{T}$, $\mathcal{T} \leftarrow \mathcal{T} \setminus \{T\}$
  4: **output** T
  5: $G \leftarrow$ COEFFICIENTS(res$(h_{T_{>U_d}}, T_{>U_d}), U)$
  6: $\mathcal{T} \leftarrow \mathcal{T} \cup$ **Triangularize**(G, T)
  7: **end while**

Proposition 7. *Algorithm 4 computes a pre-comprehensive triangular decomposition of* $\mathbf{V}(F)$.

PROOF. The loop satisfies the following invariant: the union of all $\mathbf{W}(T)$, where T ranges over $\mathcal{T}$, and of the $\mathbf{W}(T')$, where T' ranges over the current output, equals $\mathbf{V}(F)$. Indeed, the invariant holds at the beginning, when the output is empty; and for the regular chain T taken from $\mathcal{T}$ at the current iteration, we have $\mathbf{W}(T) \setminus \mathbf{W}_C(T) = \mathbf{V}(G) \cap \mathbf{W}(T)$ by Proposition 5 (1). Then, correctness of the algorithm follows from Proposition 6 and the fact that at the end $\mathcal{T} = \varnothing$.

Since polynomials in G do not involve the main variables of T, by Lemma 3 they are regular w.r.t Sat(T). Then by Lemma 1, either the output of **Triangularize**(G, T) is empty or the dimensions of the regular chains computed by **Triangularize**(G, T) are strictly less than that of T. Therefore, the algorithm terminates.

Proposition 8. *Algorithm 5 computes a comprehensive triangular decomposition of* $F \subset \mathbb{K}[U, X]$.

PROOF. Let $\mathcal{T}$ be the output of **PCTD**(F). By Proposition 6 and Proposition 5 (2), we have

$$\Pi_U(\mathbf{V}(F)) = \bigcup_{T \in \mathcal{T}} \mathbf{D}^U(T).$$

Then the conclusion follows from the definition of comprehensive triangular decomposition, Proposition 3, 7 and Remark 2.

Given a polynomial set $F \subset \mathbb{K}[U, X]$, a natural question is to describe the points u of $\overline{\mathbb{K}}^d$ for which the specialized system $F(u)$ admits a finite and positive number of solutions in $\overline{\mathbb{K}}^m$. This question is formalized by the following definition.

Definition 15. *The* discriminant set *of F is defined as the set of all points $u \in \overline{\mathbb{K}}^d$ for which $\mathbf{V}(F(u))$ is empty or infinite.*

Algorithm 5. CTD(F)

Input: A finite set $F \subset \mathbb{K}[U, X]$.
Output: A CTD of $\mathbf{V}(F)$.
1: $\mathcal{T} \leftarrow \mathbf{PCTD}(F)$
2: $\mathcal{S} \leftarrow \emptyset$
3: **for** $T \in \mathcal{T}$ **do**
4: $\quad \mathcal{S} \leftarrow \mathcal{S} \cup \mathcal{R}^U(T)$
5: **end for**
6: $\mathcal{S} \leftarrow \mathbf{SMPD}(\mathcal{S})$
7: **while** $\mathcal{S} \neq \emptyset$ **do**
8: $\quad$ let $C \in \mathcal{S}, \mathcal{S} \leftarrow \mathcal{S} \setminus C$
9: $\quad \mathcal{T}_C \leftarrow$ regular chains in $\mathcal{T}$ associated to C
10: $\quad$ **output** $(C, \mathcal{T}_C)$
11: **end while**

Theorem 4. *If $\mathcal{T}$ is a pre-comprehensive triangular decomposition of $\mathbf{V}(F)$, then the following is the discriminant set of F:*

$$\left(\bigcup_{\substack{T \in \mathcal{T} \\ X \not\subseteq \mathrm{mvar}(T)}} \mathbf{D}^U(T) \right) \cup \left(\bigcap_{\substack{T \in \mathcal{T} \\ X \subseteq \mathrm{mvar}(T)}} \overline{\mathbb{K}}^d \setminus \mathbf{D}^U(T) \right).$$

PROOF. By Proposition 4, for every parameter value $u \in \overline{\mathbb{K}}^d$, the set $\{T(u) \mid T \in \mathcal{T}$ and $u \in \mathbf{D}^U(T)\}$ is a triangular decomposition of $\mathbf{V}(F(u))$ into regular chains. In particular, if there exists no $T \in \mathcal{T}$ such that $u \in \mathbf{D}^U(T)$ holds, then $\mathbf{V}(F(u)) = \emptyset$.

Therefore, u yields finitely many solutions (and at least one) if and only if the following conditions hold:

– u belongs to at least one $\mathbf{D}^U(T)$ such that $X \subseteq \mathrm{mvar}(T)$, i.e., $T(u)$ is a zero-dimensional regular chain.
– u does not belong to any $\mathbf{D}^U(T)$ such that $X \not\subseteq \mathrm{mvar}(T)$, i.e., $T(u)$ is a positive-dimensional regular chain.

Remark 3. *By Theorem 4 and Proposition 8, we have completely answered the two problems proposed in the introduction.*

6 Implementation

We have implemented the algorithm for computing comprehensive triangular decompositions (CTD) based on *RegularChains* library in Maple 11. Our main function CTD calls essentially three functions

– Triangularize, computing a triangular decomposition of the input system F,
– PCTD, deducing a pre-comprehensive triangular decomposition of F,
– SMPD, obtaining a comprehensive triangular decomposition of F.

Table 1. Solving timings and number of cells of CTD

Sys	Name	Triangularize	PCTD	SMPD	CTD	#Cells
1	MontesS1	0.089	0.002	0.031	0.122	3
2	MontesS2	0.031	0.002	0	0.033	1
3	MontesS3	0.103	0.006	0.005	0.114	2
4	MontesS4	0.101	0.016	0	0.117	1
5	MontesS5	0.383	0.022	0.465	0.870	11
6	MontesS6	0.395	0.019	0.121	0.535	4
7	MontesS7	0.416	0.215	0.108	0.739	4
8	MontesS8	0.729	0.001	0.016	0.746	2
9	MontesS9	0.945	0.116	3.817	4.878	23
10	MontesS10	5.325	0.684	1.138	7.147	10
11	MontesS11	0.757	0.208	12.302	13.267	28
12	MontesS12	14.199	2.419	10.114	26.732	10
13	MontesS13	0.415	0.143	1.268	1.826	9
14	MontesS14	41.167	31.510	0.303	72.980	4
15	MontesS15	6.919	0.579	1.123	8.621	5
16	MontesS16	6.963	0.083	2.407	9.453	21
17	AlkashiSinus	0.716	0.191	0.574	1.481	6
18	Bronstein	2.526	0.017	0.548	3.091	6
19	Gerdt	3.863	0.006	0.733	4.602	5
20	Hereman-2	1.826	0.019	0.020	1.865	2
21	Lanconelli	2.056	0.336	3.430	5.822	14
22	genLinSyst-3-2	1.624	0.275	25.413	27.312	32
23	genLinSyst-3-3	9.571	1.824	1097.291	1108.686	116
24	Wang93	6.795	37.232	11.828	55.855	8
25	Maclane	12.955	0.403	54.197	67.555	21
26	Neural	15.279	19.313	0.530	35.122	4
27	Leykin-1	1261.751	86.460	27.180	1375.391	57
28	Lazard-ascm2001	60.698	2817.801	–	–	–
29	Pavelle	–	–	–	–	–
30	Cheaters-homotopy	–	–	–	–	–

We provide comparative benchmarks with MAPLE implementations of related methods for solving parametric polynomial systems, namely: *decomposition into regular systems* by Wang [19] and *discussing parametric Gröbner bases* by Montes [14]. Corresponding MAPLE functions are RegSer and DISPGB, respectively.

Note that the specifications of these three methods are different. The outputs of CTD and DISPGB depend on the choice of the parameter sets, whereas RegSer does not require to specify parameters. RegSer decomposes the input system into pairwise disjoint constructible sets given by regular systems. CTD computes a comprehensive triangular decomposition, and thus a family of triangular decompositions with a partition of the parameter space. DISPGB computes a family of comprehensive Gröbner bases with a partition of the parameter space.

We run CTD in Maple 11 using an Intel Pentium 4 processor (3.20GHz CPU, 2.0GB total memory, and Red Hat 4.0.0-9); we set the time-out to 1 hour. Due to the current availability of RegSer and DISPGB, the timings obtained by these two functions are

Table 2. Solving timings and number of components/cells in three algorithms

	DISPGB		RegSer		CTD	
Sys	Time (s)	# Cells	Time (s)	# Components	Time (s)	# Cells
1	0.509	2	0.021	3	0.122	3
2	0.410	2	0.021	1	0.033	1
3	0.550	2	0.060	3	0.114	2
4	1.511	2	0.070	1	0.117	1
5	1.030	3	0.099	4	0.870	11
6	1.350	4	0.049	5	0.535	4
7	1.609	2	0.180	4	0.739	4
8	2.181	3	0.150	4	0.746	2
9	10.710	5	0.171	7	4.878	23
10	9.659	5	0.329	5	7.147	10
11	0.489	3	0.260	9	13.267	28
12	259.730	5	2.381	23	26.732	10
13	5.830	9	0.199	9	1.826	9
14	–	–	–	–	72.980	4
15	30.470	7	0.640	10	8.621	5
16	61.831	7	6.060	22	9.453	21
17	4.619	6	0.150	5	1.481	6
18	8.791	5	0.319	6	3.091	6
19	20.739	5	3.019	10	4.602	5
20	101.251	2	0.371	7	1.865	2
21	43.441	4	0.330	7	5.822	14
22	–	–	0.350	18	27.312	32
23	–	–	2.031	61	1108.686	116
24	–	–	4.040	6	55.855	8
25	83.210	11	–	–	67.555	21
26	–	–	–	–	35.122	4
27	–	–	–	–	1375.391	57
28	–	–	–	–	–	–
29	–	–	–	–	–	–
30	–	–	–	–	–	–

performed in Maple 8 on Intel Pentium 4 machines (1.60GHz CPU, 513MB memory and Red Hat Linux 3.2.2-5); and the time-out is 2 hours. The 30 test-systems used in our experimentation are chosen from [13,18,21].

As shown in the above two tables, our implementation of the CTD algorithm can solve all problems which can be solved by the other methods. In addition, the CTD can solve 4 test-systems which are out of reach of the other two methods, generally due to memory consumption.

7   Conclusion

Comprehensive triangular decomposition is a powerful tool for the analysis of parametric polynomial systems: its purpose is to partition the parameter space into regions, so that within each region the "geometry" of the algebraic variety of the specialized system is the same for all values of the parameters.

As the main technical tool, we proposed an algorithm that represents the difference of two constructible sets as finite unions of regular systems. From there, we have deduced an algorithmic solution for a set theoretical instance of the coprime factorization problem: refining a family of constructible sets into a family of pairwise disjoint constructible sets.

We have reported on an implementation of our algorithm computing CTDs, based on the RegularChains library in MAPLE. Our comparative benchmarks, with MAPLE implementations of related methods for solving parametric polynomial systems, illustrate the good performances of our CTD code.

References

1. Aubry, P., Lazard, D., Moreno Maza, M.: On the theories of triangular sets. J. Symb. Comp. 28(1-2), 105–124 (1999)
2. Bernstein, D.J.: Factoring into coprimes in essentially linear time. J. Algorithms 54(1), 1–30 (2005)
3. Boulier, F., Lemaire, F., Moreno Maza, M.: Well known theorems on triangular systems and the D5 principle. In: Proc. of Transgressive Computing 2006, Granada, Spain (2006)
4. Caviness, B., Johnson, J. (eds.): Quantifier Elimination and Cylindical Algebraic Decomposition, Texts and Mongraphs in Symbolic Computation. Springer, Heidelberg (1998)
5. Chen, F., Wang, D. (eds.): Geometric Computation. Lecture Notes Series on Computing, vol. 11. World Scientific Publishing Co., Singapore, New Jersey (2004)
6. Chou, S.C., Gao, X.S.: Computations with parametric equations. In: Proc. ISAAC'91, Bonn, Germany, pp. 122–127 (1991)
7. Chou, S.C., Gao, X.S.: Solving parametric algebraic systems. In: Proc. ISSAC'92, Berkeley, California, pp. 335–341 (1992)
8. Dahan, X., Moreno Maza, M., Schost, É., Xie, Y.: On the complexity of the D5 principle. In: Proc. of Transgressive Computing 2006, Granada, Spain (2006)
9. Duval, D.: Algebraic Numbers: an Example of Dynamic Evaluation. J. Symb. Comp. 18(5), 429–446 (1994)
10. Gómez Díaz, T.: Quelques applications de l'évaluation dynamique. PhD thesis, Université de Limoges (1994)
11. Kalkbrener, M.: A generalized euclidean algorithm for computing triangular representations of algebraic varieties. J. Symb. Comp. 15, 143–167 (1993)
12. Lazard, D., Rouillier, F.: Solving parametric polynomial systems. Technical Report 5322, INRIA (2004)
13. Manubens, M., Montes, A.: Improving dispgb algorithm using the discriminant ideal (2006)
14. Montes, A.: A new algorithm for discussing gröbner bases with parameters. J. Symb. Comput. 33(2), 183–208 (2002)
15. Moreno Maza, M.: On triangular decompositions of algebraic varieties. Technical Report TR 4/99, NAG Ltd, Oxford, UK (1999), http://www.csd.uwo.ca/~moreno
16. Samuel, P., Zariski, O.: Commutative algebra. D. Van Nostrand Company, INC (1967)
17. Suzuki, A., Sato, Y.: A simple algorithm to compute comprehensive Gröbner bases. In: Proc. ISSAC'06, pp. 326–331. ACM Press, New York (2006)
18. The SymbolicData Project (2000–2006), http://www.SymbolicData.org
19. Wang, D.M.: Computing triangular systems and regular systems. Journal of Symbolic Computation 30(2), 221–236 (2000)

20. Wang, D.M.: Decomposing polynomial systems into simple systems. J. Symb. Comp. 25(3), 295–314 (1998)
21. Wang, D.M.: Elimination Methods. Springer, Wein, New York (2000)
22. Weispfenning, V.: Comprehensive grobner bases. J. Symb. Comp. 14, 1–29 (1992)
23. Weispfenning, V.: Canonical comprehensive grobner bases. In: ISSAC 2002, pp. 270–276. ACM Press, New York (2002)
24. Wu, W.T.: A zero structure theorem for polynomial equations solving. MM Research Preprints 1, 2–12 (1987)
25. Wu, W.T.: On a projection theorem of quasi-varieties in elimination theory. MM Research Preprints 4, 40–53 (1989)
26. Yang, L., Hou, X.R., Xia, B.C.: A complete algorithm for automated discovering of a class of inequality-type theorem. Science in China, Series E 44(6), 33–49 (2001)

Stability Investigation of a Difference Scheme for Incompressible Navier-Stokes Equations

Dmytro Chibisov, Victor Ganzha, Ernst W. Mayr[1], and Evgenii V. Vorozhtsov[2]

[1] Institute of Informatics, Technische Universität München, 85748 Garching, Boltzmannstr. 3, Germany
chibisov@in.tum.de, ganzha@in.tum.de, mayr@in.tum.de
[2] Institute of Theoretical and Applied Mechanics, Russian Academy of Sciences, Novosibirsk 630090, Russia
vorozh@itam.nsc.ru

Abstract. We investigate the stability of the modified difference scheme of Kim and Moin for numerical integration of two-dimensional incompressible Navier–Stokes equations by the Fourier method and by the method of discrete perturbations. The obtained analytic-form stability condition gives the maximum time steps allowed by stability, which are by factors from 2 to 58 higher than the steps obtained from previous empirical stability conditions. The stability criteria derived with the aid of CAS Mathematica are verified by numerical solution of two test problems one of which has a closed-form analytic solution.

1   Introduction

It is well known (see, for example, [13]) that the influence of the compressibility of a gas or a liquid may be neglected if the flow Mach number does not exceed the value 0.3. In such cases, it is reasonable to use the Navier-Stokes equations governing the viscous incompressible fluid flows. These equations are somewhat simpler than the system of Navier-Stokes equations for compressible media. The incompressible Navier-Stokes equations are widely used when investigating such applied problems as the buoyancy-driven convection of air in rooms, the propagation of pollutants in the atmosphere, the water flow around a moving ship or submarine, etc.

The numerical solution of Navier-Stokes equations is simplified greatly if they are discretized on a uniform rectangular spatial grid in Cartesian coordinates. It is natural and convenient to use such grids at the solution of problems in regions of rectangular shape. Many applied problems are, however, characterized by the presence of curved boundaries. In such cases, other grid types are often used: curvilinear grids, structured and unstructured triangular and polygonal grids. Although such grids simplify the implementation of boundary conditions, their use leads to new difficulties, such as the extra (metric) terms in equations, extra interpolations, larger computational molecules, etc. [9].

During the last decade, a new method for numerical solution of the Navier-Stokes equations in regions with complex geometry has enjoyed a powerful development: the immersed boundary method (IBM). In this method, the computation

V.G. Ganzha, E.W. Mayr, and E.V. Vorozhtsov (Eds.): CASC 2007, LNCS 4770, pp. 102–117, 2007.
© Springer-Verlag Berlin Heidelberg 2007

of gas motion is carried out on a rectangular grid, and the curved boundary is interpreted as an interface. The grid cells lying outside the region occupied by the fluid are classified as the ghost cells in which the Navier-Stokes equations are, however, also solved numerically. A survey of different recent realizations of the IBM may be found in [10,14,16]. The immersed boundary method has extended significantly the scope of applicability of the rectangular Cartesian grids at the numerical solution of applied problems of the incompressible fluid dynamics.

The difference scheme proposed in [8] is often used within the IBM framework. The convective terms are approximated in this scheme with the aid of the explicit three-level Adams–Bashforth second-order scheme, and the viscous terms are approximated by the implicit second-order Crank–Nicolson scheme. Despite the popularity of scheme [8], its stability was not investigated even in the case of two spatial variables.

The purpose of the present work is the stability investigation of a modified scheme from [8]. This investigation is carried out at first by the Fourier method. Since this analysis method is applicable only to linear difference schemes with constant coefficients we employ one more method for stability analysis of nonlinear difference equations approximating the Navier-Stokes equations. This method was proposed in [11] and reduces to the investigation of the behaviour of solution of difference equations in the case when the oscillating velocity profiles are specified on two lower time levels. The obtained stability conditions have been verified by computations of two test problems one of which is the lid-driven cavity problem.

2   Governing Equations and Difference Method

The Navier-Stokes equations governing two-dimensional unsteady flows of an incompressible viscous fluid may be written in the vector form as follows:

$$\operatorname{div} \boldsymbol{v} = 0, \tag{1}$$

$$\frac{\partial \boldsymbol{v}}{\partial t} + (\boldsymbol{v}\nabla)\boldsymbol{v} + \frac{1}{\rho}\nabla p = \nu \Delta \boldsymbol{v}, \tag{2}$$

where $\boldsymbol{v} = (u, v)^T$ is the velocity vector (the superscript T denotes the transposition operation), u and v are the vector components along the x, y axes of Cartesian coordinates, p is the pressure, ρ is the density, $\nu = \mu/\rho$, μ is the dynamic viscosity ($\nu = \mathrm{const} > 0$), Δ is the Laplace operator.

Following [8] we will discretize the momentum equation (2) in time by using a hybrid second-order scheme:

$$\frac{\boldsymbol{v}^* - \boldsymbol{v}^n}{\tau} + \frac{3}{2}H(\boldsymbol{v}^n) - \frac{1}{2}H(\boldsymbol{v}^{n-1}) + \frac{1}{\rho}Gp^n = \frac{\nu}{2}\left[L(\boldsymbol{v}^*) + L(\boldsymbol{v}^n)\right]. \tag{3}$$

Here τ is the time step, $H(\boldsymbol{v}^n)$ is the difference operator approximating the operator $(\boldsymbol{v}\nabla)\boldsymbol{v}$, G is the discrete gradient, L is the discrete Laplace operator, n is the time level. Thus, the convective terms in (3) are approximated explicitly by the second-order Adams–Bashforth scheme, and the diffusion terms $\nu\Delta\boldsymbol{v}$

are treated implicitly using second-order Crank–Nicolson scheme. The implicit approximation of viscous terms is applied according to [8] in order to eliminate a restriction for time step τ dictated by the computational stability.

At the second fractional step, the field of intermediate velocities v^* is corrected to ensure the mass conservation:

$$(v^{n+1} - v^*)/\tau_n = -Gp', \tag{4}$$

The pressure correction p' is computed in such a way that a divergence-free velocity field is obtained at the $(n+1)$th time step. To this end, let us apply the divergence operator to the both sides of equation (4):

$$(Dv^{n+1} - Dv^*)/\tau_n = -Lp', \tag{5}$$

where D is a discrete analog of the divergence operator. Since it is required that $Dv^{n+1} = 0$, we obtain from (5) the Poisson equation for the pressure correction:

$$Lp' = (1/\tau_n)Dv^*. \tag{6}$$

The correction p' found as the solution of equation (6) is then used for the correction of the velocity field according to (4): $v^{n+1} = v^* - \tau_n Gp'$ and of the pressure field: $p^{n+1} = p^n + p'$. The Poisson equation (6) was solved by the BiCGSTAB method [15]. As was pointed out in [9], the pressure correction method was found to be the fastest of the methods tested by Armfield and Street [1] and is the method used here.

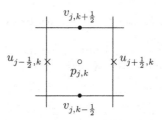

Fig. 1. The staggered grid in two dimensions

Following [8] we will approximate all spatial derivatives by second-order central differences on a staggered grid (see Fig. 1). The advantages of staggered grid at the numerical integration of the Navier-Stokes equations for incompressible fluid are discussed in detail in [8,9]. For example, the term $\partial^2 v/\partial y^2$ is approximated on the staggered grid as follows:

$$\left(\partial^2 v/\partial y^2\right)_{j,k+1/2} = (v_{j,k+3/2} - 2v_{j,k+1/2} + v_{j,k-1/2})/(h_2^2),$$

where h_1, h_2 are the steps of uniform rectangular grid along the x- and y-axes, respectively; the subscripts j, k refer to the cell center. To approximate the convective terms $H(v^n)$ we use in (3) the difference formulas of the MAC-method [6,12,8]. These formulas are applied to the divergence form of motion equations:

$$\frac{\partial u}{\partial t} + \frac{\partial (u^2)}{\partial x} + \frac{\partial uv}{\partial y} + \frac{1}{\rho}\frac{\partial p}{\partial x} = \nu \Delta u; \quad \frac{\partial v}{\partial t} + \frac{\partial uv}{\partial x} + \frac{\partial (v^2)}{\partial y} + \frac{1}{\rho}\frac{\partial p}{\partial y} = \nu \Delta v.$$

For example, $(\partial u^2/\partial x)_{j+1/2,k} = (u^2_{j+1,k} - u^2_{j,k})/h_1$, where $u_{j,k} = (1/2)(u_{j-1/2,k} + u_{j+1/2,k})$.

We now mention several stability conditions, which were used previously at the computation of time step τ entering the difference scheme (3). Roache [12] discussed the stability of the Adams–Bashforth scheme at its application for approximation of the one-dimensional advection-diffusion equation

$$\partial \zeta/\partial t + \partial(u\zeta)/\partial x = \nu \partial^2 \zeta/\partial x^2. \tag{7}$$

This scheme proved to be unconditionally unstable, and it has a weak divergence caused by the fact that the scheme amplification factor G obtained by the Fourier method has the form $G = 1 + O(\tau^2)$. It is, however, to be noted that the above scheme from [12] for equation (7) is explicit, whereas there are in scheme (3) also the implicit operators, which stabilize the numerical computation. It is to be noted here that since $\nu = O(1/\text{Re})$, where Re is the Reynolds number, then at high Reynolds numbers, the stabilizing effect of the implicit term in (3) becomes insignificant. The computation nevertheless remains stable at the solution of practical problems by scheme (3) also for the value Re = 25 000, as this was shown in [7]. It was proposed in [7] to compute the time step τ at the computation by scheme (3) by using the formula

$$\tau = \min_{j,k} \left[\tau_{conv}^{-1}/C_{conv} + \tau_{diff}^{-1}/C_{diff} \right]^{-1}, \tag{8}$$

where the items are computed in each (j, k) cell as follows:

$$\tau_{conv}^{-1} = |u|/h_1 + |v|/h_2, \quad \tau_{diff}^{-1} = \nu \cdot \left(1/h_1^2 + 1/h_2^2\right).$$

For the diffusion component in (8) the Courant number $C_{diff} = 0.25$ according to [7], and for the convective component the values of C_{conv} were taken from 0.5 to 1. Note that formula (8) is similar to the one used in [2], but in [2], the common Courant number $C_{conv} = C_{diff} = 0.25$ was used. Owing to the application of formula (8) with different values of C_{conv} and C_{diff} the authors of [7] were able to reduce the required CPU time at the computations of unsteady flows by a factor of nearly four.

The stability analysis results were presented in [5] for the schemes of Runge–Kutta type with the stage numbers three and five for the two-dimensional advection-diffusion equation

$$\partial f/\partial t + u\partial f/\partial x + v\partial f/\partial y = \nu(\partial^2 f/\partial x^2 + \partial^2 f/\partial y^2).$$

It turned out that for the both studied schemes, the stability condition has the form

$$\left(\frac{|\kappa_1| + |\kappa_2|}{a}\right)^2 + \left(\frac{\kappa_3'}{b}\right)^2 \leq 1, \tag{9}$$

where $\kappa_1 = u\tau/h_1$, $\kappa_2 = v\tau/h_2$, $\kappa_3' = v\tau(1/h_1^2 + 1/h_2^2) = \kappa_3(1 + \kappa_4^2)$, $\kappa_3 = v\tau/(h_1^2)$, $\kappa_4 = h_1/h_2$, a and b are certain constants depending on the specific method of the Runge–Kutta type. Despite the fact that condition (9) as well as the empirical stability condition (8) were obtained for different difference schemes their structure is similar. Formula (8) can indeed be written in terms of dimensionless quantities κ_1, κ_2 and κ_3' as

$$\frac{|\kappa_1| + |\kappa_2|}{C_{conv}} + \frac{\kappa_3'}{C_{diff}} \leq 1.$$

3   Fourier Symbol

The stability analysis of difference schemes by the Fourier method is known to be applicable only to linear schemes with constant coefficients. Difference scheme (3) is nonlinear, therefore, prior to the Fourier method application it is necessary to linearize the scheme. Linearization may be implemented in two different ways. One of them consists of that the original differential equations (in our case these are equations (2)) are at first linearized, and the difference scheme (3) is then applied to linearized differential equations. Another technique reduces to a direct linearization of difference equations (3). We use the first of the above techniques because it involves a slightly shorter calculation.

Thus, let us assume that $U(x, y, t)$, $V(x, y, t)$, $P(x, y, t)$ is an exact solution of equation (2), where U and V are the components of the velocity vector along the x- and y-axes, respectively, P is the pressure. According to difference equation (3), only the velocity components are varied at a passage from the nth time level to the $(n+1)$th time level. We can, therefore, present solution v of system (2) as

$$u = U + \delta u, \quad v = V + \delta v, \quad p = P, \tag{10}$$

where δu and δv are the errors, which are small in their absolute values and which are caused by the approximation error, machine roundoff errors, etc. Since the "big" quantities U, V, P satisfy equation (2), as a result of substituting formulas (10) in (2) and neglecting the second-order terms with respect to δu and δv we obtain the following linear differential equations:

$$\frac{\partial \delta u}{\partial t} + U\frac{\partial \delta u}{\partial x} + V\frac{\partial \delta u}{\partial y} = \nu\left(\frac{\partial^2 \delta u}{\partial x^2} + \frac{\partial^2 \delta u}{\partial y^2}\right);$$

$$\frac{\partial \delta v}{\partial t} + U\frac{\partial \delta v}{\partial x} + V\frac{\partial \delta v}{\partial y} = \nu\left(\frac{\partial^2 \delta v}{\partial x^2} + \frac{\partial^2 \delta v}{\partial y^2}\right). \tag{11}$$

Let us now approximate system (11) by difference scheme (3) on a staggered grid. Since this difference scheme is a three-level scheme we introduce two auxiliary dependent variables δr^n and δs^n by formulas [4]: $\delta r^n = \delta u^{n-1}$, $\delta s^n = \delta v^{n-1}$ before the investigation of its stability. Let $\boldsymbol{V} = (U, V)^T$, $\delta \boldsymbol{v}^n = (\delta u^n, \delta v^n)^T$, $\delta \boldsymbol{r}^n = (\delta r^n, \delta s^n)^T$. We can then write difference scheme (3) as applied to system (11) in the form:

$$\frac{\delta v^* - \delta v^n}{\tau} + \frac{3}{2}(V^n \nabla)\delta v^n - \frac{1}{2}(V^{n-1}\nabla)\delta r^n = \frac{\nu}{2}\left[L(\delta v^*) + L(\delta v^n)\right]. \quad (12)$$

Thus, (12) is a two-layer difference scheme. Upon "freezing" its coefficients V^n, V^{n-1} we can apply the von Neumann stability analysis [3,4] to obtain the necessary stability condition. According to the procedure of this analysis we substitute into the system of difference equations

$$\frac{\delta v^* - \delta v^n}{\tau} + \frac{3}{2}(v^n \nabla)\delta v^n - \frac{1}{2}(V^{n-1}\nabla)\delta r^n = \frac{\nu}{2}\left[L(\delta v^*) + L(\delta v^n)\right];$$
$$\delta r^{n+1} = \delta u^n; \qquad\qquad\qquad\qquad\qquad\qquad\qquad\qquad (13)$$
$$\delta s^{n+1} = \delta v^n$$

the solution of the form

$$\delta w_{j,k}^n = \delta w_0 \lambda^n \exp[i(jm_1 h_1 + km_2 h_2)], \quad (14)$$

where $\delta w^n = (\delta u^n, \delta v^n, \delta r^n, \delta s^n)^T$, δw_0 is a constant vector, m_1 and m_2 are real components of the wave vector, λ is a complex number, $i = \sqrt{-1}$. As a result of the substitution of particular solution of the form (14) into system (13) we obtain the system

$$A\delta w_{j,k}^{n+1} = B\delta w_{j,k}^n, \quad (15)$$

where

$$A = \begin{pmatrix} a & 0 & 0 & 0 \\ 0 & a & 0 & 0 \\ 0 & 0 & 1 & 0 \\ 0 & 0 & 0 & 1 \end{pmatrix}, \quad B = \begin{pmatrix} b & 0 & c & 0 \\ 0 & b & 0 & c \\ 1 & 0 & 0 & 0 \\ 0 & 1 & 0 & 0 \end{pmatrix},$$

$$a = 1 + \kappa_3(1 - \cos\xi) + \kappa_4(1 - \cos\eta), \quad c = (1/2)i(\kappa_1 \sin\xi + \kappa_2 \sin\eta), \quad (16)$$
$$b = 1 - 3c - \kappa_3(1 - \cos\xi) - \kappa_4(1 - \cos\eta),$$

$$\kappa_1 = \frac{U\tau}{h_1}, \quad \kappa_2 = \frac{V\tau}{h_2}, \quad \kappa_3 = \frac{\nu\tau}{h_1^2}, \quad \kappa_4 = \frac{\nu\tau}{h_2^2}, \quad (17)$$

$\xi = m_1 h_1$, $\eta = m_2 h_2$. The quantities κ_3 and κ_4 are nonnegative by virtue of their physical meaning, therefore, $a \geq 1$, and, hence, matrix A is invertible. Multiplying the both sides of equation (15) from the left by A^{-1} we obtain the system

$$\delta w_{j,k}^{n+1} = G\,\delta w_{j,k}^n, \quad (18)$$

where matrix $G = A^{-1}B$ is called the amplification matrix of the difference scheme with constant coefficients. But in our case, the coefficients depend on x, y, and t with regard for (17). Therefore, we will consider in the following the matrix G in (18) for fixed values of x, y, t and will term the corresponding matrix G the Fourier symbol of the difference scheme.

All analytic formulas presented in this section and in the next section were obtained with the aid of the computer algebra system (CAS) Mathematica. In particular,

$$G = A^{-1}B = \begin{pmatrix} \frac{b}{a} & 0 & \frac{c}{a} & 0 \\ 0 & \frac{b}{a} & 0 & \frac{c}{a} \\ 1 & 0 & 0 & 0 \\ 0 & 1 & 0 & 0 \end{pmatrix}. \tag{19}$$

Denote by $\lambda_1, \lambda_2, \lambda_3, \lambda_4$ the eigenvalues of matrix G. The von Neumann necessary stability conditions then have the form [3]

$$|\lambda_m| \le 1 + O(\tau), \quad m = 1, \ldots, 4. \tag{20}$$

Let $\kappa = (\kappa_1, \kappa_2, \kappa_3, \kappa_4)$. We have found the expression for the characteristic polynomial $f(\lambda, \kappa, \xi, \eta) = \text{Det}(G - \lambda I)$ of matrix G, where I is the identity matrix, with the aid of the Mathematica command `charpol = Det[G - λ*IdentityMatrix[4]]` Application of the Mathematica function `Factor[charpol]` yields

$$f(\lambda, \kappa, \xi, \eta) = \frac{(a\lambda^2 - b\lambda - c)^2}{a^2}. \tag{21}$$

This equation has two roots λ_1, λ_2, and the multiplicity of each of these roots is equal to two:

$$\lambda_1 = \frac{b - \sqrt{b^2 + 4ac}}{2a}, \quad \lambda_2 = \frac{b + \sqrt{b^2 + 4ac}}{2a}. \tag{22}$$

4   Analytic Investigation of Eigenvalues

We first consider the particular case of creeping fluid flows when $U \approx 0$, $V \approx 0$. Assuming then $\kappa_1 = \kappa_2 = 0$ we obtain the following expression for λ_2: $\lambda_2 = (1 - \sigma)/(1 + \sigma)$, where $\sigma = 2[\kappa_3 \sin^2(\xi/2) + \kappa_4 \sin^2(\eta/2)] \ge 0$. It is easy to be sure of the fact that $|\lambda_2| \le 1$ for any $\kappa_3, \kappa_4, \xi, \eta$. That is there are no limitations for κ_3 and κ_4. This is not surprising because for $\kappa_1 = \kappa_2 = 0$ scheme (3) is implicit, therefore, it is absolutely stable [4].

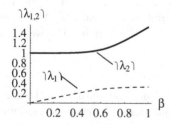
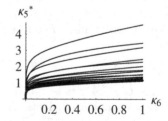

Fig. 2. The graphs of $|\lambda_{1,2}|$ vs. β **Fig. 3.** The graphs of the root $(\kappa_5^*)_2$ vs. κ_6 for different ξ

We now consider the particular case when $\kappa_3 = \kappa_4 = 0$, $\kappa_1 \geq 0$, $\kappa_2 \geq 0$. It is clear that the coefficient c in (16) reaches its maximum over ξ, η at $\xi = \eta = \pi/2$. If $\kappa_1 < 0$, $\kappa_2 < 0$, then this maximum is reached at $\sin \xi = \operatorname{sgn} \kappa_1$, $\sin \eta = \operatorname{sgn} \kappa_2$. Then in the general case it is obvious that $\max_{\xi,\eta} |c| = (1/2)(|\kappa_1| + |\kappa_2|)$. The graphs of the quantities $|\lambda_1|$, $|\lambda_2|$ are shown in Fig. 2 as the functions of the quantity $\beta = |\kappa_1| + |\kappa_2|$. It is seen that $|\lambda_2|$ exceeds unity by a small value in the interval $0 \leq \beta < 0.5$. That is scheme (3) is weakly unstable in this interval.

It follows from the above consideration of particular cases that the necessary stability condition of scheme (3) for values $\kappa_1, \kappa_2, \kappa_3, \kappa_4$ different from zero must have the following form: $|\kappa_1| + |\kappa_2| \leq \varphi(\kappa_3, \kappa_4)$, where the function $\varphi(\kappa_3, \kappa_4)$ should satisfy the following properties:

- $\varphi(0,0) = 0$;
- $\varphi(\kappa_3, \kappa_4) > 0$, $|\kappa_3| + |\kappa_4| > 0$.

The property $\varphi(0,0) = 0$ ensures the presence of the above revealed instability of scheme (3) for $\kappa_3 = \kappa_4 = 0$.

In the case when $\kappa_1 \neq 0, \kappa_2 \neq 0, \kappa_3 \neq 0, \kappa_4 \neq 0$ the derivation of stability condition in an analytic form from (22) is difficult because of the availability of square roots of complex numbers. In this connection, we use in the following the concept of the resultant, to which one can reduce the problem of determining the stability region boundary. The corresponding procedure was described in [3], therefore, we present it only briefly here. Thus, let $f(\lambda, \kappa, \xi, \eta)$ be the characteristic polynomial of a difference scheme, and let its degree in λ be equal to m $(m \geq 1)$. Following [3] let us perform the Möbius transformation $\lambda = (\omega + 1)/(\omega - 1)$. Then we obtain the polynomial

$$g(\omega, \kappa, \xi, \eta) = (\omega - 1)^m f((\omega + 1)/(\omega - 1), \kappa, \xi, \eta).$$

Let $\omega_1, \ldots, \omega_m$ be the roots of polynomial g. The condition $\operatorname{Re} \omega_j \leq 0$, $j = 1, \ldots, m$, corresponds to condition $|\lambda_j| \leq 1$, $j = 1, \ldots, m$. Then at the boundary Γ of the stability region the polynomial g must have at least one purely imaginary zero. Set $\omega = i\sigma$ and consider the polynomial $\psi(\sigma, \kappa, \xi, \eta) = g(i\sigma, \kappa, \xi, \eta)$. It is clear that the boundary Γ is determined by those values of quantities κ, ξ, η, at which the polynomial ψ has a real zero σ. Zeroes of polynomial ψ are determined by the system of two equations with real coefficients $\operatorname{Re} \psi = 0$, $\operatorname{Im} \psi = 0$. This system has the solution if and only if the resultant of equations $\operatorname{Re} \psi = 0$, $\operatorname{Im} \psi = 0$ equals zero:

$$\operatorname{Res}(\operatorname{Re} \psi, \operatorname{Im} \psi) = 0. \tag{23}$$

We now present a fragment of the Mathematica program which enables us to obtain the analytic expression for the resultant in (23) for the case of scheme (3).

```
A = {{a, 0, 0, 0}, {0, a, 0, 0}, {0, 0, 1, 0}, {0, 0, 0, 1}};
B = {{b, 0, c, 0}, {0, b, 0, c}, {1, 0, 0, 0}, {0, 1, 0, 0}};
G = Inverse[A].B; charpol = Det[G - lam*IdentityMatrix[4]];
poly = Factor[charpol]; poly=PowerExpand[Sqrt[poly]];
poly2 = poly/.{b-> b1+I*b2,c-> I*c1};
```

```
g = ComplexExpand[(w-1)^2*poly2/.{lam-> (w+1)/(w-1)}];
g1 = Expand[Simplify[g]]; g2 = g1/.{w->I*sig};
reg = ComplexExpand[Re[g2]]; img = ComplexExpand[Im[g2]];
resul = a^4*Resultant[reg,img,sig]/4;
```

As a result, we obtain the following formula for $\mathrm{Res}(\mathrm{Re}\,\psi, \mathrm{Im}\,\psi)$:

$$R(\kappa, \xi, \eta) = \mathrm{Res}(\mathrm{Re}\,\psi, \mathrm{Im}\,\psi) = -a^4 + a^2 b_1^2 + a^2 b_2^2 + 4ab_1 b_2 c_1 + 2a^2 c_1^2$$
$$+ b_1^2 c_1^2 + b_2^2 c_1^2 - c_1^4, \tag{24}$$

where in accordance with (16)

$$a = 1 + \kappa_3(1 - \cos\xi) + \kappa_4(1 - \cos\eta), \quad b_1 = 1 - \kappa_3(1 - \cos\xi) - \kappa_4(1 - \cos\eta),$$
$$b_2 = -(3/2)(\kappa_1 \sin\xi + \kappa_2 \sin\eta), \quad c_1 = (1/2)(\kappa_1 \sin\xi + \kappa_2 \sin\eta). \tag{25}$$

The substitution of expressions (25) in (24) leads to a bulky formula, which we do not present here for the sake of brevity.

As we have shown above in this section, in the particular case when $\kappa_3 = \kappa_4 = 0$ the most restrictive stability condition is obtained for $\sin\xi = \sin\eta = 1$. In this connection, we will investigate in the following the case $\xi = \eta$ in more detail. We use the following Mathematica commands:

```
resul2 = Simplify[resul/.{ξ → η}];
resz = resul2/.{κ1^2 →z,κ1^4 →z^2};
```

As a result, we obtain a quadratic equation in $z = \kappa_1^2$ to determine the roots of equation $R(\kappa, \xi, \xi) = 0$. Using the Mathematica function Solve[...] we have obtained the analytic expressions for the both roots. For the sake of brevity we present only the second root z_2. We introduce the notation $\kappa_5 = |\kappa_1| + |\kappa_2|$, $\kappa_6 = \kappa_3 + \kappa_4$, $z = \kappa_5^2$. Denote by κ_5^* the value of quantity κ_5 at the stability region boundary. Then

$$z_2 = (\kappa_5^{*2})_2 = \frac{1}{2}\mathrm{Csc}^4\xi \left(-10\kappa_6 \sin^2\xi - 12\kappa_6^2 \sin^2\xi + 10\kappa_6 \cos\xi \sin^2\xi \right.$$
$$+ 24\kappa_6^2 \cos\xi \sin^2\xi - 12\kappa_6^2 \cos^2\xi \sin^2\xi + 2\sqrt{\kappa_6}\sqrt{-1 + \cos\xi} \times$$
$$\left. (-1 - 2\kappa_6 + 2\kappa_6 \cos\xi)\sqrt{-8 - 9\kappa_6 + 9\kappa_6 \cos\xi} \sin^2\xi\right). \tag{26}$$

In particular, at $\xi = \eta = \pi/2$ we obtain the following expressions for the both roots $(\kappa_5^*)_1$ and $(\kappa_5^*)_2$:

$$(\kappa_5^*)_1 = (-5\kappa_6 - 6\kappa_6^2 - \sqrt{\kappa_6}(1 + 2\kappa_6)\sqrt{8 + 9\kappa_6})^{1/2},$$
$$(\kappa_5^*)_2 = (-5\kappa_6 - 6\kappa_6^2 + \sqrt{\kappa_6}(1 + 2\kappa_6)\sqrt{8 + 9\kappa_6})^{1/2}.$$

The radicand in formula for $(\kappa_5^*)_1$ is negative because it is the sum of negative items. Therefore, it is worthwhile considering only the root z_2 given by (26). In order to be sure that the values $\xi = \eta = \pi/2$ yield the most restrictive stability condition we have constructed twenty curves of the family $(\kappa_5^*)_2(\xi, \xi)$ with the step $\Delta\xi = 0.045\pi$. These curves are shown in Fig. 3, in which the curve for the particular pair $\xi = \eta = \pi/2$ is shown as a thick line. We can see that this line is the lowest one in Fig. 3. Thus, we have obtained an approximate form of the necessary stability condition:

$$|\kappa_1| + |\kappa_2| \leq (-5\kappa_6 - 6\kappa_6^2 + \sqrt{\kappa_6}(1 + 2\kappa_6)\sqrt{8 + 9\kappa_6})^{1/2}. \tag{27}$$

For $\xi = \eta = \pi/2$, the expression for the resultant becomes especially simple:

$$R(\kappa, \pi/2, \pi/2) = (1/2)z^2 - 4\kappa_6 + 5z\kappa_6 - 8\kappa_6^2 + 6z\kappa_6^2 - 4\kappa_6^3. \tag{28}$$

Substituting the expressions for $\kappa_1, \kappa_2, \kappa_3, \kappa_4$ from (17) into (28) we obtain a fourth-degree polynomial equation for determining the time step τ. Its solution is efficiently found with the aid of the Mathematica function $\mathtt{Solve[...]}$, and it turns out that equation $R = 0$ has two real roots and two complex conjugate roots. The real root $\tau = 0$ is of no practical value. The other real root is as follows:

$$\tau = -\frac{2(5ab - 4b^3)}{3(a^2 + 12ab^2)} - \left(2^{1/3}(-148a^2b^2 - 416ab^4 - 64b^6)\right)/(3(a^2 + 12ab^2) \times$$
$$(216a^4b + 1744a^3b^3 + 19776a^2b^5 + 9984ab^7 + 1024b^9$$
$$+ 24\sqrt{3}a^{3/2}b\sqrt{27a + 4b^2}(a^2 + 8ab^2 - 48b^4))^{1/3})$$
$$+ \frac{1}{3 \cdot 2^{1/3}(a^2 + 12ab^2)}\left((216a^4b + 1744a^3b^3 + 19776a^2b^5 + 9984ab^7\right.$$
$$\left.+ 1024b^9 + 24\sqrt{3}a^{3/2}b\sqrt{27a + 4b^2}(a^2 + 8ab^2 - 48b^4)\right)^{1/3}, \tag{29}$$

where

$$a = \left(\frac{|U|}{h_1} + \frac{|V|}{h_2}\right)^2, \quad b = \frac{\nu}{h_1^2} + \frac{\nu}{h_2^2}. \tag{30}$$

Note that after the non-dimensionalization of the Navier-Stokes equations, the value ν is usually replaced with $\nu = 1/\mathrm{Re}$.

We show in Fig. 4 the surface $\tau = \tau(a, b)$. We can draw the following conclusions from this figure: (i) for sufficiently large values of $|U|$ and $|V|$, such

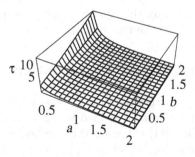

Fig. 4. The surface $\tau = \tau(a, b)$

that $a > 0.5$, the time steps are smaller than for $a < 0.5$; (ii) for low Reynolds numbers, when b is sufficiently high, the maximum time step becomes higher and higher with increasing b for sufficiently low a. This may be explained by the well-known fact that with decreasing Re, the dissipative effects become more pronounced, and right these effects are known to stabilize the difference solution.

Let us consider the case when $0 < \kappa_6 \ll 1$ (high Reynolds numbers). Using the Mathematica command Series[τ,b,0,1}] we find:

$$\tau = \frac{2(a^4)^{1/3}b^{1/3}}{a^2} - \frac{10b}{3a} + O(b^{4/3}). \tag{31}$$

If, for example, Re $= 1/\nu = 10^4$, then $\tau = O(10^{-4/3})$. This consideration explains why the computations by scheme (3) are stable also for such high Reynolds numbers.

Note that formula (29) for the maximum time step allowed by stability is approximate because for $\xi \neq \eta$ one may expect, in principle, a somewhat more restrictive stability condition. Therefore, it is advisable to compute the time step τ_n in computer code implementing scheme (3) from the known difference solution at the nth time level by formula

$$\tau_n = \theta \cdot \min_{j,k} \tau(a_{j,k}, b), \tag{32}$$

where τ is computed by (29) at each grid cell (j, k), and θ is the user-specified safety factor, $0 < \theta \leq 1$ (for example, $\theta = 0.98$).

On the other hand, although the stability condition (29) is approximate, it has a correct analytic form obtained from the von Neumann stability condition with the aid of the algebra of resultants. This enables the obtaining of information on the stability properties of a numerical method under the variation of such important physical parameters as the Reynolds number and the gas velocity.

A shortcoming of symbolic-numerical methods for stability investigation consists of the fact that although it is possible to obtain with their aid a finite set of the stability region boundary points these methods do not give information about the structure of the analytic form of the stability region boundary. Although one can obtain the analytic approximation for the maximum time step τ with the aid of the method of least squares the resulting analytic formulas have a shortcoming that they specify this analytic form in a user-predefined class of forms, which may be far from the true analytic dependence.

5   The Method of Discrete Perturbations

The method of discrete perturbations as a method for stability investigation of difference schemes was previously described in [12,3] as applied to scalar linear difference schemes. Minion [11] extended this method for the case of two grid functions u^n and v^n sought for. Another important peculiarity of the extension of the method of discrete perturbations described in [11] is that this method was applied to *nonlinear* difference equations.

Following [11] we consider the following oscillatory velocity field at the nth time level:

$$u_{j+1/2,k}^n = 1 - \varepsilon \cdot (-1)^{j+k}; \quad v_{j,k+1/2}^n = 1 + \varepsilon \cdot (h_2/h_1) \cdot (-1)^{j+k}, \qquad (33)$$

where $0 < \varepsilon \ll 1$. Besides, we assume that the discrete pressure is constant. The second-order approximation of equation (1) has the form

$$(u_{j+1/2,k}^n - u_{j-1/2,k}^n)/h_1 + (v_{j,k+1/2}^n - v_{j,k-1/2}^n)/h_2 = 0. \qquad (34)$$

The substitution of formulas (33) in (34) shows that the velocity field (33) satisfies equation (34).

The difference scheme (3) is a three-level scheme. This gives rise to the problem of computing by the scheme for $n = 0$. This problem is usually solved by using the explicit Euler method for $n = 0$:

$$\frac{\boldsymbol{v}^* - \boldsymbol{v}^n}{\tau} + H(\boldsymbol{v}^n) + \frac{1}{\rho}Gp^n = \frac{\nu}{2}\left[L(\boldsymbol{v}^*) + L(\boldsymbol{v}^n)\right]. \qquad (35)$$

The substitution of the velocity distribution (33) in (35) for $n = 0$ was carried out by us with the aid of symbolic computations in CAS Mathematica. To this end, we have at first introduced the following two functions:

```
u[j_,k_]:= 1 - eps*(-1)^(j+k); v[j_,k_]:= 1+ eps*s*(-1)^(j+k);
```

where the first function corresponds to $u_{j+1/2,k}^n$ in (33), and the second function corresponds to $v_{j,k+1/2}^n$.

We further assume following [11] that the quantities u^* and v^* obtained as the solution of difference equation (35) for $n = 0$ have the form

$$u_{j+1/2,k}^* = 1 - \alpha\varepsilon \cdot (-1)^{j+k}; \quad v_{j,k+1/2}^* = 1 + \beta\varepsilon \cdot (h_2/h_1) \cdot (-1)^{j+k}, \qquad (36)$$

where the real constants α and β are to be determined. Substituting formulas (36) in (35), we have found with the aid of CAS Mathematica the following expressions for α and β:

$$\alpha = \frac{s^2 - 2(1 + s^2)\kappa_3}{s^2 + 2(1 + s^2)\kappa_3}, \quad \beta = \frac{1 - 2(1 + s^2)\kappa_4}{1 + 2(1 + s^2)\kappa_4}, \qquad (37)$$

where $s = h_2/h_1$.

Now consider the case when $n = 1$ in (3). We can then implement the computations by three-level scheme (3) in order to find $\boldsymbol{v}^*$. Let us specify the velocity components u^1 and v^1 by the same formulas as u^*, v^* in (36). Then we find from (3) for $n = 1$:

$$u_{j+1/2,k}^* = 1 - \alpha^2\varepsilon \cdot (-1)^{j+k}; \quad v_{j,k+1/2}^* = 1 + \beta^2\varepsilon \cdot (h_2/h_1) \cdot (-1)^{j+k}. \qquad (38)$$

Formulas (37) and (38) imply the following positive property of the approximation of convective terms by the MAC method: it is insensitive to sawtooth perturbations of the form (33). In order for the oscillating part of the solution (37), (38) to be damped, it it necessary that $|\alpha| < 1$, $|\beta| < 1$. We first consider the case when $\kappa_3 = \kappa_4 = 0$. Then $\alpha = \beta = 1$, and there is no damping of oscillations.

If $\kappa_3 > 0$ and $\kappa_4 > 0$ it is easy to be sure of the fact that $0 < |\alpha| < 1$, $0 < |\beta| < 1$ for any κ_1, κ_2. That is there is the damping of oscillations (the stability). This result agrees with the result obtained above within the framework of the stability analysis by the Fourier method for the case of creeping flows when $u \approx 0$, $v \approx 0$.

6   Verification of Stability Conditions

6.1   The Taylor–Green Vortex

The Taylor–Green vortex is one of few analytical solutions of the two-dimensional Navier-Stokes equations. The solution, with $\nu = 1$ and $\rho = 1$, is given by formulas [8]

$$u = -e^{-2t}\cos x \sin y, \quad v = e^{-2t}\sin x \cos y, \quad p = -e^{-4t}(\cos 2x + \cos 2y)/4. \quad (39)$$

The flow is represented by periodic counter-rotating vortices that decay in time. The computational domain is over $\pi/2 \leq x, y \leq 5\pi/2$, which corresponds to homogeneous Dirichlet boundary conditions for the velocity component normal to the boundary and homogeneous Neumann boundary condition for the velocity component tangent to the boundary. The pressure boundary condition is homogeneous Neumann everywhere.

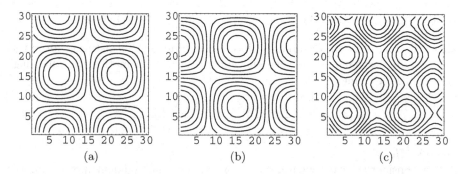

Fig. 5. The contours of u (a), v (b), and p (c) for $n = 20$ ($t = 0.96157$)

We have carried several computations of this test problem using formula (29) for the time step τ. It turns out that this formula gives the τ values, which are by factors from 2 to 6 higher than those obtained from formula (8). These factors varied depending on the grid step sizes h_1 and h_2 and the number of executed time steps, that is on the local values of the velocity components. The

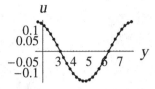

Fig. 6. The profile of $u = u(x_0, y)$, $x_0 = 3.45575$ ($\approx 1.1\pi$); solid line is the exact solution, dotted line is the numerical solution for $n = 20$ ($t = 0.96157$)

computation by the above described difference method nevertheless remained stable when using (29) with safety factor $\theta = 0.5$.

Since the amplitudes of the velocity components decay exponentially with time in the given task, we present in Figures 5 and 6 the results for the case of using formula (8) with $C_{conv} = 0.5$, $C_{diff} = 0$ and the 30×30 grid to show that our computer code works correctly also after executing several dozens of time steps.

6.2 Lid-Driven Cavity Problem

This problem is frequently used as a test of numerical methods for the incompressible Navier-Stokes equations, although it has no known exact analytic solution. In this problem the no-slip boundary conditions are imposed on the left, bottom, and right walls of the cavity, and the x-component U_0 of the velocity is specified at the upper boundary (the moving "lid"). Let B be the horizontal cavity size. Then the dimensional lengths are non-dimensionalized with respect to B, and the Reynolds number Re has the form Re $= U_0 B \rho / \mu$. The dimensionless velocity component $u = 1$ at the lid. The pressure boundary condition is homogeneous Neumann everywhere.

We have done numerous computations by the difference method of Section 2 for the purpose of elucidating the validity of formula (29) for the maximum time step allowed by stability. We at first consider the case when the Reynolds number Re $= 1$. It turns out that the computation remains stable even if the actual time step exceeds the value given by (29) by a factor of three, that is $\theta = 3$ in (32). But, on the other hand, for $\theta > 1$ the convergence to the stationary solution of the lid-driven cavity problem slows down with increasing θ.

Another interesting fact revealed by our computations in the low Reynolds number case is that the actual time step computed with the aid of (32) was by factors from 33 to 58 higher than in the case of using the known empirical formula (8), in which we specified the values $C_{conv} = 0.5$, $C_{diff} = 0$. This result means that in the case of numerical solution of more complex stationary flow problems with low Reynolds numbers it is possible to have very significant savings in CPU times (by a factor of up to 58).

And the final observation, which we have drawn from our numerical experiments involving (29) is that it ensures the fastest convergence to the stationary solution in the case of Re $= 1$ when the value on the right-hand side of (29) is multiplied by a safety factor of about 0.6.

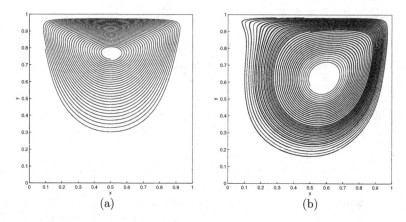

Fig. 7. Streamlines in the lid-driven cavity problem: (a) Re = 1; (b) Re = 400

In the case of a higher Reynolds number, namely Re = 400, the computation using (32) with $\theta = 1$ proves to be unstable. In order to ensure the stability for Re = 400, one must take the value $\theta < 0.1$ in (32). But even in this case, the actual "stable" time step exceeded the value given by (8) by a factor of about five.

Although the computation using (32) may remain stable also for $\theta > 1$, in the case of large time steps one should ensure the needed accuracy of the results. For this purpose, one can use the known test problems for which the exact analytic solutions are available.

We show in Fig. 7 some numerical results obtained with the use of formula (29), in which the right-hand side was multiplied by the safety factor $\theta = 0.6$ in the case of Re = 1. One can verify that these two figures are very similar to Figs. 4, (a) and (b) from [8]. Figure 7 was obtained on a mesh of 30×30 cells.

References

1. Armfield, S.W., Street, R.: The fractional-step method for the Navier-Stokes equations on staggered grids: the accuracy of three variations. J. Comp. Phys. 153, 660–665 (1999)
2. Boersma, B.J., Brethower, G., Nieuwstadt, F.T.M.: A numerical investigtion on the effect of the inflow conditions on the self-similar region of a round jet. Phys. Fluids 10, 899–909 (1998)
3. Ganzha, V.G., Vorozhtsov, E.V.: Computer-Aided Analysis of Difference Schemes for Partial Differential Equations. Wiley-Interscience, New York (1996)
4. Ganzha, V.G., Vorozhtsov, E.V.: Numerical Solution for Partial Differential Equations: Problem Solving Using Mathematica. CRC Press, Boca Raton, Ann Arbor (1996)
5. Ganzha, V.G., Vorozhtsov, E.V.: Symbolic-numerical computation of the stability regions for Jameson's schemes. Mathematics and Computers in Simulation 42, 607–615 (1996)

6. Harlow, F.H., Welch, J.E.: Numerical calculation of time-dependent viscous incompressible flow of fluid with free surface. Phys. Fluids 8, 2182–2189 (1965)
7. Ilyushin, B.B., Krasinsky, D.V.: Large eddy simulation of the turbulent round jet dynamics. Thermophysics and Aeromechanics 13(1), 43–54 (2006)
8. Kim, J., Moin, P.: Application of a fractional-step method to incompressible Navier-Stokes equations. J. Comp. Phys. 59, 308–323 (1985)
9. Kirkpatrick, M.P., Armfield, S.W., Kent, J.H.: A representtion of curved boundaries for the solution of the Navier-Stokes equations on a staggered three-dimensional Cartesian grid. J. Comp. Phys. 184, 1–36 (2003)
10. Marella, S., Krishnan, S., Liu, H., Udaykumar, H.S.: Sharp interface Cartesian grid method I: An easily implemented technique for 3D moving boundary computations. J. Comp. Phys. 210, 1–31 (2005)
11. Minion, M.L.: On the stability of Godunov-projection methods for incompressible flow. J. Comp. Phys. 123, 435–449 (1996)
12. Roache, P.J.: Computational Fluid Dynamics. Hermosa, Albuquerque, New Mexico (1976)
13. Schlichting, H., Truckenbrodt, E.: Aerodynamics of the Airplane. McGraw-Hill, New York (1979)
14. Uhlmann, M.: An immersed boundary method with direct forcing for the simulation of particulate flows. J. Comp. Phys. 209, 448–476 (2005)
15. Vorst, van der: Bi-CGSTAB: A fast and smoothly converging variant of Bi-CG for the solution of nonsymmetric linear systems. SIAM J. Sci. Statist. Comput. 13, 631–644 (1992)
16. Yang, J., Balaras, E.: An embedded-boundary formulation for large-eddy simulation of turbulent flows interacting with moving boundaries. J. Comp. Phys. 215, 12–40 (2006)

A Symbolic-Numerical Algorithm for Solving the Eigenvalue Problem for a Hydrogen Atom in the Magnetic Field: Cylindrical Coordinates

Ochbadrakh Chuluunbaatar[1], Alexander Gusev[1], Vladimir Gerdt[1],
Michail Kaschiev[2], Vitaly Rostovtsev[1], Valentin Samoylov[1],
Tatyana Tupikova[1], and Sergue Vinitsky[1]

[1] Joint Institute for Nuclear Research, Dubna, Moscow Region, Russia
vinitsky@theor.jinr.ru
[2] Institute of Mathematics and Informatics, BAS, Sofia, Bulgaria

Abstract. The boundary problem in cylindrical coordinates for the Schrödinger equation describing a hydrogen-like atom in a strong homogeneous magnetic field is reduced to the problem for a set of the longitudinal equations in the framework of the Kantorovich method. The effective potentials of these equations are given by integrals over transversal variable of a product of transverse basis functions depending on the longitudinal variable as a parameter and their first derivatives with respect to the parameter. A symbolic-numerical algorithm for evaluating the transverse basis functions and corresponding eigenvalues which depend on the parameter, their derivatives with respect to the parameter and corresponded effective potentials is presented. The efficiency and accuracy of the algorithm and of the numerical scheme derived are confirmed by computations of eigenenergies and eigenfunctions for the low-excited states of a hydrogen atom in the strong homogeneous magnetic field.

1 Introduction

To solve the problem of photoionization of low-lying excited states of a hydrogen atom in a strong magnetic field [1,2] symbolic-numerical algorithms (SNA) and the Finite Element Method (FEM) code have been elaborated [3,4,5,6]. Next investigations are shown that to impose on boundary conditions for the scattering problem in spherical coordinates (r, θ, φ), one needs to consider solution of this problem in cylindrical coordinates (z, ρ, φ) and to construct an asymptotics of solutions for both small and large values of the longitudinal variable [2,7].

With this end in view we consider a SNA for evaluating the transverse basis functions and eigenvalues depending on a longitudinal parameter, $|z|$, for their derivatives with respect to the $|z|$ and for the effective potentials depended on $|z|$ of the 1-D problem for a set of second order differential equations in the frame of the Kantorovich method (KM) [8]. For solving the above problems on a grid of the longitudinal parameter, $|z|$, from a finite interval, we elaborate the SNA to reduce a transverse eigenvalue problem for a second order ordinary

V.G. Ganzha, E.W. Mayr, and E.V. Vorozhtsov (Eds.): CASC 2007, LNCS 4770, pp. 118–133, 2007.

differential equation to algebraic one applied the FEM [9,10] or some expansions of the solution over an appropriate basis such that corresponded integrals over transversal variable will be calculated analytically [11,12]. A symbolic algorithm for evaluating the asymptotic effective potentials with respect to the $|z|$, using a series expansion in the Laguerre polynomials, is implemented in MAPLE and is used to continue the calculated numerical values of effective potentials to large values of $|z|$.

The main goal of this paper is to develop a symbolic algorithm for generation of algebraic eigenvalue problem to calculate economically the transverse basis on a grid points of finite interval of the longitudinal parameter, $|z|$, and its continuation from matching point to large $|z|$. The obtained asymptotic of effective potentials at large values of the longitudinal variable are used as input file for an auxiliary symbolic algorithm of evaluation in analytical form the asymptotics of solutions of a set of the second order differential equations with respect to the longitudinal variable, $|z|$, in the KM. The algorithms are explicitly presented and implemented in MAPLE. The developed approach is applied to numerical calculation of effective potentials for the Schrödinger equation describing a hydrogen-like atom in a strong magnetic field. A region of applicability versus a strength of the magnetic field, efficiency and accuracy of the developed algorithms and accompanying numerical schemes is confirmed by computation of eigenenergies and eigenfunctions of a hydrogen atom in the strong homogeneous magnetic field.

The paper is organized as follows. In section 2 we briefly describe a reduction of the 2D-eigenvalue problem to the 1D-eigenvalue problem for a set of the closed longitudinal equations by means of the KM. In section 3 algorithm of generation of an algebraic problem by means of the FEM. We examine the algorithm for evaluating the transverse basis functions on a grid of the longitudinal parameter from a finite interval. In section 4 the algorithm for asymptotic calculation of matrix elements at large values of the longitudinal variable is presented. In section 5 the auxiliary algorithm of evaluation the asymptotics of the longitude solutions at large $|z|$ in the KM. In section 6 the method is applied to calculating the low-lying states of a hydrogen atom in a strong magnetic field. The convergence rate is explicitly demonstrated for typical examples. The obtained results are compared with the known ones obtained in the spherical coordinates to establish of an applicability range of the method. In section 7 the conclusions are made, and the possible future applications of the method are discussed.

2   Statement of the Problem in Cylindrical Coordinates

The wave function $\hat{\Psi}(\rho, z, \varphi) = \Psi(\rho, z) \exp(\imath m \varphi)/\sqrt{2\pi}$ of a hydrogen atom in an axially symmetric magnetic field $\boldsymbol{B} = (0, 0, B)$ in cylindrical coordinates (ρ, z, φ) satisfies the 2D Schrödinger equation

$$-\frac{\partial^2}{\partial z^2}\Psi(\rho, z) + \hat{A}_c\Psi(\rho, z) = \epsilon\Psi(\rho, z), \tag{1}$$

$$\hat{A}_c = \hat{A}_c^{(0)} - \frac{2Z}{\sqrt{\rho^2 + z^2}}, \quad \hat{A}_c^{(0)} = -\frac{1}{\rho}\frac{\partial}{\partial \rho}\rho\frac{\partial}{\partial \rho} + \frac{m^2}{\rho^2} + m\gamma + \frac{\gamma^2\rho^2}{4},$$

in the region Ω_c: $0 < \rho < \infty$ and $-\infty < z < \infty$. Here $m = 0, \pm 1, \ldots$ is the magnetic quantum number, $\gamma = B/B_0$, $B_0 \cong 2.35 \times 10^5\,T$ is a dimensionless parameter which determines the field strength B. We use the atomic units ($a.u.$) $\hbar = m_e = e = 1$ and assume the mass of the nucleus to be infinite. In these expressions $\epsilon = 2E$, E is the energy (expressed in Rydbergs, $1\,Ry = (1/2)\,a.u.$) of the bound state $|m\sigma\rangle$ with fixed values of m and z-parity $\sigma = \pm 1$, and $\Psi(\rho, z) \equiv \Psi^{m\sigma}(\rho, z) = \sigma\Psi^{m\sigma}(\rho, -z)$ is the corresponding wave function. Boundary conditions in each $m\sigma$ subspace of the full Hilbert space have the form

$$\lim_{\rho \to 0} \rho \frac{\partial \Psi(\rho, z)}{\partial \rho} = 0, \quad \text{for} \quad m = 0, \quad \text{and} \quad \Psi(0, z) = 0, \quad \text{for} \quad m \neq 0, \quad (2)$$

$$\lim_{\rho \to \infty} \Psi(\rho, z) = 0. \quad (3)$$

The wave function of the discrete spectrum obeys the asymptotic boundary condition. Approximately this condition is replaced by the boundary condition of the second and/or first type at small and large $|z|$, but finite $|z| = z_{\max} \gg 1$,

$$\lim_{z \to 0} \frac{\partial \Psi(\rho, z)}{\partial z} = 0, \quad \sigma = +1, \quad \Psi(\rho, 0) = 0, \quad \sigma = -1, \quad (4)$$

$$\lim_{z \to \pm \infty} \Psi(\rho, z) = 0 \quad \to \quad \Psi(\rho, \pm|z_{\max}|) = 0. \quad (5)$$

These functions satisfy the additional normalization condition

$$\int_{-z_{\max}}^{z_{\max}} \int_0^\infty |\Psi(\rho, z)|^2 \rho d\rho dz = 2 \int_0^{z_{\max}} \int_0^\infty |\Psi(\rho, z)|^2 \rho d\rho dz = 1. \quad (6)$$

2.1 Kantorovich Expansion

Consider a formal expansion of the partial solution $\Psi_i^{Em\sigma}(\rho, z)$ of Eqs. (1)–(3), corresponding to the eigenstate $|m\sigma i\rangle$, expanded in the finite set of one-dimensional basis functions $\{\hat{\Phi}_j^m(\rho; z)\}_{j=1}^{j_{\max}}$

$$\Psi_i^{Em\sigma}(\rho, z) = \sum_{j=1}^{j_{\max}} \hat{\Phi}_j^m(\rho; z) \hat{\chi}_j^{(m\sigma i)}(E, z). \quad (7)$$

In Eq. (7) the functions $\hat{\boldsymbol{\chi}}^{(i)}(z) \equiv \hat{\boldsymbol{\chi}}^{(m\sigma i)}(E, z)$, $(\hat{\boldsymbol{\chi}}^{(i)}(z))^T = (\hat{\chi}_1^{(i)}(z), \ldots, \hat{\chi}_{j_{\max}}^{(i)}(z))$ are unknown, and the surface functions $\hat{\boldsymbol{\Phi}}(\rho; z) \equiv \hat{\boldsymbol{\Phi}}^m(\rho; z) = \hat{\boldsymbol{\Phi}}^m(\rho; -z)$, $(\hat{\boldsymbol{\Phi}}(\rho; z))^T = (\hat{\Phi}_1(\rho; z), \ldots, \hat{\Phi}_{j_{\max}}(\rho; z))$ form an orthonormal basis for each value of the variable z which is treated as a parameter.

In the KM the wave functions $\hat{\Phi}_j(\rho; z)$ and the potential curves $\hat{E}_j(z)$ (in Ry) are determined as the solutions of the following eigenvalue problem

$$\hat{A}_c \hat{\Phi}_j(\rho; z) = \hat{E}_j(z) \hat{\Phi}_j(\rho; z), \quad (8)$$

with the boundary conditions

$$\lim_{\rho \to 0} \rho \frac{\partial \hat{\Phi}_j(\rho; z)}{\partial \rho} = 0, \quad \text{for} \quad m = 0, \quad \text{and} \quad \hat{\Phi}_j(0; z) = 0, \quad \text{for} \quad m \neq 0, \quad (9)$$

$$\lim_{\rho \to \infty} \hat{\Phi}_j(\rho; z) = 0. \quad (10)$$

Since the operator in the left-hand side of Eq. (8) is self-adjoint, its eigenfunctions are orthonormal

$$\left\langle \hat{\Phi}_i(\rho; z) \middle| \hat{\Phi}_j(\rho; z) \right\rangle_\rho = \int_0^\infty \hat{\Phi}_i(\rho; z)\hat{\Phi}_j(\rho; z)\rho d\rho = \delta_{ij}, \qquad (11)$$

where δ_{ij} is the Kronecker symbol. Therefore we transform the solution of the above problem into the solution of an eigenvalue problem for a set of j_{max} ordinary second-order differential equations that determines the energy ϵ and the coefficients $\hat{\chi}^{(i)}(z)$ of the expansion (7)

$$\left(-\mathbf{I}\frac{d^2}{dz^2} + \hat{\mathbf{U}}(z) + \hat{\mathbf{Q}}(z)\frac{d}{dz} + \frac{d\hat{\mathbf{Q}}(z)}{dz} \right) \hat{\chi}^{(i)}(z) = \epsilon_i \mathbf{I}\hat{\chi}^{(i)}(z). \qquad (12)$$

Here $\mathbf{I}$, $\hat{\mathbf{U}}(z) = \hat{\mathbf{U}}(-z)$ and $\hat{\mathbf{Q}}(z) = -\hat{\mathbf{Q}}(-z)$ are the $j_{max} \times j_{max}$ matrices whose elements are expressed as

$$\hat{U}_{ij}(z) = \left(\frac{\hat{E}_i(z) + \hat{E}_j(z)}{2} \right) \delta_{ij} + \hat{H}_{ij}(z), \quad I_{ij} = \delta_{ij},$$

$$\hat{H}_{ij}(z) = \hat{H}_{ji}(z) = \int_0^\infty \frac{\partial \hat{\Phi}_i(\rho; z)}{\partial z} \frac{\partial \hat{\Phi}_j(\rho; z)}{\partial z} \rho d\rho, \qquad (13)$$

$$\hat{Q}_{ij}(z) = -\hat{Q}_{ji}(z) = -\int_0^\infty \hat{\Phi}_i(\rho; z)\frac{\partial \hat{\Phi}_j(\rho; z)}{\partial z} \rho d\rho.$$

The discrete spectrum solutions obey the asymptotic boundary condition and the orthonormality conditions

$$\lim_{z\to 0}\left(\frac{d}{dz} - \hat{\mathbf{Q}}(z) \right) \hat{\chi}^{(i)}(z) = 0, \quad \sigma = +1, \quad \hat{\chi}^{(i)}(0) = 0, \quad \sigma = -1, \quad (14)$$

$$\lim_{z\to\pm\infty} \hat{\chi}^{(i)}(z) = 0 \quad \to \quad \hat{\chi}^{(i)}(\pm z_{max}) = 0, \qquad (15)$$

$$\int_{-z_{max}}^{z_{max}} \left(\hat{\chi}^{(i)}(z) \right)^T \hat{\chi}^{(j)}(z)dz = 2\int_0^{z_{max}} \left(\hat{\chi}^{(i)}(z) \right)^T \hat{\chi}^{(j)}(z)dz = \delta_{ij}. \qquad (16)$$

3   Algorithm 1 of Generation of Parametric Algebraic Problems by the Finite Element Method

To solve eigenvalue problem for equation (8) the boundary conditions (9), (10) and the normalization condition (11) with respect to the space variable ρ on an infinite interval are replaced with appropriate conditions (9), (11) and $\hat{\Phi}(\rho_{max}; z) = 0$ on a finite interval $\rho \in [\rho_{min} \equiv 0, \rho_{max}]$.

We consider a discrete representation of solutions $\hat{\Phi}(\rho; z)$ of the problem (8) by means of the FEM on the grid, $\Omega_{h(\rho)}^p = (\rho_0 = \rho_{min}, \rho_j = \rho_{j-1} + h_j, \rho_{\bar{n}} = \rho_{max})$, in a finite sum in each $z = z_k$ of the grid $\Omega_{h(z)}^p[z_{min}, z_{max}]$:

$$\hat{\Phi}(\rho; z) = \sum_{\mu=0}^{\bar{n}p} \Phi_\mu^h(z) N_\mu^p(\rho) = \sum_{r=0}^{\bar{n}} \sum_{j=1}^{p} \Phi_{r+p(j-1)}^h(z) N_{r+p(j-1)}^p(\rho), \qquad (17)$$

where $N_\mu^p(\rho)$ are local functions and $\Phi_\mu^h(z)$ are node values of $\hat{\Phi}(\rho_\mu; z)$. The local functions $N_\mu^p(\rho)$ are piece-wise polynomial of the given order p equals one only in the node ρ_μ and equals zero in all other nodes $\rho_\nu \neq \rho_\mu$ of the grid $\Omega_{h(\rho)}^p$, i.e., $N_\nu^p(\rho_\mu) = \delta_{\nu\mu}$, $\mu, \nu = 0, 1, \ldots, \bar{n}p$. The coefficients $\Phi_\nu(z)$ are formally connected with solution $\hat{\Phi}(\rho_{j,r}^p; z)$ in a node $\rho_\nu = \rho_{j,r}^p$, $r = 1, \ldots, p$, $j = 0, \ldots, \bar{n}$:

$$\Phi_\nu^h(z) = \Phi_{r+p(j-1)}^h(z) \approx \hat{\Phi}(\rho_{j,r}^p; z), \quad \rho_{j,r}^p = \rho_{j-1} + \frac{h_j}{p} r.$$

The theoretical estimate for the $\mathbf{H^0}$ norm between the exact and numerical solution has the order of

$$|\hat{E}_m^h(z) - \hat{E}_m(z)| \leq c_1 |\hat{E}_m(z)| h^{2p}, \quad \left\| \Phi_m^h(z) - \Phi_m(z) \right\|_0 \leq c_2 |\hat{E}_m(z)| h^{p+1},$$

where $h = \max_{1 < j < \bar{n}} h_j$ is maximum step of grid [9]. It has been shown that we have a possibility to construct schemes with high order of accuracy comparable with the computer one [14]. Let us consider the reduction of differential equations (8) on the interval $\Delta: \rho_{\min} < \rho < \rho_{\max}$ with boundary conditions in points $\rho_{\min}$ and $\rho_{\max}$ rewriting in the form

$$\mathbf{A}(z)\hat{\Phi}(\rho; z) = \hat{E}(z)\mathbf{B}(z)\hat{\Phi}(\rho; z), \qquad (18)$$

where $\mathbf{A}$ and $\mathbf{B}$ are differential operators. Substituting expansion (17) to (18) and integration with respect to ρ by parts in the interval $\Delta = \cup_{j=1}^{\bar{n}} \Delta_j$, we arrive to a system of the linear algebraic equations

$$\mathbf{a}_{\mu\nu}^p \Phi_\mu^h(z) = \hat{E}(z)\mathbf{b}_{\mu\nu}^p \Phi_\mu^h(z), \qquad (19)$$

in framework of the briefly described FEM. Using p-order Lagrange elements [9], we present below an algorithm 1 for construction of algebraic problem (19) by the FEM in the form of conventional pseudocode. It MAPLE realization allow us show explicitly recalculation of indices μ, ν and test of correspondent modules in FORTRAN code.

In order to solve the generalized eigenvalue problem (19), the subspace iteration method [9,10] elaborated by Bathe [10] for the solution of large symmetric banded matrix eigenvalue problems has been chosen. This method uses a skyline storage mode, which stores components of the matrix column vectors within the banded region of the matrix, and is ideally suited for banded finite element matrices. The procedure chooses a vector subspace of the full solution space and iterates upon the successive solutions in the subspace (for details, see [10]). The iterations continue until the desired set of solutions in the iteration subspace converges to within the specified tolerance on the Rayleigh quotients for the eigenpairs. Generally, 10-16 iterations are required for the subspace iterations to converge the subspace to within the prescribe tolerance. If matrix $\mathbf{a}^p$ in Eq. (19) is not positively defined, problem (19) is replaced by the following problem:

$$\tilde{\mathbf{a}}^p \, \boldsymbol{\Phi}^h = \check{E}^h \, \mathbf{b}^p \, \boldsymbol{\Phi}^h, \quad \tilde{\mathbf{a}}^p = \mathbf{a}^p - \alpha \mathbf{b}^p. \tag{20}$$

The number α (the shift of the energy spectrum) is chosen in such a way that matrix $\tilde{\mathbf{a}}^p$ is positive. The eigenvector of problem (20) is the same, and $\hat{E}^h = \check{E}^h + \alpha$.

Algorithm 1

Input:
$\Delta = \cup_{j=1}^{\bar{n}} \Delta_j = [\rho_{\min}, \rho_{\max}]$, is interval of changing of space variable ρ;
$h_j = \rho_j - \rho_{j-1}$ is a grid step;
$\bar{n}$ is a number of subintervals $\Delta_j = [\rho_{j-1}, \rho_j]$;
p is a order of finite elements;
$\mathbf{A}(z), \mathbf{B}(z)$ are differential operators in Eq. (18);
Output:
N_μ^p is a basis functions in (17);
$\mathbf{a}_{\mu\nu}^p, \mathbf{b}_{\mu\nu}^p$ are matrix elements in system of algebraic equations (19);
Local:
$\rho_{j,r}^p$ are nodes;
$\phi_{j,r}^p(\rho)$ are Lagrange elements;
$\mu, \nu = 0, 1, \ldots, \bar{n}p$;

1: for $j:=1$ to $\bar{n}$ do
    for $r:=0$ to p do
        $\rho_{j,r}^p = \rho_{j-1} + \frac{h_j}{p} r$
    end for;
    end for;
2: $\phi_{j,r}^p(\rho) = \prod_{k \neq r}[(\rho - \rho_{j,k}^p)(\rho_{j,r}^p - \rho_{j,k}^p)^{-1}]$
3: $N_0^p(\rho) := \{\phi_{1,0}^p(\rho), \rho \in \Delta_1; 0, \rho \notin \Delta_1\}$;
    for $j:=1$ to $\bar{n}$ do
        for $r:=1$ to $p-1$ do
            $N_{r+p(j-1)}^p(\rho) := \{\phi_{j,r}^p(\rho), \rho \in \Delta_j; 0, \rho \notin \Delta_j, \}$
        end for;
        $N_{jp}^p(\rho) := \{\phi_{j,p}^p(\rho), \rho \in \Delta_j; \phi_{j+1,0}^p(\rho), \rho \in \Delta_{j+1}; 0, \rho \notin \Delta_j \cup \Delta_{j+1}\}$;
    end for;
    $N_{\bar{n}p}^p(\rho) := \{\phi_{\bar{n},p}^p(\rho), \rho \in \Delta_{\bar{n}}; 0, \rho \notin \Delta_{\bar{n}}\}$;
4: for $\mu, \nu:=0$ to $\bar{n}p$ do
    $\mathbf{a}_{\mu\nu}^p := \int_\Delta N_\mu^p(\rho) \mathbf{A}(z) N_\nu^p(\rho) \rho d\rho; \quad \mathbf{b}_{\mu\nu}^p := \int_\Delta N_\mu^p(\rho) \mathbf{B}(z) N_\nu^p(\rho) \rho d\rho;$
    end for;

Remarks

1. For equation (8) matrix elements of the operator,

$$\hat{A}_c = -\frac{1}{\rho}\frac{\partial}{\partial\rho}\rho\frac{\partial}{\partial\rho} + V(\rho; z), \quad V(\rho; z) = -\frac{2Z}{\sqrt{\rho^2 + z^2}} + \frac{m^2}{\rho^2} + m\gamma + \frac{\gamma^2\rho^2}{4},$$

between local functions N_μ and N_ν defined in same interval Δ_j calculated by formula

$$(\mathbf{a}(z_k))_{q+p(j-1),r+p(j-1)} = \int\limits_{-1}^{+1}\left\{\frac{4}{h_j^2}(\phi_{j,q}^p)'(\phi_{j,r}^p)' + V(\rho; z_k)\phi_{j,q}^p\phi_{j,r}^p\right\}\frac{h_j}{2}\rho d\eta,$$

$$(\mathbf{b}(z_k))_{q+p(j-1),r+p(j-1)} = \int\limits_{-1}^{+1}\phi_{j,q}^p\phi_{j,r}^p\frac{h_j}{2}\rho d\eta.$$

2. If integrals do not calculated analytically, for example, like in [11,12], then they have been calculated by numerical methods [9], by means of the Gauss quadrature formulae of the order $p + 1$.

3. For calculations matrix elements (19) and the corresponded derivatives of eigenfunctions by z we used algorithm described in [3]. Starting from matching point $z_m < z_{\max}$ of the grid $\Omega^p_{h(z)}[z_{\min}, z_{\max}]$ the calculation has been performed using an asymptotic expansion from next section ($z_m \sim 20$, $z_{\max} \sim 100$).

4. The problem (8)–(10) has been solved using a grid $\Omega^p_{h(\rho)}[\rho_{\min}, \rho_{\max}] = 0(500)4(500)30$ (the number in parentheses denotes the number of finite elements of order $p = 4$ in each interval). As an example, at $m = -1$ and $\gamma = 10$ the calculated the potential curves $\hat{E}_j(z)$, effective potentials $\hat{Q}_{ij}(z)$, $\hat{H}_{ij}(z)$ are shown in Fig. 1.

4   Algorithm 2 of Evaluation the Asymptotics of Effective Potentials at Large $|z|$ in Kantorovich Method

Step 1. In (8) apply the transformation to a scaled variable x

$$x = \frac{\gamma\rho^2}{2}, \quad \rho = \frac{\sqrt{x}}{\sqrt{\gamma/2}}, \tag{21}$$

and put $\lambda = \hat{E}_j(z)/(2\gamma) = \lambda^{(0)} + m/2 - Z/(\gamma|z|) + \delta\lambda$. Eigenvalue problem reads

$$\left(-\frac{\partial}{\partial x}x\frac{\partial}{\partial x} + \frac{m^2}{4x} + \frac{x}{4} + \frac{m}{2} - \frac{Z}{\gamma\sqrt{\frac{2x}{\gamma} + z^2}} - \lambda\right)\hat{\Phi}_j(x; z) = 0, \tag{22}$$

with a normalization condition

$$\frac{1}{\gamma}\int_0^\infty \hat{\Phi}_j(x; z)^2 dx = 1. \tag{23}$$

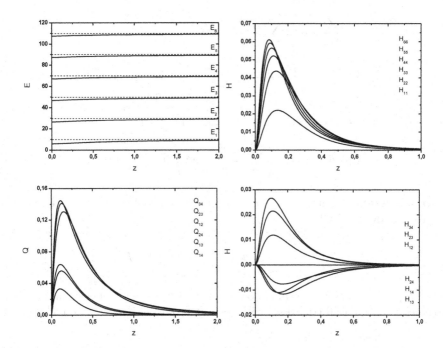

Fig. 1. The behaviour of potential curves $\hat{E}_j(z)$, effective potentials $\hat{Q}_{ij}(z)$ and $\hat{H}_{ij}(z)$ for $\gamma = 10$, $m = -1$

At $Z = 0$ Eq. (22) takes the form

$$L(n)\Phi_{nm}^{(0)}(x) = 0, \qquad L(n) = -\frac{\partial}{\partial x}x\frac{\partial}{\partial x} + \frac{m^2}{4x} + \frac{x}{4} - \lambda^{(0)}, \tag{24}$$

and has the regular and bounded solutions at

$$\lambda^{(0)} = n + (|m| + 1)/2, \tag{25}$$

where transverse quantum number $n \equiv N_\rho = j - 1 = 0, 1, \ldots$ determines the number of nodes of the solution $\Phi_{nm}^{(0)}(x)$ with respect to the variable x. Normalized solutions of Eq. (24), take the form

$$\Phi_{nm}^{(0)}(x) = C_{n|m|}e^{-\frac{x}{2}}x^{\frac{|m|}{2}}L_n^{|m|}(x), \qquad C_{n|m|} = \left[\gamma\frac{n!}{(n+|m|)!}\right]^{\frac{1}{2}}, \tag{26}$$

$$\frac{1}{\gamma}\int_0^\infty \Phi_{nm}^{(0)}(x)\Phi_{n'm}^{(0)}(x)dx = \delta_{nn'}, \tag{27}$$

where $L_n^{|m|}(x)$ are Laguerre polynomials [13].

Step 2. Substituting notation $\delta\lambda = \lambda - \lambda^{(0)} - m/2 + Z/(\gamma|z|) \equiv \hat{E}_j(z)/(2\gamma) - (n + (m + |m| + 1)/2) + Z/(\gamma|z|)$, and decomposition

$$\frac{Z}{\gamma|z|} - \frac{Z}{\gamma\sqrt{\frac{2x}{\gamma} + z^2}} = \sum_{k=1}^{j_{\max}} \frac{V^{(k)}}{|z|^k},$$

$$V^{(k)} = \begin{cases} -(-1)^{k'} \frac{(2k'-1)!!}{k'!} \frac{Z x^{k'}}{\gamma^{k'+1}}, & k = 2k'+1, \quad k' = 1, 2, \ldots, \\ 0, & \text{otherwise,} \end{cases}$$

to Eq. (22) at $Z \neq 0$, transform it in the following form

$$L(n)\hat{\varPhi}_j(x; z) + \left(\sum_{k=1}^{j_{\max}} \frac{V^{(k)}}{|z|^k} - \delta\lambda\right)\hat{\varPhi}_j(x; z) = 0. \tag{28}$$

Step 3. Solution of equation (28) is found in the form of the perturbation series by inverse powers of $|z|$

$$\delta\lambda = \sum_{k=0}^{k_{\max}} |z|^{-k}\lambda^{(k)}, \quad \varPhi_j(x; z) = \sum_{k=0}^{k_{\max}} |z|^{-k}\varPhi_n^{(k)}(x). \tag{29}$$

Equating coefficients at the same powers of $|z|$, we arrive to the system of inhomogeneous differential equations with respect to corrections $\lambda^{(k)}$ and $\varPhi^{(k)}$

$$L(n)\varPhi^{(0)}(x) = 0 \equiv f^{(0)},$$

$$L(n)\varPhi^{(k)}(x) = \sum_{p=0}^{k-1} (\lambda^{(k-p)} - V^{(k-p)})\varPhi^{(p)}(x) \equiv f^{(k)}, \quad k \geq 1. \tag{30}$$

For solving the Eqs. (28) the unnormalized orthogonal basis

$$\varPhi_{n+s}(x) = C_{n|m|} e^{-\frac{x}{2}} x^{\frac{|m|}{2}} L_{n+s}^{|m|}(x) = C_{n|m|} C_{n+s|m|}^{-1} \varPhi_{n+s,m}^{(0)}(x), \tag{31}$$

$$\langle s|s'\rangle = \int_0^\infty \varPhi_{n+s}(x)\varPhi_{n+s'}(x)dx = \delta_{ss'}\gamma\frac{n!}{(n+|m|)!}\frac{(n+s+|m|)!}{(n+s)!},$$

has been applied. The operators $L(n)$ and x on the functions $\varPhi_{n+s}(x)$ are defined by the relations without fractional powers of quantum numbers n and m

$$L(n)\varPhi_{n+s}(x) = s\varPhi_{n+s}(x), \tag{32}$$

$$x\varPhi_{n+s}(x) = -(n+s+|m|)\varPhi_{n+s-1}(x) + (2(n+s)+|m|+1)\varPhi_{n+s}(x)$$
$$-(n+s+1)\varPhi_{n+s+1}(x).$$

Step 4. Applying relations (32), the right-hand side $f^{(k)}$ and solutions $\varPhi^{(k)}(x)$ of the system (30) are expanded over basis states $\varPhi_{n+s}(x)$

$$\varPhi_n^{(k)}(x) = \sum_{s=-k}^{k} b_s^{(k)} \varPhi_{n+s}(x), \quad f^{(k)} = \sum_{s=-k}^{k} f_s^{(k)} \varPhi_{n+s}(x). \tag{33}$$

Then a recurrent set of linear algebraic equations for unknown coefficients $b_s^{(k)}$ and corrections $\lambda^{(k)}$ is obtained

$$sb_s^{(k)} - f_s^{(k)} = 0, \quad s = -k, \ldots, k.$$

that is solved sequentially for $k = 1, 2, \ldots, k_{max}$:

$$f_0^{(k)} = 0 \quad \to \lambda^{(k)}; \quad b_s^{(k)} = f_s^{(k)}/s, \quad s = -k, \ldots, k, \quad s \neq 0.$$

The initial conditions (25) and $b_s^{(0)} = \delta_{s0}$ are followed from (24) and (27).

Step 5. To obtain the normalized wave function $\hat{\Phi}_j(x; z)$ up to the k-th order, the coefficient $b_0^{(k)}$ are defined by the following relation:

$$b_0^{(k)} = -\frac{1}{2\gamma} \sum_{p=1}^{k-1} \sum_{s'=p-k}^{k-p} \sum_{s=-p}^{p} b_s^{(k-p)} \langle s|s' \rangle b_{s'}^{(p)}, \quad b_0^{(k=1,\ldots,5)} = 0.$$

As an example of output file at steps 1–5, we display nonzero coefficients $\lambda^{(k)}$, $b_s^{(k)}$ of the inverse power series (29), (33) up to $O(|z|^{-5})$:

$$\lambda^{(0)} = n + (|m| + 1)/2, \quad \lambda^{(3)} = Z(2n + |m| + 1)/\gamma^2,$$
$$b_0^{(0)} = 1, \quad b_{-1}^{(3)} = -Z(n + |m|)/\gamma^2, \quad b_1^{(3)} = Z(n + 1)/\gamma^2. \tag{34}$$

Step 6. In scaled variable x the relations of effective potentials $\hat{H}_{ij}(z) = \hat{H}_{ji}(z)$ and $\hat{Q}_{ij}(z) = -\hat{Q}_{ji}(z)$ takes form

$$\hat{H}_{ij}(z) = \frac{1}{\gamma} \int_0^\infty dx \frac{\partial \hat{\Phi}_i(x; z)}{\partial z} \frac{\partial \hat{\Phi}_j(x; z)}{\partial z}, \quad \hat{Q}_{ij}(z) = -\frac{1}{\gamma} \int_0^\infty dx \hat{\Phi}_i(x; z) \frac{\partial \hat{\Phi}_j(x; z)}{\partial z}. \tag{35}$$

For their evaluation the decomposition of solution Eqs. (24) over the normalized orthogonal basis $\Phi_{n+s}^{(0)}$ with the normalized coefficients $b_{n;n+s}^{(k)}$,

$$\Phi_n^{(k)}(x) = \sum_{s=-k}^{k} b_{n;n+s}^{(k)} \Phi_{n+s}^{(0)}, \tag{36}$$

has been applied. The normalized coefficients $b_{n;n+s}^{(k)}$ are calculated via $b_s^{(k)}$,

$$b_{n;n+s}^{(k)} = b_s^{(k)} \sqrt{\frac{n!}{(n+|m|)!} \frac{(n+s+|m|)!}{(n+s)!}} \tag{37}$$

as follows from (33), (36) and (31).

Step 7. In a result of substitution (29), (36) in (35), matrix elements takes form

$$\hat{Q}_{jj+t}(z) = -\sum_{k=0}^{k_{max}-1} |z|^{-k-1} \sum_{k'=0}^{k} \sum_{s=\max(-k,k'-k-t)}^{\min(k,k-k'-t)} (k-k') b_{n;n+s}^{(k')} b_{n+t;n+s}^{(k-k')},$$

$$\hat{H}_{jj+t}(z) = \sum_{k=0}^{k_{max}-2} |z|^{-k-2} \sum_{k'=0}^{k} \sum_{s=\max(-k,k'-k-t)}^{\min(k,k-k'-t)} k'(k-k') b_{n;n+s}^{(k')} b_{n+t;n+s}^{(k-k')}. \tag{38}$$

Collecting of coefficients of (38) at equal powers of $|z|$, algorithm leads to final expansions of eigenvalues and effective potentials of output file

$$\hat{E}_j(z) = \sum_{k=0}^{k_{\max}} |z|^{-k} E_j^{(k)}, \quad \hat{H}_{ij}(z) = \sum_{k=8}^{k_{\max}} |z|^{-k} H_{ij}^{(k)}, \quad \hat{Q}_{ij}(z) = \sum_{k=4}^{k_{\max}} |z|^{-k} Q_{ij}^{(k)}. \tag{39}$$

The successful run of the above algorithm was occurs up to $k_{\max} = 16$ (Run time is 95s on Intel Pentuim IV, 2.40 GHz, 512 MB). The some first nonzero coefficients takes form $(j = n + 1)$

$$E_j^{(0)} = 2\gamma(n + (m + |m| + 1)/2),$$

$$E_j^{(1)} = -2Z,$$

$$E_j^{(3)} = 2Z(2n + |m| + 1)/\gamma,$$

$$E_j^{(5)} = -3Z(2 + 3|m| + 6n^2 + |m|^2 + 6n|m| + 6n)/\gamma^2,$$

$$E_j^{(6)} = -2Z^2(2n + |m| + 1)/\gamma^3,$$

$$Q_{jj+1}^{(4)} = 3Z\sqrt{n+1}\sqrt{n+|m|+1}/\gamma^2,$$

$$Q_{jj+1}^{(6)} = -15Z\sqrt{n+1}\sqrt{n+|m|+1}(2n + |m| + 2)/\gamma^3,$$

$$Q_{jj+2}^{(6)} = 15Z\sqrt{n+1}\sqrt{n+2}\sqrt{n+|m|+1}\sqrt{n+|m|+2}/(4\gamma^3),$$

$$H_{jj}^{(8)} = 9Z^2(2n^2 + 2n|m| + 2n + |m| + 1)/\gamma^4,$$

$$H_{jj}^{(10)} = -90Z^2(2n + |m| + 1)(2n^2 + 2n|m| + 2n + |m| + 2)/\gamma^5,$$

$$H_{jj+1}^{(10)} = 45Z^2\sqrt{n+1}\sqrt{n+|m|+1}(n^2 + n|m| + 2n + |m| + 2)/(2\gamma^5),$$

$$H_{jj+2}^{(8)} = -9Z^2\sqrt{n+1}\sqrt{n+2}\sqrt{n+|m|+1}\sqrt{n+|m|+2}/\gamma^4,$$

$$H_{jj+2}^{(10)} = 90Z^2\sqrt{n+1}\sqrt{n+2}\sqrt{n+|m|+1}\sqrt{n+|m|+2}(2n + |m| + 3)/\gamma^5,$$

$$H_{jj+3}^{(10)} = -45Z^2\sqrt{n+1}\sqrt{n+2}\sqrt{n+3}\sqrt{n+|m|+1}\sqrt{n+|m|+2}\sqrt{n+|m|+3}/(2\gamma^5).$$

As an example, in Table 1 we show true convergence of partial sums of asymptotic expansions (39) of effective potentials $\hat{Q}_{ij}(z)$ to the corresponding numerical values calculated by algorithm 1, described in section 3.

5   Algorithm 3 of Evaluation the Asymptotics of Solutions at Large $|z|$ in Kantorovich Method

Step 1. We write the set of differential equations (12) at fixed values m, and ϵ in the explicit form for $\chi_{ji_o}(z) \equiv \hat{\chi}_j^{(i_o)}(z)$ and $j = 1, 2, \ldots, j_{\max}$, $i_o = 1, 2, \ldots, N_o$

$$-\frac{d^2\chi_{ji_o}(z)}{dz^2} - \frac{2Z}{|z|}\chi_{ji_o}(z) - \left(\epsilon - \hat{E}_j(z) - \frac{2Z}{|z|}\right)\chi_{ji_o}(z) + \hat{H}_{jj}(z)\chi_{ji_o}(z)$$

$$= \sum_{j'=1, j'\neq j}^{j_{\max}} \left(-\hat{Q}_{jj'}(z)\frac{d}{dz} - \hat{H}_{jj'}(z) - \frac{d\hat{Q}_{jj'}(z)}{dz}\right)\chi_{j'i_o}(z), \tag{40}$$

Table 1. Values of the partial sums (39) depending on $k_{\max}$ for $m = -1$, $Z = 1$, $z = 10$, $\gamma = 10$. The last row contains the corresponding numerical values (n.v.).

i,j	$\hat{Q}_{12}, 10^{-6}$	$\hat{Q}_{23}, 10^{-6}$	$\hat{Q}_{34}, 10^{-5}$	$\hat{Q}_{13}, 10^{-8}$	$\hat{Q}_{24}, 10^{-8}$	$\hat{Q}_{14}, 10^{-11}$
$z^{-4}Q_{ij}^{(4)}$	4.24264069	7.34846923	1.03923048	0	0	0
$+z^{-6}Q_{ij}^{(6)}$	4.17900108	7.16475750	1.00285742	1.29903811	3.18198052	0
$+z^{-7}Q_{ij}^{(7)}$	4.17883137	7.16446356	1.00281585	1.29903811	3.18198052	0
$+z^{-8}Q_{ij}^{(8)}$	4.17972233	7.16857870	1.00394341	1.26266504	3.04833733	7.0000
$+z^{-9}Q_{ij}^{(9)}$	4.17972824	7.16859579	1.00394680	1.26260268	3.04818460	7.0000
$+z^{-10}Q_{ij}^{(10)}$	4.17971489	7.16850321	1.00391243	1.26342108	3.05252800	6.6850
$+z^{-11}Q_{ij}^{(11)}$	4.17971474	7.16850253	1.00391224	1.26342451	3.05254060	6.6846
$+z^{-12}Q_{ij}^{(12)}$	4.17971496	7.16850469	1.00391330	1.26340651	3.05240830	6.6950
$+z^{-13}Q_{ij}^{(13)}$	4.17971496	7.16850471	1.00391331	1.26340638	3.05240762	6.6950
$+z^{-14}Q_{ij}^{(14)}$	4.17971496	7.16850466	1.00391328	1.26340679	3.05241163	6.6947
$+z^{-15}Q_{ij}^{(15)}$	4.17971496	7.16850466	1.00391327	1.26340679	3.05241166	6.6947
$+z^{-16}Q_{ij}^{(16)}$	4.17971496	7.16850466	1.00391328	1.26340678	3.05241154	6.6947
(n.v.)	4.17971496	7.16850466	1.00391328	1.26340678	3.05241154	6.6947

where matrix elements $\hat{Q}_{jj'}(z)$ and $\hat{H}_{jj'}(z)$ have of the form (39).

Note, that at large z, $E_i^{(2)} = H_{ii}^{(2)} = 0$, i.e., the centrifugal terms are eliminated and the longitudinal solution has the asymptotic form corresponding to zero angular momentum solutions, or to the one-dimensional problem on a semi-axis:

$$\chi_{ji_o}(z) = \frac{\exp(w(z))}{\sqrt{p_{i_o}}}\phi_{ji_o}(z), \quad \phi_{ji_o}(z) = \sum_{k=0}^{k_{\max}} \phi_{ji_o}^{(k)}|z|^{-k}, \qquad (41)$$

where $w(z) = \imath p_{i_o}|z| + \imath\zeta\ln(2p_{i_o}|z|) + \imath\delta_{i_o}$, p_{i_o} is the momentum in the channel, ζ is the characteristic parameter, and δ_{i_o} is the phase shift. The components $\phi_{ji_o}^{(k)}$ satisfy the system of ordinary differential equations

$$(p_{i_o}^2 - 2E + E_j^{(0)})\phi_{ji_o}^{(k)} = f_{ji_o}^{(k)}(\phi_{j'i_o}^{(k'=0,\ldots,k-1)}, p_{i_o})$$

$$\equiv -2(\zeta p_{i_o} + \imath(k-1)p_{i_o} - Z)\phi_{ji_o}^{(k-1)} - (\zeta + \imath(k-2))(\zeta + \imath(k-1))\phi_{ji_o}^{(k-2)}$$

$$- \sum_{k'=3}^{k}(E_j^{(k')} + H_{jj}^{(k')})\phi_{ji_o}^{(k-k')} + \sum_{j'=1}^{j_{\max}}\sum_{k'=4}^{k}(-2\imath Q_{jj'}^{(k')}p_{i_o} - H_{jj'}^{(k')})\phi_{j'i_o}^{(k-k')}$$

$$+ \sum_{j'=1}^{j_{\max}}\sum_{k'=5}^{k}(2k - 1 - k' - 2\imath\zeta)Q_{jj'}^{(k'-1)}\phi_{j'i_o}^{(k-k')},$$

$$k = 0, 1, \ldots, k_{\max}, \quad \phi_{ji_o}^{(-1)} \equiv 0, \quad \phi_{ji_o}^{(-2)} \equiv 0, \quad k_{\max} \le j_{\max} - i_o. \qquad (42)$$

Here index of summation, j', takes integer values, except i_o and j, ($j' = 1, \ldots, j_{\max}$, $j' \ne i_o$, $j' \ne j$).

Step 2. From first two equations ($k = 0, 1$) of set (42) we have the leading terms of eigenfunction $\phi_{ji_o}^{(0)}$, eigenvalue $p_{i_o}^2$ and characteristic parameter ζ, i.e initial data for solving recurrence sequence,

$$\phi_{ji_o}^{(0)} = \delta_{ji_o}, \quad p_{i_o}^2 = 2E - E_{i_o}^{(0)} \rightarrow p_{i_o} = \sqrt{2E - E_{i_o}^{(0)}}, \quad \zeta = Z/p_{i_o}. \quad (43)$$

Open channels have $p_{i_o}^2 \geq 0$, and close channels have $p_{i_o}^2 < 0$. Lets there are $N_o \leq j_{\max}$ open channels, i.e., $p_{i_o}^2 \geq 0$ for $i_o = 1, \ldots N_o$ and $p_{i_o}^2 < 0$ for $i_o = N_o + 1, \ldots j_{\max}$.

Step 3. Substituting (43) in (42), we obtain the following recurrent set of algebraic equations for the unknown coefficients $\phi_{ji_o}(z)$ for $k = 1, 2, \ldots, k_{\max}$:

$$(E_{i_o}^{(0)} - E_j^{(0)})\phi_{ji_o}^{(k)} = f_{ji_o}^{(k)}(\phi_{j'i_o}^{(k'=0,\ldots,k-1)}, p_{i_o}) \quad (44)$$

that is solved sequentially for $k = 1, 2, \ldots, k_{\max}$:

$$\phi_{ji_o}^{(k)} = f_{ji_o}^{(k)}(\phi_{j'i_o}^{(k'=0,\ldots,k-1)}, p_{i_o})/(E_{i_o}^{(0)} - E_j^{(0)}), \quad j \neq i_o,$$

$$f_{i_o i_o}^{(k+1)}(\phi_{j'i_o}^{(k'=0,\ldots,k)}, p_{i_o}) = 0 \rightarrow \phi_{i_o i_o}^{(k)}. \quad (45)$$

The successful run of the above algorithm was occurs up to $k_{\max} = 16$ (Run time is 167s on Intel Pentuim IV, 2.40 GHz, 512 MB). The some first nonzero coefficients takes form ($j = n + 1$)

$$\phi_{ji_o}^{(0)} = \delta_{ji_o},$$

$$\phi_{ji_o}^{(1)} = \delta_{ji_o} \imath Z(Z + \imath p_{i_o})/(2p_{i_o}^3),$$

$$\phi_{ji_o}^{(2)} = \delta_{ji_o}[\imath E_j^{(3)}/(4p_{i_o}) - Z(Z + \imath p_{i_o})^2(Z + 2\imath p_{i_o})/(8p_{i_o}^6)],$$

$$\phi_{ji_o}^{(3)} = \delta_{ji_o}[-E_j^{(3)}(3Z^2 + 7\imath p_{i_o}Z - 6p_{i_o}^2)/(24p_{i_o}^4)$$
$$-\imath Z(Z + \imath p_{i_o})^2(Z + 2\imath p_{i_o})^2(Z + 3\imath p_{i_o})/(48p_{i_o}^9)],$$

$$\phi_{ji_o}^{(4)} = \delta_{ji_o}[\imath E_j^{(5)}/(8p_{i_o}) - (E_j^{(3)})^2/(32p_{i_o}^2)$$
$$-\imath E_j^{(3)}(3Z^4 + 20\imath p_{i_o}Z^3 - 53p_{i_o}^2 Z^2 - 66\imath p_{i_o}^3 Z + 36p_{i_o}^4))/(96p_{i_o}^7)$$
$$+Z(Z + \imath p_{i_o})^2(Z + 2\imath p_{i_o})^2(Z + 3\imath p_{i_o})^2(Z + 4\imath p_{i_o})/(384p_{i_o}^{12})]$$
$$+2\imath p_{i_o} Q_{ji_o}^{(4)}/(E_{i_o}^{(0)} - E_j^{(0)}).$$

Remarks

1. Expansion (41) holds true for $|z_m| \gg \max(Z^2/(2p_{i_o}^3), 2Z(2i_o + |m| - 1)/(8\gamma p_{i_o}^2))$. The choice of a new value of $z_{\max}$ for the constructed expansions of the linearly independent solutions for $p_{i_o} > 0$ is controlled by the fulfillment of the Wronskian condition with a long derivative $D_z \equiv \mathbf{I}d/dz - \mathbf{Q}(z)$

$$Wr(\mathbf{Q}(z); \boldsymbol{\chi}^*(z), \boldsymbol{\chi}(z)) = (\boldsymbol{\chi}^*(z))^T D_z\boldsymbol{\chi}(z) - (D_z\boldsymbol{\chi}^*(z))^T \boldsymbol{\chi}(z) = 2\imath \mathbf{I}_{oo}$$

up to the prescribed accuracy. Here $\mathbf{I}_{oo}$ is the N_o-by-N_o identity matrix.

2. This algorithm can be applied also for evaluation asymptotics of solutions in closed channels $p_{i_o} = \imath \kappa_{i_o}$, $\kappa_{i_o} > 0$.

Table 2. Convergence of the method for the binding energy $\mathcal{E} = \gamma/2 - E$ (in a.u.) of even wave functions $m = -1$, $\gamma = 10$ and $\gamma = 5$ versus the number j_{max} of coupled equations (40)

j_{max}	$2p_{-1}$ $(\gamma = 10)$	$3p_{-1}$ $(\gamma = 10)$	$2p_{-1}$ $(\gamma = 5)$	$3p_{-1}$ $(\gamma = 5)$
1	1.123 532 554 (3)	0.182 190 992 (2)	0.857 495 336 (9)	0.165 082 403 (4)
2	1.125 069 513 (1)	0.182 282 868 (7)	0.859 374 058 (2)	0.165 234 428 (1)
3	1.125 280 781 (8)	0.182 294 472 (5)	0.859 641 357 (6)	0.165 253 152 (9)
4	1.125 343 075 (2)	0.182 297 825 (6)	0.859 721 942 (4)	0.165 258 606 (4)
6	1.125 381 347 (9)	0.182 299 867 (7)	0.859 772 441 (3)	0.165 261 973 (6)
8	1.125 392 776 (1)	0.182 300 474 (6)	0.859 787 833 (7)	0.165 262 991 (9)
10	1.125 397 502 (9)	0.182 300 725 (2)	0.859 794 289 (0)	0.165 263 418 (0)
12	1.125 399 854 (7)	0.182 300 849 (8)	0.859 797 533 (8)	0.165 263 631 (9)
[6]	1.125 422 341 (8)	0.182 301 494 (7)	0.859 832 622 (6)	0.165 264 273 (1)

6 Applications Algorithms for Solving the Eigenvalue Problem

The efficiency and accuracy of the elaborated SNA and of the corresponded numerical scheme derived are confirmed by computations of eigenenergies and eigenfunctions for the low-excited states of a hydrogen atom in the strong homogeneous magnetic field. These algorithms are used to generate an input file of effective potentials in the Gaussian points $z = z_k$ of the FEM grid $\Omega^p_{h(z)}[z_{min} = 0, z_{max}]$ and asymptotic of solutions of a set of longitudinal equations (12)–(16) for the KANTBP code [5]. In Table 2 we show convergence of the method for the binding energy $\mathcal{E} = \gamma/2 - E$ (in a.u.) of the even wave functions at $m = -1$, $\gamma = 10$ and $\gamma = 5$ versus the number j_{max} of coupled equations (40). The calculations was performed on a grid $\Omega^p_{h(z)} = \{0(200)2(600)150\}$ (the number in parentheses denotes the number of finite elements of order $p = 4$ in each interval). Comparison with corresponding calculations given in spherical coordinates from [1,6] is shown that elaborated method in cylindrical coordinates is applicable for strength magnetic field $\gamma > 5$ and magnetic number m of order of ~ 10. The main goal of the method consists in the fact that for states having preferably a cylindrical symmetry a convergence rate is increased at fixed m with growing values of $\gamma \gg 1$ or the high-$|m|$ Rydberg states at $|m| > 150$ in laboratory magnetic fields $B = 6.10\mathrm{T}$ ($\gamma = 2.595 \cdot 10^{-5}$ a.u.), such that several equations are provide a given accuracy [7].

7 Conclusion

A new effective method of calculating wave functions of a hydrogen atom in a strong magnetic field is developed. The method is based on the Kantorovich approach to parametric eigenvalue problems in cylindrical coordinates. The rate of convergence is examined numerically and illustrated by a set of typical examples. The results are in a good agreement with calculations executed in spherical coordinates at fixed m for $\gamma > 5$. The elaborated SNA for calculating effective potentials and asymptotic solutions allows us to generate effective approximations for a finite set of longitudinal equations describing an open channel. The developed approach yields a useful tool for calculation of threshold phenomena in formation and ionization of (anti)hydrogen like atoms and ions in magnetic traps [2,7] and channeling of ions in thin films [15].

This work was partly supported by the Russian Foundation for Basic Research (grant No. 07-01-00660) and by Grant I-1402/2004-2007 of the Bulgarian Foundation for Scientific Investigations.

References

1. Dimova, M.G., Kaschiev, M.S., Vinitsky, S.I.: The Kantorovich method for high-accuracy calculations of a hydrogen atom in a strong magnetic field: low-lying excited states. Journal of Physics B: At. Mol. Phys. 38, 2337–2352 (2005)
2. Chuluunbaatar, O., et al.: Application of Kantorovich method for calculations of a hydrogen atom photoionization in a strong magnetic field. In: Proc. SPIE 6537, 653706–1–18 (2007)
3. Gusev, A.A., et al.: A symbolic-numerical algorithm for solving the eigenvalue problem for a hydrogen atom in magnetic field. In: Ganzha, V.G., Mayr, E.W., Vorozhtsov, E.V. (eds.) CASC 2006. LNCS, vol. 4194, pp. 205–218. Springer, Heidelberg (2006)
4. Vinitsky, S.I., et al.: A symbolic-numerical algorithm for the computation of matrix elements in the parametric eigenvalue problem. Programming and Computer Software 33, 105–116 (2007)
5. Chuluunbaatar, O., et al.: KANTBP: A program for computing energy levels, reaction matrix and radial wave functions in the coupled-channel hyperspherical adiabatic approach. submitted to Comput. Phys. Commun. (2007)
6. Chuluunbaatar, O., et al.: POTHMF: A program for computing potential curves and matrix elements of the coupled adiabatic radial equations for a Hydrogen-like atom in a homogeneous magnetic field. submitted to Comput. Phys. Commun. (2007)
7. Guest, J.R., Choi, J.-H., Raithel, G.: Decay rates of high-$|m|$ Rydberg states in strong magnetic fields. Phys. Rev. A 68, 022509–1–9 (2003)
8. Kantorovich, L.V., Krylov, V.I.: Approximate Methods of Higher Analysis. Wiley, New York (1964)
9. Strang, G., Fix, G.J.: An Analysis of the Finite Element Method. Prentice-Hall, Englewood Cliffs, New York (1973)
10. Bathe, K.J.: Finite Element Procedures in Engineering Analysis. Prentice Hall, Englewood Cliffs, New York (1982)
11. Simola, J., Virtamo, J.: Energy levels of hydrogen atoms in a strong magnetic field. J. Phys. B: At. Mol. Phys. 11, 3309–3322 (1978)
12. Fridrich, H.: Bound-state spectrum of the hydrogen atom in strong magnetic fields. Phys. Rev. A 26, 1827–1838 (1982)

13. Abramovits, M., Stegun, I.A.: Handbook of Mathematical Functions. Dover, New York (1965)
14. Abrashkevich, A.G., Kaschiev, M.S., Vinitsky, S.I.: A New Method for Soling an Eigenvalue Problem for a System Of Three Coulomb Particles within the Hyperspherical Adiabatic Representation. J. Comp. Phys. 163, 328–348 (2000)
15. Demkov, Yu.N., Meyer, J.D.: A sub-atomic microscope, superfocusing in channeling and close encounter atomic and nuclear reactions. Eur. Phys. J. B 42, 361–365 (2004)

An Algorithm for Construction of Normal Forms

Victor F. Edneral*

Skobeltsyn Institute of Nuclear Physics
of Moscow State University
Leninskie Gory, Moscow, 119991, Russia
edneral@theory.sinp.msu.ru

Abstract. The normal form method is widely used in the theory of
nonlinear ordinary differential equations (ODEs). But in practice it is
impossible to evaluate the corresponding transformations without computer algebra packages. Here we describe an algorithm for normalization
of nonlinear autonomous ODEs. Some implementations of these algorithms are also discussed.

Keywords: resonant normal form, nonlinear ordinary differential equations, computer algebra.

1 Introduction

The normal form method is based on a transformation of an ODEs system to a
simpler set called the normal form. The importance of this method for analyzing
the ODEs near stationary point has been recognized for a long time. A resonant
normal form was introduced in the fundamental paper of H. Poincaré [1], where
he considered the linear case of the form. A polynomial case of this form was
discussed by H. Dulac [2] and the infinite case by A.D. Bruno [3]. We will call
the resonant form of this kind by the Poincaré–Dulac–Bruno (PDB) normal
form. For the history of this subject see, for instance, [4]. Definitions of normal
form and normalizing transformation can be formulated in different ways for
some special cases, such as Hamiltonian systems but we restrict ourselves in this
consideration by the general case only.

Since the system MAO [5] by which the Delaney's theory of the Moon motion
was checked many algorithms (and their implementations) were developed for
creating normal forms and corresponding transformations.

In this paper we will use the algorithm based on the approach, which was
developed by A.D. Bruno [6,7,8,9] for the PDB normal form. The important advantage of this approach is a possibility of considering a wide class of autonomous
systems in a single, easily algorithmized frame.

Another advantage of the used approach is an algorithmic simplicity of the
creation of that normal form and the corresponding transformations. We have a
direct recurrence formula for this procedure. The usage does not demand keeping

* This research was supported by grant of President RF Sc.Sch. - 4476.06.02.

V.G. Ganzha, E.W. Mayr, and E.V. Vorozhtsov (Eds.): CASC 2007, LNCS 4770, pp. 134–142, 2007.

of some large intermediate results as it is in other algorithms. The approach is free from a necessity to solve intermediate systems of equations and from any restrictions on low resonance cases.

The discussed implementation was originally described in [10,11] for the RE-DUCE system and in [12] for the MATHEMATICA system. Unfortunately the implementation was not presented in detail there, so this paper closes this gap. About other implementations see, for example, [13,14].

A normal form usage, in particular, provides a constructive way for obtaining the approximations of local families of periodic and conditionally periodic solutions in the form of power/Fourier series for real families and in the form of power series in time dependent exponents for complex ones. It is especially important that the problem of convergence of used transformations was considered in [6,7]. This circumstance allows us to hope that approximations of frequencies and corresponding periodic solutions families near stationary points by finite formulas can be done with acceptable precision [15] – [20]. Except solutions themselves we can find also approximations of initial conditions, which initiate such periodic solutions. I.e., we can produce some elements of a phase analysis.

It is also possible to approximate by the proposed method the non-periodic families of solutions ("crude" case). The results are close to the results of the Carleman linearization method. For periodic and conditionally periodic cases, the method [21] is a generalization of the Poincare–Lindstedt approach [22], chapter 10. The approach was also used in the center-focus problem [23].

The normal form method is widely used for bifurcation analysis. About methods of such an investigation see, for example, [24,25]. You can see in these books that the numerical bifurcation analysis is indeed based on the normal form method. We can make from the lowest not vanishing coefficients of the normal form the qualitative conclusions about the behavior of the original system. It is sufficient to know only the lowest orders of the normal form for such an analysis. Sometimes this job can be done by hand, but rather by computer algebra systems [26,27,17].

By the normal form method it is possible to study the structure of the normal form and the first integrals. By an example of the Euler–Poisson system of equations describing the motion of a rigid body with a fixed point it was shown that there is a sequence of necessary conditions of existence of an additional formal first integral at different values of the parameter. Violation of any of these conditions is enough for absence of a formal integrability of the system, and so a local, and thus a global integrability [28,29,30,31].

Below we describe the creation of the resonant normal form and corresponding transformations.

2 Problem Formulation

Consider the system of autonomous ordinary differential equations:

$$\dot{\mathbf{x}} = \mathbf{\Phi}(\mathbf{x}), \tag{1}$$

where $\mathbf{x} = (x_1, \ldots, x_n)$ is a vector function of time, $\dot{\mathbf{x}} \overset{\text{def}}{=} d\mathbf{x}/dt$ is the time derivative, $\mathbf{\Phi} = (\Phi_1, \ldots, \Phi_n)$ is a vector which is a function of $\mathbf{x}$ and probably of some parameters. Such a type of equations originates from many scientific and engineering problems where oscillations, vibrations or wave processes take place. The main idea here is in replacing system (1) with some "model" system having finite order polynomial right-hand sides and transforming them to the canonical (normal) form.

The study of systems of type (1) in the neighborhood of *stationary* point $\mathbf{x}^0$, where $\mathbf{\Phi}(\mathbf{x}^0) = \mathbf{0}$, typically includes three preliminary steps. Firstly $\mathbf{x}$ is shifted by $-\mathbf{x}^0$ so that $\mathbf{\Phi}(\mathbf{0}) = \mathbf{0}$, i.e. $\mathbf{0}$ is the stationary point to be studied. Each stationary point of the system is considered separately.

The second step is a reduction of the system to a model form where the vector $\mathbf{\Phi}(\mathbf{x})$ is approximated by a vector of polynomials. If in some neighborhood of the stationary point $\mathbf{\Phi}$ is an analytic function of $\mathbf{x}$ then its power series can be used to obtain a smooth approximation with desired precision. Often this step is made simultaneously with a reduction of the system to its central manifold. In any case, the right-hand sides of the model system will be polynomials without constant terms.

The third step is the transformation of the linear parts matrix to Jordan's form by a complex linear change of $\mathbf{x}$ variables.

After these steps, system (1) has the form:

$$\dot{y}_i = \lambda_i y_i + \sigma_i y_{i-1} + \tilde{\Phi}_i(\mathbf{y}), \qquad \sigma_1 = 0, \qquad i = 1, \ldots, n, \tag{2}$$

where $\mathbf{\Lambda} = (\lambda_1, \ldots, \lambda_n)$ is the vector of eigenvalues of the matrix of the linear part of the system and $\tilde{\mathbf{\Phi}} = (\tilde{\Phi}_1, \ldots, \tilde{\Phi}_n)$ is a vector of polynomials of finite degree without constant and linear terms.

For this paper, we assume that system (2) satisfies the following assumptions:

- the system is autonomous and has polynomial nonlinearities;
- $\mathbf{0}$ is a stationary point, and the system will be studied near $\mathbf{y} = \mathbf{0}$;
- the linear part of the right hand side is diagonal, and not all eigenvalues are zero, i.e. $\mathbf{\Lambda} \neq \mathbf{0}$.

Remark that the last assumption is a restriction of a current implementation rather the approach itself. But on the other hand it is assumed that neither the system is Hamiltonian, nor that it preserves the phase volume nor that it has any internal symmetry.

3 The Normal Form Method

Equations (2) can be written in the form:

$$\dot{y}_i = \lambda_i y_i + y_i \sum_{\mathbf{q} \in N_i} f_{i,\mathbf{q}} \mathbf{y}^{\mathbf{q}}, \qquad i = 1, \ldots, n, \tag{3}$$

where we use the multi-index notation:

$$\mathbf{y^q} = \prod_{j=1}^{n} y_j^{q_j},$$

with the power exponent vector $\mathbf{q} = (q_1, \ldots, q_n)$
and the sets:

$$N_i = \{\mathbf{q} \in \mathbb{Z}^n : q_i \geq -1 \text{ and } q_j \geq 0, \text{ if } j \neq i, \quad j = 1, \ldots, n\},$$

because the factor y_i has been moved out of the sum in (3).

The normalization is done with a near-identity transformation:

$$y_i = z_i + z_i \sum_{\mathbf{q} \in N_i} h_{i,\mathbf{q}} \mathbf{z^q}, \quad i = 1, \ldots, n \tag{4}$$

and then we will have system (3) in the normal form:

$$\dot{z}_i = \psi_i(\mathbf{z}) \stackrel{\text{def}}{=} \lambda_i z_i + z_i \sum_{\substack{\langle \mathbf{q}, \mathbf{\Lambda} \rangle = 0 \\ \mathbf{q} \in N_i}} g_{i,\mathbf{q}} \mathbf{z^q}, \quad i = 1, \ldots, n. \tag{5}$$

The important difference between (3) and (5) is a restriction on the range of the summation, which is defined by the equation:

$$\langle \mathbf{q}, \mathbf{\Lambda} \rangle \stackrel{\text{def}}{=} \sum_{j=1}^{n} q_j \lambda_j = 0. \tag{6}$$

The h and g coefficients in (4) and (5) are found by using the recurrence formula:

$$g_{i,\mathbf{q}} + \langle \mathbf{q}, \mathbf{\Lambda} \rangle \cdot h_{i,\mathbf{q}} = -\sum_{j=1}^{n} \sum_{\substack{\mathbf{p} + \mathbf{r} = \mathbf{q} \\ \mathbf{p}, \mathbf{r} \in \bigcup_i N_i \\ \mathbf{q} \in N_i}} (p_j + \delta_{ij}) \cdot h_{i,\mathbf{p}} \cdot g_{j,\mathbf{r}} + \tilde{\Phi}_{i,\mathbf{q}}, \tag{7}$$

where the second summation on the right-hand side is over all integer vectors satisfying the constraint $\mathbf{p} + \mathbf{r} = \mathbf{q}$, and $\tilde{\Phi}_{i,\mathbf{q}}$ is a coefficient of the factor $z_i \mathbf{z^q}$ in the polynomial $\tilde{\Phi}_i$ in (2), arguments of which have been transformed by (4). Here $||\mathbf{p}||$ and $||\mathbf{r}|| < ||\mathbf{q}||$, where $||\mathbf{q}|| \stackrel{\text{def}}{=} q_1 + \ldots + q_n$. So (7) is a recurrence formula.

The ambiguity in (7) is usually fixed by the conventions:

$$\begin{aligned} h_{i,\mathbf{q}} &= 0, & \text{if} \quad \langle \mathbf{q}, \mathbf{\Lambda} \rangle = 0, \\ g_{i,\mathbf{q}} &= 0, & \text{if} \quad \langle \mathbf{q}, \mathbf{\Lambda} \rangle \neq 0, \end{aligned} \tag{8}$$

and then the normalizing transformation is called a "basic" one.

4 Main Algorithm

The algorithm of the calculation of g and h in (4), (5) is based on (7) and (8). It is convenient to choose the representation of sets of coefficients $g_{i,\mathbf{q}}$ and $h_{i,\mathbf{q}}$ in such a way that they would be combined in homogeneous subgroups where each subgroup has the same order n, i.e. contains only terms with such vector-indexes $\mathbf{q} = \{q_j\}$ that $\|\mathbf{q}\| = n$ for each i. You can calculate the sets g and h of the next order by using sets of g and h with smaller order only, i.e. (7) is a recurrence formula.

The algorithm:
Let n be a dimension of the system. For its normalization till order m we are to do:

(i). for $i = 1, 2, \ldots, n$ do:
Calculate all squared in $\mathbf{y}$ elements in the right-hand side nonlinearity $\tilde{\Phi}_i(\mathbf{y})$ in (2), i.e., calculate the subgroup of the first-order ($\|\mathbf{q}\| = 1$) elements of the set $f_{i,\mathbf{q}}$ in (3) and sort it into two subsets depending on the value of scalar product (6). The first set where this product is zero will be the first order subgroup of g_i, and the second set after a division by the value of the corresponding scalar product will be the first order subgroup of h_i

(ii). for $k = 2, 3, \ldots, m$ do:
(a) for $i = 1, 2, \ldots, n$ do:
calculate the subgroup of order k of the nonlinear terms $\tilde{\Phi}_i(\mathbf{y})$ in (2) for which the substitution $\mathbf{y}$ is evaluated by (4) till order $k - 1$ and define coefficients at monomials $z_i \mathbf{z}^{\mathbf{q}}$ as $f_{i,\mathbf{q}}$;
(b) for $i = 1, 2, \ldots, n$ do:
Calculate the subgroups of g_i and h_i of order k by a subdivision of set $f_{i,\mathbf{q}}$ into two subsets as in step (i). After that you can supplement the set g_i till full order k and a part of the set h_i without a contribution from the first term of the right-hand side in (7).
(c) for $i = 1, 2, \ldots, n$ do:
for $j = 1, 2, \ldots, n$ do:
supplement the preliminary set of order k of h_i with properly sorted multiplications of *all* elements of such subgroups of $h_{i,\mathbf{p}}$ and $g_{j,\mathbf{r}}$ that their total order, i.e. $\|\mathbf{p} + \mathbf{r}\| = k$. Not all these multiplications should be really calculated because of the factor $(p_j + \delta_{i,j})$ is zero at some values of j index. Before the supplement all elements above are to be divided on the corresponding scalar products too.

A cost of the above algorithm is low in comparison with a cost of evaluation of the right-hand side of the nonlinear system. Under such circumstances it is very important to calculate the right-hand sides very economically, using so much as possible the fact that we need to calculate at each step of (ii) the homogeneous terms of $\tilde{\Phi}_i$ of order k only, and all terms of lower orders are not changed during the later operations. The problem of optimization of this evaluation is one of the main limitations for an automation of generating codes for the right-hand side calculation.

5 Computer Algebra Implementation of the Normal Form Method

The calculation of the coefficients of the normal form (5) and corresponding transformation (4) with respect to (7) and (8) was implemented as the NORT package. Earlier attempts of the author to compute sufficiently high orders of the normal form using high level of the REDUCE language were not successful. Because of this, the NORT package [10,11,15] was created. The NORT is written in Standard LISP and contains about 2000 operators. The NORT is a package of procedures to treat truncated multivariate formal power series in arbitrary dimensions. In addition to procedures for arithmetic operations with series, there are special procedures for the creation of normal forms and procedures for substitutions, for calculations of some elementary functions (when it is possible), for differentiating, for printing, and for inverting multivariate power series, etc. It contains also special procedures for the calculation of Lyapunov's values [23]. The NORT can be used as a separate program or as a REDUCE package.

Besides series, expressions in NORT can contain also non-negligible variables (parameters). There is implemented multivariate series-polynomial arithmetic. The complex-valued numerical coefficients of the truncated power series-polynomials may be treated in three different arithmetics: rational, modular, floating point, and approximate rational. There are also several options for the output form of these numbers, the output is in a REDUCE readable form. The program uses an internal recurrence representation for its objects. Remark that a garbage collection time for examples below was smaller than 3 % of evaluation time. This can characterize the NORT package as a program with a good enough internal organization. Many important results described in references were obtained by a computer with 1 Mbyte RAM only.

Unfortunately at this moment the NORT package has no friendly user interface yet. So we create a package for usage with MATHEMATICA package [12]. This package works with truncated multivariate formal power series. The *PolynomialSeries* package can be accessed at *www.mathsource.com* site. The existing version is enough for a support of a normal form method. The comparison of MATHEMATICA package with an earlier version of normal form package NORT written in LISP demonstrates that the calculations within the MATHEMATICA system are more flexible and convenient but are considerably slower than under the LISP.

A key moment for a realization of the Main algorithm above and both implementations is an internal representation of formal power series. We group terms of series in homogeneous sums in variable order, and we store the value of this order with the corresponding sum. For example, if we have a truncated series:

$$y_2\,y_3 + y_1\,y_2\,y_3 + y_2^2\,y_3 + y_1\,y_2^2\,y_3 + y_2\,y_3^2 + y_1\,y_2\,y_3^2 + y_2^2\,y_3^2 + y_1\,y_2^2\,y_3^2,$$

then the internal representation of the above series with respect to y_1, y_2, y_3 is

$$\{\{2, y_2\,y_3\}, \{3, y_1\,y_2\,y_3 + y_2^2\,y_3 + y_2\,y_3^2\}, \{4, y_1\,y_2^2\,y_3 + y_1\,y_2\,y_3^2 + y_2^2\,y_3^2\}, \{5, y_1\,y_2^2\,y_3^2\}\}$$

It is obvious that this form is very convenient for summation. And objects in this representation can very efficiently be multiplied in the sense of truncated series – for excluding from results negligible for corresponding order of truncation terms, it is enough to eliminate from the multiplied groups the terms with common orders which are over the negligible one. For example, if we wish to calculate a square of the above series till the 5th order we need to square only the sum of the first two homogeneous groups above (with 2 and 3 common orders), no more. One more very important advantage of such representation is that an implementation of formulae (7) is a realization of some kind of tensor production as you can see from the main algorithm.

6   Conclusions

Here we can conclude that the obtaining of PDB normal forms of high order is very useful for analysis of autonomous nonlinear ODEs.

The special choice of an internal representation allows us to build an efficient algorithm for evaluation of the PDB normal form.

References

1. Poincaré, H.: Les Méthodes Nouvelles de la Méchanique Celeste. vol. 1,2,3. Gauthier-Villars, Paris (1872–1879) Reprint: Dover, New York, 1957. National Aeronautics and Space Administration, Washington (1967)
2. Dulac, H.: Solution d'une système d'équations différentielles dans le voisinage des valeurs singulieres. Bull. Soc. Math. France 40, 324–384 (1912)
3. Bruno(Brjuno), A.D.: The normal form of differential equations. Dokl. Akad. Nauk SSSR 157, 1276–1279 (1964) [Russian]. Sov. Math. Dokl. 5, 1105–1108 (1964) [English]
4. Arnold, V.I., Anosov, D.V. (eds.): Dynamical Systems, I. Encyclopaedia of Mathematical Sciences. Springer, New York (1988)
5. Rom, A.: Mechanized algebraic operations (MAO). Celestial Mechanics 1, 301–319 (1970)
6. Bruno(Brjuno), A.D.: Analytical form of differential equations. Trans. Mosc. Mat. Soc. 25, 131–288 (1971), 26, 199–239 (1972)
7. Bruno, A.D.: Local method in nonlinear differential equations. Springer Series in Soviet Mathematics, p. 370 (1989), ISBN 3-540-18926-2
8. Bruno, A.D.: Normal forms. J. Mathematics and Computers in Simulation 45, 413–427 (1998)
9. Bruno, A.D.: The power geometry in algebraic and differential equations, p. 385. Elsevier, Amsterdam (2000)
10. Edneral, V.F., Khrustalev, O.A.: The normalizing transformation for nonlinear systems of ODEs. The realization of the algorithm (Russian). In: Proc. Int. Conf. on Computer Algebra and its Application in Theoretical Physics, USSR, Dubna, pp. 219–224. JINR publ., Dubna (1985)
11. Edneral, V.F., Khrustalev, O.A.: Program for recasting ODE systems in normal form. Sov. J. Programmirovanie (5), 73–80 (1992) [Russian]

12. Edneral, V.F., Khanin, R.: Multivariate power series and normal form calculation in MATHEMATICA. In: Ganzha, V., et al. (eds.) CASC 2002. Proc. Fifth Workshop on Computer Algebra in Scientific Computing, Big Yalta, Ukraine, pp. 63–70. Tech. Univ. München, Munich (2002)

13. Vallier, L.: An Algorithm for the computation of normal forms and invariant manifolds. In: Bronstein, M. (ed.) Proc. 1993 Int. Symp. on Symbolic and Algebraic Computation, Kiev, Ukraine, pp. 225–233. ACM Press, New York (1993)

14. Mikram, J., Zinoun, F.: Normal form methods for symbolic creation of approximate solutions of nonlinear dynamical systems. J. Mathematics and Computers in Simulation 57, 253–290 (2001)

15. Edneral, V.F.: Computer generation of normalizing transformation for systems of nonlinear ODE. In: Bronstein, M. (ed.) Proc. 1993 Int. Symp. on Symbolic and Algebraic Computation, Kiev, Ukraine, pp. 14–19. ACM Press, New York (1993)

16. Edneral, V.F.: A symbolic approximation of periodic solutions of the Henon–Heiles system by the normal form method. J. Mathematics and Computers in Simulation 45, 445–463 (1998)

17. Edneral, V.F.: Bifurcation analysis of low resonant case of the generalized Henon - Heiles system. In: Ganzha, et al. (eds.) CASC 2001. Proc. Fourth Workshop on Computer Algebra in Scientific Computing, Konstanz, Germany, pp. 167–176. Springer, Heidelberg (2001)

18. Edneral, V.F.: Periodic solutions of a cubic ODE system. In: Ganzha, et al. (eds.) CASC 2003. Proc. Fifth Workshop on Computer Algebra in Scientific Computing, Passau, Germany, pp. 77–80. Tech. Univ. München, Munich (2003)

19. Edneral, V.F., Khanin, R.: Application of the resonant normal form to high order nonlinear ODEs using MATHEMATICA. Nuclear Inst. and Methods in Physics Research A 502, 643–645 (2003)

20. Edneral, V.F., Khanin, R.: Investigation of the double pendulum system by the normal form method in MATHEMATICA. Programming and Computer Software 30, 115–117 (2004)

21. Edneral, V.F.: Looking for periodic solutions of ode systems by the normal form method. In: Wang, D., Zheng, Z. (eds.) Differential Equations with Symbolic Computation, Birkhauzer, Basel, Boston, Berlin, pp. 173–200 (2005)

22. Verhulst, F.: Nonlinear differential equations and dynamical systems, p. 227. Springer, Heidelberg (1990)

23. Edneral, V.F.: Computer evaluation of cyclicity in planar cubic system. In: Küchlin, W. (ed.) Proc. ISSAC'97, Hawaii, USA, pp. 305–309. ACM, New York (1997)

24. Hassard, B.D., Kazarinoff, N.D., Wan, Y.H.: Theory and Applications of Hopf Bifurcation, p. 280. Cambridge Univ. Press, Cambridge (1981)

25. Guckenheimer, J., Holmes, P.: Nonlinear Oscillations, Dynamical Systems and Bifurcations of Vector Fields. Springer, New York (1986)

26. Rand, R., Armbruster, D.: Perturbation Methods, Bifurcation Theory and Computer Algebra. Springer, New York (1987)

27. Bruno, A.D.: Bifurcation of the periodic solutions in the case of a multiple pair of imaginary eigenvalues. Selecta Mathematica *formerly* Sovietica 12, 1–12 (1993)

28. Bruno, A.D.: Local integrability of the Euler–Poisson equations. Doklady Math. 74, 512–516 (2006)
29. Bruno, A.D., Edneral, V.F.: Normal forms and integrability of ODE systems. Programming and Computer Software 32, 139–144 (2006)
30. Bruno, A.D., Edneral, V.F.: On integrability of the Euler–Poisson equations. Fundamental and Applied Mathematics 13, 45–59 (2007)
31. Bruno, A.D., Edneral, V.F.: Computation of normal forms of the Euler–Poisson equations. Preprint No. 1 of the Keldysh Institute of Applied Mathematics of RAS. Moscow, p. 17 (2007) [Russian]

Computer Algebra: A 'Classical' Path to Explore Decoherence and Entanglement Phenomena in Quantum Information Theory

Stephan Fritzsche

Max–Planck–Institut für Kernphysik, D–69117 Heidelberg, Germany
s.fritzsche@gsi.de

(Plenary Talk)

Abstract. During the past decade, quantum information theory has attracted a lot of interest because of its promise for solving problems that are *intractable* otherwise. Despite of the recent advancements in understanding the basic principles of quantum information systems, however, there are still a large number of difficulties to be resolved. One of the great challenges concerns for instance the decoherence in quantum systems and how entanglement is lost or transfered between the subsystems, if they are coupled to their enviroment. — To overcome these difficulties, several schemes for studying the decay of quantum states and their interaction with an environment have been developed during recent years, including a large variety of separability and entanglement measures, decoherence-free subspaces as well as (quantum) error correction codes.

To support the investigation of entanglement and decoherence phenomena in general $N-$qubit quantum systems, we recently developed the FEYNMAN program [1], a computer-algebraic approach within the framework of MAPLE, which facilitates the symbolic and numerical manipulation of quantum registers and quantum transformations. This program has been designed for studying the dynamics of quantum registers owing to their interaction with external fields and perturbations. In a recent addition to this program [2], moreover, we now implemented also various noise models as well as a number of entanglement measures (and related quantities). In this lecture, I shall display the interactive use of the program by a number of simple but intuitive examples.

To make quantum information theory *alive*, an active (re-) search has been initiated during the past decade to find and explore physical systems that are suitable to produce and control the entanglement in course of their time evolution. In atomic photoionization, for instance, we have shown how the polarization can be transfered from the incoming photons to the emitted photoelectrons, giving rise to a (spin-spin) entanglement between the photoelectron and the remaining (photo-) ion. Detailed computations on the entanglement as function of the energy and polarization of the incoming light have been carried out along various isoelectronic sequences [3]. For the two-photon decay of atomic hydrogen, moreover, we analyzed the geometrical control of the polarization entanglement of the emitted photons.

V.G. Ganzha, E.W. Mayr, and E.V. Vorozhtsov (Eds.): CASC 2007, LNCS 4770, pp. 143–144, 2007.
© Springer-Verlag Berlin Heidelberg 2007

References

1. Radtke, T., Fritzsche, S.: Comput. Phys. Commun. 173, 91 (2006), ibid. 175, 145 (2005)
2. Radtke, T., Fritzsche, S.: Comput. Phys. Commun. 176, 617 (2007)
3. Radtke, T., Fritzsche, S., Surzhykov, A.: Phys. Lett. A347, 73 (2005), Phys. Rev. A74, 032709 (2005)

Deducing the Constraints in the Light-Cone $SU(3)$ Yang-Mills Mechanics Via Gröbner Bases

Vladimir Gerdt[1], Arsen Khvedelidze[1,2], and Yuri Palii[1,3]

[1] Laboratory of Information Technologies, Joint Institute for Nuclear Research,
Dubna, Moscow Region, 141980, Russia
gerdt@jinr.ru
[2] Department of Theoretical Physics, A.Razmadze Mathematical Institute,
Tbilisi, GE-0193, Georgia
akhved@jinr.ru
[3] Institute of Applied Physics, Moldova Academy of Sciences,
Chisinau, MD-2028, Republic of Moldova
palii@jinr.ru

Abstract. The algorithmic methods of commutative algebra based on the Gröbner bases technique are briefly sketched out in the context of an application to the constrained finite dimensional polynomial Hamiltonian systems. The effectiveness of the proposed algorithms and their implementation in *Mathematica* is demonstrated for the light-cone version of the $SU(3)$ Yang-Mills mechanics. The special homogeneous Gröbner basis is constructed that allow us to find and classify the complete set of constraints the model possesses.

1 Introduction

The basic procedure, *completion to involution* [1,2,3,4,5,6], of systems of differential equations represents a highly nontrivial issue in view of its practical application. Particularly, a manipulation with functions modulo a set of algebraic relations requires an efficient algorithmization and implementation in a proper computer algebra software. For the practical purposes of wide class of theories and models of the contemporary theoretical and mathematical physics and especially of the degenerate Hamiltonian systems [7]-[9] the problem of completion to involution being very topical became nowadays feasible due to the progress in computer technologies. Our attempts to implement such an algorithmic description for the degenerate polynomial Hamiltonian mechanical models have been summarized in the recent papers [10]-[14], where the method based on the most universal algorithmic tool of commutative algebra, the well-known *Gröbner bases* theory [15]-[17], has been elaborated. Since this technique provides an effective algorithmic instrument to verify whether a polynomial vanishes on the manifold defined by a set of other polynomials, the Gröbner bases plays the principal role in algorithmic implementation of the basic operations of the Dirac constraint formalism: computation and separation of constrains.

V.G. Ganzha, E.W. Mayr, and E.V. Vorozhtsov (Eds.): CASC 2007, LNCS 4770, pp. 145–159, 2007.
© Springer-Verlag Berlin Heidelberg 2007

Here we briefly sketch out this very central element of the *Dirac-Bergmann-Gröbner* algorithmic procedure suggested in [10] to deal with the practically important case of finite-dimensional degenerate polynomial Lagrangian system. Afterwards we apply this algorithm to examine the mechanical system with a rich set of constraints, the so-called light-cone $SU(3)$ Yang-Mills mechanics, where computation of constraints has not been done before.

The outline of the article is as follows. In section 2 the basic elements of the Dirac-Bergmann-Gröbner algorithm to compute and classify the constraints are given. Then the formulation of the mechanical model, the light-cone $SU(3)$ Yang-Mills mechanics is presented. In section 3 the results of computation of the complete set of constraints are given. The section 4 is devoted to the discussion of the specially constructed homogeneous Gröbner basis that provides our calculations and categorization of the constraints.

2   Elements of the Dirac-Bergmann-Gröbner Algorithm

The Dirac method to determine and classify constraints for degenerate Hamiltonian systems is *easy formulate but difficult to implement* at practical level of computation when the both, number of degrees of freedom as well as the number of free parameters of the model are sufficiently large. Here we describe a possible way to make this procedure computationally effective. We start with the discussion of the Dirac constraint formalism for a finite dimensional degenerate Lagrangian system aiming its algorithmic reformulation.

Consider an n-dimensional mechanical system whose configuration space is $\mathbf{R}^n$ and the Lagrangian $L(q, \dot{q})$ is defined on a tangent space as a function of the coordinates $q := q_1, q_2, \ldots, q_n$ and velocities $\dot{q} := \dot{q}_1, \dot{q}_2, \ldots, \dot{q}_n$.

The Lagrangian system is *regular* if the rank $r := rank\|H_{ij}\|$ of the corresponding Hessian function $H_{ij} := \partial^2 L / \partial \dot{q}_i \partial \dot{q}_j$ is maximal ($r = n$). In this case the Euler-Lagrange equations

$$\frac{\mathrm{d}}{\mathrm{d}t} \left(\frac{\partial L}{\partial \dot{q}_i} \right) - \frac{\partial L}{\partial q_i} = 0, \qquad 1 \leq i \leq n \tag{1}$$

rewritten explicitly as

$$H_{ij} \ddot{q}_j + \frac{\partial^2 L}{\partial q_j \partial \dot{q}_i} \dot{q}_j - \frac{\partial L}{\partial q_i} = 0$$

can be resolved with respect to the accelerations ($\ddot{q}$) and there are no *hidden constraints*. Otherwise, if $r < n$, the Euler-Lagrange equations (and, thus, the Lagrangian system itself) are *degenerate* or *singular*. In this case not all differential equations (1) are of second order, namely there are $n - r$ independent equations, Lagrangian constraints, containing only coordinates and velocities. Passing to the Hamiltonian description via a Legendre transformation

$$p_i := \frac{\partial L}{\partial \dot{q}_i} \tag{2}$$

the degeneracy of the Hessian results in the existence of $n - r$ relations between coordinates and momenta, the *primary constraints*

$$\varphi_a^{(1)}(p,q) = 0, \quad 1 \le a \le n - r. \tag{3}$$

Equations (3) define the so-called *primary constraints subset* (manifold, if certain regularity conditions assumed) Σ_1.[1] This definition is implicit and therefore it is necessary to provide an effective algorithm to compute *all primary constraints describing the subset* Σ_1.

From (3) the dynamics is constrained by the set Σ_1 and by the Dirac prescription is governed by the *total* Hamiltonian

$$H_T := H_C + U_a \, \varphi_a^{(1)}, \tag{4}$$

which differs from the *canonical* Hamiltonian $H_C(p,q) = p_i \dot{q}_i - L$ by a linear combination of the primary constraints with the *Lagrange multipliers* U_a.

The next step is to analyze the dynamical requirement that classical trajectories remain in Σ_1 during the evolution

$$\dot{\varphi}_a^{(1)} = \{H_T, \varphi_a^{(1)}\} \overset{\Sigma_1}{=} 0. \tag{5}$$

In (5) the evolutional changes are generated by the canonical *Poisson brackets* with the total Hamiltonian (4) and the abbreviation $\overset{\Sigma_1}{=}$ stands for *a week equality*, i.e., the right-hand side of (5) vanishes modulo the primary constraints (3).

The consistency condition (5), unless it is satisfied identically, may lead either to a contradiction or to a determination of the Lagrange multipliers U_a or to new constraints. The former case indicates that the given Hamiltonian system is inconsistent.

In the latter case when (5) is not satisfied identically and is independent of the multipliers U_a the left-hand side of (5) defines the new constraints. Otherwise, if the left-hand side depends on some Lagrange multipliers U_a the consistency condition determines these multipliers, and, therefore, the constraints set is not enlarged by new constraints. The subsequent iteration of this consistency check ends up with the complete set of constraints and/or determination of some/or all Lagrange multipliers.

The number of Lagrange multipliers U_a which can be found is determined by the rank of the so-called *Poisson bracket matrix*

$$\mathbf{M}_{\alpha\beta} := \{\phi_\alpha, \phi_\beta\}, \tag{6}$$

where Σ denotes the subset of a phase space defined by the complete set of constraints $\Phi := (\phi_1, \phi_2, \dots, \phi_k)$

---

[1] Everywhere in this paper we suppose that all constraints satisfy the so-called *regularity conditions* (see explanations in §1.1.2 of [9]).

$$\Sigma : \quad \phi_\alpha(p, q) = 0, \qquad 1 \le \alpha \le k. \tag{7}$$

including all primary $\varphi^{(1)}$, *secondary* $\varphi^{(2)}$, *ternary* $\varphi^{(3)}$, etc., constraints,

If $rank\,(\mathbf{M}) = m$, then $s := k - m$ linear combinations of constraints ϕ_α

$$\psi_\alpha(p, q) = \sum_\beta c_{\alpha\beta}(p, q)\,\phi_\beta, \tag{8}$$

define the *first-class constraints*, whose Poisson brackets are weakly zero

$$\{\psi_\alpha(p, q), \psi_\beta(p, q)\} \overset{\Sigma}{=} 0 \qquad 1 \le \alpha, \beta \le s. \tag{9}$$

The remaining functionally independent constraints form the subset of *second-class constraints*.

It is worth to note here that the described method to find constraints within the Dirac formalism represent the reformulation of *completion* of the initial Hamiltonian equations to *involution* in another words and constraints corresponds to a set of the *integrability conditions* [18,19,20].

Now the algorithmic reformulation of the above stated scheme will be described using the ideas and the terminology of the Gröbner bases theory. In doing so, we restrict our consideration to an arbitrary dynamical system with finitely many degrees of freedom whose Lagrangian is a polynomial in coordinates and velocities with rational (possibly parametric) coefficients $L(q, \dot{q}) \in \mathbf{Q}[q, \dot{q}]$. Thereafter we use the standard notions and definitions of commutative algebra (see, e.g., [15,16,17]).

Algorithm to determine the primary constraints

The primary constraints (3) are consequences of the polynomial relations (2). These relations generate the polynomial ideal in $\mathbf{Q}[p, q, \dot{q}]$

$$I_{p,q,\dot{q}} \equiv \mathrm{Id}(\cup_{i=1}^n \{p_i - \partial L/\partial \dot{q}_i\}) \subset \mathbf{Q}[p, q, \dot{q}]. \tag{10}$$

Thereby, primary constraints (3) belong to the radical $\sqrt{I_{p,q}}$ of the elimination ideal

$$I_{p,q} = I_{p,q,\dot{q}} \cap \mathbf{Q}[p, q]. $$

Correspondingly, for an appropriate term ordering which eliminates $\dot{q}$, a Gröbner basis of $I_{p,q}$ (denotation: $GB(I_{p,q})$) is given by [15,16,17]

$$GB(I_{p,q}) = GB(I_{p,q,\dot{q}}) \cap \mathbf{Q}[p, q]. $$

This means that construction of the Gröbner basis for the ideal (10) with omitting elements in the basis depending on velocities and then constructing of $GB(I_{p,q})$ allows to compute the set of primary constraints. If $GB(I_{p,q}) = \emptyset$, then the dynamical system is regular. Otherwise, the algebraically independent set Φ_1 of (effective) primary constraints can be found as the subset $\Phi_1 \subset GB(\sqrt{I_{p,q}})$ such that

$$\forall \phi(p, q) \in \Phi_1 : \quad \phi(p, q) \notin \mathrm{Id}(\Phi_1 \setminus \{\phi(p, q)\}). \tag{11}$$

Verification of (11) is algorithmically done by computing the following *normal form*: $NF(\phi, GB(\mathrm{Id}(\Phi_1 \setminus \{\phi\})))$. In addition, the canonical Hamiltonian $H_c(p, q)$ is computed as $NF(p_i \dot{q}_i - L, GB(I_{p,q,\dot{q}}))$. This form of $H_c(p, q)$ is used in the next steps of the Dirac-Bergman-Gröbner algorithm.

Algorithm to determine the higher constraints and to classify them

The dynamical consequences (5) of a primary constraint can also be algorithmically analyzed by computing the normal form of the Poisson brackets of the primary constraint and the total Hamiltonian modulo $\sqrt{I_{p,q}}$). Here the Lagrange multipliers U_a in (4) are treated as time-dependent functions. If the non-vanishing normal form does not contain U_a, then it is nothing else than the secondary constraint. In this case the set of primary constraints is enlarged by the secondary constraint obtained and the process is iterated. At the end either the complete set Φ of constraints (7) is constructed or some inconsistency is detected. The detection holds when the intermediate Gröbner basis, whose computation is a part of the iterative procedure, becomes $\{1\}$.

In order to separate the set $\Phi = \{\phi_1, \ldots, \phi_k\}$ into subsets of the first and second classes constraints the entries of Poisson brackets matrix $\mathbf{M}$ are evaluated as normal forms of the Poisson brackets of the constraints modulo a Gröbner basis of the ideal generated by set Φ. Afterwards if the basis $E = \{\mathbf{e}^{(1)}, \ldots, \mathbf{e}^{(k-m)}\}$ of the null space (kernel) of this matrix $\mathbf{M}$ is known the each basis vector $\mathbf{e}^{(s)} \in E, s = 1, \ldots, k - m$ generates the first-class constraint of form $\mathbf{e}_\alpha^{(s)} \phi_\alpha$. The second class constraints are build using the basis of the m-dimensional orthogonal complement $E_\perp$, of subspace E. With the aid of these vectors $\mathbf{e}_\perp^{(l)} \in E_\perp, l = 1, \ldots, m$ the second-class constraint are constructed as $\mathbf{e}_{\perp\alpha}^{(l)} \phi_\alpha$.

Concluding we see that the constraints separation can be performed using the linear algebra operations with the matrix $\mathbf{M}$ alone. Together with the Gröbner bases technique this implies full algorithmisation for computing the complete set of algebraically independent constraints and their classification.

Implementation

The above described algorithms were implemented first in *Maple* [10,14]. However, the Gröbner bases routines built-in *Maple* are not efficient enough to perform computation needed for the light-cone $SU(3)$ Yang-Mills mechanics (Sect.3.2). We also tried recent extensions of the Maple Gröbner bases facilities with the external packages *Gb* and *Fgb* created by J.C.Faugère [21]. Unfortunately *Gb* runs for our problems even slower than the built-in package whereas *Fgb* cannot deal with the parametric coefficients. By the last reason we cannot use yet[2] the *Ginv* [22] software that is a C++ module of Python and implements the efficient involutive algorithms [6] for the construction of the involutive or/and Gröbner bases.

---

[2] The implementation in *Ginv* of multivariate GCD computation that is necessary for computation of Gröbner bases with the parametric coefficients is in progress now in collaboration with the RWTH, Aachen.

It should be emphasized that manipulation with the parametric coefficients is essential for the Dirac formalism due to the presence of physical parameters (e.g. masses, coupling constants) in the initial Lagrangian, the Lagrange multipliers in the total Hamiltonian (4). Having these needs in mind we implemented the algorithms in *Mathematica* whose built-in routine GroebnerBasis as well as Groebner in *Maple* allows to compute parametric Gröbner bases but performs computations much faster.

3   Light-Cone Yang-Mills Mechanics

Here we apply the above described scheme to a mechanical model originated from the Yang-Mills gauge field theory assuming a certain homogeneity of fields. Namely, we consider the so-called light-cone Yang-Mills mechanics which differs from the well-known instant form of Yang-Mills mechanics intensively studied during the last twenty years for a variety of reasons, both in physics and in mathematics (see e.g. [23]-[33]). The alternative light-cone Yang-Mills mechanics is formulated as the light-front form version of the $SU(n)$ Yang-Mills gauge theory when the additional supposition of the gauge potentials dependence on the light-cone time only is made.

The coordinate free representation of the $SU(n)$ Yang-Mills fields action in four-dimensional Minkowski space M_4, endowed with a metric η reads

$$S := \frac{1}{g_0^2} \int_{M_4} \operatorname{tr} F \wedge *F \,, \tag{12}$$

where g_0 is a coupling constant and the $su(n)$ algebra valued curvature two-form

$$F := dA + A \wedge A$$

is constructed from the connection one-form A. The connection and curvature, as Lie algebra valued quantities are expanded in some basis T^a

$$A = A^a T^a \,, \qquad F = F^a T^a \,. \qquad a = 1, 2, \ldots, n^2 - 1 \,.$$

The metric $\eta_{\gamma\delta}$ enters the action through the dual field strength tensor $*F_{\mu\nu} := \frac{1}{2}\sqrt{-\det\eta}\,\epsilon_{\mu\nu\alpha\beta}\,F^{\alpha\beta}$, with the totally antisymmetric tensor $\epsilon_{\mu\nu\alpha\beta}$.

To formulate the light-cone version of the $SU(n)$ mechanics we expand the one-form A in so-called light-cone basis

$$A := A_+ \, dx^+ + A_- \, dx^- + A_k \, dx^k \,, \quad k = 1, 2 \,, \tag{13}$$

where the basic one-forms $dx^\pm$ in (13) are dual to the vectors $e_\pm := \frac{1}{\sqrt{2}}\,(e_0 \pm e_3)$ tangent to the light-cone. The corresponding coordinates, light-cone coordinates $x^\mu = (x^+, x^-, x^\perp)$ are

$$x^\pm := \frac{1}{\sqrt{2}}\,(x^0 \pm x^3) \,, \qquad x^\perp := x^k \,, \quad k = 1, 2 \,,$$

and non-zero components of the metric read $\eta_{+-} = \eta_{-+} = -\eta_{11} = -\eta_{22} = 1$.

Now if the components of the connection one-form A in (13) are functions of the light-cone "time variable" x^+ only

$$A_\pm = A_\pm(x^+), \qquad A_k = A_k(x^+).$$

the classical action (12) reduces to the following form

$$S_{\text{LC}} := \frac{V^{(3)}}{2g_0^2} \int dx^+ \left(F_{+-}^a \, F_{+-}^a + 2 \, F_{+k}^a \, F_{-k}^a - F_{12}^a \, F_{12}^a \right) . \tag{14}$$

This expression can be identified with the action of a finite dimensional model named as light-cone Yang-Mills mechanics whose dynamics is governed by the Lagrangian

$$L := \frac{1}{2} \left(F_{+-}^a \, F_{+-}^a + 2 \, F_{+k}^a \, F_{-k}^a - F_{12}^a \, F_{12}^a \right) , \tag{15}$$

Deriving (15) we fix the "renormalized" coupling constant $g_0^2/V^{(3)} = 1$ in (14) to simplify formulaes and use the following expression for the light-cone components of the field-strength tensor

$$F_{+-}^a := \frac{\partial A_-^a}{\partial x^+} + f^{abc} \, A_+^b \, A_-^c \, ,$$

$$F_{+k}^a := \frac{\partial A_k^a}{\partial x^+} + f^{abc} \, A_+^b \, A_k^c \, ,$$

$$F_{-k}^a := f^{abc} \, A_-^b \, A_k^c \, ,$$

$$F_{ij}^a := f^{abc} \, A_i^b \, A_j^c \, , \quad i, j, k = 1, 2 \, .$$

The Lagrangian (15) defines the $SU(n)$ Yang-Mills light-cone mechanics with $4(n^2 - 1)$- degrees of freedom $A_\pm$, A_k evolving with respect to the light-cone time $\tau := x^+$.

However due to the gauge invariance of the initial Yang-Mills theory and because in the light-cone dynamics the instant time states are given at the light-cone characteristics the corresponding evolutionary equations degenerate (see e.g. discussion in [8],[34]): not all of them are second order with respect to the light-cone time. Some of the Euler-Lagrange equations that follow from (15) represent the constraints on the variables from the extended configuration.

In the Hamiltonian description this can be seen as follows. The Legendre transformation gives the momentum π_a^-, canonically conjugated to A_-^a

$$\pi_a^- := \frac{\partial L}{\partial \dot{A}_-^a} = \dot{A}_-^a + f^{abc} \, A_+^b \, A_-^c \, ,$$

while defining the momenta π_a^+ and π_a^k canonically conjugated to A_-^a and A_k^a we find the set of the primary constraints

$$\varphi_a^{(1)} := \pi_a^+ = 0 \, , \tag{16}$$

$$\chi_k^a := \pi_a^k - f^{abc} \, A_-^b \, A_k^c = 0 \, . \tag{17}$$

The presence of primary constraints affects the dynamics of the degenerate system. The generic evolution is governed now by the total Hamiltonian

$$H_T := H_C + U_a(\tau)\varphi_a^{(1)} + V_k^a(\tau)\chi_k^a,$$

where the canonical Hamiltonian reads

$$H_C = \frac{1}{2}\pi_a^-\pi_a^- - f^{abc} A_+^b \left(A_-^c\,\pi_a^- + A_k^c\,\pi_a^k \right) + \frac{1}{2}F_{12}^a F_{12}^a,$$

and $U_a(\tau)$ and $V_k^a(\tau)$ are the Lagrange multipliers.

Using the total Hamiltonian and the fundamental canonical Poisson brackets

$$\{A_\pm^a,\pi_b^\pm\} = \delta_b^a, \qquad \{A_k^a,\pi_b^l\} = \delta_k^l\delta_b^a,$$

the dynamical self-consistence of the primary constraints (16) should be checked out. From the requirement of conservation of the primary constraints $\varphi_a^{(1)}$ we see that

$$0 = \dot{\varphi}_a^{(1)} = \{\pi_a^+, H_T\} = f^{abc}\left(A_-^b\,\pi_c^- + A_k^b\,\pi_c^k\right), \tag{18}$$

while the same procedure for the primary constraints χ_k^a gives the following self-consistency conditions

$$0 = \dot{\chi}_k^a = \{\chi_k^a, H_C\} - 2\,f^{abc}\,A_-^b\,V_k^c. \tag{19}$$

It is straightforward to check that the consistency conditions (18) define the $n^2 - 1$ secondary constraints $\varphi_a^{(2)}$

$$\varphi_a^{(2)} := f_{abc}\left(A_-^b\,\pi_c^- + A_k^b\,\pi_c^k\right) = 0 \tag{20}$$

which obey the $su(n)$ algebra

$$\{\varphi_a^{(2)},\varphi_b^{(2)}\} = f_{abc}\,\varphi_c^{(2)}.$$

However, the further analysis of the consistency conditions (19) represents not so easy tractable issue. First of all, the number of Lagrange multipliers that can be determined from (19) depends on the rank of the structure group. This can bee seen from the non-vanishing Poisson brackets between constraints χ_i^a

$$\{\chi_i^a,\chi_j^b\} = 2\,f^{abc}A_-^c\,\delta_{ij}. \tag{21}$$

The simplest case of the special unitary group of rank 1, the $SU(2)$ group, has been analyzed in our previous papers. The constraints analysis of the $SU(2)$ model including their separation into the first and second class can be found in [11,12,13]. Below we only state these results and then discuss in more details the model with the first non-trivial rank 2 structure group, the $SU(3)$ Yang-Mills light-cone mechanics.

3.1   The $SU(2)$ Structure Group

For the $su(2)$ algebra we use the standard Pauli matrices $\sigma_1, \sigma_2, \sigma_3$ providing the structure constants as the totally antisymmetric three dimensional Levi-Civita symbol: $f^{abc} := \epsilon^{abc}$, $\epsilon^{123} = 1$.

According to the equations (16) and (17), there are $(2^2 - 1) + (2^2 - 1) \times 2 = 9$ primary constraints $\varphi_a^{(1)}$ and χ_k^a. From the consistency condition (19) for the primary constraints χ_k^a the following picture stands out

- Apart from the easy recognizable *abelian* constraints π_a^+ and *non-abelian first-class* constraints $\varphi_a^{(2)}$, (20), there are two more constraints absent in the instant form of $SU(2)$ Yang-Mills mechanics

$$\psi_k := A_-^a \chi_k^a .$$

Here A_-^a *is the null vector* of the Poisson brackets $C_{ab} = \epsilon_{abc} A_-^c$ in (21).
- The remaining four "orthogonal" constraints

$$\chi_{k\perp}^a := \chi_k^a - A_-^a \left(A_-^b \chi_k^b \right) ,$$

are the *second-class* and satisfy the relations

$$\{\chi_{i\perp}^a , \chi_{j\perp}^b\} = 2\, \epsilon^{abc}\, A_-^c\, \delta_{ij} ,$$
$$\{\varphi_a^{(2)} , \chi_{k\perp}^b\} = \epsilon^{abc}\, \chi_{k\perp}^c .$$

Further analysis shows that apart from the secondary Gauss law constraints φ_a^2 there are no new constraints. Indeed, the abelian constraints ψ_i do not create new ones

$$\{\psi_i, H_T\} = -A_i^a \varphi_a^{(2)} + \pi_a^- \chi_i^a + \epsilon_{abc} A_i^a A_k^b \chi_k^c \stackrel{\Sigma}{=} 0 . \tag{22}$$

The consistency condition (19) for the "orthogonal" constraints $\chi_{i\perp}^a$ allows to determine the corresponding four Lagrange multiplier $V_\perp(\tau)$ and therefore summarizing, the $SU(2)$ light-cone Yang-Mills mechanics possesses 8 functionally independent first-class constraints $\varphi_a^{(1)}, \psi_k, \varphi_a^{(2)}$ and 4 second-class constraints $\chi_{k\perp}^a$.

3.2   The $SU(3)$ Structure Group

The algebraic properties of the $su(3)$ algebra are encoded in the two independent set the skew-symmetric f_{abc} and symmetric d_{abc} structure constants. For the basis usually used in physical applications–the Gell-Mann basis–they are listed in the Appendix.

Since the rank of the $su(3)$ algebra is two, the null space of the matrix $C_{ab} = f_{abc} A_-^c$ is 2-dimensional. It can be spanned by two null-vectors, one linear and another one quadratic in the coordinates

$$e_a^{(1)} := A_-^a , \qquad e_a^{(2)} := d_{abc}\, A_-^b\, A_-^c .$$

Using vectors $e_a^{(1,2)}$ we decompose the set of $2 \times (3^2 - 1) = 16$ primary constraints χ_k^a as

$$\chi_i^a = (\chi_{i\perp}^a, \psi_i, \varsigma_i), \tag{23}$$

where

$$\psi_i := e_a^{(1)} \chi_i^a, \qquad \varsigma_i := e_a^{(2)} \chi_i^a. \tag{24}$$

The decomposition (23) turn to be very useful owing to the special Poisson brackets relations for the decomposition components

$$\{\chi_k^a, \psi_i\} = 0, \quad \{\chi_k^a, \varsigma_i\} = 0, \quad \{\psi_i, \varsigma_k\} = 0, \quad \{\psi_i, \psi_j\} = 0, \quad \{\varsigma_i, \varsigma_k\} = 0.$$

The consistency conditions (19) allow to find the corresponding Lagrange multipliers $V_{k\perp}^a$ and to get the expressions modulo primary constraints

$$\{\psi_i, H_T\} = -A_i^a \varphi_a^{(2)} + \text{primary constraints}, \tag{25}$$
$$\{\varsigma_i, H_T\} = \mathrm{d}_{abc} A_i^a F_{-k}^b F_{-k}^c - 2\, \mathrm{d}_{abc} A_-^a A_i^b \varphi_c^{(2)} + \text{primary constraints}.$$

According to the upper equalities (25), the constraints ψ_i do not give rise to new secondary constraints. However, the second equation (26) states that there are two more *new secondary constraints*

$$\zeta_i = \mathrm{d}_{abc} A_i^a F_{-k}^b F_{-k}^c. \tag{26}$$

The new constraints ζ_i obey the following relations:

$$\{\zeta_i, \zeta_j\} = 0,$$
$$\{\psi_i, \zeta_j\} = \delta_{ij}\, \mathrm{d}_{abc} A_-^a \left(F_{-k}^b \chi_k^c - \frac{1}{2} A_-^b \varphi_-^{(2)} \right),$$
$$\{\varsigma_i, \zeta_j\} = -\delta_{ij}\, \mathrm{d}_{abc} \mathrm{d}_{cpq} A_-^a A_-^b F_{-k}^p F_{-k}^q. \tag{27}$$

Evaluation of the right hand side in the last equations (27) by using the Gröbner basis technique (details of the basis used are given in the subsequent Sect. 4) modulo all known constraints shows that the further search for the ternary constraints terminates and from the consistency condition[3]

$$\{\zeta_i, H_T\} \stackrel{\Sigma_2}{=} \{\zeta_i, H_C\} + \{\zeta_i, \varsigma_k\} V_k^\varsigma \tag{28}$$

one can fix two unknown functions V_k^ς entering the decomposition for the Lagrange multipliers $V_k^a = \left(V_{k\perp}^a, V_k^\psi, V_k^\varsigma \right)$.

Therefore we can now finally conclude with the statement about the complete set of constraints for the light-cone $SU(3)$ Yang-Mills mechanics. The complete set of constraints consists of 34 constraints, and among them there are

- $8 + 8 + 2 = 18$ *first-class* constraints: π_+^a, $\varphi_a^{(2)}$ and ψ_k,
- $2 \times 6 + 2 + 2 = 16$ *second-class* constraints: $\chi_{k\perp}^a$, ς_k and ζ_k.

---

[3] The Σ_2 stands here for the constraint manifold defined by the primary and secondary constraints.

It is worth to note here that these results are based on the tedious calculation of the Poisson bracket relations and their subsequent evaluation modulo the constraint functions using the specially constructed Gröbner basis. At the present moment, to the best of our knowledge, there is no way to overpass these straightforward calculations with a high computational complexity.

4   Computation of the Gröbner Basis

The goal of this section is to discuss certain properties of a Gröbner basis used in the calculation of the light-cone $SU(3)$ Yang-Mills mechanics and describe some computational aspects of its construction.

The actual calculations were performed using the the computer algebra system *Mathematica* (version 5.0) running on the machine 2xOpteron-242 (1.6 Ghz) with 6Gb of RAM. For the simplest nontrivial case of the $SU(n)$ light-cone mechanics having the structure group $SU(2)$ we used the built-in-function `GroebnerBasis` with monomial order `DegreeReverseLexicographic`. However, for the $SU(3)$ group due to substantial increase of the number of variables as well as the number of non-vanishing structure constants f_{abc} and d_{abc} the memory of the above computer turns to be insufficient. To overcome this problem a special *Mathematica* program has been written in order to calculate the homogeneous Gröbner bases ([15] §10.2) allowing to use step by step the partially constructed Gröbner bases.

In doing so we built a homogeneous (grading compatible) Gröbner bases for the $SU(3)$ structure group using the *grading* Γ determined by the following weights of the variables:

$$\Gamma(\pi_\mu^a) = 2\,, \quad \text{for all momenta} \quad a = 1, 2, \ldots, 8\,, \quad \mu = -, 1, 2\,,$$
$$\Gamma(A_\mu^a) = 1\,, \quad \text{for all coordinates} \quad a = 1, 2, \ldots, 8\,, \quad \mu = -, 1, 2\,.$$

As a monomial ordering we used a grading (degree) one with breaking ties by the following pure lexicographical order on the variables.

The order on the variables with different spatial indices was chosen as

$$\pi_a^- \succ \pi_b^1 \succ \pi_c^2 \succ A_-^a \succ A_1^b \succ A_2^c \qquad \text{for all} \quad a, b, c = 1, 2, \ldots, 8\,,$$

whereas for the variables the equal spatial indices our choice was

$$\pi_a^k \succ \pi_b^k \succ A_k^a \succ A_k^b \qquad \text{if} \quad a < b \qquad \text{for any} \quad a, b = 1, 2, \ldots, 8\,.$$

According to the introduced grading the Γ-degrees of the basic constraints (17), (20), (26) are given by

Γ – degree	Constraints
2	$\chi_k^a = \pi_a^k - f^{abc} A_-^b A_k^c\,,$
3	$\varphi_a^{(2)} = f_{abc}\left(A_-^b \pi_c^- + A_k^b \pi_c^k\right),$
5	$\zeta_i = d_{abc} A_i^a F_{-k}^b F_{-k}^c\,.$

With such a choice of grading the constraints χ_k^a and $\varphi^{(2)}$ are the lowest degree homogeneous Gröbner basis elements G_2 and G_3 of the order 2 and 3, respectively. Higher degree elements of the basis are constructed step by step by doing the following manipulations:

(i) formation of all S−polynomials (G_i, G_j) ;
(ii) elimination of some superfluous S−polynomials according to the Buchberger's criteria [15,16,17];
(iii) computation of the normal forms of S-polynomials modulo the lower order elements with respect to the grading chosen.

The results of computation of the Gröbner basis elements of different orders n are shown in the following table where we explicitly indicated only S−polynomials with non-vanishing normal form.

G_n	Polynomials #	Constraints and S-polynomials
G_2	16	χ_k^a
G_3	8	$\varphi_a^{(2)}$
G_4	15	(G_3, G_3)
G_5	14	$\zeta_i,\ (\zeta_i, G_j)_{j=2,3,4}$ $(G_2, G_4), (G_3, G_4), (G_4, G_4)$ (G_3, G_3)
G_6	13	$(G_2, G_5), (G_3, G_5), (G_4, G_5), (G_5, G_5)$ $(G_3, G_4), (G_4, G_4)$

Our attempts to compute G_n for $n > 6$ were failed because of enormous computational expenses. It should be noted that in comparison to the $SU(3)$ Gröbner computation, the case of the $SU(2)$ structure group is computationally much more simple. In this case the construction of the *complete homogeneous Gröbner basis* of 64 elements takes about 60 seconds for the following lexicographic order

$$\{\pi_1^1, \pi_1^2, \pi_2^1, \pi_2^2, \pi_3^1, \pi_3^2, \pi_1^-, \pi_2^-, \pi_3^-, A_1^1, A_2^1, A_1^2, A_2^2, A_1^3, A_2^3, A_-^1, A_-^2, A_-^3\} .$$

5   Concluding Remarks

By applying the Dirac-Bergmann-Gröbner algorithm to the light-cone $SU(3)$ Yang-Mills mechanics we found the complete set of constraints and identified them in accordance with the Dirac classification as the first and the second class constraints. This was achieved by means of the exploiting the special homogeneous Gröbner basis whose components G_n with $n \leq 6$ where computed. Though we were not able to determine the elements G_n $(n > 6)$ of the Gröbner basis, but the knowledge of these partial components computed suffices (Sect.3.2) to deduce the complete set of constraints and categorize them.

Performing the computation we observed that the timings in construction of polynomials of a given order may considerably vary. For instance, the element in G_4 obtained from the S-polynomial $(\varphi_2^{(2)}, \varphi_3^{(2)})$ depends on the coordinates A_-^a, A_1^a, A_2^a only and contains 286 terms. Its reduction requires a large number of monomial divisions at the elementary reduction steps. In contrast to that polynomial, the other polynomials in G_4 turn to be an irreducible.

Final remark, most of the total calculation time (about a month) was spend performing useless zero reductions. This is in agreement with the well-known experimental facts in practical computation of Gröbner bases for large polynomial systems [35].

Acknowledgments

We would like to thank T.Heinzl, D. McMullan, D. Mladenov and S. Krivonos for fruitful conversions on topics related to this work.

This work was supported in part by the grant No. 07-01-00660 from the Russian Foundation for Basic Research and by the grant No. 5362.2006.2. from the Ministry of Education and Science of the Russian Federation.

References

1. Pommaret, J.F.: Partial Differential Equations and Group Theory. New Perspectives for Applications. Kluwer, Dordrecht (1994)
2. Reid, G., Wittkopf, A., Boulton, A.: Reduction of systems of nonlinear partial differential equations to simplified involutive forms. Euro. J. Appl. Maths. 7, 604–635 (1996)
3. Gerdt, V.P., Blinkov, Y.A.: Involutive bases of polynomial ideals. Math. Comp. Simul. 45, 519–542 (1998), arXiv:math.AC/9912027.
4. Gerdt, V.P.: Completion of linear differential systems to involution. In: Ganzha, V.G., Mayr, E.W., Vorozhtsov, E.V. (eds.) Computer Algebra in Scientific Computing / CASC'99, pp. 115–137. Springer, Berlin (1999)
5. Calmet, J., Hausdorf, M., Seiler, W.M.: A constructive introduction to involution. In: Akerkar, R. (ed.) International Symposium on Applications of Computer Algebra/ISACA'2000, pp. 33–50. Allied Publishers, New Delhi (2001)
6. Gerdt, V.P.: Involutive algorithms for computing Gröbner bases. In: Cojocaru, S., Pfister, G., Ufnarovski, V. (eds.) Computational Commutative and Non-Commutative algebraic geometry. NATO Science Series, pp. 199–225. IOS Press, Amsterdam (2005), arXiv:math.AC/0501111
7. Dirac, P.A.M.: Generalized Hamiltonian dynamics, Canad. J. Math. 2, 129–148 (1950), Lectures on Quantum Mechanics, Belfer Graduate School of Science, Monographs Series, Yeshiva University, New York, 1964.
8. Sundermeyer, K.: Constrained Dynamics. Lecture Notes in Physics, vol. 169. Spinger, Heidelberg (1982)
9. Henneaux, M., Teitelboim, C.: Quantization of Gauge Systems. Princeton University Press, Princeton, New Jersey (1992)
10. Gerdt, V.P., Gogilidze, S.A.: Constrained Hamiltonian systems and Gröbner bases. In: Ganzha, V.G., Mayr, E.W., Vorozhtsov, E.V. (eds.) Computer Algebra in Scientific Computing/CASC 1999, pp. 138–146. Springer, Berlin (1999)

11. Gerdt, V.P., Khvedelidze, A.M., Mladenov, D.M.: Analysis of constraints in light-cone version of $SU(2)$ Yang-Mills mechanics. In: Gerdt, V.P. (ed.) Computer Algebra and its Applications to Physics/CAAP'2001, Dubna, JINR, pp. 83–92 (2002), arXiv:hep-th/0209107

12. Gerdt, V.P., Khvedelidze, A.M., Mladenov, D.M.: Light-cone SU(2) Yang-Mills theory and conformal mechanics, arXiv:hep-th/0222100

13. Gerdt, V., Khvedelidze, A., Mladenov, D.: On application of involutivity analysis of differential equations to constrained dynamical systems. In: Sissakian, A.N. (ed.)Symmetries and Integrable Systems, Selected papers of the seminar, 2000-2005, Dubna, JINR, vol. I, pp. 132–150 (2006), arXiv:hep-th/0311174

14. Gerdt, V., Khvedelidze, A., Palii, Y.: Towards an algorithmisation of the Dirac constraint formalism. In: Calmet, J., Seiler, W.M., Tucker, R.W. (eds.) Global Integrabilty of Field Theories/GIFT' 2006, Cockroft Institute, Daresbury, UK, pp. 135–154 (2006), arXiv:math-ph/0611021

15. Becker, T., Weispfenning, V.: Gröbner Bases. A Computational Approach to Commutative Algebra. In: Graduate Texts in Mathematics, vol. 141, Springer, New York (1993)

16. Cox, D., Little, J., O'Shea, D.: Ideals, Varieties and Algorithms, 2nd edn. Springer, New York (1996)

17. Buchberger, B., Winkler, F. (eds.): Gröbner Bases and Applications. Cambridge University Press, Cambridge (1998)

18. Hurtley, D.H., Tucker, R.W., Tuckey, P.: Constrained Hamiltonian dynamics and exterior differential systems. J. Phys. A. 24, 5252–5265 (1991)

19. Seiler, W.M., Tucker, R.W.: Involution and constrained dynamics I: the Dirac approach. J. Phys. A. 28, 4431–4451 (1995)

20. Seiler, W.M.: Involution and constrained Dynamics. II: the Faddeev-Jackiw approach. J. Phys. A. 28, 7315–7331 (1995)

21. http://fgbrs.lip6.fr/salsa/Software/

22. http://invo.jinr.ru

23. Baseian, G.Z., Matinyan, S.G., Savvidy, G.K.: Nonlinear plane waves in massless Yang-Mills theory. Pisma Zh. Eksp. Teor. Fiz. 29, 641 (1979)

24. Asatryan, H.M., Savvidy, G.K.: Configuration manifold of Yang-Mills classical mechanics. Phys. Lett. A 99, 290–292 (1983)

25. Lüscher, M.: Some analytic results cocerning the mass spectrum of Yang-Mills gauge theories on a torus. Nucl. Phys. B219, 233–261 (1983)

26. Simon, B.: Some quantum operators with discrete spectrum but classically continuos spectrum. Annals of Phys. 146, 209–220 (1983)

27. Matinyan, S.G.: Dynamical chaos of nonabelian gauge fields. Fiz. Elem. Chast. Atom. Yadra 16, 522–550 (1985)

28. Soloviev, M.A.: On the geometry of classical mechanics with nonabelian gauge symmetry. Teor. Mat. Fiz. 73, 3–15 (1987)

29. Dahmen, B., Raabe, B.: Unconstrained $SU(2)$ and $SU(3)$ Yang-Mills classical mechanics. Nucl. Phys. B384, 352–380 (1992)

30. Gogilidze, S.A., Khvedelidze, A.M., Mladenov, D.M., Pavel, H.-P.: Hamiltonian reduction of $SU(2)$ Dirac-Yang-Mills mechanics. Phys. Rev. D 57, 7488–7500 (1998)

31. Khvedelidze, A.M., Pavel, H.P.: On the groundstate of Yang-Mills quantum mechanics. Phys. Lett. A 267, 96 (2000), arXiv:hep-th/9905093

32. Khvedelidze, A.M., Mladenov, D.M.: Euler-Calogero-Moser system from SU(2) Yang-Mills theory. Phys. Rev. D 62, 125016 (1-9) (2000)

33. Pavel, H.P.: SU(2) Yang-Mills quantum mechanics of spatially constant fields, arXiv:hep-th/0701283

34. Heinzl, T.: Light-cone quantization: Foundations and applications. Lecture Notes in Physics, vol. 572, p. 55. Spinger, Heidelberg (2001)
35. Faugère, J.-C.: A new efficient algorithm for computing Gröbner bases without reduction to zero (F5). In: Mora, T. (ed.) International Symposium on Symbolic and Algebraic Computation/ISSAC' 2002, pp. 75–83. ACM Press, New York (2002)

6 Appendix

The eight traceless 3×3 Hermitian Gell-Mann matrices λ_a :

$$\lambda_1 = \begin{pmatrix} 0 & 1 & 0 \\ 1 & 0 & 0 \\ 0 & 0 & 0 \end{pmatrix}, \quad \lambda_2 = \begin{pmatrix} 0 & -i & 0 \\ i & 0 & 0 \\ 0 & 0 & 0 \end{pmatrix}, \quad \lambda_3 = \begin{pmatrix} 1 & 0 & 0 \\ 0 & -1 & 0 \\ 0 & 0 & 0 \end{pmatrix},$$

$$\lambda_4 = \begin{pmatrix} 0 & 0 & 1 \\ 0 & 0 & 0 \\ 1 & 0 & 0 \end{pmatrix}, \quad \lambda_5 = \begin{pmatrix} 0 & 0 & -i \\ 0 & 0 & 0 \\ i & 0 & 0 \end{pmatrix}, \quad \lambda_6 = \begin{pmatrix} 0 & 0 & 0 \\ 0 & 0 & 1 \\ 0 & 1 & 0 \end{pmatrix}, \tag{29}$$

$$\lambda_7 = \begin{pmatrix} 0 & 0 & 0 \\ 0 & 0 & -i \\ 0 & i & 0 \end{pmatrix}, \quad \lambda_8 = \frac{1}{\sqrt{3}} \begin{pmatrix} 1 & 0 & 0 \\ 0 & 1 & 0 \\ 0 & 0 & -2 \end{pmatrix}.$$

provide a basis for the $su(3)$ algebra

$$[\lambda_a, \lambda_b] = 2i \sum_{c=1}^{8} f_{abc} \lambda_c. \tag{30}$$

The nonzero structure constants f_{abc}, antisymmetric in all indices are listed below:

$$f_{123} = 1, \quad f_{147} = f_{246} = f_{257} = f_{345} = f_{516} = f_{637} = 1/2, \quad f_{458} = f_{678} = \sqrt{3}/2.$$

The matrices λ_a obey the product law

$$\lambda_a \lambda_b = \frac{2}{3} \delta_{ab} I + \sum_{c=1}^{8} (d_{abc} + i f_{abc}) \lambda_c,$$

with the following non-vanishing values of the symmetric constants d_{abc}

$$d_{118} = d_{228} = d_{338} = \frac{1}{\sqrt{3}}, \quad d_{146} = d_{157} = d_{256} = d_{344} = d_{355} = \frac{1}{2},$$

$$d_{247} = d_{366} = d_{377} = -\frac{1}{2}, \quad d_{448} = d_{558} = d_{668} = d_{778} = -\frac{1}{2\sqrt{3}}, \quad d_{888} = -\frac{1}{\sqrt{3}}.$$

On the Weight Spectra of Conway Matrices Related to the Non-transitive Head-or-Tail Game

Nikita Gogin and Aleksandr Mylläri

University of Turku, Finland

Abstract. Our paper is devoted to the computation of the weight spectra of Conway matrices related to the non-transitive head-or-tail game. We obtained explicit formulas for the spectra containing partial binomial sums. These sums are rather hard to deal with when the methods of classical algebra and number theory are used; but when we used methods of computer algebra, we were able to handle them quite efficiently and could easily produce visualizations. We suggest a recurrence algorithm for efficient calculation of the weight-spectrum matrices, including, as a special case, integer matrices modulo m. The algorithm is implemented with MATHEMATICA and visualizations for some interesting examples are shown.

1 Introduction

The problem in question emerges from the old problem of binary string overlapping in connection with the so-called "non-transitive head-or-tail game" (cf. [1,2]):

Two players, (1) and (2), agree on some integer $n \geq 2$. Then both of them select a binary $n-$word ("head"$=0$, "tail"$=1$), say u and v, and begin flipping a coin until either u or v appears as a block of n consecutive outcomes. Player (1) wins if u appears before v does. The problem is to find the probability $P(u, v)$ that player (1) will win for the chosen u and v.

The solution of the problem is represented by *Conway's formula* (cf. [1])

$$P(u, v) = \frac{c(v, v) - c(v, u)}{(c(u, u) - c(u, v)) + (c(v, v) - c(v, u))} \qquad (1)$$

where $c(u, v)$ is the *Conway number* (or *"correlation"*) of two binary $n-$words u and v.

Definition 1. *Let V_n be the space of the binary $n-$words, i.e. the $n-$dimensional vector space over the field $Z_2 = \{0, 1\}$ and k be an integer, $0 \leq k \leq n$. We say that binary vector $u = (\varepsilon_0, \varepsilon_1, \ldots, \varepsilon_{n-1})$ $k-$overlaps vector $v = (\eta_0, \eta_1, \ldots, \eta_{n-1})$, if $(\varepsilon_k, \varepsilon_{k+1}, \ldots, \varepsilon_{n-1}) = (\eta_0, \eta_1, \ldots, \eta_{n-k-1})$, and write this as $u =_k v$.*

V.G. Ganzha, E.W. Mayr, and E.V. Vorozhtsov (Eds.): CASC 2007, LNCS 4770, pp. 160–168, 2007.

Definition 2. *The Conway number of the pair (u, v) is defined as a sum*

$$c(u, v) = \sum_{k=0}^{n-1} 2^{n-1-k} \|u =_k= v\|, \qquad (2)$$

where $\|u =_k= v\|$ equals to 1 (respectively 0) if $u =_k= v$ is "true" (respectively "false").

For every n the set of all Conway numbers constitutes the $2^n \times 2^n$ *Conway matrix* $C_n = (c(u, v))$ with the natural ordering of binary vectors as binary expansions the indexes 0 , 1 , ... , $2^n - 1$.

In other words, we have an integer-valued function

$$C_n : \quad W_n = V_n \bigoplus V_n \to \{0, 1, \ldots, 2^n - 1\} \qquad (3)$$

on the Abelian group W_n and our aim is to find the weight-spectrum (cf. [3]) of this function, i.e. the $(n + 1) \times (n + 1)$ integer matrix $S(C_n) = (S_{p,q})$, where

$$S_{p,q} = \sum_{wt(u)=p, wt(v)=q} c(v, u) \qquad (4)$$

and $wt()$ stands for the Hamming weight.

To do this let us first consider the Fourier (called also in this case the Hadamard) transform $\hat{C}_n$ of the function C_n defined by formula (2) on the finite Abelian group W_n, i.e. we consider the matrix product

$$\hat{C}_n = H_n C_n H_n \qquad (5)$$

where H_n is the $2^n \times 2^n$ Hadamard matrix defined by the well-known (cf. [3]) recurrent formula

$$H_n = \begin{pmatrix} H_{n-1} & H_{n-1} \\ H_{n-1} & -H_{n-1} \end{pmatrix}, \quad n \geq 2, \quad H_1 = \begin{pmatrix} 1 & 1 \\ 1 & -1 \end{pmatrix}. \qquad (6)$$

The reason for this move from C_n to its Hadamard transform $\hat{C}_n$ is that it has been shown [4] that the latter has the very simple matrix structure:

Theorem 1

$$(\hat{C}_n)_{ij} = 2^{2n-1} \begin{cases} n, & \text{if } i = j = 0 \\ 1, & \text{for some } k, \ 0 \leq k \leq n-1 \end{cases} \quad 0 \leq i, j \leq 2^n - 1. \qquad (7)$$

The Proof can be found in [4].

Example 1. For $n = 3$ it can be found straightforwardly by formulas (2) and (5) that

$$C_3 = \begin{pmatrix} 7 & 3 & 1 & 1 & 0 & 0 & 0 & 0 \\ 0 & 4 & 2 & 2 & 1 & 1 & 1 & 1 \\ 1 & 1 & 5 & 1 & 2 & 2 & 0 & 0 \\ 0 & 0 & 0 & 4 & 1 & 1 & 3 & 3 \\ 3 & 3 & 1 & 1 & 4 & 0 & 0 & 0 \\ 0 & 0 & 2 & 2 & 1 & 5 & 1 & 1 \\ 1 & 1 & 1 & 1 & 2 & 2 & 4 & 0 \\ 0 & 0 & 0 & 0 & 1 & 1 & 3 & 7 \end{pmatrix}, \hat{C}_3 = 2^{2 \times 3 - 1} \begin{pmatrix} 3 & 0 & 0 & 0 & 0 & 0 & 0 & 0 \\ 0 & 1 & 1 & 0 & 1 & 0 & 0 & 0 \\ 0 & 0 & 1 & 0 & 1 & 0 & 0 & 0 \\ 0 & 0 & 0 & 1 & 0 & 0 & 1 & 0 \\ 0 & 0 & 0 & 0 & 1 & 0 & 0 & 0 \\ 0 & 0 & 0 & 0 & 0 & 1 & 0 & 0 \\ 0 & 0 & 0 & 0 & 0 & 0 & 1 & 0 \\ 0 & 0 & 0 & 0 & 0 & 0 & 0 & 1 \end{pmatrix}. \qquad (8)$$

2 An Explicit Formula for the Weight-Spectrum Matrix $S(C_n)$

Given the spectra $S(\hat{C}_n)$, we can easily find the spectrum $S(C_n)$ by the MacWilliams formula for the dual spectrums (cf. [3]) :

$$2^{2n} S_{k_1,k_2}(C_n) = \sum_{m_1,m_2=0}^{n} S_{m_1,m_2}(\hat{C}_n) P_{k_1}(m_1) P_{k_2}(m_2), \quad 0 \le k_1, k_2 \le n, \quad (9)$$

where $P_k(m)$ are the Krawtchouk polynomials (of order n) defined by the generating function

$$(1+t)^{n-m}(1-t)^m = \sum_{k=0}^{n} P_k(m) t^k, \tag{10}$$

$$P_k(m) = \sum_{l=0}^{k} (-1)^l \binom{n-m}{k-l} \binom{m}{l} \tag{11}$$

(cf. [3]).

So, our first step is to calculate the numbers $S_{m_1,m_2}(\hat{C}_n)$.

Lemma 1

$$S_{m_1,m_2}(\hat{C}_n) = 2^{2n-1} \cdot \begin{cases} n, & m_1 = m_2 = 0 \\ \binom{n+1}{m+1} \cdot \delta_{m_1,m_2}, & m_1 + m_2 \ne 0 \end{cases} \tag{12}$$

Proof. Identifying the binary expansions of the matrix indices with the binary n–vectors and using formula (7), we easily find that, if $m_1 \ne 0$ or $m_2 \ne 0$ (or both), the number $S_{m_1,m_2}(\hat{C}_n)$ is not equal to zero only when $m_1 = m_2 = m$ in which case

$$S_{m_1,m_2}(\hat{C}_n) = \sum_{wt(i)=m_1, wt(j)=m_2} (\hat{C}_n)_{ij} = 2^{2n-1} \cdot \sum_{k=0}^{n} \sum_{u \in V_{v-k}, wt(u)=m} 1$$

$$= 2^{2n-1} \cdot \sum_{k=0}^{n} \binom{n-k}{m} = 2^{2n-1} \cdot \binom{n+1}{m+1}, \tag{13}$$

so, again due to (7), we get the claimed result.

Theorem 2. *For the elements of the weight-spectrum matrix $S(C_n)$ we have*

$$S_{k_1,k_2}(C_n) = \frac{1}{2} \left(\sum_{p \le \min(k_1,k_2)} \binom{n+1}{p} \cdot \sum_{p > \max(k_1,k_2)} \binom{n+1}{p} - \binom{n}{k_1} \binom{n}{k_2} \right). \tag{14}$$

Here we need the following lemma:

Lemma 2. *Let*

$$\beta_{k_1,k_2}^{(n)} = \beta_{k_1,k_2} = \sum_{m=0}^{n} \binom{n+1}{m+1} P_{k_1}(m) P_{k_2}(m) \qquad (15)$$

(we write simply β_{k_1,k_2}, skipping the upper index n when there is no danger of confusion.) Then

$$\beta_{k_1,k_2} = \beta_{k_2,k_1} = \sum_{p \le \min(k_1,k_2)} \binom{n+1}{p} \cdot \sum_{p > \max(k_1,k_2)} \binom{n+1}{p}. \qquad (16)$$

Proof. To prove (16) we first write down formula (10) twice

$$(1+y)^{n-m}(1-y)^m = \sum_{k=0}^{n} P_k(m) y^k, \quad (1+z)^{n-m}(1-z)^m = \sum_{k=0}^{n} P_k(m) z^k$$

and then consider the following sum:

$$\sum_{m=0}^{n} \binom{n+1}{m+1}(1+y)^{n-m}(1-y)^m(1+z)^{n-m}(1-z)^m = \sum_{k_1,k_2=0}^{n} \beta_{k_1,k_2} y^{k_1} z^{k_2}. \qquad (17)$$

The left side of this equality can be easily found with the help of the elementary formula

$$\sum_{m=0}^{n} \binom{n+1}{m+1} a^{n-m} b^m = \begin{cases} (n+1)a^n, & b=0 \\ \dfrac{(a+b)^{n+1}-a^{n+1}}{b}, & b \ne 0 \end{cases}$$

with $a = (1+y)(1+z) = 1+y+z+yz$ and $b = (1-y)(1-z) = 1-y-z+yz$, and so, formula (17) can be rewritten as follows:

$$2^{n+1}(1+yz)^{n+1}-(1+y)^{n+1}(1+z)^{n+1} = (1-y-z+yz)\cdot \sum_{k_1,k_2=0}^{n} \beta_{k_1,k_2} y^{k_1} z^{k_2}. \qquad (18)$$

Denoting by γ_{k_1,k_2} the coefficients at the terms $y^{k_1} z^{k_2}$ in the left side of (18) and comparing corresponding coefficients in the left and right sides, we find that

$$\gamma_{k_1,k_2} = 2^{n+1}\binom{n+1}{k_1}\delta_{k_1,k_2} - \binom{n+1}{k_1}\binom{n+1}{k_2} \qquad (19)$$

$$= \beta_{k_1,k_2} - \beta_{k_1-1,k_2} - \beta_{k_1,k_2-1} + \beta_{k_1-1,k_2-1}$$

(understanding that $\beta_{n+1,\ldots} = \beta_{\ldots,n+1} = 0$), and now it is not difficult to verify that the numbers β_{k_1,k_2} can be expressed through the numbers γ_{k_1,k_2} as follows:

$$\beta_{k_1,k_2} = \sum_{p=0}^{k_1} \sum_{q=0}^{k_2} \gamma_{p,q}. \qquad (20)$$

So, we have

$$\beta_{k_1,k_2} = \sum_{p=0}^{k_1}\sum_{q=0}^{k_2}\left(2^{n+1}\binom{n+1}{p}\delta_{pq} - \binom{n+1}{p}\binom{n+1}{q}\right)$$

$$= 2^{n+1}\sum_{p=0}^{\min(k_1,k_2)}\binom{n+1}{p} - \sum_{p=0}^{k_1}\binom{n+1}{p}\sum_{q=0}^{k_2}\binom{n+1}{q}$$

$$= \sum_{p=0}^{\min(k_1,k_2)}\binom{n+1}{p}\cdot\left(2^{n+1} - \sum_{p=0}^{\max(k_1,k_2)}\binom{n+1}{p}\right)$$

$$= \sum_{p\leq\min(k_1,k_2)}\binom{n+1}{p}\cdot\sum_{p>>\max(k_1,k_2)}\binom{n+1}{p}$$

and this completes the proof of Lemma 2.

Remark 1. For any integers m, $m \geq 0$ and r, $0 \leq r \leq m$ let

$$b_m(r) = \sum_{x\leq r}\binom{m}{x} \tag{21}$$

be the r−th partial sum of the binomial coefficients. Then formula (16) can be written in a little more elegant form:

$$\beta_{k_1,k_2} = \beta^{(n)}_{k_1,k_2} = \beta^{(n)}_{k_2,k_1} = b_{n+1}(\min(k_1,k_2))\cdot b_{n+1}(n - \max(k_1,k_2)) \tag{22}$$

because evidently $\sum_{x>r}\binom{m}{x} = b_m(m - r - 1)$.

Now we can easily complete the proof of Theorem 2:

Proof. According to the MacWilliams formula (9) we find that

$$2^{2n}S_{k_1,k_2}(C_n) = \sum_{m_1,m_2=0}^{n} S_{m_1,m_2}(\hat{C}_n)P_{k_1}(m_1)P_{k_2}(m_2)$$

$$= 2^{2n-1}\cdot\left(n\binom{n}{k_1}\binom{n}{k_2} + \sum_{m=1}^{n}\binom{n+1}{m+1}P_{k_1}(m)P_{k_2}(m)\right)$$

$$= 2^{2n-1}\cdot\left(-\binom{n}{k_1}\binom{n}{k_2} + \sum_{m=0}^{n}\binom{n+1}{m+1}P_{k_1}(m)P_{k_2}(m)\right)$$

and due to formula (16) this concludes the proof.

Let now

$$B_n = (\beta_{k_1,k_2})_{0\leq k_1,k_2\leq n}, \quad Y_n = \binom{n}{s}_{0\leq s\leq n}. \tag{23}$$

Then formula (14) evidently can be written in a matrix form as follows:

$$S(C_n) = \frac{1}{2}(B_n - Y_n \bigotimes Y_n^T).\tag{24}$$

Since Y_n is a row of the binomial coefficients, everywhere in what follows, we shall refer to the Kronecker product $Y_n \bigotimes Y_n^T$ as a "trivial term" of the matrix $S(C_n)$ and our attention will be mainly devoted to the matrix B_n (ignoring the multiplier $\frac{1}{2}$) as a nontrivial number-theoretical, combinatorial and computational object.

Example 2. For $n = 3$ the straightforward calculations with matrix C_3 from Example 1 give the result

$$S(C_3) = \begin{pmatrix} 7 & 4 & 1 & 0 \\ 4 & 23 & 8 & 1 \\ 1 & 8 & 23 & 4 \\ 0 & 1 & 4 & 7 \end{pmatrix},$$

whereas the straightforward calculations by formulas (16) and (23) show that

$$B_3 = \begin{pmatrix} 15 & 11 & 5 & 1 \\ 11 & 55 & 25 & 5 \\ 5 & 25 & 55 & 11 \\ 1 & 5 & 11 & 15 \end{pmatrix}, \quad Y_3 \bigotimes Y_3^T = \begin{pmatrix} 1 & 3 & 3 & 1 \\ 3 & 9 & 9 & 3 \\ 3 & 9 & 9 & 3 \\ 1 & 3 & 3 & 1 \end{pmatrix},$$

so, the equality (24) $S(C_3) = \frac{1}{2}(B_3 - Y_3 \bigotimes Y_3^T)$ is evidently true.

3 A Recurrent Formula and Fast Computational Algorithm for Matrices B_n

In this section, we are going to obtain a recurrence relation between the matrix elements of B_{n+1} and B_n, using the well-known recurrent relations for the binomial coefficients.

Lemma 3. *For $n \geq 0$ and for any r and s, $0 \leq r, s \leq n + 1$, we have the following recurrent relation:*

$$\beta_{rs}^{(n+1)} = \beta_{r,s}^{(n)} + \beta_{r,s-1}^{(n)} + \beta_{r-1,s}^{(n)} + \beta_{r-1,s-1}^{(n)}.\tag{25}$$

Proof. For $r < s$ we have $\min(r, s) = r$, $\max(r, s - 1) = s - 1$, , so, we find straightforwardly from formula (16) that

$$\beta_{rs}^{(n+1)} = \sum_{x \leq r} \binom{n+1}{x} \cdot \sum_{y > s} \binom{n+1}{y}\tag{26}$$

$$= \sum_{x \leq r} \binom{n}{x} \cdot \sum_{y > s} \binom{n}{y} + \sum_{x \leq r} \binom{n}{x} \cdot \sum_{y > s} \binom{n}{y - 1}$$

$$+ \sum_{x \le r} \binom{n}{x-1} \cdot \sum_{y > s} \binom{n}{y} + \sum_{x \le r} \binom{n}{x-1} \cdot \sum_{y > s} \binom{n}{y-1}$$

$$= \beta_{r,s}^{(n)} + \sum_{x \le r} \binom{n}{x} \cdot \sum_{y > s-1} \binom{n}{y}$$

$$+ \sum_{x \le r-1} \binom{n}{x} \cdot \sum_{y > s} \binom{n}{y} + \sum_{x \le r-1} \binom{n}{x} \cdot \sum_{y > s-1} \binom{n}{y}$$

$$= \beta_{r,s}^{(n)} + \beta_{r,s-1}^{(n)} + \beta_{r-1,s}^{(n)} + \beta_{r-1,s-1}^{(n)}.$$

For $r = s$ we have $\min(r, s) = r$, but $\max(r, s-1) = r$, so we need to change the above calculations only in the single term:

$$\sum_{x \le r} \binom{n}{x} \cdot \sum_{y > s} \binom{n}{y-1} = \sum_{x \le r} \binom{n}{x} \cdot \sum_{y > s-1} \binom{n}{y} \qquad (27)$$

$$= \sum_{x \le r} \binom{n}{x} \cdot \sum_{y > r-1} \binom{n}{y} = \left(2^n - \sum_{x > r} \binom{n}{x} \right) \cdot \left(2^n - \sum_{y \le r-1} \binom{n}{y} \right)$$

$$= 2^{2n} + \beta_{r-1,r}^{(n)} - 2^n \left(2^n - \binom{n}{r} \right) = \beta_{r-1,r}^{(n)} + 2^n \binom{n}{r}$$

$$= \beta_{r,r-1}^{(n)} + 2^n \binom{n}{r},$$

So, in this case we get

$$\beta_{r,r}^{(n+1)} = \beta_{r,r}^{(n)} + \beta_{r,r-1}^{(n)} + \beta_{r-1,r}^{(n)} + \beta_{r-1,r-1}^{(n)} + 2^n \binom{n}{r} \qquad (28)$$

and the general formula for any r and s coincides with (25).

In order to rewrite the recurrence relations (25) in matrix form we introduce the following notations:

Let Δ_{n+1} be the diagonal matrix: $\Delta_{n+1} = 2^n \cdot \operatorname{diag} \left(\binom{n}{r} \right)_{0 \le r \le n}$.

Furthermore, for any matrix A let $|\overline{A}$ stand for the zero-padding of A from its left and upper sides and let the notations $\overline{A}|$, $|\underline{A}$ and $\underline{A}|$ have analogous meaning.

Then it's easy to see that the relations (25) can be written in matrix form as follows:

$$B_{n+1} = |\overline{B_n} + \overline{B_n}| + |\underline{B_n} + \underline{B_n}| + \Delta_{n+1}, \quad \Delta_{n+1} = 2 \cdot \left(|\overline{\Delta_n} + \underline{\Delta_n}| \right), \quad n \ge 1 \quad (29)$$

with initial values $B_0 = \Delta_0 = (1)$.

Formulae (29) provide us with a recursive construction of matrices B_n "without multiplications" (ignoring the multiplication by 2), so we refer to them as a "fast algorithm" for B_n.

Example 3

$$B_1 = \begin{pmatrix} 0\,0 \\ 0\,1 \end{pmatrix} + \begin{pmatrix} 0\,0 \\ 1\,0 \end{pmatrix} + \begin{pmatrix} 0\,1 \\ 0\,0 \end{pmatrix} + \begin{pmatrix} 1\,0 \\ 0\,0 \end{pmatrix} + 2\left(\begin{pmatrix} 0\,0 \\ 0\,1 \end{pmatrix} + \begin{pmatrix} 1\,0 \\ 0\,0 \end{pmatrix} \right) = \begin{pmatrix} 3\,1 \\ 1\,3 \end{pmatrix},$$

$$B_2 = \begin{pmatrix} 0\,0\,0 \\ 0\,3\,1 \\ 0\,1\,3 \end{pmatrix} + \begin{pmatrix} 0\,0\,0 \\ 3\,1\,0 \\ 1\,3\,0 \end{pmatrix} + \begin{pmatrix} 0\,3\,1 \\ 0\,1\,3 \\ 0\,0\,0 \end{pmatrix} + \begin{pmatrix} 3\,1\,0 \\ 1\,3\,0 \\ 0\,0\,0 \end{pmatrix}$$
$$+ 2\left(\begin{pmatrix} 0\,0\,0 \\ 0\,2\,0 \\ 0\,0\,2 \end{pmatrix} + \begin{pmatrix} 2\,0\,0 \\ 0\,2\,0 \\ 0\,0\,0 \end{pmatrix} \right) = \begin{pmatrix} 7\ \ 4\ \ 1 \\ 4\ 16\ 4 \\ 1\ \ 4\ \ 7 \end{pmatrix}$$

$$B_3 = \begin{pmatrix} 0\,0\ \ 0\ \ 0 \\ 0\,7\ \ 4\ \ 1 \\ 0\,4\ 16\ 4 \\ 0\,1\ \ 4\ \ 7 \end{pmatrix} + \begin{pmatrix} 0\ \ 0\ \ 0\,0 \\ 7\ \ 4\ \ 1\,0 \\ 4\ 16\ 4\,0 \\ 1\ \ 4\ \ 7\,0 \end{pmatrix} + \begin{pmatrix} 0\,7\ \ 4\ \ 1 \\ 0\,4\ 16\ 4 \\ 0\,1\ \ 4\ \ 7 \\ 0\,0\ \ 0\ \ 0 \end{pmatrix} + \begin{pmatrix} 7\ \ 4\ \ 1\,0 \\ 4\ 16\ 4\,0 \\ 1\ \ 4\ \ 7\,0 \\ 0\ \ 0\ \ 0\,0 \end{pmatrix}$$
$$+ 2\left(\begin{pmatrix} 0\,0\,0\,0 \\ 0\,4\,0\,0 \\ 0\,0\,8\,0 \\ 0\,0\,0\,4 \end{pmatrix} + \begin{pmatrix} 4\,0\,0\,0 \\ 0\,8\,0\,0 \\ 0\,0\,4\,0 \\ 0\,0\,0\,0 \end{pmatrix} \right) = \begin{pmatrix} 15\ 11\ \ 5\ \ \ 1 \\ 11\ 55\ 25\ \ 5 \\ \ 5\ 25\ 55\ 11 \\ \ 1\ \ 5\ 11\ 15 \end{pmatrix}$$

and so on.

Formulae (29) show that in fact B_n is the sum of the results of the n–th steps of two cellular automata, where B_n is a 2D cellular automata and its diagonal Δ_n is one-dimensional.

We found most interesting the behavior of B_n modulo some integer m that we observed with the help of WSCONW.nb, a MATHEMATICA program for visualizing formulae (29) that was written by the authors of this paper. The structure of the resulting images reflects the difficulties related to studying these partial binomial sums, difficulties which have been mentioned by many authors (see, e.g. [5,6]).

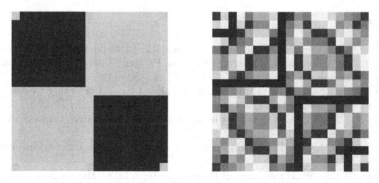

Fig. 1. WSCONW[18,3], WSCONW[18,7]

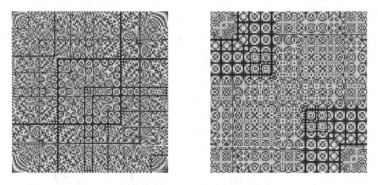

Fig. 2. WSCONW[227,23], WSCONW[193,21]

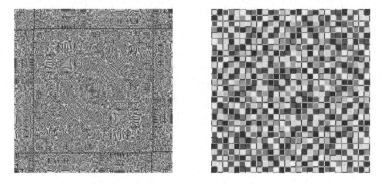

Fig. 3. WSCONW[361,121], WSCONW[196,49]

Some examples of the visualizations of matrices are given below. Notice the effect of the optical illusions of "looking through the frosted glass" in the case $n = 361, m = 121$ and askew deformations in the case $n = 196, m = 49$ on the Figure 3.

References

1. Gardner, M.: Time Travel and Other Mathematical Bewilderments. NY (1988)
2. Guibas, L.J., Odlyzko, A.M.: String Overlaps, Pattern Matching, and Nontransitive Games. Journal of Combinatorial Theory 30, 183–208 (1981)
3. MacWilliams, F.J., Sloane, N.J.A.: The Theory of Error-Correcting Codes. North-Holland P.C., Amsterdam (1977)
4. Gogin, N., Mylläri, A.: Choosing the Strategy in non-Transitive Games: Visualization (to be published)
5. Graham, R.L., Knuth, D.E., Patashnik, O.: Concrete Mathematics. Addison-Wesley, Reading (1989)
6. Ollerton, R.L.: Partial row-sums of Pascal's triangle. International Journal of Mathematical Education in Science and Technology 38(1/15), 124–127 (2007)

Properties of the Liapunov Stability Zones of the Lagrange Triangle

E.A. Grebenikov[1], D. Kozak-Skoworodkin[2], and N.I. Zemtsova[1]

[1] Dorodnicyn Computing Center of RAS
Vavilova str. 40, 119991 Moscow, Russia
greben@ccas.ru
[2] University of Podlasie, Poland
kdorota@ap.siedlce.pl

Abstract. We derived the quantitative estimates for geometrical parameters of the stability domains of the Lagrange triangle in the restricted three-body problem. We have shown that these domains are plane ellipse-similar figures, extended along a tangent to a circle, on which the Lagrange triangular solutions are located. We have proposed the heuristic algorithm for finding the maximal sizes of the stability domains.

History of the restricted three-body problem has been starting since famous works of Euler and Lagrange [1,2]. Remind that the problem is to study motion of the body of infinitesimal mass in the Newtonian gravitational field generated by two bodies, having finite masses and moving in three-dimensional space on Kepler orbits about their common center of mass. Many outstanding mathematicians and mechanicians investigated this problem in the 19th and 20th centuries. One should mention such names as K. Gauss, K. Jacobi, G. Hill, H. Poincaré, A. Liapunov, T. Levi–Civita, G. Birkhoff, E. Whittaker, J. Chazy, N.D. Moiseev and many other scientists who studied analytical, qualitative, and numerical properties of the restricted 3-body problem. A comprehensive review of this problem was done in the famous book of V. Szebehely [3].

The Liapunov stability of the Lagrange triangle was investigated on the basis of the KAM-theory [4,5,6], which was developed by A. Kolmogorov, V. Arnold, and J. Moser. It is the metric theory of the existence of quasi-periodic solutions of the Hamiltonian systems determined on many-dimensional toruses. Linearization of Hamiltonian systems in the vicinity of stationary solutions (the equilibrium solutions) always generates linear systems with symplectic matrix[7,8], i.e., any equilibrium solutions of the Hamiltonian system can be linearly stable only in such cases when all eigenvalues of the matrix of first approximation are purely imaginary numbers. Therefore, the theorems of the Liapunov First Method [9] cannot be applied. Just this situation stimulated development of the KAM-theory. Also the book of A.P. Markeev [10] and article of A.G. Sokol'sky [11] should be pointed out who investigated a problem of Liapunov stability of the Lagrange solutions in the "resonant" case, when pure imaginary eigenvalues of symplectic matrix of linear approximation are rationally commensurable.

V.G. Ganzha, E.W. Mayr, and E.V. Vorozhtsov (Eds.): CASC 2007, LNCS 4770, pp. 169–180, 2007.
© Springer-Verlag Berlin Heidelberg 2007

In this article we investigate geometrical form and dimensions of the Liapunov stability zones of the Lagrange solutions in the restricted circular three-body problem. As the analytical methods are unsuitable for this, we have made a computing experiment with the system *Mathematica* [12].

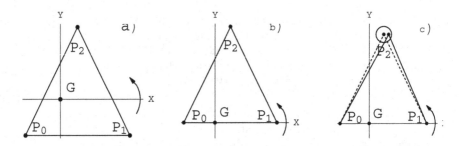

Fig. 1. Gxy is a rotating Cartesian coordinate system; a) equilateral Lagrange triangle with masses $m_0 \neq 0$, $m_1 \neq 0$, $m_2 \neq 0$; b) equilateral Lagrange triangle, representing the restricted 3-body problem ($m_2 = 0$); c) perturbed Lagrange triangle

Let two particles P_0 and P_1 having non-zero masses m_0 and m_1, respectively, and point P_2 of "zero" mass be situated in the Cartesian coordinate plane (see Fig. 1). Triangle $P_0P_1P_2$ represents an elementary model of the Newton three-body problem for which Lagrange showed that the differential equations of motion in the three-dimensional space have exact particular solution. This solution is geometrically represented by the equilateral triangle rotating about the center of mass of the system G. Obviously, in the case of $m_2 = 0$ the center of mass G is situated on the side P_0P_1 of the triangle that connects masses $1 - \mu, \mu$. Triangle $P_0P_1P_2$ rotates with constant angular velocity being equal to angular velocity of gravitating masses rotating about the point G on circular orbits according to Kepler's laws [13].

Does such triangular configuration in the restricted three-body exist that is stable in the first approximation and in Liapunov's sense? This, apparently, simple problem has been solved completely only 200 years later after its formulation.

In 1875 Routh [14] found necessary conditions of linear stability of the Lagrange triangle in general case of planar three-body problem ($m_0 \neq 0$, $m_1 \neq 0$, $m_2 \neq 0$, and a force of their mutual attraction is given by $F \sim \frac{1}{r^n}$, the case $n = 2$ corresponds to the force of Newton's attraction) in the form

$$3 \left(\frac{n+1}{n-3} \right)^2 < \frac{(m_0 + m_1 + m_2)^2}{m_0 m_1 + m_0 m_2 + m_1 m_2}, \quad n < 3.$$

For $n \geq 3$ and any m_0, m_1, m_2 the Lagrange triangle is unstable. The case $n = 2$, obviously, satisfies the Routh condition.

A.M. Liapunov studied the generalized spatial three-body problem [9] and proved correctness of the Routh conditions in this case.

Using notations $m_0 = 1 - \mu$, $m_1 = \mu$, $m_2 = 0$, $n = 2$, we can rewrite the Routh inequality for *the restricted circular three-body problem* in a simpler form [13]:

$$27\mu(1 - \mu) < 1. \tag{1}$$

Consequently, if condition (1) is satisfied, the Lagrange triangle in the restricted circular three-body problem is *stable in the first approximation*. Inequality (1) takes place for any value of μ from an interval

$$0 < \mu < \frac{1}{2}\left(1 - \sqrt{\frac{23}{27}}\right) \approx 0.0385209. \tag{2}$$

For the values μ satisfying the inequality

$$0.0385209 \approx \frac{1}{2}\left(1 - \sqrt{\frac{23}{27}}\right) < \mu \leq \frac{1}{2},$$

the Lagrange triangle is unstable in the first approximation. We can write two numbers playing a role of frequencies in linear model of this problem:

$$\begin{cases} \sigma_1 = \frac{\sqrt{2}}{2}\sqrt{1 + \sqrt{1 - 27\mu(1 - \mu)}}, \\ \sigma_2 = \frac{\sqrt{2}}{2}\sqrt{1 - \sqrt{1 - 27\mu(1 - \mu)}}. \end{cases} \tag{3}$$

It is supposed, obviously, that parameter μ in (3) satisfies inequality (2).

Arnold – Moser theorem [10] contains *"a condition of the absence of frequency resonances"*

$$k_1\sigma_1 + k_2\sigma_2 \neq 0, \tag{4}$$

where k_1, k_2 are integers satisfying the inequality $0 < |k_1| + |k_2| \leq 4$.

Here the following problem arises: *find all values of the parameter μ from the interval (2) for which the resonant equality is precisely satisfied.* This problem has two solutions [10]

$$\begin{cases} \mu_1 = \frac{45 - \sqrt{1833}}{90} \approx 0.024294 < 0.0385209, \\ \mu_2 = \frac{15 - \sqrt{213}}{30} \approx 0.013516 < 0.0385209. \end{cases} \tag{5}$$

For them

$$\begin{cases} \sigma_1^{(1)} = \frac{2}{\sqrt{5}}, & \sigma_2^{(1)} = \frac{1}{\sqrt{5}}, \\ \sigma_1^{(2)} = \frac{3}{2}\sqrt{\frac{2}{5}}, & \sigma_1^{(2)} = \frac{1}{2}\sqrt{\frac{2}{5}}. \end{cases} \tag{6}$$

One can readily see from (6) that frequencies σ_1 and σ_2 are *irrational numbers and, nevertheless, frequency resonances present,* the first for $k_1 = 1$, $k_2 = 2$, the second for $k_1 = 1$, $k_2 = 3$. In other words, for $\mu = \mu_1$ we have a frequency resonance $\sigma_1^{(1)} = 2\sigma_2^{(1)}$, and for $\mu = \mu_2$ we have a resonance $\sigma_1^{(2)} = 3\sigma_2^{(2)}$. The resonance $\sigma_1^{(1)} = 2\sigma_2^{(1)}$ *refers to as the third order resonance*, and the resonance $\sigma_1^{(2)} = 3\sigma_2^{(2)}$ is *the 4th order resonance* [10].

Except for condition (4), Arnold–Moser theorem contains also the inequality

$$c_{20}\sigma_2^2 + c_{11}\sigma_1\sigma_2 + c_{02}\sigma_1^2 \neq 0, \tag{7}$$

where c_{20}, c_{11}, c_{02} are the coefficients in the fourth-order term in the expansion of the Hamiltonian in the neighborhood of equilibrium points reduced to the normal Birkhoff form [15].

Calculations of A.P. Markeev have shown that in interval (2) there is only one value

$$\mu_3 \approx 0.01091367, \tag{8}$$

which transforms inequality (7) into equality.

Thus, all over again by means of the theorems of Arnold [5] and Deprit [15] it has been proved that the Lagrange triangle is stable not only in the first approximation, but also in Liapunov's sense at all values of μ from the interval (2), except for three specified values μ_1, μ_2, and μ_3. More precisely, A.M.Leontovich in the article [16] based on Arnold theorem [5] proved stability in Liapunov's sense for all μ satisfying inequality (2), except for some set of a zero measure theoretically generating frequency resonances.

At first A.M.Leontovich in article [16] based on the Arnold theorem [5] proved stability in Liapunov sense for all μ satisfying inequality (2), except for some set of a zero measure, theoretically generating frequency resonances. Afterwards A. Deprit showed [15] that this set consists of only three values μ_1, μ_2, and μ_3.

Research of stability of the Lagrange triangle in Liapunov sense for these three values of μ was executed by A.P. Markeev [10]. He proved that the Lagrange triangle for $\mu = \mu_1$, $\mu = \mu_2$ is unstable in Liapunov sense and stable for $\mu = \mu_3$.

In our article we investigate properties of the stability domain of stationary solutions of the restricted three-body problem using for this purpose the algorithms and methods of the Computer Algebra System *Mathematica* [12].

The differential equations determining motion of the body P of mass $m = 0$ in the steadily revolving Cartesian coordinates P_0xyz are [3,13]:

$$\begin{cases} x'' - 2y' = \frac{\partial \Omega}{\partial x}, \\ y'' + 2x' = \frac{\partial \Omega}{\partial y}, \\ z'' = \frac{\partial \Omega}{\partial z}, \end{cases} \tag{9}$$

where

$$\begin{aligned} \Omega &= \tfrac{1}{2}(x^2 + y^2) + \tfrac{1-\mu}{r_1} + \tfrac{\mu}{r_2}, \\ \mu &= \tfrac{m_2}{m_1+m_2}, \ \left(0 < \mu \leq \tfrac{1}{2}\right), \\ r_1^2 &= (x + \mu)^2 + y^2 + z^2, \\ r_2^2 &= (x - 1 + \mu)^2 + y^2 + z^2. \end{aligned}$$

The program written in the system "Mathematica" codes [12] supposes graphic representation of equilibrium points in the restricted 3-body problem for different values of parameter μ. It is simultaneously possible to obtain the coordinates of

pic[0.5]   **pic[0.01]**

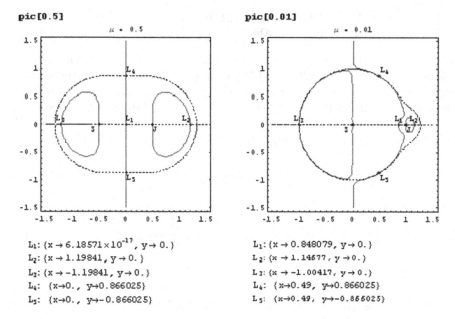

L_1: {x → 6.18571×10^{-17}, y → 0.} L_1: {x → 0.848079, y → 0.}
L_2: {x → 1.19841, y → 0.} L_2: {x → 1.14677, y → 0.}
L_3: {x → -1.19841, y → 0.} L_3: {x → -1.00417, y → 0.}
L_4: {x→0., y→0.866025} L_4: {x→0.49, y→0.866025}
L_5: {x→0., y→-0.866025} L_5: {x→0.49, y→-0.866025}

Fig. 2.

equilibrium points of the model under consideration. For example, for $\mu = 0.5$ and $\mu = 0.01$, we have obtained the results presented in fig. 2.

Let us consider a specific "triangular" stationary point L_4 (similar results are obtained for the point L_5) with coordinates:

$$L_4 = \left(\frac{1 - 2\mu}{2}, \frac{\sqrt{3}}{2}, 0 \right). \tag{10}$$

Using the system "Mathematica" [12], we solve system (9) for different values of parameter μ and on large enough interval of time in the form of interpolation functions.

Let us consider the plane restricted 3-body problem ($z = 0$), for example, for $\mu = 0.01$ when the Lagrange triangle is stable in the Liapunov sense.

Let $x(0) = x_1$, $y(0) = y_1$, $x'(0) = 0$, $y'(0) = 0$, where $x_1 = \frac{1-2\mu}{2}$, $y_1 = \frac{\sqrt{3}}{2}$.

Then solving equations (9) for $0 < t < 10000$, for example, we obtain by means of the following instruction (s11 and s22 are the right-hand sides of equations (9)):

```
r1=NDSolve[{ x"[t]-2y'[t]==s11,x[0]==x1 ,x'[0]==0,
y"[t]+2x'[t]==s22,y[0]==y1 ,y'[0]==0},{x,y},{t,0,10000}]

{{x→InterpolatingFunction[{{0.,10000.}},¿¿],
y→InterpolatingFunction[{{0.,10000.}},¿¿]}}
```

The graphs of the obtained functions have been constructed by means of the following instruction:

ParametricPlot[Evaluate[{x[t],y[t]}/.r1],{t,0,t1},
AxesLabel→{"x[t]","y[t]"},AxesOrigin→{x1,y1}],

where solutions are represented in the form of interpolation functions, which can be presented graphically for different integration intervals by replacing $t1$ with specific values (coordinate axes go through the point L_4). Figure 3 presents the graphs for intervals : $0 < t < 250$, $0 < t < 1000$, $0 < t < 5000$, and $0 < t < 10000$.

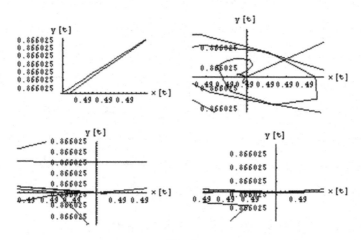

Fig. 3.

Considering scales of the coordinate axes one can easily see from Fig. 3 that the trajectory does not go far away from the initial point.

Let $\Delta r(t)$ be the local distance of point on the trajectory from the stationary point L_4 for t. Figure 4 shows the behavior of function $\Delta r(t)$ for intervals of t.

Let us now change initial conditions by perturbing a little bit the initial coordinates. Let

$$x(0) = x_1 + \alpha\text{Cos}[\varphi], \quad y(0) = y_1 + \alpha\text{Sin}[\varphi],$$
$$x'(0) = 0, \quad y'(0) = 0. \tag{11}$$

Then, as a result of numerical integration, we obtain, for example, for $\varphi = \pi/4$ and $\alpha = 0.014$ that

r2=NDSolve[{x"[t]-2y'[t]==s11,x[0]==x1+αCos[φ] ,x'[0]==0,
y"[t]+2x'[t]==s22,y[0]==y1+αSin[φ] ,y'[0]==0},{x,y},{t,0,1000}]

{{x→InterpolatingFunction[{{0.,1000.}},¿¿],
y→InterpolatingFunction[{{0.,1000.}},¿¿]}}

The graphs of obtained functions for $0 < t < 250$ and $0 < t < 1000$ and corresponding graphs of function $\Delta r(t)$ are shown in Fig. 5.

Let now $\varphi = \pi/4$ and $\alpha = 0.015$. Then, as a result of numerical integration, we obtain

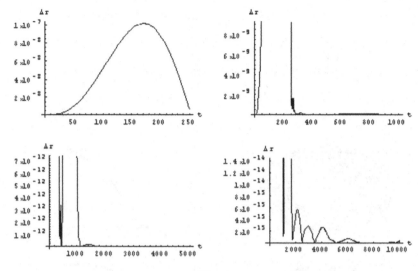

Fig. 4.

r3=NDSolve[{x″[t]-2y′[t]==s11,x[0]==x1+αCos[φ] ,x′[0]==0,

y″[t]+2x′[t]==s22,y[0]==y1+αSin[φ] ,y′[0]==0},{x,y},{t,0,700}]

{{x→InterpolatingFunction[{{0.,700.}},¡¿],
y→InterpolatingFunction[{{0.,700.}},¡¿]}}

and the corresponding graphs for $0 < t < 100$, $0 < t < 400$, and $0 < t < 700$ are presented in Fig. 6.

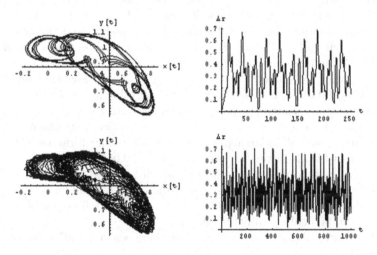

Fig. 5.

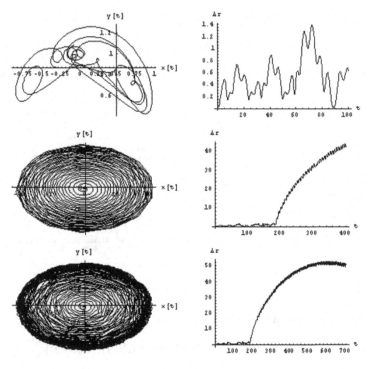

Fig. 6.

Let us compare the results presented in Figs. 4 and 5. We see that in the first case the trajectory does not abandon the neighborhood of origin during a sufficiently long period of time. In the second case, for values $\varphi = \pi/4$ and $\alpha = 0.015$, the trajectory quickly enough "leaves" the origin (Fig. 6).

Really, the value Δx for $t \geq 200$ grows quickly enough, and the graphs disposed at the left in Fig. 6, show that the value Δy also increases not only as function of Δx, but also as function of time t.

The computing experiments presented in figures 2 - 5 show that on the parameter interval $0.014 < \alpha < 0.015$, there is some value $\bar{\alpha}$, that at $\alpha < \bar{\alpha}$, the trajectory "rotates" around the position of equilibrium L_4.

It is possible to assert, on the basis of a big numerical experiment, that for specific value, $\varphi = \pi/4$, changing initial conditions under formulas (11) up to some value α, which is in the interval $0.014 < \alpha < 0.015$, (we shall designate it through α_{max})), we obtain the trajectory, which "rotates" around the position of equilibrium L_4 and does not move "far" from point L_4, and for $\alpha > \alpha_{max}$ the trajectory leaves from a point L_4. The angle φ shows the direction of an initial vector-position in local coordinate system.

In Table 1 intervals are presented in which there is a "critical" value α_{max} for other φ values.

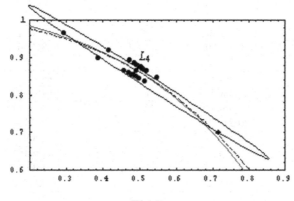

Fig. 7.

These results allow us to present the domain of attraction (domain of stability) of point L_4 (Fig. 7). Calculations show that the "real" domain of stability of point L_4 for $\mu = 0.01$ is the figure being similar to ellipse. It is prolate along a trajectory of point L_4 in a fixed barycentric coordinate system.

From Table 1 we see also that for $\alpha \cong 0.013$ we have stability, and the domain of stability is a circle with such radius.

Table 1.

φ	α_{max}
0	$0.027 < \alpha_{max} < 0.028$
$\pi/10$	$0.018 < \alpha_{max} < 0.019$
$2\pi/10$	$0.015 < \alpha_{max} < 0.016$
$3\pi/10$	$0.013 < \alpha_{max} < 0.014$
$4\pi/10$	$0.014 < \alpha_{max} < 0.015$
$5\pi/10$	$0.016 < \alpha_{max} < 0.017$
$6\pi/10$	$0.020 < \alpha_{max} < 0.021$
$7\pi/10,$	$0.033 < \alpha_{max} < 0.034$
$8\pi/10$	$0.093 < \alpha_{max} < 0.094$
$9\pi/10$	$0.11 < \alpha_{max} < 0.12$
π	$0.033 < \alpha_{max} < 0.034$
$11\pi/10$	$0.021 < \alpha_{max} < 0.022$
$12\pi/10$	$0.017 < \alpha_{max} < 0.018$
$13\pi/10$	$0.015 < \alpha_{max} < 0.016$
$14\pi/10$	$0.015 < \alpha_{max} < 0.016$
$15\pi/10$	$0.017 < \alpha_{max} < 0.018$
$16\pi/10$	$0.022 < \alpha_{max} < 0.023$
$17\pi/10$	$0.037 < \alpha_{max} < 0.038$
$18\pi/10$	$0.28 < \alpha_{max} < 0.29$
$19\pi/10$	$0.059 < \alpha_{max} < 0.060$
$17\pi/20$	$0.22 < \alpha_{max} < 0.23$

Table 2.

φ	$\Delta r(t)_{max}$	φ	$\Delta r(t)_{max}$
0	0.234896	$\pi/10$	0.379857
$2\pi/10$	0.519831	$3\pi/10$	0.607559
$4\pi/10$	0.586242	$5\pi/10$	0.474076
$6\pi/10$	0.330625	$7\pi/10$	0.189032
$8\pi/10$	0.0525061	$9\pi/10$	0.0868816
π	0.216706	$11\pi/10$	0.33713
$12\pi/10$	0.435889	$13\pi/10$	0.506919
$14\pi/10$	0.497495	$15\pi/10$	0.420164
$16\pi/10$	0.306633	$17\pi/10$	0.179645
$18\pi/10$	0.0475908	$19\pi/10$	0.0942737

Let us denote by $\Delta r(t)$ the local distance of a point on a trajectory from the stationary point L_4 for t. We shall calculate the maximal distance, on which the trajectory for $\alpha = 0.013$ and for different values of parameter φ escapes. These results are presented in table 2. They are obtained for interval of time $0 < t < 1000$.

We see from Table 2 that the maximal distance for $\varphi \in (0, 2\pi)$, is $\Delta r = 0.607559$. It means that if we shall change initial conditions under formulas (11) for $\alpha = 0.013$, the trajectory will not leave essentially the considered position of equilibrium L_4, and the local distance of a point on a trajectory from the stationary point L_4 will not be more than 0.607559 (Fig. 8)

In case of $\mu_3 \approx 0.01091367$, when the following equality takes place

$$c_{20}\sigma_2^2 + c_{11}\sigma_1\sigma_2 + c_{02}\sigma_1^2 0 = 0,$$

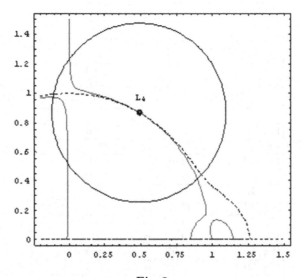

Fig. 8.

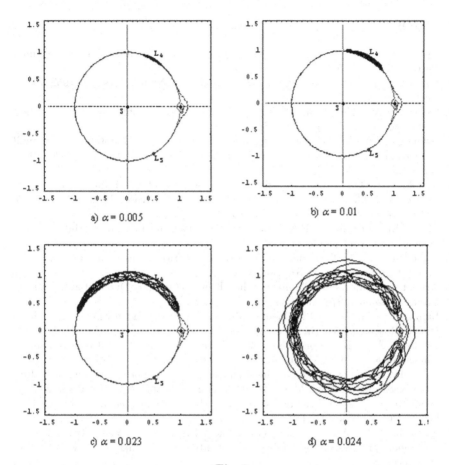

Fig. 9.

inverse to condition of Arnold–Moser (7), our results essentially do not differ from results for value $\mu = 0.01$.

Let us consider also the plane restricted 3-body problem for $\mu = 0.001$. We shall change initial conditions under formulas (11), for example, for $\varphi = 0$. Besides we shall represent in figures the results of direct numerical integration of equations (9). The following graphs are obtained for different values of parameter α.

it can be seen from Figure 9 that for $\alpha = 0.005$ the trajectory "rotates" around the position of equilibrium L_4 and does not leave it essentially. For $\alpha = 0.01$ and $\alpha = 0.023$, the trajectory "rotates" around the position of equilibrium L_4 as well, but together with the growth of α begins more and more to move away from it.

For $\alpha = 0.023$, the trajectory "flees" from point L_4.

Thus, the computing procedures implemented in system "Mathematica" allow us to draw some qualitative conclusions on estimations of the sizes and forms of domains of stability on large enough intervals of time (of the order of thousand turns of the Lagrange triangle in a non-rotating Cartesian coordinate system).

References

1. Euler, L.: De motu rectilineo trium corporum se mutuo attrahentium. Nov. Comm. Petrop. 11, 144 (1765)
2. Lagrange, G.: The Analytical Mechanics. vol. 1–2. Nauka, Moscow (1952–1953), (in Russian)
3. Szebehely, V.: Theory of Orbits. In: The Restricted Problem of Three Bodies, Academic Press, New York, London (1967)
4. Kolmogorov, A.N.: On the conservation of quasi-periodic motions for a small change in the Hamiltonian function. Dokl. Akad. Nauk USSR 98(4), 527–530 (1954) (in Russian)
5. Arnold, V.I.: On stability of equilibrium positions of the Hamiltonian system of ordinary differential equations in general elliptic case. Dokl. Akad. Nauk USSR 137(2), 255–257 (1961) (in Russian)
6. Moser, J.K.: Lectures on Hamiltonian Systems. Courant Institute of Mathematical Science, New York (1968)
7. Samoilenko, A.M.: To a question on structure of trajectories on a tour. Ukrainian Math. J. 16(6) (1964) (in Russian)
8. Arnold, V.I.: Additional Chapters of the Theory of Ordinary Differential Equations. Nauka, Moscow (1978) (in Russian)
9. Liapunov, A.M.: General Problem of Motion Stability. Academy of Sciences of the USSR, Moscow (1954) (in Russian)
10. Markeev, A.P.: Libration Points in Celestial Mechanics and Cosmodynamics. Nauka, Moscow (1978) (in Russian)
11. Sokol'sky, A.G.: On stability of the Hamiltonian autonomous system with the first order resonance. J. Appl. Math. & Mech. 39(2), 366–369 (1975) (in Russian)
12. Wolfram, S.: The Mathematica Book, 4th edn. Wolfram Media/Cambridge University Press, Cambridge (1999)
13. Abalakin, V.K., Aksenov, E.P., Grebenikov, E.A., Demin, V.G., Ryabov, Y.A.: Handbook on Celestial Mechanics and Astrodynamics. Nauka, Moscow (1976) (in Russian)
14. Routh, E.I.: Proceedings of the London Mathematical Society 6 (1875)
15. Deprit, A.: Limiting orbits around the equilateral centers of libration. Astron. J. 71(2), 77–87 (1966)
16. Leontovich, A.M.: On stability of Lagrange periodic solutions in the restricted three-body problem. Dokl. Akad. Nauk USSR 143(3), 525–528 (1962) (in Russian)
17. Grebenicov, E.A., Kozak-Skovorodkin, D., Diarova, D.M.: Numerical study of stability domains of Hamiltonian equation solutions. In: Ganzha, V.G., Mayr, E.W., Vorozhtsov, E.V. (eds.) CASC 2006. LNCS, vol. 4194, pp. 178–191. Springer, Heidelberg (2006)

Studying the Stability of the Second Order Non-autonomous Hamiltonian System

Evgenii A. Grebenikov[1], Ersain V. Ikhsanov[2], and Alexander N. Prokopenya[3]

[1] Dorodnicyn Computing Center of RAS
Vavilova str. 40, 119991 Moscow, Russia
`greben@ccas.ru`
[2] Engineer-Humane Institute
Atyrau, Kazakhstan
`unatatyrau@nursat.kz`
[3] Brest State Technical University
Moskowskaya str. 267, 224017 Brest, Belarus
`prokopenya@brest.by`

Abstract. The problem of studying the stability of equilibrium solution of the second order non-autonomous Hamiltonian system, containing a small parameter, is considered. The main steps in solving this problem and application of the computer algebra systems for doing necessary calculations are discussed. As an example, we analyze stability of some equilibrium solutions in the elliptic restricted $(2n + 1)$-body problem. The problem is solved in a strict nonlinear formulation. All calculations are done with the computer algebra system Mathematica.

1 Introduction

The theory of stability of the Hamiltonian systems gives quite general methods for solving many problems of motion stability [1]. But usually application of this theory requires quite cumbersome analytical calculations and so the stability problem can be totally solved only with a computer and modern software. Besides, in every concrete case one needs both to adopt already known algorithms of calculations and to develop some new ones.

A typical example is the restricted $(n+1)$-body problem [2] when a body P_n of infinitesimal mass moves in the gravitational field generated by the system of n massive bodies $P_0, \ldots, P_{n-1}$ whose motion is determined by some exact solution of the corresponding n-body problem. Although differential equations of motion of the body P_n can be written in the Hamiltonian form, they are essentially nonlinear and are not integrable in general. But some equilibrium solutions of these equations can be usually found and the problem is to investigate their stability. It should be emphasized that this problem is very complicated and can be solved only in a strict nonlinear formulation on the basis of the KAM-theory [3,4,5]. Application of this theory implies construction of a sequence of the Birkhoff canonical transformations reducing the Hamiltonian function to the normal form. In the most cases this can be done only with some modern computer

V.G. Ganzha, E.W. Mayr, and E.V. Vorozhtsov (Eds.): CASC 2007, LNCS 4770, pp. 181–194, 2007.

algebra system. The case of planar circular restricted four-body problem [6], for example, when perturbed motion of the body P_3 is described by the fourth order autonomous Hamiltonian system, was analyzed in detail only owing to using the system *Mathematica* [7].

If the system considered is non-autonomous the stability problem becomes much more complicated, even in the simplest case of the second order Hamiltonian system. In the present paper we investigate stability of some equilibrium solutions in the elliptic restricted $(2n+1)$-body problem. In section 2 we obtain equations of the perturbed motion in the form of the Hamiltonian system of two differential equations. Section 3 is devoted to studying the stability of equilibrium solution in linear approximation. In sections 4 and 5 we construct the canonical transformation reducing the Hamiltonian function to its normal form. In section 6 we analyze the influence of the fourth order resonance on stability of the equilibrium solution. And we conclude the paper in the last section. Note that all symbolic and numerical calculations are done with the computer algebra system *Mathematica* [7].

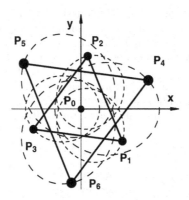

Fig. 1. Geometrical configuration of the system for $n = 3$

2   Equations of the Perturbed Motion

Let us consider the system of $2n$ bodies moving in the Oxy plane of the barycentric inertial frame on similar elliptic orbits about their common center of mass (see Fig. 1). The bodies $P_1, \ldots, P_n$ having equal masses m_1 and $P_{n+1}, \ldots, P_{2n}$ having equal masses m_2 at any instant of time form two regular concentric polygons with n sides. The polygons $P_1 P_2 \ldots P_n$ and $P_{n+1} P_{n+2} \ldots P_{2n}$ may be homothetic when the bodies P_k, P_{n+k} are situated on the same half-line $P_0 P_{n+k}$ or polygon $P_{n+1} P_{n+2} \ldots P_{2n}$ is rotated with respect to $P_1 P_2 \ldots P_n$ about Oz axis at the angle π/n (this case is shown in Fig. 1 for $n = 3$). Existence of the corresponding solutions in the $2n$-body problem was proved in [8].

Using cylindrical coordinates (r, φ, z), we can write equations determining the corresponding trajectories of the bodies in the form

$$r_j = \frac{p}{1 + e \cos \nu} \ , \quad r_{n+j} = \frac{pR}{1 + e \cos \nu} \ , \quad \frac{d\nu}{dt} = w(1 + e \cos \nu)^2 \ ,$$

$$\varphi_j = \nu + \frac{2\pi}{n} j \ , \quad \varphi_{n+j} = \nu + \frac{2\pi}{n} j + \beta \ , \quad z_j = z_{n+j} = 0 \ , \quad (j = 1, \ldots, n) \quad (1)$$

where p, pR, e are parameters and eccentricity of elliptic orbits of the bodies, respectively, ν is a true anomaly, $\beta = 0, \frac{\pi}{n}$, and parameter w is given by

$$w^2 = \frac{Gm_1}{p^3} (S_1 + \mu f(R, n, \beta)) \ . \tag{2}$$

Parameter μ in (2) gives a ratio of masses $\mu = m_2/m_1$, G is a gravity constant,

$$S_1 = \sum_{k=1}^{n-1} \left(\sin \left(\frac{\pi k}{n} \right) \right)^{-1} \ ,$$

$$f(R, n, \beta) = \sum_{k=1}^{n} \frac{1 - R \cos \left(\frac{2\pi k}{n} - \beta \right)}{\left(1 + R^2 - 2R \cos \left(\frac{2\pi k}{n} - \beta \right) \right)^{3/2}} \ ,$$

and parameter R is determined as a root of the equation

$$\frac{1}{4} S_1(\mu - R^3) + f\left(\frac{1}{R}, n, \beta \right) - \mu R^3 f(R, n, \beta) = 0 \ . \tag{3}$$

Let us consider motion of the body P_0 of infinitesimal mass in the gravitational field generated by the bodies $P_1, \ldots, P_{2n}$. Obviously, the origin is an equilibrium point of the body P_0 and under the appropriate initial conditions it will move only along the Oz axis that is just an axis of symmetry of the system. Takin into account (1) and using true anomaly ν as an independent variable, we can write equations of motion in the Hamiltonian form

$$\frac{dz}{d\nu} = \frac{\partial H}{\partial p_z} \ , \quad \frac{dp_z}{d\nu} = -\frac{\partial H}{\partial z} \ , \tag{4}$$

where the Hamiltonian function H is given by

$$H = \frac{p_z^2}{2} + \frac{1}{1 + e \cos \nu} \left(\frac{e \cos \nu}{2} z^2 - \frac{Gm_1 n}{w^2 p^3} \left(\frac{1}{\sqrt{1 + z^2}} + \frac{\mu}{\sqrt{R^2 + z^2}} \right) \right) \ . \tag{5}$$

One can readily verify that equations (4) have a solution $z = p_z = 0$. The Hamiltonian (5) is analytic function with respect to coordinate z and momentum p_z and in the neighborhood of the equilibrium solution it can be represented in the form

$$H = H_2 + H_4 + \ldots \ , \tag{6}$$

where H_k is the kth order homogeneous polynomial with respect to z and p_z and its coefficients are 2π-periodic functions of the independent variable ν. The quadratic form H_2 and the fourth order form H_4 are given by

$$H_2 = \frac{p_z^2}{2} + \frac{a + e\cos\nu}{1 + e\cos\nu}\frac{z^2}{2} \ , \quad H_4 = -\frac{3bz^4}{8(1 + e\cos\nu)} \ , \tag{7}$$

where

$$a = \frac{Gm_1n}{\omega^2 p^3}\left(1 + \frac{\mu}{R^3}\right) \ , \quad b = \frac{Gm_1n}{\omega^2 p^3}\left(1 + \frac{\mu}{R^5}\right) \ . \tag{8}$$

Thus, in order to investigate stability of the equilibrium solution we have to normalize the terms $H_2, H_4, \ldots$ in (6) successively and to apply some general theorems on stability of solutions of the Hamiltonian systems (see [1]).

3   Stability of the Linearized System

Stability analysis of any Hamiltonian system is started from studying the linearized equations of the perturbed motion. They are obtained from (4) if only quadratic part H_2 of the Hamiltonian (6) is taken into account and can be written in the form

$$\frac{dz}{d\nu} = p_z \ , \quad \frac{dp_z}{d\nu} = -\frac{a + e\cos\nu}{1 + e\cos\nu}z \ . \tag{9}$$

The system (9) is reduced to the Hill equation that was investigated in detail in [9]. It was shown there that domains of instability of its equilibrium solution exist only in the neighborhood of the points $a = (2k-1)^2/4$ $(k = 1, 2, \ldots)$ in the plane of parameters Oea. Note that equation (3), determining possible values of R for any $\mu \geq 0$, $n \geq 2$, can be solved only numerically. But using the system *Mathematica*, we can find R with arbitrary precision. Taking into account (2), (3), (8), we can easily visualize the function $a = a(\mu)$ for any n. In the case of $n = 3$, $\beta = \pi/3$, for example, the corresponding dependence $a = a(\mu)$ is shown in Fig. 2. As for small values of the parameter μ equation (3) has three roots, three corresponding values of a exist.

Fig. 2 shows that there are only three values of the parameter μ for which the condition $a = (2k-1)^2/4$ $(k = 5, 7, 9)$ is fulfilled. Note that boundaries $a = a(e)$

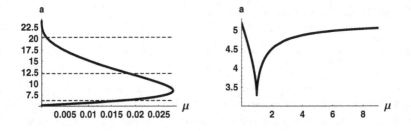

Fig. 2. Dependence $a = a(\mu)$ for $n = 3$, $\beta = \pi/3$

of the corresponding domains of instability were found in [9] in the form of power series in e. Using (2), (3), (8), we can find the corresponding curves $\mu = \mu(e)$ in the $Oe\mu$ plane. They are

$$\mu = 0.0174023 - 0.0242406e^2 - 0.0240934e^4 \pm 2.95904 \times 10^{-6}e^5 - $$
$$-0.0159875e^6 \pm 7.82116 \times 10^{-6}e^7 - 0.00894437e^8 \pm 1.09707 \times 10^{-5}e^9 , \quad (10)$$

$$\mu = 0.0187383 + 0.0141168e^2 + 0.00408211e^4 - 0.000495739e^6 \pm$$
$$\pm 1.34629 \times 10^{-8}e^7 - 0.002313e^8 \pm 2.50386 \times 10^{-8}e^9 , \quad (11)$$

$$\mu = 0.00107148 + 0.00585103e^2 + 0.0109524e^4 + 0.0078372e^6 +$$
$$+0.00247988e^8 \pm 5.67624 \times 10^{-11}e^9 . \quad (12)$$

Expressions (10), (11), (12) show that a bandwidth of the domain of instability in the neighborhood of the point $a = (2k - 1)^2/4$ $(k = 5, 7, 9)$ is $O(e^k)$ and decreases very fast if the number k is growing up. As coefficients of the corresponding terms e^k in (10), (11), (12) are very small we can conclude that equilibrium solution of equations (4) is stable in linear approximation for $n = 3$, $\beta = \pi/3$ for almost all values of the mass parameter μ. The same conclusion is true in the case of $n = 3$, $\beta = 0$.

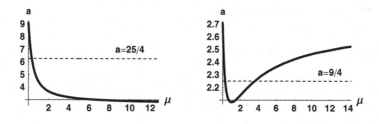

Fig. 3. Dependence $a = a(\mu)$ for $n = 9$, $\beta = \pi/3$

Note that parameter a decreases if the number n is growing up. For $n = 9$, $\beta = \pi/3$, for example, possible values of a are shown in Fig. 3. Now in addition to the domain of instability in the neighborhood of the point $a = 25/4$ two additional domains arise in the neighborhood of the point $a = 9/4$, their boundaries are given in (13)–(15). These domains are a little bit larger, their bandwidth is $O(e^3)$ but it remains quite small because of smallness of the corresponding coefficients of e^3.

$$\mu = 0.332199 + 0.385372e^2 + 0.301524e^4 \pm 0.000047043e^5 +$$
$$+0.27319e^6 \pm 0.00010444e^7 + 0.25835e^8 \pm 0.00016834e^9 , \quad (13)$$

$$\mu = 0.277417 + 0.679105e^2 \pm 0.028296e^3 + 1.18549e^4 \pm 0.0992603e^5 +$$
$$+2.47899e^6 \pm 0.309623e^7 + 5.78482e^8 \pm 0.95982e^9 , \quad (14)$$

$$\mu = 3.60468 - 8.82408e^2 \mp 0.36767e^3 + 6.19711e^4 \pm 0.510322e^5 -$$
$$-9.63604e^6 \mp 1.17605e^7 + 0.871407e^8 \pm 0.126048e^9 , \tag{15}$$

With further growth of the number n the interval of parameter a changing decreases. Numerical analysis of the expressions (2), (3), (8) shows that for $900 < n < 50000$ it belongs to the interval $\frac{1}{4} < a < \frac{9}{4}$ and, hence, there are not any domains of instability of the equilibrium solution, at least, if parameter e is sufficiently small. This conclusion is true in both cases $\beta = 0$ and $\beta = \pi/n$. Further we'll consider only such values of parameters of the system for which the equilibrium solution is linearly stable.

4   Normalization of the Quadratic Part of the Hamiltonian

It is well-known (see [1]) that equilibrium solution of the Hamiltonian system of differential equations, being stable in linear approximation, may become unstable if nonlinear terms in the expansion (6) are taken into account. In order to investigate its stability in a strict nonlinear formulation we must construct the Birkhoff canonical transformation [10] reducing the Hamiltonian function (6) to its normal form.

Let us start from normalizing the quadratic part H_2 of the Hamiltonian. One can readily see that H_2 is an analytic function of parameter e in the domain $|e| < 1$ and can be represented in the form

$$H_2 = \frac{1}{2}(p_z^2 + az^2) + \frac{z^2}{2}(a - 1)\sum_{k=1}^{\infty}(-e\cos\nu)^k . \tag{16}$$

Hence, for its normalization we can apply an algorithm developed in [11]. Doing the corresponding symbolic calculations with the system *Mathematica*, we construct the canonical transformation

$$z \to Z_{11}q + Z_{12}p , \quad p_z \to Z_{21}q + Z_{22}p , \tag{17}$$

where the coefficients $Z_{11}, Z_{12}, Z_{21}, Z_{22}$ are represented in the form of power series in e and are given by

$$Z_{11} = \frac{1}{\sqrt{a}} + \frac{(a-1)e\cos\nu}{\sqrt{a}(4a-1)} + \frac{\sqrt{a}e^2(10 - 2a - 8a^2 + (5 - 22a + 8a^2)\cos(2\nu))}{8(4a-1)^2} +$$
$$+ \frac{\sqrt{a}(a-1)e^3}{16(4a-1)^3(4a-9)}((-441 + 52a - 80a^2 + 64a^3)\cos\nu -$$
$$-(4a-1)^2(4a-11)\cos(3\nu)) + \frac{\sqrt{a}e^4}{768(4a-1)^4(4a-9)}(-11664 + 783a +$$
$$+37299a^2 - 133968a^3 + 148896a^4 - 44544a^5 + 768a^6 - 8(804 - 6563a +$$
$$+21067a^2 - 36338a^3 + 23816a^4 - 5344a^5 + 128a^6)\cos(2\nu) +$$
$$+(4a-1)^3(114 - 149a + 31a^2 + 4a^3)\cos(4\nu)) ,$$

$$Z_{12} = -\frac{2(a-1)e\sin\nu}{4a-1} + \frac{(a+2)e^2\sin(2\nu)}{8(4a-1)} - \frac{(a-1)e^3\sin\nu}{24(4a-1)^3(4a-9)}(-234 +$$
$$+205a - 2580a^2 + 1296a^3 - 64a^4 + (4a-1)^2(18 - 13a + 4a^2)\cos(2\nu)) -$$
$$-\frac{e^4\sin\nu}{384(4a-1)^3(4a-9)}((1656 - 1493a + 233a^2 + 11796a^3 - 6160a^4 +$$
$$+448a^5)\cos\nu + (4a-1)^2(72 - 157a + 113a^2 - 28a^3)\cos(3\nu)) \; ,$$

$$Z_{21} = \frac{(1 - 3a + 2a^2)e\sin\nu}{\sqrt{a}(4a-1)} + \frac{\sqrt{a}(5a-8)e^2\sin(2\nu)}{8(4a-1)} +$$
$$+\frac{\sqrt{a}(a-1)e^3}{48(4a-1)^3(4a-9)}((189 + 3939a - 9348a^2 + 3792a^3 - 192a^4)\sin\nu -$$
$$-(4a-1)^2(81 - 23a - 4a^2)\sin(3\nu)) + \frac{\sqrt{a}e^4\sin(2\nu)}{384(4a-1)^3(4a-9)}(-4992 +$$
$$+23192a - 60308a^2 + 53460a^3 - 15296a^4 + 704a^5 -$$
$$-(4a-1)^2(384 - 439a + 11a^2 + 44a^3)\cos(2\nu)) \; ,$$

$$Z_{22} = 1 - \frac{(a-1)e\cos\nu}{4a-1} + \frac{e^2}{8(4a-1)^2}(2(a-1)^2(3-4a) - (4 - 19a +$$
$$+14a^2 - 8a^3)\cos(2\nu)) + \frac{(a-1)e^3\cos\nu}{8(4a-1)^3(4a-9)}(36 - 105a + 860a^2 - 640a^3 +$$
$$+128a^4 - 2(3 - 14a + 8a^2)^2\cos(2\nu)) + \frac{e^4}{768(4a-1)^4(4a-9)}(-4536 +$$
$$+24372a - 133245a^2 + 361011a^3 - 456528a^4 + 259488a^5 - 53760a^6 +$$
$$+768a^7 + 8(396 - 2983a + 10738a^2 - 27839a^3 + 34386a^4 - 22600a^5 +$$
$$+5600a^6 - 128a^7)\cos(2\nu) - (4a-1)^3(288 - 742a + 601a^2 -$$
$$-143a^3 - 4a^3)\cos(4\nu)) \; . \tag{18}$$

Note that transformation (17) exists only if the condition $a \neq (2k-1)^2/4$ ($k = 1, 2, \ldots$) is fulfilled. This just corresponds to the considered case of linearly stable equilibrium solution. And normal form obtained for the quadratic part H_2 of the Hamiltonian is

$$\tilde{H}_2 = \frac{\omega}{2}(p^2 + q^2) \; , \tag{19}$$

where

$$\omega = \sqrt{a}\left(1 + \frac{3(a-1)e^2}{4(4a-1)} - \frac{3(14 - 69a + 195a^2 - 140a^3)e^4}{64(4a-1)^3}\right) \; . \tag{20}$$

5   Normalization of the Fourth Order Term

On substituting (17), (18) into (7) we can rewrite the fourth order term H_4 in the form

$$\tilde{H}_4 = \sum_{j=0}^{4} \tilde{h}_j^{(4)} q^{4-j} p^j \ . \tag{21}$$

Coefficients $\tilde{h}_j^{(4)}$ in (21) are periodic functions of ν and are given by

$$\tilde{h}_0 = -\frac{3b}{8a^{3/2}} + \frac{9be\cos\nu}{8a^{3/2}(4a-1)} - \frac{3be^2}{32a^{3/2}(4a-1)^2}(9 + 5a + 20a^2 - 16a^3 +$$

$$+2(3 + 5a - 16a^2 + 8a^3)\cos(2\nu)) - \frac{9be^3}{32a^{3/2}(4a-1)^3(4a-9)}(2(18 + 73a - 45a^2 +$$

$$+31a^3 - 48a^4 + 16a^5)\cos\nu + (3 + 25a - 78a^2 - 14a^3 + 96a^4 - 32a^5)\cos(3\nu)) +$$

$$+\frac{be^4}{512a^{3/2}(4a-1)^4(4a-9)}(3(1026 + 7833a + 7935a^2 - 39517a^3 + 90220a^4 -$$

$$-85296a^5 + 28864a^6 - 2560a^7) + 8(243 + 3276a - 8423a^2 + 23056a^3 - 43112a^4 +$$

$$+35648a^5 - 11968a^6 + 1280a^7)\cos(2\nu) + a(666 + 1036a - 17189a^2 + 28588a^3 -$$

$$-16048a^4 + 9152a^5 - 2560a^6)\cos(4\nu)) \ ,$$

$$\tilde{h}_1 = \frac{3(a-1)be\sin\nu}{a(4a-1)} + \frac{9be^2\sin(2\nu)}{16a(4a-1)^2}(6 - 5a - 4a^2) +$$

$$+\frac{be^3\sin\nu}{16a(4a-1)^3(4a-9)}(882 - 1375a + 4936a^2 - 8592a^3 + 4384a^4 -$$

$$-640a^5 + (306 + 517a - 4252a^2 + 5424a^3 - 3040a^4 + 640a^5)\cos(2\nu)) -$$

$$-\frac{3be^4\sin(2\nu)}{512a(4a-1)^4(4a-9)}(96(75 + 15a + 119a^2 + 118a^3 - 1134a^4 + 832a^5 -$$

$$-160a^6) + 2(72 + 2293a - 7627a^2 - 7404a^3 + 35216a^4 - 31040a^5 + 7680a^6)\cos(2\nu)) \ ,$$

$$\tilde{h}_2 = -\frac{9(a-1)^2be^2\sin^2\nu}{(4a-1)^2\sqrt{a}} + \frac{27(2 - 3a - 3a^2 + 4a^3)be^3\cos\nu\sin^2\nu}{4\sqrt{a}(4a-1)^3} -$$

$$-\frac{3be^4\sin^2\nu}{128\sqrt{a}(4a-1)^4(4a-9)}(-7308 + 24772a - 86939a^2 + 173620a^3 -$$

$$-149584a^4 + 48320a^5 - 4096a^6 - (684 + 4060a - 29957a^2 + 50860a^3 -$$

$$-39088a^4 + 18752a^5 - 4096a^6)\cos(2\nu)) \ ,$$

$$\tilde{h}_3 = \frac{12(a-1)^3be^3\sin^3\nu}{(4a-1)^3} + \frac{9(a-1)^2be^4\cos\nu\sin^3\nu}{2(4a-1)^4}(2 + a - 12a^2) \ ,$$

$$\tilde{h}_4 = -\frac{6\sqrt{a}(a-1)^4 be^4 \sin^4 \nu}{(4a-1)^4} . \tag{22}$$

The second step in normalization of the Hamiltonian is to find such canonical transformation that reduces the fourth order term H_4 to its the simplest form. Following to Birkhoff [10], we seek the corresponding generating function in the form of the fourth order polynomial

$$S(\tilde{p}, q, \nu) = \tilde{p}q + \sum_{j=0}^{4} s_j(\nu) q^{4-j} \tilde{p}^j , \tag{23}$$

where coefficients $s_j(\nu)$ are 2π-periodic functions. Then new canonical variables $\tilde{q}, \tilde{p}$ and old ones q, p are connected by the following relationships

$$\tilde{q} = \frac{\partial S}{\partial \tilde{p}} = q + \sum_{j=1}^{4} j s_j(\nu) q^{4-j} \tilde{p}^{j-1} , \quad p = \frac{\partial S}{\partial q} = \tilde{p} + \sum_{j=0}^{3} (4-j) s_j(\nu) q^{3-j} \tilde{p}^j . \tag{24}$$

Obviously, relationships (24) are just the equations with respect to old canonical variables q, p which are analytic functions of new canonical variables $\tilde{q}$, $\tilde{p}$ in the neighborhood of the point $\tilde{q} = \tilde{p} = 0$. Solving (24), we obtain

$$q = \tilde{q} - \sum_{j=1}^{4} j s_j(\nu) \tilde{q}^{4-j} \tilde{p}^{j-1}, \quad p = \tilde{p} + \sum_{j=0}^{3} (4-j) s_j(\nu) \tilde{q}^{3-j} \tilde{p}^j . \tag{25}$$

Then new Hamiltonian function $\hat{H}(\tilde{q}, \tilde{p}, \nu)$ is determined as

$$\hat{H}(\tilde{q}, \tilde{p}, \nu) = H(q, p, \nu) + \frac{\partial S(\tilde{p}, q, \nu)}{\partial \nu} , \tag{26}$$

where p, q in the right-hand side must be replaced by the corresponding expressions (25). As a result, we obtain

$$\hat{H} = \frac{\omega}{2}(\tilde{p}^2 + \tilde{q}^2) + \left(\frac{ds_0}{d\nu} - \omega s_1 + \tilde{h}_0\right) \tilde{q}^4 + \left(\frac{ds_1}{d\nu} + 4\omega s_0 - 2\omega s_2 + \tilde{h}_1\right) \tilde{q}^3 \tilde{p} +$$

$$+ \left(\frac{ds_2}{d\nu} + 3\omega s_1 - 3\omega s_3 + \tilde{h}_2\right) \tilde{q}^2 \tilde{p}^2 + \left(\frac{ds_3}{d\nu} + 2\omega s_2 - 4\omega s_4 + \tilde{h}_3\right) \tilde{q} \tilde{p}^3 +$$

$$+ \left(\frac{ds_4}{d\nu} + \omega s_3 + \tilde{h}_4\right) \tilde{p}^4 + O((\tilde{p}^2 + \tilde{q}^2)^{5/2}) . \tag{27}$$

Remind that coefficients $s_j(\nu)$ in (23) are 2π-periodic functions and they should be chosen in such a way that the Hamiltonian (27) takes the simplest form. Obviously, they can be represented as the following Fourier series

$$s_j(\nu) = b_0^{(j)} + \sum_{k=1}^{\infty} \left(b_{1k}^{(j)} \cos(k\nu) + b_{2k}^{(j)} \sin(k\nu)\right) . \tag{28}$$

As coefficients of $\tilde{q}^{4-j}\tilde{p}^j$ $(j = 0, \ldots, 4)$ in (27) are linear functions of $s_j(\nu)$, we can analyze each harmonic in the series (28) separately. First of all, one can readily show that we can choose constant terms $b_0^{(j)}$ in such a way that coefficients of $\tilde{q}^3\tilde{p}$, $\tilde{q}\tilde{p}^3$ become equal to zero while coefficients of $\tilde{q}^4$, $\tilde{p}^4$, $\tilde{q}^2\tilde{p}^2$ are equal to $c/4$, $c/4$, $c/2$, respectively, where

$$c = -\frac{9b}{16a^{3/2}} - \frac{27(3 + 7a - 4a^2)be^2}{64a^{3/2}(4a - 1)^2} -$$

$$- \frac{27be^4}{1024a^{3/2}(4a - 1)^4}(38 + 479a - 27a^2 + 472a^3 - 368a^4) . \qquad (29)$$

On substituting (28) into (27) and equating coefficients of $\cos(k\nu)$, $\sin(k\nu)$ to zero, we obtain the following linear system of equations

$$kb_{2k}^{(0)} - \omega b_{1k}^{(1)} + h_{1k}^{(0)} = 0 ,$$
$$-kb_{1k}^{(0)} - \omega b_{2k}^{(1)} + h_{2k}^{(0)} = 0 ,$$
$$4\omega b_{1k}^{(0)} - 2\omega b_{1k}^{(2)} + kb_{2k}^{(1)} + h_{1k}^{(1)} = 0 ,$$
$$-kb_{1k}^{(1)} + 4\omega b_{2k}^{(0)} - 2\omega b_{2k}^{(2)} + h_{2k}^{(1)} = 0 ,$$
$$3\omega b_{1k}^{(1)} - 3\omega b_{1k}^{(3)} + kb_{2k}^{(2)} + h_{1k}^{(2)} = 0 ,$$
$$-kb_{1k}^{(2)} + 3\omega b_{2k}^{(1)} - 3\omega b_{2k}^{(3)} + h_{2k}^{(2)} = 0 ,$$
$$2\omega b_{1k}^{(2)} - 4\omega b_{1k}^{(4)} + kb_{2k}^{(3)} + h_{1k}^{(3)} = 0 ,$$
$$-kb_{1k}^{(3)} + 2\omega b_{2k}^{(2)} - 4\omega b_{2k}^{(4)} + h_{2k}^{(3)} = 0 ,$$
$$\omega b_{1k}^{(3)} + kb_{2k}^{(4)} + h_{1k}^{(4)} = 0 ,$$
$$-kb_{1k}^{(4)} + \omega b_{2k}^{(3)} + h_{2k}^{(4)} = 0 , \qquad (30)$$

where $h_{1k}^{(j)}, h_{2k}^{(j)}$ are coefficients of $\cos(k\nu)$, $\sin(k\nu)$ which are obtained if we rewrite (22) in the form

$$\tilde{h}_j^{(4)} = h_0^{(j)} + \sum_{k=1}^{\infty} \left(h_{1k}^{(j)} \cos(k\nu) + h_{2k}^{(j)} \sin(k\nu) \right) .$$

The system (30) determines coefficients $b_{1k}^{(j)}, b_{2k}^{(j)}$ of the expansion (28). Its determinant is equal to

$$D = k^2(k - 4\omega)^2(k - 2\omega)^2(k + 2\omega)^2(k + 4\omega)^2 . \qquad (31)$$

Obviously, determinant (31) is not equal to zero if the conditions

$$2\omega \neq N, \quad 4\omega \neq N, \quad (N = 1, 2, \ldots) \qquad (32)$$

are fulfilled. In this case the system (30) has a unique solution and, hence, the functions $s_j(\nu)$ can be chosen in such a way that all coefficients in the Hamiltonian (27), depending on ν, will be equal to zero and it will take a form

$$\hat{H} = \frac{\omega}{2}(\tilde{p}^2 + \tilde{q}^2) + \frac{c}{4}(\tilde{p}^2 + \tilde{q}^2)^2 + O((\tilde{p}^2 + \tilde{q}^2)^{5/2})) . \qquad (33)$$

Doing the last canonical transformation

$$\tilde{q} = \sqrt{2r}\sin\varphi, \ \tilde{p} = \sqrt{2r}\cos\varphi \ , \tag{34}$$

where r, φ are new canonically conjugate variables of the "action-angle" kind, we can rewrite the Hamiltonian function (33) in the form

$$H^* = \omega r + cr^2 + O(r^{5/2}) \ . \tag{35}$$

Coefficients ω and c in (35) are determined in (20), (29), respectively, and can be calculated for any values of parameters n, β, μ and sufficiently small values of e. If ω satisfies inequalities (32), which means the absence of the second and the fourth order resonances in the system, and parameter c is not equal to zero then we can apply Arnold-Moser's theorem [4,12,5] and conclude that equilibrium solution is stable in Liapunov sense.

6   The Case of the Fourth Order Resonance

Let us suppose that for some values of the system parameters the second order resonances are absent but there is the fourth order resonance $4\omega = N$. Then determinant (31) is equal to zero for $k = N$ and the system (30) has not any solutions. It means that terms being proportional to $\cos(N\nu), \sin(N\nu)$ in the Hamiltonian (27) can not be removed. Following to Markeev [13], we seek such coefficients $b_{1N}^{(j)}, b_{2N}^{(j)}$ for which the Hamiltonian (27) takes the form

$$\hat{H} = \frac{\omega}{2}(\tilde{p}^2 + \tilde{q}^2) + \frac{c}{4}(\tilde{p}^2 + \tilde{q}^2)^2 + (K_2\cos(N\nu) - K_3\sin(N\nu))(\tilde{p}^4 - 6\tilde{p}^2\tilde{q}^2 + \tilde{q}^4) +$$

$$+ 4(K_2\sin(N\nu) + K_3\cos(N\nu))(\tilde{q}\tilde{p}^3 - \tilde{q}^3\tilde{p}) + O((\tilde{p}^2 + \tilde{q}^2)^{5/2})) \ , \tag{36}$$

where K_2, K_3 are some unknown constant. Extracting coefficients of $\tilde{q}^{4-j}\tilde{p}^j$ in (27) and (36) for $j = 0, 1, \ldots, 4$ and equating the corresponding coefficients of $\cos(N\nu), \sin(N\nu)$, we obtain the following system of equations

$$Nb_{2N}^{(0)} - \frac{N}{4}b_{1N}^{(1)} + h_{1N}^{(0)} = K_2 \ ,$$

$$-Nb_{1N}^{(0)} - \frac{N}{4}b_{2N}^{(1)} + h_{2N}^{(0)} = -K_3 \ ,$$

$$Nb_{1k}^{(0)} - \frac{N}{2}b_{1N}^{(2)} + Nb_{2N}^{(1)} + h_{1N}^{(1)} = -4K_3 \ ,$$

$$-Nb_{1N}^{(1)} + Nb_{2N}^{(0)} - \frac{N}{2}b_{2N}^{(2)} + h_{2N}^{(1)} = -4K_2 \ ,$$

$$\frac{3N}{4}b_{1N}^{(1)} - \frac{3N}{4}b_{1N}^{(3)} + Nb_{2N}^{(2)} + h_{1N}^{(2)} = -6K_2 \ ,$$

$$-Nb_{1N}^{(2)} + \frac{3N}{4}b_{2N}^{(1)} - \frac{3N}{4}b_{2N}^{(3)} + h_{2N}^{(2)} = 6K_3 \ ,$$

$$\frac{N}{2}b_{1N}^{(2)} - Nb_{1N}^{(4)} + Nb_{2N}^{(3)} + h_{1N}^{(3)} = 4K_3 \ ,$$

$$-Nb_{1N}^{(3)} + \frac{N}{2}b_{2N}^{(2)} - Nb_{2N}^{(4)} + h_{2N}^{(3)} = 4K_2 \, ,$$

$$\frac{N}{4}b_{1N}^{(3)} + Nb_{2N}^{(4)} + h_{1N}^{(4)} = K_2 \, ,$$

$$-Nb_{1N}^{(4)} + \frac{N}{4}b_{2N}^{(3)} + h_{2N}^{(4)} = -K_3 \, . \tag{37}$$

One can readily see that extracting coefficients of $b_{1N}^{(j)}, b_{2N}^{(j)}$ ($j = 0, 1, 2, 3, 4$) in each equation of the system (37), we obtain the matrix coinciding with the corresponding matrix of the system (30) in the case of $4\omega = N$. Hence, its determinant is equal to zero and all coefficients $b_{1N}^{(j)}, b_{2N}^{(j)}$ can not be found as solution of the system (37). Nevertheless, the system (37) can be solved with respect to $b_{1N}^{(j)}, b_{2N}^{(j)}$ ($j = 0, 1, 2, 3$) and K_2, K_3 and the corresponding solution contains coefficients $b_{1N}^{(4)}, b_{2N}^{(4)}$ as parameters. Thus, the Hamiltonian is reduced to the form (36) and coefficients K_2, K_3 are

$$K_2 = \frac{1}{16}(h_{1N}^{(0)} - h_{1N}^{(2)} + h_{1N}^{(4)} - h_{2N}^{(1)} + h_{2N}^{(3)}) \, ,$$

$$K_3 = -\frac{1}{16}(h_{1N}^{(1)} - h_{1N}^{(3)} + h_{2N}^{(0)} - h_{2N}^{(2)} + h_{2N}^{(4)}) \, . \tag{38}$$

Now we can do the following canonical transformation

$$\tilde{q} = \sqrt{2r}\sin\left(\frac{N\nu}{4} + \varphi - \theta\right) \, , \quad \tilde{p} = \sqrt{2r}\cos\left(\frac{N\nu}{4} + \varphi - \theta\right) \, , \tag{39}$$

where r, φ are new canonical variables of the "action-angle" kind and parameter θ is determined from the conditions

$$\cos(4\theta) = \frac{K_2}{\sqrt{K_2^2 + K_3^2}} \, , \quad \sin(4\theta) = \frac{K_3}{\sqrt{K_2^2 + K_3^2}} \, .$$

Then the Hamiltonian function (36) takes a form

$$H^* = r^2(c + d\cos(4\varphi)) + O(r^{5/2}) \, , \tag{40}$$

where

$$d = 4\sqrt{K_2^2 + K_3^2} \, . \tag{41}$$

Now the theorem of A.P. Markeev [13] can be applied which states that equilibrium solution is stable in Liapunov sense for $|c| > d$ and is unstable for $|c| < d$. Thus, the stability problem in the case of the fourth order resonance is solved by means of estimation and comparison of the parameters c and d.

7   Conclusion

In the present paper we have analyzed the problem of stability of the equilibrium solution in the elliptic restricted $(2n+1)$-body problem. Perturbed motion of the system is described by the second order non-autonomous Hamiltonian system of nonlinear differential equations. In such a case the stability problem is solved in several steps and each step requires quite cumbersome calculations which can be effectively done with modern computer algebra systems.

On the first step we investigate stability in linear approximation and determine domains of instability in the space of parameters of the system. As the Hamiltonian function contains a small parameter e, the boundaries of these domains can be found in the form of power series in e. Although coefficients of these series are found only numerically, the calculations can be done with any necessary precision.

Linearly stable equilibrium solutions may become unstable if nonlinear terms in the equations of perturbed motion are taken into account. In order to study the stability problem in a strict nonlinear formulation we have to normalize the Hamiltonian function and to apply theorems of the KAM-theory. We have normalized the quadratic form H_2 and the fourth order form H_4 in the Hamiltonian expansion and described the algorithms of corresponding calculations in detail. If the second and the fourth order resonance are absent in the system then the Hamiltonian function can be reduced to the form (35) and Arnold-Moser's theorem on stability of the equilibrium solution in general elliptic case is applied. In the case of the fourth order resonance the Hamiltonian function is reduced to the form (40) and the theorem of A.P. Markeev is applied. In both cases the stability problem is totally solved. It should be noted that all necessary symbolic and numeric calculations in this paper are done with the computer algebra system *Mathematica*.

References

1. Markeev, A.P.: The stability of hamiltonian systems. In: Matrosov, V.M., Rumyantsev, V.V., Karapetyan, A.V. (eds.) Nonlinear meczhanics, Moscow, Fizmatlit, pp. 114–130 (2001) (in Russian)
2. Grebenikov, E.A., Kozak-Skovorodkin, D., Jakubiak, M.: The methods of computer algebra in the many-body problem. Moscow, Ed. RUDN (2002) (in Russian)
3. Kolmogorov, A.N.: On the conservation of quasi-periodic motions for a small change in the Hamiltonian function. Dokl. Akad. Nauk USSR 98(4), 527–530 (1954) (in Russian)
4. Arnold, V.I.: Small denominators and problems of stability of motion in classical and celestial mechanics. Uspekhi Math. Nauk 18(6), 91–192 (1963) (in Russian)
5. Moser, J.: Lectures on the Hamiltonian systems. Moscow, Mir (1973)
6. Grebenikov, E.A., Gadomski, L., Prokopenya, A.N.: Studying the stability of equilibrium solutions in the planar circular restricted four-body problem. Nonlinear oscillations 10(1), 66–82 (2007)
7. Wolfram, S.: The Mathematica Book, 4th edn. Wolfram Media/Cambridge University Press (1999)
8. Grebenikov, E.A., Prokopenya, A.N.: On existence of a new class of exact solutions in the planar Newtonian many-body problem. In: The questions of modelling and analysis in the problems of making decision. Dorodnicyn Computing Center of RAS, Moscow, pp. 39–56 (2004) (in Russian)

9. Grebenikov, E.A., Prokopenya, A.N.: Determination of the boundaries between the domains of stability and instability for the Hill equation. Nonlinear Oscillations 6(1), 42–51 (2003)
10. Birkhoff, G.D.: Dynamical systems. Moscow, GITTL (1941) (in Russian)
11. Prokopenya, A.N.: Normalization of the non-autonomous linear Hamiltonian system with a small parameter. Mathematical Modelling 17(6), 33–42 (2005) (in Russian)
12. Arnold, V.I.: On stability of equilibrium positions of the Hamiltonian system of ordinary differential equations in general elliptic case. Dokl. Akad. Nauk USSR 137(2), 255–257 (1961) (in Russian)
13. Markeev, A.P.: Libration points in celestial mechanics and cosmodynamics. Moscow, Nauka (1978) (in Russian)

On the Peculiar Properties of Families of Invariant Manifolds of Conservative Systems

Valentin Irtegov and Tatyana Titorenko

Institute for Systems Dynamics and Control Theory SB RAS,
134, Lermontov str., Irkutsk, 664033, Russia
irteg@icc.ru

Abstract. The paper discusses a technique for investigation of peculiar properties of invariant manifolds of conservative systems. The technique is based on constructing the envelope for the family of first integrals of such systems. Routh–Lyapunov's method [1] has been applied for obtaining the families of invariant manifolds.

With the use of the method of envelope we have analyzed some peculiar properties of families of invariant manifolds in the problems related to rigid body dynamics and vortex theory. For the purpose of solving the computational problems arising in the process of investigations we employed the computer algebra system (CAS) *Mathematica*. This paper presents a development of our approach [2] to investigation of some qualitative properties of conservative systems.

1 Introduction

It is well known that constructing envelopes for the families of ODE's solutions allows one to find peculiar solutions [3]. In the present paper, we use the envelope for the family of first integrals of a conservative system to find out the peculiar properties of families of invariant manifolds for the system. The invariant manifolds, on which elements of the indicated family of first integrals assume stationary value, are considered. We call such manifolds the invariant manifolds of steady motions (IMSMs) [4]. In the capacity of peculiar invariant manifolds we will understand the IMSMs on which the first integrals of several problem assume a stationary value.

We have employed the method of envelope to analyze peculiar properties of the families of IMSMs in problems related to motion of a rigid body having one fixed point, a rigid body in fluid, and to vortex dynamics. Already in the problems of rigid body dynamics, the computations, which are needed for implementation of the technique applied, become rather bulky, and there appears the necessity of employment of special computational tools. We used CAS *Mathematica* for performing computations: standard tools and packages developed by the authors in the language *Mathematica* (see, e.g., [5]) for the purpose of deriving stationarity equations, investigation of stationary solutions and IMSMs for stability, constructing the envelope integral, etc. These tools have given us a possibility to analyze a number of dynamic systems on the basis of approach

V.G. Ganzha, E.W. Mayr, and E.V. Vorozhtsov (Eds.): CASC 2007, LNCS 4770, pp. 195–210, 2007.
© Springer-Verlag Berlin Heidelberg 2007

proposed. Manual investigation of the systems would be problematic because of bulky computations.

Let us demonstrate our approach to investigation of IMSMs on Brun's problem well-known in dynamics of a rigid body having one fixed point [6].

2   Brun's Problem

Equations of motion of a rigid body in this case write:

$$
\begin{aligned}
A\dot{p} &= (B-C)qr - \mu(B-C)\gamma_2\gamma_3, & \dot{\gamma}_1 &= r\gamma_2 - q\gamma_3, \\
B\dot{q} &= (C-A)rp - \mu(C-A)\gamma_3\gamma_1, & \dot{\gamma}_2 &= p\gamma_3 - r\gamma_1, \\
C\dot{r} &= (A-B)rp - \mu(A-B)\gamma_1\gamma_2, & \dot{\gamma}_3 &= q\gamma_1 - p\gamma_2.
\end{aligned}
\tag{1}
$$

Here A, B, and C are the moments of body inertia; p, q, and r are projections of the body angular rate onto the axes bound up with it; γ_1, γ_2, and γ_3 are directional cosines of angles between the vertical and the axes bound up with the body.

System (1) possesses the family of first integrals:

$$
\begin{aligned}
2K =\ & Ap^2 + Bq^2 + Cr^2 + \mu(A\gamma_1^2 + B\gamma_2^2 + C\gamma_3^2) - 2\lambda_1(Ap\gamma_1 + Bq\gamma_2 + Cr\gamma_3) \\
& -\lambda_2[A^2p^2 + B^2q^2 + C^2r^2 - \mu(BC\gamma_1^2 + CA\gamma_2^2 + AB\gamma_3^2)] \\
& -\lambda_3(\gamma_1^2 + \gamma_2^2 + \gamma_3^2),
\end{aligned}
\tag{2}
$$

which for

$$
\begin{aligned}
\lambda_1^2 &= \mu[(1 - A\lambda_2)(1 - B\lambda_2)(1 - C\lambda_2)], \\
\lambda_3 &= \mu[\lambda_2(AB + BC + CA) - \lambda_2^2 ABC]
\end{aligned}
\tag{3}
$$

assumes a stationary value on the family of invariant manifolds:

$$
A'p - \lambda_1\gamma_1 = 0, \quad B'q - \lambda_1\gamma_2 = 0, \quad C'r - \lambda_1\gamma_3 = 0.
\tag{4}
$$

Turning partial derivatives of the family of integrals K (2) with respect to variables of the problem on the IMSMs (4) to zero is obvious because conditions of stationarity for K have the form:

$$
\begin{aligned}
\frac{\partial K}{\partial p} &= A(A'p - \lambda_1\gamma_1) = 0, & \frac{\partial K}{\partial \gamma_1} &= [\mu(A + \lambda_2 BC) - \lambda_3]\gamma_1 - \lambda_1 Ap = 0, \\
\frac{\partial K}{\partial q} &= B(B'q - \lambda_1\gamma_2) = 0, & \frac{\partial K}{\partial \gamma_2} &= [\mu(B + \lambda_2 CA) - \lambda_3]\gamma_2 - \lambda_1 Bq = 0, \\
\frac{\partial K}{\partial r} &= C(C'q - \lambda_1\gamma_3) = 0, & \frac{\partial K}{\partial \gamma_3} &= [\mu(C + \lambda_2 AB) - \lambda_3]\gamma_3 - \lambda_1 Cr = 0.
\end{aligned}
\tag{5}
$$

After substituting (3) into (5), the latter assume the form:

$$
\begin{aligned}
A'\frac{\partial K}{\partial \gamma_1} &= A\lambda_1(\lambda_1\gamma_1 - A'p), & B'\frac{\partial K}{\partial \gamma_2} &= B\lambda_1(\lambda_1\gamma_2 - B'q), \\
C'\frac{\partial K}{\partial \gamma_3} &= C\lambda_1(\lambda_1\gamma_3 - C'r).
\end{aligned}
$$

For the sake of brevity, we have introduced the notations: $A' = (1 - A\lambda_2)$, $B' = (1 - B\lambda_2)$, and $C' = (1 - C\lambda_2)$. Henceforth we also intend to use the following abbreviations: $N = A + B + C$, $M = AB + BC + CA$, $L = ABC$.

2.1 Finding Peculiar Properties of Invariant Manifolds

Consider the problem of finding peculiar properties of the family of invariant manifolds (4) with the use of the envelope for the family of first integrals (2). To this end, we employ a standard procedure of constructing the envelope for the one-parameter family of integrals (2) (here λ_2 is a parameter).

Compute the derivative of K with respect to parameter λ_2 and equate it to zero:

$$\frac{\partial K}{\partial \lambda_2} = -\frac{d\lambda_1}{d\lambda_2}V_1 - \frac{1}{2}V_2 - \frac{1}{2}\frac{d\lambda_3}{d\lambda_2}V_3 = 0.$$

Here for the brevity, the multipliers of λ_1, λ_2, and λ_3 in (2) are denoted by V_1, V_2, V_3, respectively.

Having considered expressions for λ_1 and λ_3 as functions of λ_2 (3), after obvious transformations, we obtain the following equation:

$$\mu(1 - \lambda_2 N + \lambda_2^2 M - \lambda_2^3 L)[V_2^2 + \mu^2 V_3^2(M^2 - 4\lambda_2 ML + 4\lambda_2^2 L^2) + 2\mu V_2 V_3(M$$
$$-2\lambda_2 L)] - \mu^2 V_1^2(N^2 + 4\lambda_2^2 M^2 + 9\lambda_2^4 L^2 - 4\lambda_2 MN + 6\lambda_2^2 NL - 12\lambda_2^3 ML) = 0.$$

Having removed one multiplier μ and collecting the terms, which have the same powers of λ_2, we obtain:

$$-4\lambda_2^5\mu^2 V_3^2 L^3 + \lambda_2^4\mu L^2(8\mu MV_3 + 4V_2 V_3^2 - 9V_1^2) + \lambda_2^3 L[-V_2^2 - \mu^2 V_3^2(5M^2$$
$$+4LN) - 6\mu MV_2 V_3 + 12\mu MV_1^2] + \lambda_2^2[MV_2^2 + \mu^2 V_3^2(M^3 + 4L^2 + 4NML)$$
$$+2\mu V_2 V_3(M^2 + 2NL) - \mu V_1^2(4M^2 + 6NL)] + \lambda_2[-NV_2^2$$
$$-\mu^2 V_3^2 M(MN + 4L) - 2\mu V_2 V_3(MN + 2L) + 4\mu NMV_1^2]$$
$$+(V_2^2 + \mu^2 M^2 V_3^2 + 2\mu MV_2 V_3 - \mu MV_1^2) = 0.$$

Here analysis of peculiar properties is practically reduced to investigation of solutions of the equation obtained. Since it requires much space, we shall restrict our consideration to the most characteristic peculiar properties of the family of invariant manifolds (4). To this end, we transform the latter equation to the form

$$4\lambda_2^2\mu^2 V_3^2 L^2(-\lambda_2^3 L + \lambda_2^2 M - \lambda_2 N + 1) + \lambda_2^4\mu L^2[4V_3(V_2 + \mu MV_3) - 9V_1^2] + \lambda_2^3 L$$
$$\times[-(V_2 + \mu MV_3)^2 - 4\mu MV_3(V_2 + \mu MV_3) + 12\mu MV_1^2] + \lambda_2^2[M(V_2 + \mu MV_3)^2$$
$$+4\mu LNV_3(V_2 + \mu MV_3) - 2\mu V_1^2(2M^2 + 3NL)] + \lambda_2[-N(V_2 + \mu MV_3)^2$$
$$-4\mu LV_3(V_2 + \mu MV_3) + 4\mu NMV_1^2] + [(V_2 + \mu MV_3)^2 - \mu V_1^2 N^2] = 0. \qquad (6)$$

The bracket $(-\lambda_2^3 L + \lambda_2^2 M - \lambda_2 N + 1)$ represents a cubic polynomial having coefficients, which are elementary symmetric functions of A, B, C. Therefore, the following real solutions will be the roots of this polynomial: $\lambda_2 = A^{-1}$,

$\lambda_2 = B^{-1}$, $\lambda_2 = C^{-1}$. It can easily be verified that the rest of the terms of equation (6) vanish when $V_1 = 0$, $V_3 = 1$ and $V_2 = \mu M$. Direct substitution of the values of $\lambda_2 = A^{-1}$, $\lambda_2 = B^{-1}$, $\lambda_2 = C^{-1}$ into equations of IMSMs (4) gives evidence that these values of λ_2 correspond to the body pendular oscillations about its main horizontal inertial axes. For example, when $\lambda_2 = A^{-1}$ such an axis is x, and equations of the invariant manifold write:

$$q = 0, r = 0, \gamma_1 = 0. \tag{7}$$

Finally, having equated the free term of (6) to zero (then, for $V_3 = 1$ the following relation between the constants of the first integrals V_2 and V_1 : $(V_2 + \mu M)^2 - \mu N^2 V_1^2 = 0$ takes place) we obtain the zero root of equation (6), i.e. $\lambda_2 = 0$. Equations of the family of IMSMs (4) will be reduced in this case to the following ones:

$$p = \sqrt{\mu}\gamma_1, \quad q = \sqrt{\mu}\gamma_2, \quad r = \sqrt{\mu}\gamma_3. \tag{8}$$

Since no restriction has been imposed on the directional cosines γ_1, γ_2, and γ_3 for the IMSM obtained, the family of IMSMs (8) will define permanent rotations of the body about the vertical when the angular rate $\omega = \sqrt{\mu}$. Furthermore, the axis of rotation in the body can have any fixed direction.

It is possible to apply Lyapunov's second method to investigate stability of peculiar IMSMs obtained by the technique applied. Here the first integral K (2) will play the role of the Lyapunov function.

For example, in case of IMSM (7), which defines pendular oscillations of the body about its main horizontal axis x, the 2nd variation of integral K has the form

$$\delta^2 K = \mu \frac{(A-B)(A-C)}{A}\eta_1^2 + B\frac{(A-B)}{A}\xi_1^2 + C\frac{(A-C)}{A}\xi_3^2.$$

Here η_1, ξ_1, and ξ_2 are deviations from undisturbed values of the variables q, r, γ_1.

Since the latter quadratic form is sign-definite with respect to all its variables for $A > B$, $A > C$, due to Zubov's theorem [7], the IMSM (7) is stable. In other words, pendular oscillations of the body about its main horizontal axis x are stable when the moment of inertia of the body with respect to this axis is the largest.

3   Euler's Equations in Lie Algebras

Consider the dynamic system [8] in Lie algebra in the capacity of the second problem. Differential equations of the system write:

$$
\begin{aligned}
\dot{s}_1 &= -\alpha^2 r_2 r_3 + \alpha r_1 s_2 - (\beta r_3 - s_2)(\beta r_2 + s_3), \\
\dot{s}_2 &= (\alpha^2 + \beta^2) r_1 r_3 - (\alpha r_1 + \beta r_2) s_1 + (\alpha r_3 - s_1) s_3, \\
\dot{s}_3 &= (\beta r_1 - \alpha r_2) s_3, \\
\dot{r}_1 &= r_2(\alpha r_1 + \beta r_2 + 2s_3) - r_3 s_2 - ((\alpha^2 + \beta^2) r_3 s_2 + \beta s_3^2)x, \\
\dot{r}_2 &= r_3 s_1 - r_1(\alpha r_1 + \beta r_2 + 2s_3) + ((\alpha^2 + \beta^2) r_3 s_1 + \alpha s_3^2)x, \\
\dot{r}_3 &= r_1 s_2 - r_2 s_1 + (\beta s_1 - \alpha s_2) s_3 x.
\end{aligned}
\tag{9}
$$

Equations (9), besides quadratic first integrals,

$$2V_0 = (s_1^2 + s_2^2 + 2s_3^2) + 2(\alpha r_1 + \beta r_2)s_3 - (\alpha^2 + \beta^2)r_3^2 = 2h,$$
$$V_1 = s_1 r_1 + s_2 r_2 + s_3 r_3 = c_1, \quad V_2 = x(s_1^2 + s_2^2 + s_3^2) + r_1^2 + r_2^2 + r_3^2 = c_2, \quad (10)$$

assume the 4th-order additional first integral

$$V_3 = (r_1 s_1 + r_2 s_2)((\alpha^2 + \beta^2)(r_1 s_1 + r_2 s_2) + 2(\alpha s_1 + \beta s_2)s_3) + s_3^2(s_1^2 + s_2^2$$
$$+(\alpha r_1 + \beta r_2 + s_3)^2) + x s_3^2(\beta s_1 - \alpha s_2)^2 = c_3. \quad (11)$$

Here s_i and r_i are components of two three-dimensional vectors, α, β, and x are arbitrary constants.

The cases when $x > 0$ and $x < 0$ correspond to Euler's equations in Lie algebras $so(4)$ and $so(3,1)$, respectively. When $x = 1$, equations (9) coincide with the Poincare–Zhukovsky equations, which describe the motion of a rigid body having an ellipsoidal cavity filled with vortex incompressible fluid, and when $x = 0$, the system under scrutiny corresponds to the integrable case in Kirchhoff's problem [9].

System (9)-(11) has been investigated in [10]–[12]. These equations describe the bifurcation analysis of the given system in Lie algebra $e(3)$ and $so(4)$. We will consider the problem of finding the families of IMSMs of equations (9) and investigation of their peculiar properties.

3.1   Finding Invariant Manifolds

For the purpose of finding the families of IMSMs of equations (9) we shall use Routh–Lyapunov's method as it was said above. According to this method, some combinations are formed from the problem first integrals – families of first integrals K. We will restrict here our consideration to linear combinations of first integrals (to the end of complete analysis of the problem stated it is necessary to use also nonlinear combinations of the integrals (see [13])):

$$K = \lambda_0 V_0 - \lambda_1 V_1 - \frac{\lambda_2}{2}V_2 - \frac{\lambda_3}{2}V_3 \quad (\lambda_i = const). \quad (12)$$

Here $\lambda_0, \lambda_1, \lambda_2$, and λ_3 are some constants, which may assume also zero values.

The conditions of stationarity of K with respect to variables s_1, s_2, s_3, r_1, r_2, and r_3 write:

$$\partial K/\partial s_1 = \lambda_0 s_1 - \lambda_1 r_1 - \lambda_3[(\alpha^2 + \beta^2)r_1(r_1 s_1 + r_2 s_2) + s_1 s_3(\alpha r_2 + \beta r_1)$$
$$+ s_2 s_3(2\alpha r_1 + s_3)] - x(\lambda_3 \beta s_3^2(\beta s_1 - \alpha s_2) + \lambda_2 s_1) = 0,$$
$$\partial K/\partial s_2 = \lambda_0 s_2 - \lambda_1 r_2 - \lambda_3[(\alpha^2 + \beta^2)r_2(r_1 s_1 + r_2 s_2) + s_1 s_3(\alpha r_2 + \beta r_1)$$
$$+ s_2 s_3(2\beta r_2 + s_3)] - x(\lambda_3 \alpha s_3^2(\alpha s_2 - \beta s_1) + \lambda_2 s_2) = 0,$$
$$\partial K/\partial s_3 = \lambda_0(\alpha r_1 + \beta r_2 + 2s_3) - \lambda_1 r_3 - \lambda_3[(r_1 s_1^2 \alpha s_1 + \beta s_2)(s_1 r_1 + s_2 r_2)$$
$$+ s_3(\alpha r_1 + \beta r_2)^2 + s_3(s_1^2 + s_2^2 + 2s_3^2) + 3s_3^2(\alpha r_1 + \beta r_2)]$$
$$- x s_3(\lambda_3(\beta s_1 - \alpha s_2)^2 + \lambda_2) = 0,$$

$$\partial K/\partial r_1 = \lambda_0 \alpha s_3 - \lambda_1 s_1 - \lambda_2 r_1 - \lambda_3[(\alpha^2 + \beta^2)s_1(r_1 s_1 + r_2 s_2) + \alpha s_3^2$$
$$\times(\beta r_2 + \alpha r_1) + \alpha s_3(s_1^2 + s_2^2) + \beta\lambda_3 s_1 s_2 s_3] = 0,$$
$$\partial K/\partial r_2 = -\lambda_0\beta s_3 + \lambda_1 s_2 + \lambda_2 r_2 + \lambda_3[(\alpha^2 + \beta^2)s_2(s_1 r_1 + r_2 s_2) + s_2 s_3$$
$$\times(\alpha s_1 + \beta s_2) + \beta s_3^2(\alpha r_1 + \beta r_2) + \beta s_3^3] = 0,$$
$$\partial K/\partial r_3 = -((\alpha^2 + \beta^2)\lambda_0 + \lambda_2)r_3 + \lambda_1 s_3 = 0. \tag{13}$$

Conditions of existence of IMSMs for equations (9) may be obtained by equating the Jacobian of system (13) to zero. The solutions of system (13) obtained under these conditions will, generally speaking, be the desired IMSMs. Unfortunately, in this problem the expression of the Jacobian is rather bulky, so its complete analysis is problematic even with the use of computer algebra tools. Some particular conditions of turning the Jacobian to zero, which have allowed us to obtain IMSMs of equations (9), are considered below.

When $\lambda_1 = 0, \lambda_2 = -(\alpha^2 + \beta^2)\lambda_0$ the last equation of system (13) turns to zero, i.e., the system undoubtedly becomes degenerate, and, consequently, the system Jacobian is zero. Having substituted the above values for λ_1 and λ_2 into (13), and using the method of Gröbner bases, we have found a series of families of IMSMs. Some of these families are given below:

$$r_1 = \frac{1}{a_1 a_2(\alpha^2 + \beta^2)\lambda_3}\Big(a_3\alpha\lambda_3 s_3 + \sqrt{a_1(x(\alpha^2 + \beta^2) + 1)\lambda_3}\ ((\alpha^2 + \beta^2)$$
$$\times \sqrt{a_1(\alpha^2 + \beta^2)(\lambda_0 - \lambda_3 s_3^2)\lambda_0\lambda_3}\ s_1 \mp a_1\alpha\beta\lambda_3 s_3^2)\Big),$$
$$r_2 = -\frac{1}{(\alpha^2 + \beta^2)\lambda_3}\Big(\beta\lambda_3 s_3 \mp \sqrt{a_1(x(\alpha^2 + \beta^2) + 1)\lambda_3}\Big),$$
$$s_2 = \frac{1}{a_2\lambda_3}\Big(\alpha\beta\lambda_3^2 s_1 s_3^2 \pm \sqrt{a_1(\alpha^2 + \beta^2)(\lambda_0 - \lambda_3 s_3^2)\lambda_0\lambda_3}\Big). \tag{14}$$

Here for the sake of brevity, we have used the following notations: $a_1 = (\alpha^2 + \beta^2)(\lambda_0 - \lambda_3 s_1^2) - \alpha^2\lambda_3 s_3^2$, $a_2 = \alpha^2\lambda_3 s_3^2 - (\alpha^2 + \beta^2)\lambda_0$, $a_3 = (\alpha^2 + \beta^2)^2(\lambda_0 - \lambda_3 s_1^2)\lambda_0 + \alpha^2(\alpha^2 + \beta^2)(\lambda_3 s_1^2 - 2\lambda_0)\lambda_3 s_3^2 + \alpha^4\lambda_3^2 s_3^4$.

We have also analyzed the set of solutions of system (13) with the use of the method of Gröbner bases. It has also enabled us to obtain additional families of the IMSMs. For this purpose, we have constructed the Gröbner basis for the equations (13) (the variable r_3 has been preliminarily removed from the equations with the use of the system last equation). The basis obtained under elimination monomial order writes:

$$(a_{139}s_1 + a_{138}s_2 + a_{141}s_3)(a_{135} + a_{136}s_1^2 + a_{136}s_2^2 + a_{106}s_1 s_3 + a_{105}s_2 s_3$$
$$+a_{132}s_3^2) = 0,$$
$$s_3(a_{140}s_1 + a_{142}s_3)(a_{135} + a_{136}s_1^2 + a_{136}s_2^2 + a_{106}s_1 s_3 + a_{105}s_2 s_3 + a_{132}s_3^2) = 0,$$
$$s_3(a_{133}s_1^2 + a_{133}s_2^2 + a_{108}s_1 s_3 + a_{107}s_2 s_3 + a_{110}s_3^2 + a_{128}s_1^2 s_3^2 + a_{72}s_1 s_2 s_3^2$$
$$+a_{130}s_2^2 s_3^2 + a_{79}s_1 s_3^3 + a_{78}s_2 s_3^3 + a_{127}s_3^4) = 0,$$
$$a_{118}s_1 + a_{117}s_2 + a_{119}s_3 + a_{81}s_1^2 s_3 + a_{39}s_1 s_2 s_3 + a_{80}s_2^2 s_3 + a_{62}s_1 s_3^2 + a_{124}s_1^3 s_3^2$$

$$+a_{71}s_2s_3^2 + a_{101}s_1^2s_2s_3^2 + a_{121}s_2^3 + a_{125}s_1^2s_3^3 + a_{34}s_1s_2s_3^3 + a_{102}s_1s_3^4 + a_{77}s_2s_3^4$$
$$+a_{126}s_3^5 = 0,$$

$$a_{48}s_1 + a_5s_1^3 + a_{63}s_1^5 + a_{18}s_2 + a_6s_1s_2^2 + a_{63}s_1^3s_2^2 + a_{25}s_3 + a_{86}s_1^2s_3 + a_8s_1^4s_3$$
$$+a_4s_1s_2s_3 + a_7s_1^3s_2s_3 + a_{85}s_2^2s_3 + a_{111}s_1s_3^2 + a_{92}s_1^3s_3^2 + a_{47}s_2s_3^2 + a_{84}s_3^3$$
$$+a_{49}s_1^2s_3^3 + a_2s_1s_2s_3^3 + a_{50}s_1s_3^4 + a_3s_2s_3^4 + a_{88}s_3^5 = 0,$$

$$s_3(a_{55}s_1 + a_{56}s_2 + a_{57}s_3 + a_{21}s_1^2s_3 + a_{37}s_1s_2s_3 + a_{38}s_2^2s_3 + a_{120}s_1s_3^2 + a_{123}s_1^3s_3^2$$
$$+a_{33}s_2s_3^2 + a_{98}s_3^3 + a_{74}s_1^2s_3^3 + a_{76}s_1s_3^4 + a_{104}s_3^5) = 0,$$

$$a_{94}s_1^2 + a_{27}s_1s_2 + a_{54}s_2^2 + a_{53}s_1s_3 + a_{36}s_1^3s_3 + a_{42}s_2s_3 + a_{35}s_1^2s_2s_3 + a_{93}s_3^2$$
$$+a_{95}s_1^2s_3^2 + a_{131}s_1^4s_3^2 + a_{29}s_1s_2s_3^2 + a_{99}s_2^2s_3^2 + a_{28}s_1s_3^3 + a_{23}s_2s_3^3 + a_{97}s_3^4$$
$$+a_{73}s_1^2s_3^4 + a_{75}s_1s_3^5 + a_{103}s_3^6 = 0,$$

$$s_3(a_{122} + a_{114}s_1^2 + a_{114}s_2^2 + a_{41}s_1s_3 + a_{40}s_2s_3 + a_{129}s_3^2 + a_{22}s_1^2s_3^2 + a_{15}s_1s_2s_3^2$$
$$+a_{17}s_1s_3^3 + a_{16}s_2s_3^3 + a_{51}s_3^4 + a_{67}s_1^2s_3^4 + a_9s_1s_2s_3^4 + a_{10}s_1s_3^5 + a_1s_2s_3^5$$
$$+a_{32}s_3^6) = 0,$$

$$a_{69}r_2 + a_{43}s_1 + a_{30}s_1^3 + a_{52}s_2 + a_{30}s_1s_2^2 + a_{44}s_3 + a_{109}s_1^2s_3 + a_{24}s_1s_2s_3$$
$$+a_{116}s_2^2s_3 + a_{87}s_1s_3^2 + a_{68}s_1^3s_3^2 + a_{113}s_2s_3^2 + a_{89}s_3^3 + a_{96}s_1^2s_3^3 + a_{11}s_1s_2s_3^3$$
$$+a_{60}s_1s_3^4 + a_{13}s_2s_3^4 + a_{46}s_3^5 = 0,$$

$$a_{70}r_1 + a_{83}s_1 + a_{65}s_1^3 + a_{19}s_2 + a_{65}s_1s_2^2 + a_{20}s_3 + a_{91}s_1^2s_3 + a_{45}s_1s_2s_3$$
$$+a_{82}s_2^2s_3 + a_{112}s_1s_3^2 + a_{100}s_1^3s_3^2 + a_{90}s_2s_3^2 + a_{26}s_3^3 + a_{59}s_1^2s_3^3 + a_{12}s_1s_2s_3^3$$
$$+a_{58}s_1s_3^4 + a_{14}s_2s_3^4 + a_{61}s_3^5 = 0, \tag{15}$$

where a_i $(i = 1, \ldots, 142)$ are some polynomials in λ_i, α, β, and x. The timing for constructing the basis, as measured on 1100 MHz Pentium with 256 MB RAM running under Windows XP, is 5.71 minutes.

As obvious from (15), the given system of equations is factored, i.e., decomposed into a number of subsystems, which may be analyzed separately. A lexicographic Gröbner basis has been constructed for each of the subsystems. Analysis of these bases has given evidence that the system of equations (15) (and, consequently, 13)) has an infinite set of solutions (variable s_3 being free). We have found some solutions of system (13), which contain the free variable. These represent the families of IMSMs for the system of differential equations (9). Some of these solutions are given below:

$$\left\{s_1 = -\frac{\alpha\lambda_1 s_3}{(\alpha^2 + \beta^2)\lambda_0}, \quad s_2 = -\frac{\beta\lambda_1 s_3}{(\alpha^2 + \beta^2)\lambda_0}, \quad r_1 = \frac{\lambda_0 - \lambda_3 s_3(\beta r_2 + s_3)}{\alpha\lambda_3 s_3}, \right.$$
$$\left. r_3 = -\frac{\lambda_1 s_3}{(\alpha^2 + \beta^2)\lambda_0}\right\} \quad \text{when } \lambda_2 = 0; \tag{16}$$

$$\left\{r_1 = \frac{x\alpha s_3\sqrt{x\lambda_1^2 + a\lambda_0^2}}{\sqrt{x\lambda_1^2 + a\lambda_0^2} \mp a\lambda_0}, r_2 = \frac{x\beta s_3\sqrt{x\lambda_1^2 + a\lambda_0^2}}{\sqrt{x\lambda_1^2 + a\lambda_0^2} \mp a\lambda_0}, r_3 = -\frac{x\lambda_1 s_3}{a\lambda_0 \mp \sqrt{x\lambda_1^2 + a\lambda_0^2}}, \right.$$
$$\left. s_1 = -\frac{x\alpha\lambda_1 s_3}{a\lambda_0 \mp \sqrt{x\lambda_1^2 + a\lambda_0^2}}, \quad s_2 = -\frac{x\beta\lambda_1 s_3}{a\lambda_0 \mp \sqrt{x\lambda_1^2 + a\lambda_0^2}}\right\},$$

when $\lambda_2 = \dfrac{\lambda_0 \mp \sqrt{x\lambda_1^2 + a\lambda_0^2}}{x}$, $\lambda_3 = 0.$ \hfill (17)

Here $a = (\alpha^2 + \beta^2)x + 1$. It can be seen that the solutions have been obtained under some restrictions imposed on parameters λ_2, λ_3.

From the geometrical viewpoint, each of the elements of the family of IMSMs (17) – for fixed values of parameters $\lambda_0, \lambda_1, \lambda_2$, and λ_3 – describes the straight lines, which lie in R^6 at the intersection of 5 hyperplanes. Furthermore, since there is a vector field of form $\dot{s}_3 = 0$ defined on each family of IMSMs (17), each point of the given lines is a degenerate stationary solution of the initial differential equations.

Similarly, each of the families of IMSMs (14), (16) (for fixed values of parameters λ_i) describes in R^6 a surface of dimension of 3 or 2, respectively. In the first case, the surface lies at the intersection of a hyperplane and two 4th-order and 6th-order hypersurfaces; in the second case, – three hyperplanes and the 2nd-order surface.

3.2   Analysis of Peculiar Properties of Invariant Manifolds

Now we use the technique of enveloping integral to analyze peculiar properties of IMSMs for the given problem. Consider one of the families of IMSMs (17), for example:

$$\{r_1 = \frac{x\alpha s_3 \sqrt{x\lambda_1^2 + a\lambda_0^2}}{\sqrt{x\lambda_1^2 + a\lambda_0^2} + a\lambda_0}, \quad r_2 = \frac{x\beta s_3 \sqrt{x\lambda_1^2 + a\lambda_0^2}}{\sqrt{x\lambda_1^2 + a\lambda_0^2} + a\lambda_0},$$

$$r_3 = -\frac{x\lambda_1 s_3}{a\lambda_0 + \sqrt{x\lambda_1^2 + a\lambda_0^2}},$$

$$s_1 = -\frac{x\alpha\lambda_1 s_3}{a\lambda_0 + \sqrt{x\lambda_1^2 + a\lambda_0^2}}, \quad s_2 = -\frac{x\beta\lambda_1 s_3}{a\lambda_0 + \sqrt{x\lambda_1^2 + a\lambda_0^2}}\},$$

$$\lambda_2 = \frac{\lambda_0 + \sqrt{x\lambda_1^2 + a\lambda_0^2}}{x}, \quad \lambda_3 = 0.$$ \hfill (18)

Substitute the values of λ_2, λ_3, which correspond to this family of IMSMs, into the integral K (12). As a result, we obtain

$$K = \lambda_0 V_0 - \lambda_1 V_1 - \frac{\lambda_0 + \sqrt{x\lambda_1^2 + a\lambda_0^2}}{2x} V_2.$$

Compute the derivative of the latter expression with respect to λ_1 and equate it to zero:

$$\frac{\partial K}{\partial \lambda_1} = -V_1 - \frac{\lambda_1}{2\sqrt{x\lambda_1^2 + a\lambda_0^2}} V_2 = 0.$$

After some obvious transformations and substitution of the respective expressions for the integrals, the latter expression writes:

$$2\sqrt{x\lambda_1^2 + a\lambda_0^2}(s_1 r_1 + s_2 r_2 + s_3 r_3) + \lambda_1(x(s_1^2 + s_2^2 + s_3^2) + r_1^2 + r_2^2 + r_3^2) = 0. \quad (19)$$

Find the value of this expression on the family of IMSMs (18). For this purpose, substitute values of the variables r_1, r_2, r_3, s_1, s_2 from (18) into the latter equality. As a result, the equality shall turn into an identity. This means that the family of IMSMs (18) is peculiar. The latter is valid for all the values of the parameter λ_1 for which there exists the IMSMs under scrutiny.

Analysis of expansion of integral K (12) in the neighbourhood of IMSMs (18) has revealed that elements of the family of IMSMs under scrutiny are stable in the sense of Lyapunov. Their stability conditions write: $\lambda_0 > 0 \wedge \lambda_1 \neq 0 \wedge x > 0$.

Similar results have been also obtained for the 2nd family of IMSMs (17).

As far as the families of IMSMs (14), (16) are concerned, the corresponding families of first integrals will be linear with respect to parameters λ_i. In such cases, the procedure of constructing the envelope for the family leads to the requirement of turning the constant of one of the integrals V_i, which is contained in K (12), to zero. If we add the equation of this integral to equations of the IMSMs, then we obtain peculiar invariant manifolds. From the geometric viewpoint, these manifolds represent sections of the hypersurface $V_i = 0$ with the elements of the family of manifolds under scrutiny. If the family of IMSMs under investigation is multiparametric (what takes place in our case) then finding the peculiarities of one-parameter subfamilies of such IMSMs is of interest.

Consider, for example, the family of IMSMs (16).

Like in the previous cases, substitute into K (12) the values of λ_i (in the given case $\lambda_2 = 0$) corresponding to the family of IMSMs under investigation. As a result, we have the family of first integrals:

$$K = \lambda_0 V_0 - \lambda_1 V_1 - \frac{1}{2}\lambda_3 V_3.$$

Find a one-parameter subfamily of the family of IMSMs (16). To this end, we consider λ_0 and λ_3 as functions of λ_1. Compute the derivative of the latter expression with respect to λ_1 and equate it to zero:

$$\frac{\partial K}{d\lambda_1} = \frac{d\lambda_0}{d\lambda_1}V_0 - V_1 - \frac{1}{2}\frac{d\lambda_3}{d\lambda_1}V_3 = 0. \tag{20}$$

Having substituted the expressions of the integrals into (20) and having then excluded variables with the use of equations (16), equality (20) assumes the form:

$$\frac{2\lambda_3(\lambda_1)(\lambda_1 + (\alpha^2 + \beta^2)\lambda_0(\lambda_1)\dot{\lambda}_0(\lambda_1)) - (\lambda_1^2 + (\alpha^2 + \beta^2)\lambda_0^2(\lambda_1))\dot{\lambda}_3(\lambda_1)}{2(\alpha^2 + \beta^2)\lambda_3^2(\lambda_1)} = 0, \tag{21}$$

where $\dot{\lambda}_i = d\lambda_i/d\lambda_1$.

Integration of expression (21) gives the following relationship between the parameters:

$$\lambda_3(\lambda_1) = \bar{C}(\lambda_1^2 + (\alpha^2 + \beta^2)\lambda_0^2(\lambda_1)). \tag{22}$$

Here $\bar{C}$ is a constant of integration.

Obviously, equations of family (16), after excluding, for example, parameter λ_3 from them with the aid of relationship (22), define a subfamily of peculiar IMSMs. A similar result has been obtained for the family of IMSMs (14).

4   The Problem Related to Motion of a System of Vortices

Finally, let us consider the problem of motion of N parallel direct vortex lines (having intensities α_i) in an unbounded volume of an ideal fluid. The points of intersection of vortex lines with the plane perpendicular to them have the Cartesian coordinates (x_i, y_i). Motion of such a system is described by Kirchhoff's equations [14]. Consider the case when $N = 3$. For $N = 3$ equations of motion write:

$$
\begin{aligned}
\dot{x}_1 &= -\frac{\alpha_2(y_1 - y_2)}{4\pi((x_1 - x_2)^2 + (y_1 - y_2)^2)} - \frac{\alpha_3(y_1 - y_3)}{4\pi((x_1 - x_3)^2 + (y_1 - y_3)^2)}, \\
\dot{x}_2 &= \frac{\alpha_1(y_1 - y_2)}{4\pi((x_1 - x_2)^2 + (y_1 - y_2)^2)} - \frac{\alpha_3(y_2 - y_3)}{4\pi((x_2 - x_3)^2 + (y_2 - y_3)^2)}, \\
\dot{x}_3 &= \frac{\alpha_1(y_1 - y_3)}{4\pi((x_1 - x_3)^2 + (y_1 - y_3)^2)} + \frac{\alpha_2(y_2 - y_3)}{4\pi((x_2 - x_3)^2 + (y_2 - y_3)^2)}, \\
\dot{y}_1 &= \frac{\alpha_2(x_1 - x_2)}{4\pi((x_1 - x_2)^2 + (y_1 - y_2)^2)} + \frac{\alpha_3(x_1 - x_3)}{4\pi((x_1 - x_3)^2 + (y_1 - y_3)^2)}, \\
\dot{y}_2 &= -\frac{\alpha_1(x_1 - x_2)}{4\pi((x_1 - x_2)^2 + (y_1 - y_2)^2)} + \frac{\alpha_3(x_2 - x_3)}{4\pi((x_2 - x_3)^2 + (y_2 - y_3)^2)}, \\
\dot{y}_3 &= -\frac{\alpha_1(x_1 - x_3)}{4\pi((x_1 - x_3)^2 + (y_1 - y_3)^2)} - \frac{\alpha_2(x_2 - x_3)}{4\pi((x_2 - x_3)^2 + (y_2 - y_3)^2)}
\end{aligned}
\tag{23}
$$

The differential equations (23) possess the following first integrals:

$$
\begin{aligned}
V_0 &= -\frac{1}{8\pi}\Big(\alpha_1\alpha_2 \ln\big((x_1 - x_2)^2 + (y_1 - y_2)^2\big) \\
&\quad + \alpha_1\alpha_3 \ln\big((x_1 - x_3)^2 + (y_1 - y_3)^2\big) \\
&\quad + \alpha_2\alpha_3 \ln\big((x_2 - x_3)^2 + (y_2 - y_3)^2\big)\Big) = h = const, \\
V_1 &= \alpha_1 x_1 + \alpha_2 x_2 + \alpha_3 x_3 = c_1 = const, \\
V_2 &= \alpha_1 y_1 + \alpha_2 y_2 + \alpha_3 y_3 = c_2 = const, \\
V_3 &= \alpha_1(x_1^2 + y_1^2) + \alpha_2(x_2^2 + y_2^2) + \alpha_3(x_3^2 + y_3^2) = c_3 = const,
\end{aligned}
\tag{24}
$$

As far as the given problem is concerned, the families of IMSMs and their peculiar properties are – likewise in problems considered above – of interest for us.

4.1   Finding the Invariant Manifolds

To the end of finding the invariant manifolds for the system (23) we introduce the function

$$
K = 4\lambda_0 V_0 - \frac{1}{2}\lambda_1 V_1^2 - \frac{1}{2}\lambda_2 V_2^2 - \frac{1}{2}\lambda_3 V_3,
\tag{25}
$$

where $\lambda_0, \lambda_1, \lambda_2$, and λ_3 are some constants.

The conditions of stationarity of K with respect to variables $x_1, x_2, x_3, y_1, y_2,$ and y_3 write:

$$\partial K/\partial x_1 = -\alpha_1\lambda_0\left(\frac{\alpha_2(x_1 - x_2)}{\pi((x_1 - x_2)^2 + (y_1 - y_2)^2)} + \frac{\alpha_3(x_1 - x_3)}{\pi((x_1 - x_3)^2 + (y_1 - y_3)^2)}\right)$$
$$-\alpha_1\lambda_1(\alpha_1 x_1 + \alpha_2 x_2 + \alpha_3 x_3) - \alpha_1\lambda_3 x_1 = 0,$$

$$\partial K/\partial x_2 = \alpha_2\lambda_0\left(\frac{\alpha_1(x_1 - x_2)}{\pi((x_1 - x_2)^2 + (y_1 - y_2)^2)} - \frac{\alpha_3(x_2 - x_3)}{\pi((x_2 - x_3)^2 + (y_2 - y_3)^2)}\right)$$
$$-\alpha_2\lambda_1(\alpha_1 x_1 + \alpha_2 x_2 + \alpha_3 x_3) - \alpha_2\lambda_3 x_2 = 0,$$

$$\partial K/\partial x_3 = \alpha_3\lambda_0\left(\frac{\alpha_1(x_1 - x_3)}{\pi((x_1 - x_3)^2 + (y_1 - y_3)^2)} + \frac{\alpha_2(x_2 - x_3)}{\pi((x_2 - x_3)^2 + (y_2 - y_3)^2)}\right)$$
$$-\alpha_3\lambda_1(\alpha_1 x_1 + \alpha_2 x_2 + \alpha_3 x_3) - \alpha_3\lambda_3 x_3 = 0,$$

$$\partial K/\partial y_1 = -\alpha_1\lambda_0\left(\frac{\alpha_3(y_1 - y_3)}{\pi((x_1 - x_3)^2 + (y_1 - y_3)^2)} + \frac{\alpha_2(y_1 - y_2)}{\pi((x_1 - x_2)^2 + (y_1 - y_2)^2)}\right)$$
$$-\alpha_1\lambda_2(\alpha_1 y_1 + \alpha_2 y_2 + \alpha_3 y_3) - \alpha_1\lambda_3 y_1 = 0,$$

$$\partial K/\partial y_2 = \alpha_2\lambda_0\left(\frac{\alpha_1(y_1 - y_2)}{\pi((x_1 - x_2)^2 + (y_1 - y_2)^2)} - \frac{\alpha_3(y_2 - y_3)}{\pi((x_2 - x_3)^2 + (y_2 - y_3)^2)}\right)$$
$$-\alpha_2\lambda_2(\alpha_1 y_1 + \alpha_2 y_2 + \alpha_3 y_3) - \alpha_2\lambda_3 y_2 = 0,$$

$$\partial K/\partial y_3 = \alpha_3\lambda_0\left(\frac{\alpha_1(y_1 - y_3)}{\pi((x_1 - x_3)^2 + (y_1 - y_3)^2)} + \frac{\alpha_2(y_2 - y_3)}{\pi((x_2 - x_3)^2 + (y_2 - y_3)^2)}\right)$$
$$-\alpha_3\lambda_2(\alpha_1 y_1 + \alpha_2 y_2 + \alpha_3 y_3) - \alpha_3\lambda_3 y_3 = 0. \tag{26}$$

Using not very complex linear transformations, it is possible to reduce equations (26) to the form

$$\alpha_1\lambda_0\left(\frac{\alpha_2(x_1 - x_2)}{\pi((x_1 - x_2)^2 + (y_1 - y_2)^2)} + \frac{\alpha_3(x_1 - x_3)}{\pi((x_1 - x_3)^2 + (y_1 - y_3)^2)}\right) + \alpha_1\lambda_1(\alpha_1 x_1$$
$$+\alpha_2 x_2 + \alpha_3 x_3) + \alpha_1\lambda_3 x_1 = 0,$$

$$\alpha_2\lambda_0\left(\frac{\alpha_1(x_1 - x_2)}{\pi((x_1 - x_2)^2 + (y_1 - y_2)^2)} - \frac{\alpha_3(x_2 - x_3)}{\pi((x_2 - x_3)^2 + (y_2 - y_3)^2)}\right) - \alpha_2\lambda_1(\alpha_1 x_1$$
$$+\alpha_2 x_2 + \alpha_3 x_3) - \alpha_2\lambda_3 x_2 = 0,$$

$$(\alpha_1 x_1 + \alpha_2 x_2 + \alpha_3 x_3)((\alpha_1 + \alpha_2 + \alpha_3)\lambda_1 + \lambda_3) = 0,$$

$$\alpha_1\lambda_0\left(\frac{\alpha_3(y_1 - y_3)}{\pi((x_1 - x_3)^2 + (y_1 - y_3)^2)} + \frac{\alpha_2(y_1 - y_2)}{\pi((x_1 - x_2)^2 + (y_1 - y_2)^2)}\right) + \alpha_1\lambda_2(\alpha_1 y_1$$
$$+\alpha_2 y_2 + \alpha_3 y_3) + \alpha_1\lambda_3 y_1 = 0,$$

$$\alpha_2\lambda_0\left(\frac{\alpha_1(y_1 - y_2)}{\pi((x_1 - x_2)^2 + (y_1 - y_2)^2)} - \frac{\alpha_3(y_2 - y_3)}{\pi((x_2 - x_3)^2 + (y_2 - y_3)^2)}\right) - \alpha_2\lambda_2(\alpha_1 y_1$$
$$+\alpha_2 y_2 + \alpha_3 y_3) - \alpha_2\lambda_3 y_2 = 0,$$

$$(\alpha_1 y_1 + \alpha_2 y_2 + \alpha_3 y_3)((\alpha_1 + \alpha_2 + \alpha_3)\lambda_2 + \lambda_3) = 0. \tag{27}$$

As obvious from (27), the third and (or) the last of the equations turn to zero, when one of the following conditions hold:

i) $\lambda_1 = -\dfrac{\lambda_3}{\alpha_1 + \alpha_2 + \alpha_3}$, $\lambda_2 = -\dfrac{\lambda_3}{\alpha_1 + \alpha_2 + \alpha_3}$; ii) $\lambda_1 = -\dfrac{\lambda_3}{\alpha_1 + \alpha_2 + \alpha_3}$;

iii) $\lambda_2 = -\dfrac{\lambda_3}{\alpha_1 + \alpha_2 + \alpha_3}$ $\hfill(28)$

and hence the Jacobian of system (26) turns to zero. Conditions (28) will be considered in the capacity of necessary conditions of existence of invariant manifolds for the system of differential equations (23).

After substituting one of the conditions (28) into equations (27), these equations (with the use of simple linear transformations) are reduced to a cubic equation. Solutions of the equation are rather easy to obtain, while using the tools of CAS *Mathematica*, but these will be rather bulky. We have obtained solutions of the cubic equation under various restrictions imposed on the parameters α_i. Consider below some of the families of solutions obtained. These represent the families of IMSMs for the equations (23):

$$\frac{1}{2}a_1\alpha_3^2(x_1 - x_2)^2 - a_0(y_2 - y_3)^2 - \frac{\alpha_1\alpha_3^2\lambda_0(a_2\alpha_1 + \alpha_3^2)}{\pi\lambda_3} = 0,$$

$$(x_3 - x_2)^2 - (y_2 - y_3)^2 + \frac{\alpha_1\alpha_3^2(\alpha_3 + \sqrt{4\alpha_1^2 - 3\alpha_3^2})\lambda_0}{\pi\lambda_3(\alpha_1 - \alpha_3)a_1} = 0,$$

$$2\alpha_3 y_1 + (a_2 - \alpha_3)y_2 - (a_2 + \alpha_3)y_3 = 0 \text{ when } \lambda_1 = \lambda_2 = -\frac{\lambda_3}{\alpha_1}, \ \alpha_2 = -\alpha_3; \quad (29)$$

$$\frac{\alpha_2(x_1 - x_2)^2}{2\alpha_2 + \alpha_3} + \frac{2\alpha_2 + \alpha_3}{5\alpha_2 + 4\alpha_3}y_3^2 + \frac{\alpha_2^2\lambda_0}{\pi\lambda_3(\alpha_2 + \alpha_3)} = 0,$$

$$\frac{\alpha_2^2 a_4(x_3 - x_2)^2}{2\alpha_2 + \alpha_3} + \frac{a_3(2\alpha_2 + \alpha_3)(\alpha_2 + \alpha_3)^2}{5\alpha_2 + 4\alpha_3}y_3^2 + \frac{a_3(\alpha_2 + \alpha_3)\alpha_2^2\lambda_0}{\lambda_3\pi} = 0,$$

$$2\alpha_2\sqrt{5\alpha_2 + 4\alpha_3}y_1 + a_5 y_3 = 0, \ 2\alpha_2\sqrt{5\alpha_2 + 4\alpha_3}y_2 + a_4 y_3 = 0$$

when $\lambda_1 = -\dfrac{\lambda_3}{2\alpha_2 + \alpha_3}$, $\alpha_1 = \alpha_2$; $\hfill(30)$

$$\alpha_3^2 x_3^2 + \alpha_2^2(y_1 - y_2)^2 + \frac{\alpha_2^4\alpha_3(\alpha_2 - a_6)\lambda_0}{\pi\lambda_3 a_7} = 0,$$

$$\alpha_3^2 a_7 x_3^2 + \frac{\alpha_2^4((a_6 + \alpha_2)\alpha_3 - 2\alpha_2^2)}{\alpha_2 + \alpha_3}(y_3 - y_2)^2 + \frac{\alpha_2^4\alpha_3(\alpha_2 - a_6)\lambda_0}{\lambda_3\pi} = 0,$$

$$2\alpha_2^2 x_1 + ((a_6 - \alpha_2)\alpha_3 - 2\alpha_2^2)x_3 = 0, \ 2\alpha_2^3 x_2 + \alpha_2((a_6 + \alpha_2)\alpha_3 - 2\alpha_2^2)x_3 = 0$$

when $\lambda_2 = -\dfrac{\lambda_3}{\alpha_3}$, $\alpha_1 = -\alpha_2$. $\hfill(31)$

The following notations are used here: $a_0 = (2\alpha_1\alpha_2 + \alpha_3^2)(\alpha_3^2 - 2\alpha_1^2) - \alpha_1^2\alpha_3^2$, $a_1 = 2\alpha_1(\sqrt{4\alpha_1^2 - 3\alpha_3^2} - \alpha_3) + \alpha_3(\sqrt{4\alpha_1^2 - 3\alpha_3^2} + \alpha_3) - 4\alpha_1^2$, $a_2 = \sqrt{4\alpha_1^2 - 3\alpha_3^2} - 2\alpha_1$, $a_3 = 3\sqrt{\alpha_2} + \sqrt{5\alpha_2 + 4\alpha_3}$, $a_4 = ((2\alpha_2 + \alpha_3)\sqrt{\alpha_2} + \alpha_3\sqrt{5\alpha_2 + 4\alpha_3})$, $a_5 = \alpha_3\sqrt{5\alpha_2 + 4\alpha_3} - (2\alpha_2 + \alpha_3)\sqrt{\alpha_2}$, $a_6 = 2\alpha_3 + \sqrt{4\alpha_3^2 - 3\alpha_2^2}$, $a_7 = (\alpha_2 - 3\alpha_3)(\alpha_2^2 - \alpha_3 a_6) + a_6(\alpha_3^2 - \alpha_2^2)$.

As far as the family of IMSMs (29) is concerned, let us consider the problem of existence of such motions as triangular configurations of vortices [15] to this family.

In this connection, we shall compute the distances between the centers of the vortices on the family of IMSMs under scrutiny, while assuming that centers of the vortices are placed in the corners of the triangle. Substitute values of the variables from (29) into the formulas, which describe the distances between the corners of the triangle. As a result, we have:

$$(x_1 - x_3)^2 + (y_1 - y_3)^2 = -\frac{\alpha_1\lambda_0}{2\pi\lambda_3(\alpha_1 - \alpha_3)}(\alpha_3 + \sqrt{4\alpha_1^2 - 3\alpha_3^2}),$$

$$(x_1 - x_2)^2 + (y_1 - y_2)^2 = \frac{\alpha_1\lambda_0}{2\pi\lambda_3(\alpha_1 + \alpha_3)}(\alpha_3 - \sqrt{4\alpha_1^2 - 3\alpha_3^2}),$$

$$(x_2 - x_3)^2 + (y_2 - y_3)^2 = \frac{\alpha_1\lambda_0}{\pi\lambda_3(\alpha_1^2 - \alpha_3^2)}(\alpha_3^2 - \alpha_1(2\alpha_1 + \sqrt{4\alpha_1^2 - 3\alpha_3^2})).(32)$$

Right-hand sides of expressions (32) are positive, when

$$\alpha_1 > \wedge\left(-\frac{2\alpha_1}{\sqrt{3}} \leq \alpha_3 < -\alpha_1 \vee \alpha_1 < \alpha_3 \leq \frac{2\alpha_1}{\sqrt{3}}\right).$$

The latter is some restriction imposed on vortex intensities. When it holds, the motions, which correspond to triangular configurations of the vortices, belong to the family of IMSMs (29). Similar conditions may be obtained for the families of IMSMs (30), (31).

4.2   Analysis of Peculiar Properties of Invariant Manifolds

Let us again try to employ our approach for the analysis of peculiar properties of the families of IMSMs. Consider, for example, the family of IMSMs (29).

As above, we substitute into K (25) the values of parameters λ_1 and λ_2, which correspond to the family of IMSMs under scrutiny. As a result, we have

$$2\tilde{K} = -8\lambda_0 V_0 + \frac{\lambda_3}{\alpha_1}V_1^2 + \frac{\lambda_3}{\alpha_1}V_2^2 - \lambda_3 V_3.$$

The latter expression represents a combination of first integrals, which is linear with respect to the parameters. The partial derivative $\tilde{K}$ computed with respect to λ_3 (or λ_0) does not contain the parameter needed for constructing the enveloping integral:

$$2\frac{\partial\tilde{K}}{\partial\lambda_3} = \frac{1}{\alpha_1}(V_1^2 + V_2^2) - V_3 = 0. \tag{33}$$

Obviously, equality (33) itself defines the invariant manifold of initial differential equations (23) (i.e. the surface of the level of the corresponding integral).

Substitute into (33) the corresponding expressions of first integrals. As a result, we obtain the equation of the invariant manifold in the explicit form:

$$(\alpha_1 x_1 - \alpha_3 x_2 + \alpha_3 x_3)^2 + (\alpha_1 y_1 - \alpha_3 y_2 + \alpha_3 y_3)^2 - \alpha_1(\alpha_1(x_1^2 + y_1^2)$$
$$-\alpha_3(x_2^2 + y_2^2) + \alpha_3(x_3^2 + y_3^2)) = 0.$$

The latter expression may be rewritten as follows:

$$\alpha_3((x_2 - x_3)^2 + (y_2 - y_3)^2) - \alpha_1((x_2 - x_3)(x_1 - x_2)$$
$$+(y_2 - y_3)(y_1 - y_2) + (x_2 - x_3)(x_1 - x_3) + (y_2 - y_3)(y_1 - y_3)) = 0. \quad (34)$$

If the vectors, which connect the centers of vortices, are introduced

$$r_1 = \{(x_1-x_3),(y_1-y_3)\}, \quad r_2 = \{(x_2-x_3),(y_2-y_3)\}, \quad r_3 = \{(x_1-x_2),(y_1-y_2)\},$$

then equality (34) in its vector form writes:

$$\alpha_3(r_2, r_2) - \alpha_1(r_2, r_3) - \alpha_1(r_2, r_1) = 0.$$

It demonstrates more explicitly the restrictions, which are imposed by the invariant manifold (34) on the character of motion of the vortices defined by the family of IMSMs (29) when equation (34) is added to equations of this family.

Let us use the results of section 3.2 and find the relationship between the parameters λ_0, λ_3, under which the subfamily of IMSMs (29) is peculiar.

As in the previous case, we consider the parameter λ_3 as a function of λ_0. Having computed the partial derivative of $\tilde{K}$ with respect to λ_0 we obtain the expression

$$2\frac{\partial \tilde{K}}{\partial \lambda_3} = -8V_0 + \frac{d\lambda_3}{\alpha_1 d\lambda_0}V_1^2 + \frac{d\lambda_3}{\alpha_1 d\lambda_0}V_2^2 - \frac{d\lambda_3}{d\lambda_0}V_3 = 0. \quad (35)$$

After substitution of the expressions for the integrals into (35) and elimination of variables with the use of equations (29), the latter expression writes:

$$\frac{\alpha_3^2 \lambda_0 \dot{\lambda}_3(\lambda_0)}{2\pi\lambda_3(\lambda_0)} - \ln\left(2^{-\alpha_1\alpha_3/2\pi}\pi^{-\alpha_3^2/2\pi}\left(\frac{2\alpha_1^2 - \alpha_3(\alpha_3 + \sqrt{4\alpha_1^2 - 3\alpha_3^2})}{(\alpha_1 + \alpha_3)^2}\right)^{\alpha_1\alpha_3/2\pi}\right.$$

$$\left. \times \left(-\frac{\alpha_1\alpha_3^2\lambda_0}{(\alpha_3^2 - \alpha_1(2\alpha_1 - \sqrt{4\alpha_1^2 - 3\alpha_3^2}))\lambda_3(\lambda_0)}\right)^{\alpha_3^2/2\pi}\right) = 0. \quad (36)$$

The expression (36) will be considered as a differential equation, where λ_3 is a desired function, λ_0 is an independent variable, $\dot{\lambda}_3 = d\lambda_3/d\lambda_0$. We use the *Mathematica* function "DSolve" to find a solution of the equation. The obtained solution writes:

$$\lambda_3(\lambda_0) = -\frac{1}{\pi(\alpha_3^2 - \alpha_1(2\alpha_1 - \sqrt{4\alpha_1^2 - 3\alpha_3^2})}\left(e^{\frac{1}{\alpha_3\lambda_0}e^{C[1]\alpha_3}-1}\alpha_1\alpha_3^2\right.$$

$$\left. \times (\alpha_1 + \alpha_3)^{-2\alpha_1/\alpha_3}\left(\alpha_1^2 - \frac{1}{2}\alpha_3(\alpha_3 + \sqrt{4\alpha_1^2 - 3\alpha_3^2})\right)^{\alpha_1/\alpha_3}\lambda_0\right), \quad (37)$$

where $C[1]$ is a constant of integration.

So, when the relationship (37) between the parameters λ_0 and λ_3 takes place, the corresponding subfamily of the family of IMSMs (29) is peculiar. Similar results have been obtained for the families of IMSMs (30), (31).

5 Conclusion

The paper has presented an approach to investigation of peculiar properties of families of invariant manifolds for conservative systems. The approach is based on application of the envelopes for the families of first integrals of such systems. The Routh–Lyapunov's method has been used for finding the families of IMSMs. When solving the computational problems, we applied CAS *Mathematica*.

Analysis of peculiar properties of families of invariant manifolds in the problems related to motion of a rigid body having one fixed point (Brun's problem) and to motion of a system of 3 vortices, as well as an analysis of Euler's equations in Lie algebras have been conducted by the method of envelope.

For example, when solving Brun's problem, we have found peculiar families of IMSMs, which describe pendular oscillations of a body about the body main horizontal inertial axes, and a special class of degenerate permanent rotations of the body about the vertical. Analysis of stability in the sense of Lyapunov has been conducted for the family of invariant manifolds, which defines the body pendular oscillations.

In the process of investigating the Euler's equations in Lie algebras the conditions, under which the peculiar families of IMSMs exist, have been obtained.

Peculiar properties of the family of IMSMs have been revealed in the problem of motion for a system of 3 vortices. The conditions, under which the motions corresponding to triangular configurations of vortices belong to these invariant manifolds, have been obtained.

Acknowledgements. The research presented in this paper was supported by the grant 06-1000013-9019 from INTAS-SB RAS.

References

1. Lyapunov, A.M.: On Permanent Helical Motions of a Rigid Body in Fluid. Collected Works 1, USSR Acad. Sci., Moscow, Leningrad (1954)
2. Irtegov, V., Titorenko, T.: On the properties of families of first integrals. In: Proc. Fifth International Workshop on Computer Algebra in Scientific Computing. Yalta, Ukraine. Publ. by Techn. Univ. München, Garching, pp. 175–182 (2002)
3. Goursat, E.: The Course of Mathematical Analysis. vol. 1, Moscow, Leningrad (1936)
4. Irtegov, V.D.: Invariant Manifolds of Steady-State Motions and Their Stability. Nauka Publ., Novosibirsk (1985)
5. Irtegov, V.D., Titorenko, T.N.: Using the system "Mathematica" in problems of mechanics. Mathematics and Computers in Simulation 3-5(57), 227–237 (2001)
6. Appel, P.: Theoretical Mechanics. vol. 2, GIF-ML, Moscow (1960)
7. Zubov, V.I.: Oscillations and Waves. Leningrad, Leningrad State Univ. (1989)

8. Borisov, A.V., Mamaev, I.S., Sokolov, V.V.: A new integrable case on so(4). Dokl. Phys. 12(46), 888–889 (2001)
9. Sokolov, V.V.: A new integrable case for the Kirchhoff equations. Theoret. and Math. Phys. 1(129), 1335–1340 (2001)
10. Ryabov, P.E.: Bifurcations of first integrals in the Sokolov case. Theoret. and Math. Phys. 2(34), 181–197
11. Morozov, V.P.: The topology of Liouville foliations for the cases of Steklov's and Sokolov's integrability. Math. Sbornik 3(195), 69–114 (2004)
12. Haghighatdoost, Gh.: The topology of isoenergetic surfaces for the Sokolov integrable case in the Lie algebra so(4). Dokl. Math. 2(71), 256 (2005)
13. Rumyantsev, V.V.: Comparison of three methods of constructing Lyapunov functions. J. Appl. Math. Mech. 6(59), 873–877 (1995)
14. Lamb, G.: Hydrodynamics. Regular and Chaotic Dynamics I, II, Izhevsk (2003)
15. Meleshko, V.V., Konstantinov, M.Yu.: Dynamics of Vortex Structures. Naukova Dumka, Kyiv (1993)

A Unified Algorithm for Multivariate Analytic Factorization

Maki Iwami

Osaka University of Economics and Law, Japan
maki@keiho-u.ac.jp

Abstract. The expansion base algorithm, which was devised by Abhyankar, Kuo and McCallum is very efficient for analytic factorization of bivariate polynomials. The author had extended it to more than two variables but it was only for polynomials with non-vanishing leading coefficient at the expansion point. In this paper, we improve it to be able to apply to polynomials including the case of vanishing leading coefficient, that is, singular leading coefficient, which comes to a specific problem only for more than two variables[1].

1 Introduction

Analytic factorization is a factorization over the ring of formal power series, by fixing the expansion point, and is very important operation for local analysis of algebraic curves and algebraic surfaces. For example, the bivariate polynomial $x^2 - u^2 - u^3$ is irreducible in the polynomial ring, but reducible in the formal power series ring at the origin; $x^2 - u^2 - u^3 = (x + u + \frac{1}{2}u^2 - \frac{1}{8}u^3 + \cdots)(x - u - \frac{1}{2}u^2 + \frac{1}{8}u^3 - \cdots)$. As shown in Fig.1, the algebraic curve determined by $F_1 = x^2 - u^2 - u^3 = 0$ is factorized into two irreducible factors at the origin, which shows that the curve has two tangent lines at the origin, whereas $F_2 = x^2 - u^3 = 0$

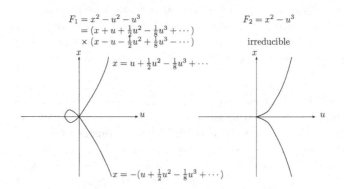

$$F_1 = x^2 - u^2 - u^3$$
$$= (x + u + \tfrac{1}{2}u^2 - \tfrac{1}{8}u^3 + \cdots)$$
$$\times (x - u - \tfrac{1}{2}u^2 + \tfrac{1}{8}u^3 - \cdots)$$

$$x = u + \tfrac{1}{2}u^2 - \tfrac{1}{8}u^3 + \cdots$$

$$x = -(u + \tfrac{1}{2}u^2 - \tfrac{1}{8}u^3 + \cdots)$$

$$F_2 = x^2 - u^3$$

irreducible

Fig. 1.

[1] A part of this work was supported by JSPS. Grant-in-Aid for Scientific Research.

V.G. Ganzha, E.W. Mayr, and E.V. Vorozhtsov (Eds.): CASC 2007, LNCS 4770, pp. 211–223, 2007.

is irreducible showing that the corresponding curve has only one tangent line at the origin.

In the case of more than two variables, the analytic factorization corresponds to the decomposition of (hyper)surface.

Let K be a number field of characteristic 0 and $\overline{K}$ be an algebraic closure of K. Let $K[u_1, \cdots, u_\ell]$, $K(u_1, \cdots, u_\ell)$ and $K\{u_1, \cdots, u_\ell\}$ be the ring of polynomials, the field of rational functions and the ring of formal power series, respectively, over K in variables $u_1, \cdots, u_\ell$. Let $(s_1, \cdots, s_\ell) \in \overline{K}^\ell$, and we abbreviate $(u_1, \cdots, u_\ell)$ and $(s_1, \cdots, s_\ell)$ to $(\boldsymbol{u})$ and $(\boldsymbol{s})$, respectively. Without loss of generality, we assume that the expansion point is $(\boldsymbol{s}) = (\boldsymbol{0})$.

Definition 1 (analytically (ir)reducible). *A nonzero polynomial $F(x, \boldsymbol{u})$ in $K[x, \boldsymbol{u}]$ which is not a unit of $K\{x, \boldsymbol{u}\}$ is said to be analytically (ir)reducible over $\overline{K}$ at the origin if $F(x, \boldsymbol{u})$ is (ir)reducible in $\overline{K}\{x, \boldsymbol{u}\}$.*

Definition 2 (analytic factorization). *Factorizing $F(x, \boldsymbol{u})$ into irreducible factors in $\overline{K}\{x, \boldsymbol{u}\}$ is called* **analytic factorization***.*

In the analytic factorization, owing to the Weierstrass preparation theorem, we can regard factors as polynomials w.r.t. one variable, let it be x. Therefore, in practice, we factorize a given polynomial $F(x, \boldsymbol{u})$ into irreducible factors in $\overline{K}\{\boldsymbol{u}\}[x]$. Then we call x the main variable, and $\boldsymbol{u}$ sub-variables.

As is well known in computer algebra, the generalized Hensel construction allows us to decompose a multivariate polynomial into factors in $\overline{K}\{\boldsymbol{u}\}[x]$ if the polynomial satisfies some conditions. Hence, we have only to consider the analytic factorization in the case that the Hensel construction breaks down. By $\deg_x(F)$, we denote the degree of F w.r.t. x. Let $F(x, \boldsymbol{u}) = f_D(\boldsymbol{u})x^D + \cdots + f_0(\boldsymbol{u})$. The generalized Hensel construction breaks down if A) $F(x, \boldsymbol{0}) = x^D$ ($D = \deg_x(F) \geq 2$) or B) $f_D(\boldsymbol{0}) = 0$; the leading coefficient disappears. The Hensel construction breaks down in the case A) because x^D cannot be decomposed into relatively prime factors.

Definition 3 (singular point, singular leading coefficient). *For $F(x, \boldsymbol{u}) = f_D(\boldsymbol{u})x^D + \cdots + f_0(\boldsymbol{u})$, we call the expansion point $(\boldsymbol{s})$ a singular point for the Hensel construction, or a singular point in short, if $F(x, \boldsymbol{s})$ is not square free. If $f_D(\boldsymbol{s}) = 0$ then we say the leading coefficient is singular at $\boldsymbol{s}$.*

Thus, in the analytic factorization, we have only to consider the case that $F(x, \boldsymbol{u})$ is singular or its leading coefficient is singular at the expansion point.

When the Hensel construction breaks down, i.e. the expansion point is at the singular point, we use the extended Hensel construction; see [11] for A), [10] for B). The procedure of the extended Hensel construction is as follows: 1) introduce the weighted total-degree variable t by $u_i \rightarrow t^{\omega_i} u_i$, 2) for each nonzero term $cx^{e_x} t^{e_t} u_1^{e_{u_1}} \cdots u_\ell^{e_{u_\ell}}$, plot a dot at the point (e_x, e_t), 3) determine so-called "Newton's polynomial F_{New}" by correcting terms on the lowest edge, 4) factorize F_{New} into relatively prime factors, and 5) perform a Hensel-like construction by treating factors of F_{New} as initial factors. In this paper, we set $\omega_i = 1$

$(i = 1, \cdots, \ell)$, and factorize F_{New} in $\overline{K}[x, \boldsymbol{u}]$. Then, problems arise when F_{New} is not square-free. Consider $F = (x^2 - u^3)^2 - u^7$ for example; $F_{\text{New}} = (x^2 - u^3)^2$ cannot be factorized into relatively prime factors in $\overline{K}[x, \boldsymbol{u}]$, but actually F is factorized analytically as $F = (x^2 - u^3 + xu^2 + \frac{1}{2}u^4 + \frac{1}{8}xu^3 + \frac{1}{8}u^5 + \cdots)(x^2 - u^3 - xu^2 + \frac{1}{2}u^4 - \frac{1}{8}xu^3 + \frac{1}{8}u^5 - \cdots)$. In this case, how can we decide whether F is analytically irreducible or not, and if F is reducible how can we decompose it?

Let $F = g^m + \cdots$, where g is irreducible in $\overline{K}[x, \boldsymbol{u}]$, $m \geq 2$ and $F_{\text{New}} = g^m$. In the bivariate case, there are two algorithms to solve the problem mentioned above. One is the "expansion-base method", which comes from ideas of Abhyankar [1,2,3,4] for calculating approximate roots, applied by Kuo [8] to bivariate analytic factorization, and implemented by McCallum [9] as the more complete algorithm. In this algorithm, we regard g as a new main variable (in the above example, $g = x^2 - u^3$ is the new main variable), and expand F w.r.t. g. New variables $G_1 = g$, $G_2, \cdots, G_s$ are generated repeatedly so long as new Newton's polynomial is not square-free over $\overline{K}$. The set $\mathcal{G} = \{G_{-1}(= u), G_0(= x), G_1(= g), \cdots, G_s\}$ is called an expansion base. Another algorithm is devised by Sasaki [12] and it utilizes the extended Hensel construction; we factorize g into linear factors as $g = x^{\hat{d}} - cu^{\hat{\delta}} = \Pi_{i=1}^{\hat{d}}(x - c^{1/\hat{d}}e^{2i\pi \mathbf{i}/\hat{d}}u^{\hat{\delta}/\hat{d}})$, $\mathbf{i} = \sqrt{-1}$, $c \in \overline{K}$ (in the above example, we factorize as $F_{\text{New}} = (x + u^{3/2})^2(x - u^{3/2})^2$), perform the extended Hensel construction by treating $(x - c^{1/\hat{d}}e^{2i\pi \mathbf{i}/\hat{d}}u^{\hat{\delta}/\hat{d}})^m$ $(i = 1, \cdots, \hat{d})$, as initial factors, and multiply the mutually conjugate extended Hensel factors. Then algebraic functions disappear, and we can obtain analytically irreducible factors.

In the multivariate case, there had been no solution to the above problem; here, by multivariate case we mean that the number of variables is more than 2. Then the author had extended the two methods mentioned above to multivariate case. In the multivariate case, another problem arises; factors obtained by the extended Hensel construction which we call extended Hensel factors, are usually not in $\overline{K}\{\boldsymbol{u}\}[x]$. In general, the extended Hensel factors are series of rational functions in sub-variables, and they belong to a ring which is larger than $\overline{K}\{\boldsymbol{u}\}[x]$; we denote it by $\overline{K}\{(\boldsymbol{u})\}[x]$.

Definition 4 (ring of the multivariate extended Hensel factors)
We define the ring of the extended Hensel factors in multivariate case as
$$\overline{K}\{(\boldsymbol{u})\} \overset{\text{def}}{=} \left\{ \sum_{k=0}^{\infty} [\frac{N_k(\boldsymbol{u})}{D_k(\boldsymbol{u})}] \,\middle|\, \begin{matrix} N_k(\boldsymbol{u}) \text{ and } D_k(\boldsymbol{u}) \text{ are homogeneous polynomials} \\ \text{in } \boldsymbol{u} \text{ s.t. } \text{tdeg}(N_k) - \text{tdeg}(D_k) = k \ (k = 0, 1, 2, \cdots) \end{matrix} \right\}$$

Therefore, in the multivariate case, we perform the factorization in $\overline{K}\{(\boldsymbol{u})\}[x]$ and we get irreducible factors in $\overline{K}\{\boldsymbol{u}\}[x]$ by combining exteded Hensel factors in $\overline{K}\{(\boldsymbol{u})\}[x]$ so as to cancel their denominators (their combinations are trivial from the type of denominators).

We perform extended Hensel construction as a preprocessing and reduce the problem. That is to say, let $g_1(x, \boldsymbol{u}), \cdots, g_r(x, \boldsymbol{u})$, $g_{r+1}(x, \boldsymbol{u}), \cdots, g_{r+r'}(x, \boldsymbol{u})$ be irreducible in $\overline{K}[x, \boldsymbol{u}]$, $m_{r+1}, \cdots, m_{r+r'}$ be natural number greater than 2, by $A \Rightarrow B$ we denote that an expression A before extended Hensel construction is transformed to an expression B after the construction;

$$F_{\text{New}} = f_D(\mathbf{0}) \; x^{n_0} \; g_1(x,\mathbf{u}) \cdots g_r(x,\mathbf{u}) \; g_{r+1}(x,\mathbf{u})^{m_{r+1}} \cdots g_{r+r'}(x,\mathbf{u})^{m_{r+r'}}$$
$$= f_D(\mathbf{0}) \; F_0^{(0)} \quad F_1^{(0)} \cdots F_r^{(0)} \qquad F_{r+1}^{(0)} \qquad \cdots \qquad F_{r+r'}^{(0)}$$

$$\Downarrow \quad \vdots \quad \Downarrow \quad \Downarrow \quad \Downarrow \quad \cdots \quad \Downarrow \qquad \Downarrow \qquad \cdots \qquad \Downarrow$$

$$F(x,\mathbf{u}) = f_D(\mathbf{u}) \; F_0^{(\infty)} \; F_1^{(\infty)} \cdots F_r^{(\infty)} \qquad F_{r+1}^{(\infty)} \qquad \cdots \qquad F_{r+r'}^{(\infty)}$$

If $n_0 = 1$ then we may include x^{n_0} in any of $g_1(x,\mathbf{u}), \cdots, g_r(x,\mathbf{u})$, else if $n_0 \geq 2$ then we perform extended Hensel construction to $F_0^{(\infty)}$ recursively. $F_1^{(\infty)}, \cdots, F_r^{(\infty)}$ are irreducible in $\overline{K}\{(\mathbf{u})\}[x]$. Therefore problems are reduced to factorization $F_{r+1}^{(\infty)}, \cdots, F_{r+r'}^{(\infty)}$ in $\overline{K}\{(\mathbf{u})\}[x]$, say, factorization of the form $F(x,\mathbf{u}) = g^m + \cdots$, where g is irreducible in $\overline{K}[x,\mathbf{u}]$, $m \geq 2$ and $F_{\text{New}} = g^m$. After all, we can say that the research for analytic factorization means how to solve this problem.

For this problem, the author extended Sasaki's method for bivariate case to multivariate case in [5]. And then the author extended "the expansion base method for bivariate case" to multivariate case in [6]. The fundamental idea which is common to both algorithms is to regard multivariate polynomials as bivariate polynomials w.r.t. x and the total-degree variable t.

The extension of Sasaki's method is as follows. First, we decompose g^m into linear factors introducing algebraic functions as $g^m = \Pi_{i=1}^{\hat{d}}(x - t^{\hat{\delta}/\hat{d}}\theta_i)^m$, and then we perform the extended Hensel construction in $\overline{K}(\theta_1, \cdots, \theta_{\hat{d}})\{(\mathbf{u})\}[x]$ by regarding the linear factors as initial factors, multiply mutually conjugate extended Hensel factors, then algebraic functions disappear, and we can obtain irreducible factorization in $\overline{K}\{(\mathbf{u})\}[x]$, and finally obtain factors in $\overline{K}\{\mathbf{u}\}[x]$ by combining the extended Hensel factors in $\overline{K}\{(\mathbf{u})\}[x]$ so as to cancel their denominators by focusing attention on their denominators. In section 2, we describe the extension of the expansion base algorithm [6]. We regard g as a new main variable, and expand F w.r.t. g. New variables are generated as $G_1(= g)$, $G_2, \cdots, G_s$ so long as new Newton's polynomial is not square-free over K, recursively. Then we obtain the expansion base $\mathcal{G} = \{G_{-1}(= t), G_0(= x), G_1(= g), \cdots, G_s\}$ over

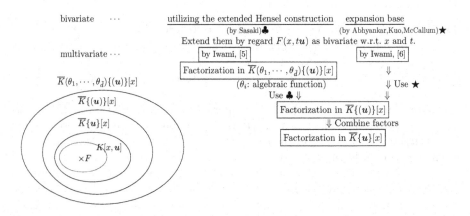

Fig. 2. Extension from Bivariate to Multivariate Case

$\overline{K}\{(\boldsymbol{u})\}$. The author also invented lifting techniques for the multivariate expansion base algorithm which we can use in the bivariate case, too.

In [6], the author did not treat the case that the leading coefficient disappears at the expansion point, say, the leading coefficient is singular. We modify [6] so as to include the case that the leading coefficient is singular, which can be seen in [7] as an extended abstract of the poster presentation of the 8th international workshop on computer algebra in scientific computing in 2005. In this paper, the theorems, proofs, algorithms and examples are given. See section **3**. Note that the problem of singular leading coefficient does not arise in the bivariate case because we can factorize F as $F = F_1 F_2$, where the leading coefficient of F_1 is not singular and F_2 is a unit in $\overline{K}\{(u)\}[x]$.

As for the implementation of multivariate analytic factorization, the author made a demonstration by Mathematica in the presentation of conference of Computer Algebra – Design of Algorithms, Implementations and Applications at RIMS in Kyoto in 2006.

These results also give us methods of factorization in the polynomial ring especially Newton's polynomial of F is not square-free.

2 Multivariate Expansion Base in the Case of Nonsingular Leading Coefficient $(f_D(0) \neq 0)$

In this section, we review [6]. We extend the expansion base method [9] to the multivariate case by regarding F as a bivariate polynomial w.r.t. x and total-degree variable t. That is to say, for $F_{\text{New}} = g^m$, we define $G_{-1} \overset{\text{def}}{=} t$, $G_0 \overset{\text{def}}{=} x$, $G_1 \overset{\text{def}}{=} g$. $\mathcal{G} \overset{\text{def}}{=} (G_{-1}, G_0, G_1)$, we call $\mathcal{G}$ expansion base. We can obtain unique expression as $F = \Sigma_{e_1=0}^m c_{(e_{-1},e_0,e_1)}(\boldsymbol{u}) G_{-1}^{e_{-1}} G_0^{e_0} G_1^{e_1}$, $c_{(e_{-1},e_0,e_1)}(\boldsymbol{u}) \in \overline{K}\{(\boldsymbol{u})\}$, by dividing G_1, G_0, G_{-1} in this order, and we call this expression $\mathcal{G}$–adic expansion of F. By w_i, we denote the weight of G_i. First, we define $w_{-1} = 1$, then w_0 is determined by the absolute value of the slope of $\mathcal{L}_{\text{New}}$. And plot terms of F on the two-dimensional plane whose horizontal line represents the exponent of G_1 and the vertical line represents the weighted total-degree of G_{-1} and G_0, say $(e_1, w_{-1}e_{-1} + w_0 e_0)$. By $\mathcal{L}_{G_i\text{New}}$ and $F_{G_i\text{New}}$, we denote the Newton's line and Newton's polynomial when horizontal line represents the exponent of G_i, respectively. Then w_1 is determined to be the absolute value of the slope of $\mathcal{L}_{G_1\text{New}}$. If $F_{G_1\text{New}}$ is such a multiple form as $F_{G_1\text{New}} = g_1^{m_1}$ then we put $G_2 = g_1$ and append to $\mathcal{G}$ as $\mathcal{G} = (G_{-1}, G_0, G_1, G_2)$. And we perform $\mathcal{G}$–adic expansion as $F = \Sigma_{e_2=0}^{m_1} c_{(e_{-1},e_0,e_1,e_2)}(\boldsymbol{u}) G_{-1}^{e_{-1}} G_0^{e_0} G_1^{e_1} G_2^{e_2}$, $c_{(e_{-1},e_0,e_1,e_2)}(\boldsymbol{u}) \in \overline{K}\{(\boldsymbol{u})\}$ by dividing G_2, G_1, G_0, G_{-1} in this order and plot terms of F as $(e_2, w_{-1}e_{-1} + w_0 e_0 + w_1 e_1)$, and determine $F_{G_2\text{New}}$. We perform these steps and generate elements of $\mathcal{G}$ as long as new Newton's polynomial cannot be decomposed into relatively prime factors over $\overline{K}\{(\boldsymbol{u})\}$ or $m_s = 1$ where $F_{G_s\text{New}} = g_s^{m_s}$. As a result, if F is reducible over $\overline{K}\{(\boldsymbol{u})\}$, Newton's polynomial is decomposed into relatively prime factors, and with a lifting technique, we can get irreducible factors in $\overline{K}\{(\boldsymbol{u})\}[x]$. Finally, we combine them so as to cancel their denominators and get analytically irreducible factors. Note that $G_i \in \mathcal{G}$ is also called **an irreducible curve germ** in algebraic geometry.

Algorithm 1. [multivariate expansion base with nonsingular leading coefficient]

INPUT   : $F(x, \boldsymbol{u}) \in \overline{K}\{(\boldsymbol{u})\}[x]$ s.t. $F_{\text{New}} = g^m$ where $m \geq 2$ and g is irreducible polynomial. $D \leftarrow \deg_x F$, $G_{-1} \leftarrow t$, $G_0 \leftarrow x$, $G_1 \leftarrow g$, $w_1 \leftarrow \deg_t F/D$, Expansion Base $\mathcal{G} \leftarrow (G_{-1}, G_0, G_1)$, Weight $\mathcal{W} \leftarrow (w_{-1}(= 1), w_0)$, $s \leftarrow 1$, $D_1 \leftarrow \deg_x F/\deg_x G_1$

OUTPUT : Irreducible factors in $\overline{K}\{\boldsymbol{u}\}[x]$

1. $\boxed{\text{If } D_s = 1 \text{ then Return } F.}$

2. Perform $\mathcal{G}$-adic expansion of F as $F = \Sigma c(\boldsymbol{u})_{e_{-1}, e_0, \cdots, e_s} G_{-1}^{e_{-1}} G_0^{e_0} \cdots G_s^{e_s}$. Plot $(e_s, \sum_{i=-1}^{s-1} e_i w_i)$, determine $F_{G_s \text{New}}$, pseudo form $F_{G_s \text{New}}^*$, satisfying $F_{G_s \text{New}}^* \equiv F_{G_s \text{New}}$ as follows and factorize it into irreducible factors over $\overline{K}\{(\boldsymbol{u})\}$;

$$F_{G_s \text{New}}^* = G_s^r \cdot ((G_s^d)^q + a_1(G_s^d)^{q-1}\Delta_s + \cdots + a_q\Delta_s^q), \quad D_s = dq + r$$
$$= G_s^r \cdot h_1(G_s^d, \Delta_s) \cdots h_R(G_s^d, \Delta_s) h_{R+1}(G_s^d, \Delta_s)^{m_{R+1}} \cdots$$
$$h_{R+R'}(G_s^d, \Delta_s)^{m_{R+R'}},$$
$$= H_0^{(0)} \cdot H_1^{(0)} \cdots H_{R+R'}^{(0)}, \quad \Delta = G_{-1}^{h_{-1}} \cdots G_{s-1}^{h_{s-1}}, \quad m_i \geq 2.$$

Append $w_s \leftarrow \sum_{i=-1}^{s-1} w_i h_i/d$ to the last element of $\mathcal{W}$.

$\boxed{\text{If } F_{G_s \text{New}}^* = h_1(G_s^d, \Delta_s) \text{ then Return } F.}$

$\boxed{\begin{array}{l} \text{If } F_{G_s \text{New}}^* = h_1(G_s^d, \Delta_s)^{m_1}, \, d = 1, \, m_1 = q \text{ then } G_s \leftarrow G_s + a_1\Delta_s/q \text{ and goto 2.} \\ \text{else if } F_{G_s \text{New}}^* = h_1(G_s^d, \Delta_s)^{m_1} \\ \quad \text{then } G_{s+1} \leftarrow h_1(G_s^d, \Delta_s), \, w_{s+1} \leftarrow D_s w_s/m_1, \text{ append } G_{s+1} \text{ to } \mathcal{G}, \\ \quad \text{append } w_{s+1} \text{ to } \mathcal{W}, \, D_{s+1} \leftarrow \deg_x(F)/\deg_x(G_{s+1}), \, s \leftarrow s+1 \text{ and goto 1.} \end{array}}$

$\boxed{\text{else } F_{G_s \text{New}}^* \text{ has more than 2 relatively prime factors hence goto 3.}}$

3. Calculate Moses-Yun's interpolation polynomials $W_i^{(j)}$ satisfying
$$W_0^{(j)} \frac{F_{G_s \text{New}}^*}{H_0^{(0)}} + \cdots + W_{R+R'}^{(j)} \frac{F_{G_s \text{New}}^*}{H_{R+R'}^{(0)}} = G_s^j, \quad \deg_{G_s}(W_i^{(j)}) < \deg_{G_s}(H_i^{(0)})$$
$(0 \leq i \leq R + R', 0 \leq j \leq D - 1)$ Let the maximum degrees of Δ_s in denominators of $W_0^{(j)}, \cdots, W_{R+R'}^{(j)}$ be $\tilde{m}_j$.

4. Calculate practical interpolation polynomials $\widetilde{W}_i^{(j)}$ $(0 \leq i \leq R + R', 0 \leq j \leq D - 1)$ s.t. $\widetilde{W}_i^{(j)} = \Delta_s^{\tilde{m}_j} W_i^{(j)} \in \overline{K}(\boldsymbol{u})[\Delta_s, G_s]$.

5. Perform lifting; Let the slope of $F_{G_s \text{New}}$ be $-\hat{\delta}/\hat{d}$ ($\hat{d}$ and $\hat{\delta}$ are positive integers s.t. $\gcd(\hat{d}, \hat{\delta}) = 1$) and $\tilde{t}$ be a weighted total-degree variable as $G_i \mapsto \tilde{t}^{w_i} G_i$ $(-1 \leq i \leq s-1)$. For $k = 1, 2, 3, \cdots$,
$$\text{ideal } S_k \overset{\text{def}}{=} (G_s \tilde{t}^{\hat{\delta}/\hat{d}})^{D_s} \times (\tilde{t}^{1/\hat{d}})^k$$
$$= (G_s^{D_s} \tilde{t}^{(k+0)/\hat{d}}, G_s^{D_s-1}\tilde{t}^{(k+\hat{\delta})/\hat{d}}, G_s^{D_s-2}\tilde{t}^{(k+2\hat{\delta})/\hat{d}}, \cdots, G_s^0 \tilde{t}^{(k+D_s\hat{\delta})/\hat{d}}).$$
$$f^{*(k)} \equiv F - H_0^{(k-1)} \cdots H_{R+R'}^{(k-1)} \pmod{S_{k+1}}$$
$$= \Sigma_{j=0}^{D_s-1} f_j^{(k)} \tilde{t}^{k/\hat{d}} \cdot \Delta_s^{\tilde{m}_j} G_s^j \, \tilde{t}^{(D_s-j)\hat{\delta}/\hat{d}}$$
$$H_i^{(k)} = H_i^{(k-1)} + \Sigma_{j=0}^{D_s-1} \widetilde{W}_i^{(j)} f_j^{(k)} \tilde{t}^{k/\hat{d}} \quad (i = 0, \cdots, R + R')$$

Then we get $F = H_0^{(\infty)} \cdots H_{R+R'}^{(\infty)}$, $H_i^{(\infty)} \in \overline{K}\{(\boldsymbol{u})\}[x]$.
We process each factor $H_0^{(\infty)} \cdots H_{R+R'}^{(\infty)}$ as follows.

$H_0^{(\infty)}$	: $F \leftarrow H_0^{(\infty)}$ and perform this algorithm recursively.
$H_1^{(\infty)}, \cdots, H_R^{(\infty)}$	: They are analytically irreducible factors in $\overline{K}\{(\boldsymbol{u})\}[x]$, because initial factors are such irreducible forms as $h_i(G_s^d, \Delta_s)$ $(i = 1, \cdots, R)$. Hence Return $H_i^{(\infty)}$.
$H_{R+1}^{(\infty)}, \cdots, H_{R+R'}^{(\infty)}$	: For each i, $F \leftarrow H_{R+i}^{(\infty)}$ $(i = 1, \cdots, R')$ and perform this algorithm recursively, because initial factors are such multiple forms as $h_{R+i}(G_s^d, \Delta_s)^{m_{R+i}}$ $(m_{R+i} \geq 2)$.

Then we obtain irreducible factors in $\overline{K}\{(\boldsymbol{u})\}[x]$.
6. Combine irreducible factors in $\overline{K}\{(\boldsymbol{u})\}[x]$ so as to eliminate their denominators. Then we can obtain irreducible factors in $\overline{K}\{\boldsymbol{u}\}[x]$.

Example 1   ◁
$F(x, u_1, u_2) = ((x^2 - (u_1 + u_2)^3)^2 + (u_1 + u_2)^7(u_1 + 2u_2)^2)$
$\qquad \times ((x^2 - (u_1 + u_2)^3)^2 - (u_1 + 2u_2)^8 - (u_1 + 3u_2)^{10}) \in \overline{K}[x, u_1, u_2]$
$F_{\text{New}} = (x^2 - t^3(u_1 + u_2)^3)^4$ cannot be decomposed into relatively prime polynomial factors. Therefore, we define $\mathcal{G} = (G_{-1}, G_0, G_1) = (t, x, x^2 - t^3(u_1 + u_2)^3)$, $\mathcal{W} = (w_{-1}, w_0) = (1, 3/2)$ as the slope of $\mathcal{L}_{\text{New}}$ is $3/2$. We perform $\mathcal{G}$-adic expansion and introduce weighted total-degree variable $\tilde{t}$ w.r.t. G_{-1}, G_0 as $G_{-1}^{e_{-1}} \mapsto \tilde{t}^1 G_{-1}^{e_{-1}}$, $G_0^{e_0} \mapsto \tilde{t}^{3/2} G_0^{e_0}$. And we plot $(e_1, 1 \cdot e_{-1} + \frac{3}{2} \cdot e_0)$ for each term of $F = \Sigma_{e_1=0}^D c_{(e_{-1}, e_0, e_1)}(\boldsymbol{u}) \tilde{t}^{1 \cdot e_{-1} + \frac{3}{2} \cdot e_0} G_{-1}^{e_{-1}} G_0^{e_0} G_1^{e_1}$, $c_{(e_{-1}, e_0, e_1)}(\boldsymbol{u}) \in \overline{K}\{(\boldsymbol{u})\}$. Then we see the slope of $\mathcal{L}_{G_1 \text{New}}$ is 4, hence we put $w_1 = 4$. Now we have $\mathcal{W} = (w_{-1}, w_0, w_1) = (1, 3/2, 4)$

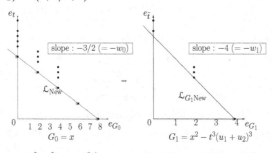

$$F_{G_0\text{New}} = (x^2 - t^3(u_1 + u_2)^3)^4, \quad F_{G_1\text{New}} = G_1^2(G_1 - \tilde{t}^4(u_1 + 2u_2)^4 G_{-1}^4)(G_1 + \tilde{t}^4(u_1 + 2u_2)^4 G_{-1}^4)$$

We decompose new Newton's polynomial over $\overline{K}\{(\boldsymbol{u})\}$ as $F_{G_1\text{New}} = G_1^2(G_1 - \tilde{t}^4(u_1 + 2u_2)^4\Delta_1)(G_1 + \tilde{t}^4(u_1 + 2u_2)^4\Delta_1)$, $\Delta_1 = G_{-1}^4$ and perform lifting as
$$H_0^{(\infty)} = G_1^2 + (u_1 + u_2)^7(u_1 + 2u_2)^2 G_{-1}^9$$
$$H_1^{(\infty)} = G_1 - (u_1 + 2u_2)^4 G_{-1}^4 - \frac{(u_1 + 3u_2)^{10}}{2(u_1 + 2u_2)^4} G_{-1}^6 + \frac{(u_1 + 3u_2)^{20}}{8(u_1 + 2u_2)^{12}} G_{-1}^8 - \cdots$$
$$H_2^{(\infty)} = G_1 + (u_1 + 2u_2)^4 G_{-1}^4 + \frac{(u_1 + 3u_2)^{10}}{2(u_1 + 2u_2)^4} G_{-1}^6 - \frac{(u_1 + 3u_2)^{20}}{8(u_1 + 2u_2)^{12}} G_{-1}^8 + \cdots.$$

Now $H_0^{(0)} = G_1^2$ is in multiple form, so we have to determine whether its irreducible or not. (As for $H_1^{(\infty)}$ and $H_2^{(\infty)}$, we cannot decompose anymore because degree w.r.t. G_1 is 1.) We put $H_0 \stackrel{-}{=} H_0^{(\infty)} = G_1^2 + (u_1 + u_2)^7(u_1 + 2u_2)^2 G_{-1}^9$

and apply the algorithm to H_0 as we have done for $F(x, u_1, u_2)$ in the previous step, recursively. In this case, $\mathcal{G} = (G_{-1}, G_0, G_1) = (t, x, x^2 - t^3(u_1 + u_2)^3)$, $\mathcal{W} = (w_{-1}, w_0, w_1) = (1, 3/2, 9/2)$. Now, $H_{0\,G_1\text{New}} = G_1^2 + (u_1 + u_2)^7(u_1 + 2u_2)^2 G_{-1}^9$.
As $G_1 = G_0^2 - (u_1 + u_2)^3 G_{-1}^3$, we obtain $G_{-1}^3 = \frac{G_0^2}{(u_1+u_2)^3} - \frac{G_1}{(u_1+u_2)^3}$. By this substitution, we get initial pseudo form as follows.

$H_{0\,G_1\text{New}} = G_1^2 + (u_1 + u_2)^7(u_1 + 2u_2)^2 G_{-1}^9$

$\quad = G_1^2 + (u_1 + u_2)^4(u_1 + 2u_2)^2 \underline{G_{-1}^6 G_0^2} - (u_1 + u_2)^4(u_1 + 2u_2)^2 \underline{G_{-1}^6 G_1}$

(weights $\quad\quad 9 \quad\quad\quad\quad\quad\quad\quad\quad\quad\quad 9 \quad\quad\quad\quad\quad\quad\quad\quad\quad\quad 10.5$)

$\quad \equiv G_1^2 + (u_1 + u_2)^4(u_1 + 2u_2)^2 G_{-1}^6 G_0^2 \pmod{S_1} \quad \overset{=}{=} \quad H_{0\,G_1\text{New}}^*$

As having the same weight 9, we can use $(u_1 + u_2)^4(u_1 + 2u_2)^2 G_{-1}^6 G_0^2$ instead of $(u_1 + u_2)^7(u_1 + 2u_2)^2 G_{-1}^9$, while $-(u_1 + u_2)^4(u_1 + 2u_2)^2 G_{-1}^6 G_1$ is put up to the higher order i.e. 10.5. Therefore,

$H_{0\,G_1\text{New}}^* = G_1^2 + (u_1 + u_2)^4(u_1 + 2u_2)^2 \Delta_{H_0}^2, \quad \Delta_{H_0} = G_{-1}^3 G_0^1$

$\quad = (G_1 + \mathbf{i}(u_1 + u_2)^2(u_1 + 2u_2)\Delta_{H_0})(G_1 - \mathbf{i}(u_1 + u_2)^2(u_1 + 2u_2)\Delta_{H_0})$

We put $H_{01}^{(0)} = (G_1 + \mathbf{i}(u_1 + u_2)^2(u_1 + 2u_2)\Delta_{H_0})$ and $H_{02}^{(0)} = (G_1 - \mathbf{i}(u_1 + u_2)^2(u_1 + 2u_2)\Delta_{H_0})$ (Let $\mathbf{i}$ be an imaginary unit). Then Moses-Yun's polynomials $W_{0i}^{(j)}$ ($i = 1, 2; j = 0, 1$), satisfying $W_{01}^{(j)} H_{02}^{(0)} + W_{02}^{(j)} H_{01}^{(0)} = G_1^j$, are as follows.

$W_{01}^{(0)} = \frac{\mathbf{i}}{2(u_1+u_2)^2(u_1+2u_2)\Delta_{H_0}}, W_{02}^{(0)} = \frac{-\mathbf{i}}{2(u_1+u_2)^2(u_1+2u_2)\Delta_{H_0}}, W_{01}^{(1)} = \frac{1}{2}, W_{02}^{(1)} = \frac{1}{2}$

Then **practical interpolation polynomials** $\widetilde{W}_{0i}^{(j)} (= \Delta_{H_0}^{\tilde{m}_j} W_i^{(j)}) \in \overline{K}(\mathbf{u})[\Delta_{H_0},$ $G_1]$ where $i = 1, 2$, and $j = 0, 1$, and $\tilde{m}_j$ is the maximum degree of denominators of $W_{0i}^{(j)}$ w.r.t. Δ_{H_0}, satisfying $\widetilde{W}_{01}^{(j)} H_{02}^{(0)} + \widetilde{W}_{02}^{(j)} H_{01}^{(0)} = \Delta_{H_0}^{\tilde{m}_j} G_1^j$, are $\widetilde{W}_{0i}^{(0)} = \Delta_{H_0} W_{0i}^{(0)}, \widetilde{W}_{0i}^{(1)} = W_{0i}^{(1)}$. (In this case, $\tilde{m}_j = 1 - j$ and $\Delta_{H_0} = G_{-1}^3 G_0$)
ideal $S_k = (G_1^2 \tilde{t}^{(k+9\cdot0)/2}, G_1^1 \tilde{t}^{(k+9\cdot1)/2}, G_1^0 \tilde{t}^{(k+9\cdot2)/2})$.

By lifting, we obtain the factorization $H_0 = H_{01}^{(\infty)} H_{02}^{(\infty)}$ in $\overline{K}\{(\mathbf{u})\}[x]$ as follows.

$H_{01}^{(\infty)} = G_1 + \mathbf{i}(u_1 + u_2)^2(u_1 + 2u_2)G_{-1}^3 G_0 - \frac{1}{2}(u_1 + u_2)^4(u_1 + 2u_2)^2 G_{-1}^6 - \cdots$

$H_{02}^{(\infty)} = G_1 - \mathbf{i}(u_1 + u_2)^2(u_1 + 2u_2)G_{-1}^3 G_0 - \frac{1}{2}(u_1 + u_2)^4(u_1 + 2u_2)^2 G_{-1}^6 + \cdots$

Therefore, we obtain factorization of F in $\overline{K}\{(\mathbf{u})\}[x]$ as $F = H_0^{(\infty)} H_1^{(\infty)} H_2^{(\infty)} = H_{01}^{(\infty)} H_{02}^{(\infty)} H_1^{(\infty)} H_2^{(\infty)}$. By focusing attention on similarities of denominators, we combine $H_1^{(\infty)}$ and $H_2^{(\infty)}$, we obtain irreducible factors in $\overline{K}\{\mathbf{u}\}[x]$ as follows.

$F = (x^2 - (u_1 + u_2)^3 + \mathbf{i}x(u_1 + u_2)^2(u_1 + 2u_2) - \frac{1}{2}(u_1 + u_2)^4(u_1 + 2u_2)^2 - \cdots)$

$\quad \times (x^2 - (u_1 + u_2)^3 - \mathbf{i}x(u_1 + u_2)^2(u_1 + 2u_2) - \frac{1}{2}(u_1 + u_2)^4(u_1 + 2u_2)^2 + \cdots)$

$\quad \times ((x^2 - (u_1 + u_2)^3)^2 - (u_1 + 2u_2)^8 - (u_1 + 3u_2)^{10}).$

3 Multivariate Expansion Base Including the Case of Singular Leading Coefficient ($f_D(0) = 0$)

In the case of singular leading coefficient, the slope of Newton's line becomes positive, 0 or negative. To avoid the case that the lifting ideal contains constant term during the lifting step, we transform Newton's line to be horizontal in a unified way, by identifying "weight of expansion base" and "slope of Newton's line". Without loss of generality, after performing the extended Hensel construction as a preprocessing, we may assume $F(x, \boldsymbol{u})$ as in the assumption of the following **Theorem 1**.

Theorem 1. *We assume* $F(x, t\boldsymbol{u})$ *s.t.* $F_{\text{New}} = f_D(\boldsymbol{u})x^D + \cdots + f_0(\boldsymbol{u}) = c(\boldsymbol{u})g^m$ *($c(\boldsymbol{u}) \in \overline{K}[\boldsymbol{u}], m \geq 2$ and g is irreducible in $\overline{K}[x, \boldsymbol{u}]$) and $\text{ord}_t f_D(t\boldsymbol{u}) \overset{def}{=} \nu \gneqq 0$ (i.e. $f_D(\boldsymbol{0}) = 0$, the leading coeffient is singular). Let the slope of the Newton's line be $\hat{\lambda}$. Let the expansion base of F be $\mathcal{G} = (G_{-1}(= t), G_0(= x), G_1(= g))$, and weights be $\mathcal{W} = \left(w_{-1}(= 1), w_0(= -\hat{\lambda}), w_1\right)$, $F(G_{-1}, G_0, G_1)$ be $\mathcal{G}$-adic expansion*

of F. If we plot terms of $\hat{F}(G_{-1}, G_0, G_1, \tilde{t}) \overset{def}{=} F(\tilde{t}^{w_{-1}}G_{-1}, \tilde{t}^{w_0}G_0, \tilde{t}^{w_1}G_1)/\tilde{t}^{\nu + Dw_0}$ on the $(e_{G_1}, e_{\tilde{t}})$-plane (whose horizontal line represents the power of G_1 and the vertical line represents the power of $\tilde{t}$) then the lattice points on the Newton's line are shifted right down to be on the horizontal axis, and a point whose horizontal component being maximal is $(m, 0)$. Then we determine Newton's polynomial $\hat{F}_{G_1\text{New}}$ and ideal $\hat{I}_k$ as $\hat{F}_{G_1\text{New}} = \hat{F}(G_{-1}, G_0, G_1, \tilde{t} = 0)$, $\hat{I}_k = \langle \tilde{t}^{k/\hat{m}} \rangle$, $k = 1, 2, 3, \cdots$, where $|$slope of Newton's line $\mathcal{L}_{G_1}| = \hat{n}/\hat{m}$ ($\hat{m}$ and $\hat{n}$ are relatively prime positive integers).

Proof. By the transformation $\hat{F}(x, \boldsymbol{u}, t) \overset{def}{=} F(t^{-\hat{\lambda}}x, t\boldsymbol{u})/t^{\nu - D\hat{\lambda}}$ in [10], all dots on the Newton's line are shifted right down to be on the horizontal axis. Then (D, ν) is shifted to $(D, 0)$, and there is no dot whose horizontal component is greater than D. Below, we extend this transformation to multivariate expansion base method by defining weights of G_i as $w_i \overset{def}{=} (-1) \times$(slope of Newton's line $\mathcal{L}_{G_i}$). By assumpsion, Newton's polynomial $F_{\text{New}} = \hat{F}(x, \boldsymbol{u}, t = 0) = c(\boldsymbol{u})g^m$ is multiple and cannot be decomposed into relatively prime nonunit factors, therefore, we may set multivariate expansion base and weights as $\mathcal{G} = (G_{-1}, G_0, G_1) = (t, x, g)$ and $\mathcal{W} = (w_{-1}, w_0) = (1, -\hat{\lambda})$, respectively. We can obtain the unique expression, $\mathcal{G}$-adic expansion of $F(x, t\boldsymbol{u})$, as $F(G_{-1}, G_0, G_1) = \sum_{e_1=0}^{m} c_{(e_{-1}, e_0, e_1)}(\boldsymbol{u}) G_{-1}^{e_{-1}} G_0^{e_0} G_1^{e_1}$, $c_{(e_{-1}, e_0, e_1)}(\boldsymbol{u}) \in \overline{K}\{(\boldsymbol{u})\}$, by dividing G_1, G_0, G_{-1} in this order. Then plot each term of $F(\tilde{t}^{w_{-1}}G_{-1}, \tilde{t}^{w_0}G_0, G_1)$ on $(e_{G_1}, e_{\tilde{t}})$-plane, we can obtain new Newton's polynomial $F_{G_1\text{New}}$, regarding G_1 as a new main variable and G_{-1}, G_0 as subvariables.

In the case of singular leading coefficient, we can transform the slope of $\mathcal{L}_{G_1}$ to be horizontal by $F(\tilde{t}^{w_{-1}}G_{-1}, \tilde{t}^{w_0}G_0, \tilde{t}^{w_1}G_1)$, where $w_1 \overset{def}{=} (-1) \times$ (slope of $\mathcal{L}_{G_1}$), and we can shift Newton's line right down to be on the horizontal axis by dividing $\tilde{t}^{\nu + Dw_0}$. Moreover, $F_{\text{New}} = c(\boldsymbol{u})g^m = c(\boldsymbol{u})G_1^m$ and obtained Newton's line is on the horizontal axis, a point whose horizontal component being maximal is $(m, 0)$. If we substitute $\tilde{t} = 0$ for $\hat{F}(G_{-1}, G_0, G_1, \tilde{t}) = F(\tilde{t}^{w_{-1}}G_{-1}, \tilde{t}^{w_0}G_0, \tilde{t}^{w_1}G_1)$, then we obtain sum of all terms corresponding to dots on the horizontal axis, i.e. Newton's polynomial $F_{G_1\text{New}}$. We may shift the Newton's line by step $1/\hat{m}$ in the $e_{\tilde{t}}$-direction successively, then the ideal is $\hat{I}_k = \langle \tilde{t}^{k/\hat{m}} \rangle$, $k = 1, 2, 3, \cdots$. □

Note that the leading coefficient of Newton's polynomial substituting $\tilde{t} = 0$, i.e. $\hat{F}_{G_1\text{New}} = \hat{F}(G_{-1}, G_0, G_1, \tilde{t} = 0)$ does not disappear. The horizontal Newton's line is easy to deal with because we can obtain Newton's polynomial just to substitute $\tilde{t} = 0$ and perfom lifting with simple ideal $\hat{I} = \langle \tilde{t}^{k/\hat{m}} \rangle$, $k = 1, 2, 3, \cdots$.

Therefore, we produce a standardization of horizontal Newton's line for multivariate expansion base $\mathcal{G} = (G_{-1}, G_0, G_1, \cdots, G_s)$ as follows.

Theorem 2. *Let the multivariate expansion base of F be $\mathcal{G} = (G_{-1}, G_0, \cdots, G_s)$ and weights be $\mathcal{W} = (w_{-1}, w_0, \cdots, w_s)$, where $w_i \overset{\text{def}}{=} (-1) \times (slope\ of\ \mathcal{L}_{G_i\text{New}})$. Let $\mathcal{G}$-adic expansion of F be $F(G_{-1}, G_0, G_1, \cdots, G_s)$. Then whether the leading coefficient is singular or not, we can obtain horizontal Newton's line by the transformation as follows.*

$$\hat{F}(G_{-1}, \cdots, G_s, \tilde{t}) \overset{\text{def}}{=} F(\tilde{t}^{w_{-1}}G_{-1}, \cdots, \tilde{t}^{w_s}G_s)/\tilde{t}^\alpha,$$

where $\alpha = \mathrm{ord}_{\tilde{t}}F(\tilde{t}^{w_{-1}}G_{-1}, \cdots, \tilde{t}^{w_s}G_s)$. Moreover, if we substitute $\tilde{t} = 0$ then the leading coefficient does not disappear. Then Newton's polynomial $\hat{F}_{G_s\text{New}}$ and ideal $\hat{I}_k$ for lifting are as follows.

$\hat{F}_{G_s\text{New}} = \hat{F}(G_{-1}, G_0, \cdots, G_s, \tilde{t} = 0)$, $\hat{I}_k = \langle \tilde{t}^{k/\hat{m}} \rangle$, $k = 1, 2, 3, \cdots$, *where $|slope\ of\ Newton's\ line\ \mathcal{L}_{G_s}| = \hat{n}/\hat{m}$ ($\hat{m}$ and $\hat{n}$ are relatively prime positive integers).*

Proof. Generation of the multivariate expansion base causes the change of the horizontal axis as $e_{G_0} \to e_{G_1} \to \cdots e_{G_s}$. Hence step by step, the value of the vertical axis, $e_{\tilde{t}}$, is accumulated as weighted sum of powers of $G_{-1}, \cdots, G_{s-1}$. To make $\mathcal{L}_{G_s}$ being horizontal, we substitute $\tilde{t}^{w_{G_s}}G_s$ for G_s. And to shift all terms on $\mathcal{L}_{G_s}$ right down to be on the horizontal axis, we divide by $\tilde{t}^{\mathrm{ord}_{\tilde{t}}F(\tilde{t}^{w_{-1}}G_{-1}, \cdots, \tilde{t}^{w_s}G_s)}$.

We can also proof that the leading coefficient does not disappear when we substitute $\tilde{t} = 0$, and $\hat{F}_{G_s\text{New}} = \hat{F}(G_{-1}, G_0, \cdots, G_s, \tilde{t} = 0)$, $\hat{I}_k = \langle \tilde{t}^{k/\hat{m}} \rangle$, $k = 1, 2, 3, \cdots$, in a similar way as proof in **Theorem 1**. □

Algorithm 2. [Factorization of $F = g^m + \cdots$ in $\overline{K}\{(u)\}[x]$ utilizing the multivariate expansion base including the case of singular leading coefficient]
INPUT 　: $F(x, u) \in \overline{K}\{(u)\}[x]$ s.t. $F_{\text{New}} = g^m$
　　　　　where $m \geq 2$ and g is irreducible in $\overline{K}[x, u]$.
OUTPUT : Irreducible factors in $\overline{K}\{(u)\}[x]$.

1. Calculate the expansion base $\mathcal{G} = (G_{-1}(= t), G_0(= x), G_1(= g), \cdots, G_s)$ and weights $\mathcal{W} = (w_{-1}(= 1), w_0, \cdots, w_s)$, and peform $\mathcal{G}$-adic expansion of F as $F(G_{-1}, G_0, \cdots, G_s) = \Sigma c(u)_{e_{-1}, e_0, \cdots, e_s} G_{-1}^{e_{-1}} G_0^{e_0} \cdots G_s^{e_s}$.
2. Plot terms of $F(\tilde{t}^{w_{-1}}G_{-1}, \tilde{t}^{w_0}G_0, \cdots, \tilde{t}^{w_{s-1}}G_{s-1}, G_s)$ on $(e_{G_s}, e_{\tilde{t}})$-plane, and define the weight of G_s as $w_s \overset{\text{def}}{=} (-1) \times (slope\ of\ Newton's\ line\ \mathcal{L}_{G_s})$
3. Calculate $\hat{F}(G_{-1}, \cdots, G_s, \tilde{t}) = F(\tilde{t}^{w_{-1}}G_{-1}, \cdots, \tilde{t}^{w_s}G_s)/\tilde{t}^{\mathrm{ord}_{\tilde{t}}F(\tilde{t}^{w_{-1}}G_{-1}, \cdots, \tilde{t}^{w_s}G_s)}$
4. Factorize $\hat{F}_{G_s\text{New}} = \hat{F}(G_{-1}, \cdots, G_s, 0)$ over $\overline{K}\{(u)\}$.
   (a) **If $\hat{F}_{G_s\text{New}}$ cannot be decomposed into relatively prime factors,** if multiplicity is one then irreducible in $\overline{K}\{(u)\}[x]$, else if muitiplicity is greater or equal to two then set $G_{s+1} \leftarrow$ (the irreducible component), and goto **1**.($s \leftarrow s + 1$).
   (b) **If $\hat{F}_{G_s\text{New}}$ can be decomposed into relatively prime factors,** then regarding them as initial factors, perform lifting with ideal $\langle \tilde{t}^{k/\hat{m}} \rangle$, $k = 1, 2, 3, \cdots$, and practical interpolation polynomials.

◁

Newton dots for $F(\tilde{t}^{w-1}G_{-1}, \cdots, \tilde{t}^{w_s-1}G_{s-1}, G_s)$

$$\Longrightarrow F(\tilde{t}^{w-1}G_{-1}, \cdots, \tilde{t}^{w_s-1}G_{s-1}, \tilde{t}^{w_s}G_s)/\tilde{t}^{\alpha}$$
$$\alpha = \mathrm{ord}_{\tilde{t}}F(\tilde{t}^{w-1}G_{-1}, \cdots, \tilde{t}^{w_s-1}G_{s-1}, \tilde{t}^{w_s}G_s)$$

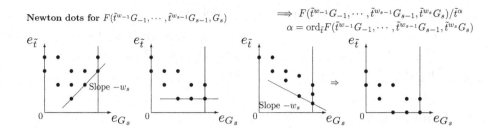

Example 2

$$F(x, u_1, u_2) \overset{\text{def}}{=} \left((u_1 + 2u_2)\left(u_2^3\,x^2 - (u_1 + u_2)\right)^2 - u_1^7 x^2 - u_1^9 \right)\left(\left(u_2^3 x^2 - (u_1 + u_2)\right)^2 - (u_1 + u_2)^3 u_2 \right).$$

For $F(x, t\boldsymbol{u})$, we plot each term on (e_x, e_t)-plane (See Fig.A-1). First, we perform the transformation $F(\tilde{t}^{-1}x, \tilde{t}\boldsymbol{u})/\tilde{t}^5$, corresponding dots on the $(e_x, e_{\tilde{t}})$-plane become as in Fig.A-2. Then Newton's polynomial F_{New} cannot be decomposed into nonunit relatively prime irreducible polynomials over $\overline{K}$ as $F_{\mathrm{New}} = (u_1 + 2u_2)t\left(u_2^3\,t^3x^2 - (u_1 + u_2)t\right)^4$, and leading coefficient disappears at $u_1 = u_2 = 0$. We put the expansion base $\mathcal{G} = (G_{-1}, G_0, G_1) = (t, x, u_2^3t^3x^2 - (u_1 + u_2)t)$. The slope of $\mathcal{L}_{\mathrm{New}}$ is 1 then we put weights as $\mathcal{W} = (w_{-1}, w_0) = (1, -1)$. Perform $\mathcal{G}$-adic expansion of F as $F(G_{-1}, G_0, G_1) = \Sigma_{e_1=0}^{m} c_{(e_{-1}, e_0, e_1)}(\boldsymbol{u})G_{-1}^{e_{-1}}G_0^{e_0}G_1^{e_1}$, $c_{(e_{-1}, e_0, e_1)}(\boldsymbol{u}) \in \overline{K}\{(\boldsymbol{u})\}$, by dividing G_1, G_0, G_{-1} in this order, and plot terms of $F(\tilde{t}^1 G_{-1}, \tilde{t}^{-1}G_0, G_1)$ on the $(e_{G_1}, e_{\tilde{t}})$ plane, then the slope of $\mathcal{L}_{G_1\mathrm{New}}$ is -2 then we put $w_1 = 2$ (See Fig B-1). Therefore, we perform translation as $\hat{F}(G_{-1}, G_0, G_1, \tilde{t}) = F(\tilde{t}^1 G_{-1}, \tilde{t}^{-1}G_0, \tilde{t}^2 G_1)/\tilde{t}^9$ (See Fig B-2). Now we get $\hat{F}_{G_1\mathrm{New}} = \hat{F}(G_{-1}, G_0, G_1, \tilde{t} = 0) = (G_1^2 - u_2(u_1 + u_2)^3 G_{-1}^4)((u_1 + 2u_2)G_{-1}G_1^2 - \frac{u_1^7}{u_2^3}(u_1 + u_2)G_{-1}^5)$. As the definition of G_1 is $(\tilde{t}^{w_1}G_1) = u_2^3(\tilde{t}^{w-1}G_{-1})^3(\tilde{t}^{w_0}G_0)^2 - (u_1 + u_2)(\tilde{t}^{w-1}G_{-1})$, we obtain $(u_1 + u_2)G_{-1} = u_2^3 G_{-1}^3 G_0^2 - \tilde{t}G_1$. And the weights of each term are

$$(u_1 + u_2)G_{-1} = u_2^3 G_{-1}^3 G_0^2 - \tilde{t}G_1.$$
$$\text{Weights} \qquad\qquad 1 \qquad\qquad 1 \qquad\qquad 2$$

Therefore, we can modify $\hat{F}_{G_1\mathrm{New}}$ to pseudo form F^* by replacing $(u_1+u_2)G_{-1} \to u_2^3 G_{-1}^3 G_0^2$ having the same weights 1 as follows.

$$F^* = (G_1^2 - u_2^4(u_1 + u_2)^2 G_{-1}^6 G_0^2)\left((u_1 + 2u_2)G_{-1}G_1^2 - u_1^7 G_{-1}^7 G_0^2\right)$$
$$= (G_1 + u_2^2(u_1 + u_2)G_{-1}^3 G_0)(G_1 - u_2^2(u_1 + u_2)G_{-1}^3 G_0)\left((u_1 + 2u_2)G_{-1}G_1^2 - u_1^7 G_{-1}^7 G_0^2\right)$$

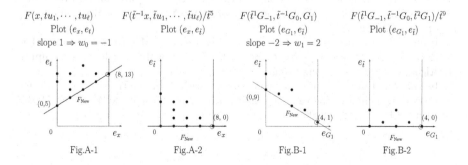

$F(x, tu_1, \cdots, tu_\ell)$
Plot (e_x, e_t)
slope $1 \Rightarrow w_0 = -1$

$F(\tilde{t}^{-1}x, \tilde{t}u_1, \cdots, \tilde{t}u_\ell)/\tilde{t}^5$
Plot $(e_x, e_{\tilde{t}})$

$F(\tilde{t}^1 G_{-1}, \tilde{t}^{-1}G_0, G_1)$
Plot $(e_{G_1}, e_{\tilde{t}})$
slope $-2 \Rightarrow w_1 = 2$

$F(\tilde{t}^1 G_{-1}, \tilde{t}^{-1}G_0, \tilde{t}^2 G_1)/\tilde{t}^9$
Plot $(e_{G_1}, e_{\tilde{t}})$

Fig.A-1   Fig.A-2   Fig.B-1   Fig.B-2

By lifting, we obtain required factors of F as follows.

$u_2^3\,x^2 - (u_1 + u_2) + u_2^2(u_1 + u_2)x + \frac{1}{2}u_2(u_1 + u_2)^2 + \frac{1}{8}u_2^2(u_1 + u_2)^3 + \cdots,$

$u_2^3\,x^2 - (u_1 + u_2) - u_2^2(u_1 + u_2)x + \frac{1}{2}u_2(u_1 + u_2)^2 - \frac{1}{8}u_2^2(u_1 + u_2)^3 - \cdots,$

$(u_1 + 2u_2)\left(u_2^3\,x^2 - (u_1 + u_2)\right)^2 - u_1^7 x^2 - u_1^9.$

4   Conclusion

In this paper, the multivariate analytic factorization algorithm in the case that the leading coefficient disappears at the expansion point was suggested. A transformation in [10] which is for the extended Hensel construction is applied to multivariate expansion base method, and we found that the analytic factorization with singular and non singular leading coefficients can be done in a unified way. This breakthrough comes from the identification of "weight of expansion base" and "slope of Newton's line". Note that weights of bivariate expansion base take only positive values, whereas weights of multivariate one can be negative, 0 or positive (slope of Newton's line can be positive, 0 or negative, respectively). We can say this unified method is a blend of techniques of "multivariate expansion base" and "the extended Hensel construction". As for factorization in $\overline{K}[x, u]$, we can treat only when Newton's polynomial is square-free in [10]. However, using [5], [6] or one in this paper, we can do even when Newton's polynomial is not square-free. As $\overline{K}[x, u] \subset \overline{K}\{u\}[x] \subset \overline{K}\{(u)\}[x]$, we can obtain irreducible factors in $\overline{K}[x, u]$ by multiplying analytically irreducible factors.

References

1. Abhyankar, S.S.: What is the difference between a parabola and a hyperbola? Math. Intelligencer 10, 37–43 (1988)
2. Abhyankar, S.S.: Irreducibility Criterion for Germs of Analytic Functions of Two Complex Variables. Advances in Math. 74, 190–257 (1989)
3. S. S. Abhyankar: Algebraic Geometry for Scientists and Engineers. In: Mathematical Surveys and Monographs. vol. 35, Amarican Mathematical Society, Providence, RI (1990)
4. Abhyankar, S.S., Moh, T.T.: Newton-Puiseux expansion and generalized Tschirnhausen transformation II. J. reine und angew. Math. 261, 29–54 (1973)
5. Iwami, M.: Analytic Factorization of the Multivariate Polynomial. In: Proceedings of the Sixth International Workshop on Computer Algebra in Scientific Computing. pp. 213–225 (2003)
6. Iwami, M.: Extension of Expansion Base Algorithm to Multivariate Analytic Factorizaion. In: Proceedings of the Seventh International Workshop on Computer Algebra in Scientific Computing, pp. 269–281 (2004)
7. Iwami, M.: Extension of expansion base algorithm for multivariate analytic factorization including the case of singular leading coefficient. ACM SIGSAM Bulletin 39(4), 122–126 (2005)

8. Kuo, T.C.: Generalized Newton-Puiseux Theory and Hensel's Lemma in C[[x,y]]. Can. J. Math., XLI(6), 1101–1116 (1989)
9. McCallum, S.: On Testing a Bivariate Polynomial for Analytic Reducibility. J. Symb. Comput. 24, 509–535 (1997)
10. Sasaki, T., Inaba, D.: Hensel Construction of $F(x, u_1, \ldots, u_l)$, $l \geq 2$, at a Singular Point and Its Applications. SIGSAM Bulletin 34, 9–17 (2000)
11. Sasaki, T., Kako, F.: Solving Multivariate Algebraic Equation by Hensel Construction. Japan J. Indus. Appl. Math. 16, 257–285 (1999)
12. Sasaki, T.: Properties of Extended Hensel Factors and Application to Approximate Factorization. Preprint, Univ. Tsukuba (2000)

On the Computation of the Defining Polynomial of the Algebraic Riccati Equation

Takuya Kitamoto[1] and Tetsu Yamaguchi[2]

[1] Yamaguchi University, 1677-1 Yoshida, Yamaguchi, Japan
[2] Cybernet Systems, Co., LTD, 2-9-3 Otsuka, Tokyo, Japan

Abstract. The algebraic Riccati equation, denoted by 'ARE' in the paper, is one of the most important equation in the post modern control theory, playing important role for solving H_2 and H_∞ optimal control problems. The solution of ARE is given in the form of a matrix, and a typical procedure of computing the solution uses eigenvalues and eigenvectors of matrix H, where H is a matrix determined by a given system. With the aid excellent numerical packages such as "LAPACK" for matrix computations, the procedure is quite efficient for the numerical systems (the systems without unknown parameters).

This paper considers a system with an unknown parameter k. In this case, the numerical procedure cannot be applied without fixing parameter k to a constant value. Let us consider some symbolic method to compute the solution of ARE which leaves parameter k symbolic. Letting entries of the solution matrix be unknown variables, ARE can be viewed as a set of m algebraic equations with m variables and parameter k, where m is the number of entries of the unknown matrix. Computing Groebner basis of the algebraic equations with lexicographic ordering, we obtain a polynomial whose roots are the solution of ARE (i.e. the defining polynomial of ARE). Although this method with Groebner basis gives us the defining polynomial of ARE, it is not practical. The method easily collapses when the size of a given system is large because of its heavy numerical complexities. This paper presents a practical algorithm to compute the defining polynomial. The proposed algorithm uses polynomial interpolations, and is easily parallelizable, implying that it is advantageous under multi-CPU environments. Numerical experiments indicate that even in the single CPU environments, the proposed algorithm is much more practical than that with Groebner basis.

1 Introduction

Computer Algebra has received an increasing attention in recent years because of the capacity to handle a parameter as an indeterminate. This is particularly advantageous in controller designs with a parameter where it is difficult to apply conventional numerical methods without substituting a numerical value into the parameter.

Various Computer Algebra techniques have been proved to be quite effective in the design and analysis of control systems. Some of the most important

V.G. Ganzha, E.W. Mayr, and E.V. Vorozhtsov (Eds.): CASC 2007, LNCS 4770, pp. 224–235, 2007.

results are produced using QE (Quantifier Elimination), a comparatively new technique in Computer Algebra. In short, QE converts mathematical formulas with quantifiers $\forall, \exists$ to an equivalent formula without a quantifier. For more on the application of QE to control system design, refer to references [1]-[4].

Another application of Computer Algebra is presented in Reference [5], where one of the authors of the paper describes an algorithm to compute the H_∞ norm of a system with a parameter under a restrictive condition (the system must be one-input or one-output). The condition is removed in [6], which can be viewed as a generalization of [5]. In [7], the same idea as the ones in [5] and [6] is used to compute the H_∞ norm of a system with a validated numerical calculation, which guarantees the accuracy of the computation.

In this paper, we treat a 'parametric' algebraic Riccati equation, which is the algebraic Riccati equation containing an unknown parameter. The algebraic Riccati equation, denote by 'ARE' in the paper, is an equation of $n \times n$ matrix P in the form of

$$PA + A^T P - PWP + Q = 0, \tag{1}$$

where A, W, Q (Q being positive semi definite and W being symmetric) are given $n \times n$ matrices. It is used for solving H_2 and H_∞ optimal control problems, and is one of the most important equations in control theory. This paper focuses on the ARE derived from H_2 optimal control problem, where $W = BR^{-1}B^T$ with R being a positive definite matrix. It is assumed that a given system is controllable, i.e.

$$\mathrm{rank}\left(\begin{bmatrix} B \ AB \ \cdots \ A^{n-1}B \end{bmatrix}\right) = n, \tag{2}$$

which guarantees the existence of solutions of the ARE.

Typically, a solution P of the equation is computed by a numerical algorithm, which utilizes the eigenvalues and eigenvectors of matrix H defined by

$$H \stackrel{\mathrm{def}}{=} \begin{bmatrix} A & -BR^{-1}B^T \\ -Q & -A^T \end{bmatrix}. \tag{3}$$

However, when matrices A, B, R, Q contain unknown parameters, the numerical algorithm cannot be applied.

Letting entries of the solution matrix P be unknown variables, ARE can be viewed as a set of m algebraic equations with m variables and parameter k, where m is the number of entries of matrix P. Computing Groebner basis of the algebraic equations with lexicographic ordering, we obtain a polynomial whose roots are the solution of ARE (i.e. the defining polynomial of ARE).

Let us give a numerical example: Let matrices A, B, R, Q in (1) be

$$A = \begin{bmatrix} k & 1 \\ 1 & -1 \end{bmatrix}, \ B = \begin{bmatrix} 1 & 0 \\ 0 & 1 \end{bmatrix}, \ R = \frac{1}{2}, \ Q = \begin{bmatrix} 1 & 0 \\ 0 & 1 \end{bmatrix}$$

and let the solution P be

$$P = \begin{bmatrix} p_{1,1} & p_{1,2} \\ p_{1,2} & p_{2,2} \end{bmatrix},$$

where $p_{1,1}, p_{1,2}, p_{2,2}$ are unknown variables. From (1,1),(1,2),(2,2)th entries of (1), we obtain a set of algebraic equations

$$
\begin{cases}
2kp_{1,1} + 2p_{1,1}^2 + 2p_{1,2} - 2p_{1,2}^2 + 1 = 0 \\
p_{1,1} - p_{1,2} + kp_{1,2} - 2p_{1,1}p_{1,2} + p_{2,2} - 2p_{1,2}p_{2,2} = 0 \\
2p_{1,2} - 2p_{1,2}^2 - 2p_{2,2} - 2p_{2,2}^2 + 1 = 0
\end{cases} .
$$

Computing Groebner basis for the equations with the lexicographic term ordering $p_{1,1} \succ p_{1,2} \succ p_{2,2} \succ k$, we obtain

$$
f_4(k)p_{2,2}^4 + f_3(k)p_{2,2}^3 + f_2(k)p_{2,2}^2 + f_1(k)p_{2,2} + f_0(k) = 0,
$$

which is the defining polynomial of (2,2)th entry p_{22} of the solution P in (1), where

$$
\begin{aligned}
&f_4(k) = 4k^3 + 4k^2 + 12k - 20, \quad f_3(k) = 8k^3 + 8k^2 + 24k - 40, \\
&f_2(k) = -4k^2 + 8k - 4, \quad f_1(k) = -4k^3 - 8k^2 - 4k + 16, \\
&f_0(k) = k^3 + 3k^2 - 3k - 1.
\end{aligned}
$$

The defining polynomials of other entries can also be computed in a similar way. Although this procedure with Groebner basis computes any defining polynomials of the entries of P theoretically, it is effective only for quite small size problems in practice because of its heavy numerical complexities. This paper presents a practical algorithm for computing the defining polynomial of the entries. The algorithm uses polynomial interpolations, and is easily parallelizable. In this paper, the following notations are used:

R: The set of real numbers.
C: The set of complex numbers.
Z: The set of integers.
N: The set of natural numbers.
$\mathbf{K}^{m,n}$: The set of matrices with entries in **K**.
$\mathbf{K}[x]$: The set of polynomials with coefficients in **K**.
$\mathbf{Res}_x(r_1(x), r_2(x))$: Resultant of polynomials $r_1(x)$ and $r_2(x)$ with respect to x.
$\mathbf{GCD}_x(r_1(x), \cdots, r_n(x))$: Polynomial GCD of $r_1(x), \cdots, r_n(x)$ with respect to x.
$\mathbf{deg}_x(r(x))$: The degree of $r(x)$ with respect to x.
$H|_{k=k_0}$: Matrix H when $k = k_0$.

2   Solution of the Algebraic Riccati Equation

In this section, we discuss the solutions of ARE. First, we explain a conventional numerical method of solving ARE. We then discuss the expression of the ARE solution using matrix determinants.

2.1   Numerical Method of Solving ARE

Given ARE (1), let matrix H be defined by (3). It is well-known that if λ is an eigenvalue of H, then so is $-\lambda$. Thus, a set of the eigenvalues of H is given in the form of $\{\mu_1, \cdots, \mu_n, -\mu_1, \cdots, -\mu_n\}$ ($\mu_i \in \mathbf{C}$). Let $\lambda_1, \cdots, \lambda_n$ be n eigenvalues among $2n$ eigenvalues of H, and $u_1, \cdots, u_n$ be corresponding eigenvectors to the eigenvalues $\lambda_1, \cdots, \lambda_n$. Then a solution P of ARE is given by $P = X_2 X_1^{-1}$, where $n \times n$ matrices X_1 and X_2 are defined by

$$\begin{bmatrix} X_1 \\ X_2 \end{bmatrix} \overset{\text{def}}{=} \begin{bmatrix} u_1 \cdots u_n \end{bmatrix}. \tag{4}$$

The above numerical method tells us that each choice of n eigenvalues among $2n$ eigenvalues constitutes a solution of ARE, implying that there exist $_{2n}C_n = \frac{(2n)!}{(n!)^2}$ solutions of ARE. In the control theory, only symmetric solutions are of interests in many cases, and our attention is only on the symmetric solutions. As we will see later, among $\frac{(2n)!}{(n!)^2}$ solutions, only 2^n solutions are symmetric.

2.2   Expression of the Solution with Matrix Determinants

In this subsection, we first describe the algorithm presented in [6], which computes solutions of ARE, leaving eigenvalues and parameter as symbol. The algorithm computes eigenvectors of matrix H defined in (3), leaving eigenvalue λ as indeterminates.

Algorithm 1 (Computation of $v(\lambda)$)

$\langle 1 \rangle$ Let x be $x = \begin{bmatrix} x_1 \cdots x_{2n} \end{bmatrix}^T$ and compose $2n$ linear equations $(H - \lambda E)x = 0$ (each entry of $(H - \lambda E)x = 0$ is a linear equation), where λ is an indeterminate.

$\langle 2 \rangle$ Select $(2n - 1)$ linear equations from $2n$ equations in $\langle 1 \rangle$, and solve the $(2n - 1)$ linear equations with respect to variables $x_1, \cdots, x_{2n-1}$.

$\langle 3 \rangle$ Substitute the solution of $x_1, \cdots, x_{2n-1}$ into x and multiply an adequate polynomial so that each entry of x is a polynomial in λ.

$\langle 4 \rangle$ Let $v(\lambda) \leftarrow x/x_{2n}$, and output $v(\lambda)$.

Vector $v(\lambda)$ computed using Algorithm 1 is the eigenvector of H corresponding to eigenvalue λ. Now let us define $n \times n$ matrix $\Lambda(y_1, \cdots, y_n) \in \mathbf{Z}[y_1, \cdots, y_n]^{n,n}$ by

$$\Lambda(y_1, \cdots, y_n) \overset{\text{def}}{=} \Gamma_2(y_1, \cdots, y_n) \Gamma_1(y_1, \cdots, y_n)^{-1}, \tag{5}$$

where $y_1, \cdots, y_n$ are variables and $n \times n$ matrices $\Gamma_1(y_1, \cdots, y_n)$ and $\Gamma_2(y_1, \cdots, y_n)$ are defined by

$$\begin{bmatrix} \Gamma_1(y_1, \cdots, y_n) \\ \Gamma_2(y_1, \cdots, y_n) \end{bmatrix} \overset{\text{def}}{=} \begin{bmatrix} v(y_1) \cdots v(y_n) \end{bmatrix}. \tag{6}$$

Then, $\Lambda(\lambda_1, \cdots, \lambda_n)$, which is a polynomial matrix, constitutes the solution of ARE corresponding to eigenvalues $\lambda_1, \cdots, \lambda_n$, and we have the following theorem:

Theorem 1. *Let P be the solution of ARE (1) corresponding to eigenvalues $\lambda_1, \cdots, \lambda_n$ of H (i.e. $P = \Lambda(\lambda_1, \cdots, \lambda_n)$). Then (i,j)th entry $p_{i,j}$ of P is given by*

$$p_{i,j} = \frac{|\bar{\Gamma}_{i,j}(\lambda_1, \cdots, \lambda_n)|}{|\Gamma_1(\lambda_1, \cdots, \lambda_n)|}, \tag{7}$$

where $\bar{\Gamma}_{i,j}(\lambda_1, \cdots, \lambda_n)$ is the matrix defined by

$$\bar{\Gamma}_{i,j}(\lambda_1, \cdots, \lambda_n) = \begin{bmatrix} v_1(\lambda_1) & \cdots & v_1(\lambda_n) \\ \vdots & \vdots & \vdots \\ v_{j-1}(\lambda_1) & \cdots & v_{j-1}(\lambda_n) \\ v_{n+i}(\lambda_1) & \cdots & v_{n+i}(\lambda_n) \\ v_{j+1}(\lambda_1) & \cdots & v_{j+1}(\lambda_n) \\ \vdots & \vdots & \vdots \\ v_n(\lambda_1) & \cdots & v_n(\lambda_n) \end{bmatrix} \tag{8}$$

($v_i(y)$ in (8) denotes i-th entry of vector $v(y)$ computed by Algorithm 1).

Proof
From $P = \Lambda(\lambda_1, \cdots, \lambda_n)$, we have

$$P = \Gamma_2(\lambda_1, \cdots, \lambda_n)\Gamma_1(\lambda_1, \cdots, \lambda_n)^{-1}.$$

Hence,

$$P\Gamma_1(\lambda_1, \cdots, \lambda_n) = \Gamma_2(\lambda_1, \cdots, \lambda_n).$$

Taking transpose of both sides, we obtain (note that P is a symmetric matrix)

$$\Gamma_1(\lambda_1, \cdots, \lambda_n)^T P = \Gamma_2(\lambda_1, \cdots, \lambda_n)^T.$$

Looking at the i-th column of both sides, we see that

$$\Gamma_1(\lambda_1, \cdots, \lambda_n)^T P_i = \begin{bmatrix} v_{n+i}(\lambda_1) & \cdots & v_{n+i}(\lambda_n) \end{bmatrix}^T,$$

where P_i denotes i-th column of P. From the Cramer's formula, we obtain that j-th entry $p_{i,j}$ of P_i is given by

$$p_{i,j} = \frac{|\bar{\Gamma}_{i,j}(\lambda_1, \cdots, \lambda_n)^T|}{|\Gamma_1(\lambda_1, \cdots, \lambda_n)^T|} = \frac{|\bar{\Gamma}_{i,j}(\lambda_1, \cdots, \lambda_n)|}{|\Gamma_1(\lambda_1, \cdots, \lambda_n)|},$$

which proves the theorem.

3   Basic Algorithm

3.1   Problem Setting

Let $A \in \mathbf{Z}[k]^{n,n}$, $B \in \mathbf{Z}[k]^{n,m}$, $R \in \mathbf{Z}[k]^{m,m}$, $Q \in \mathbf{Z}[k]^{n,n}$ be polynomial matrices in k, where R is positive definite and Q is positive semi-definite for any value of k. We assume that a given system is controllable (i.e. (2)) except at finite number of points of k. Let $p_{i,j}$ be (i,j)th entry of a symmetric solution P

of (1), and let us compute the defining polynomial of $p_{i,j}$. First, let us remark the following lemma:

Lemma 1. *Suppose that there exists numerical value k_0 of k which satisfies the two conditions:*

(C1) All eigenvalues of $H|_{k=k_0}$ are distinct.
(C2) $P(\lambda_1, \cdots, \lambda_n)$ is symmetric $\Rightarrow \forall i,j$ $(1 \le i,j \le n,\ i \ne j)$ $\lambda_i + \lambda_j \ne 0$ where $\lambda_1, \cdots, \lambda_n$ are eigenvalues of $H|_{k=k_0}$.

Then, except for finite number of points of k, ARE (1) has exactly 2^n symmetric solutions.

Proof
The fact that "Numerical ARE with $k = k_0$ satisfying the above conditions (C1) and (C2) has exactly 2^n symmetric numerical solutions" is well-known (see, for example, [9] for details), from which the lemma is deduced.

From Lemma 1, we obtain the following theorem:

Theorem 2. *Suppose that the assumption of Lemma 1 holds and let $p_{i,j}$ be (i,j)th entry of a symmetric solution P of (1). Then there exists a polynomial $f_l(k) \in \mathbf{Z}[k]$ $(l = 0, \cdots, 2^n)$ that satisfies*

$$GCD_k(f_0(k), \cdots, f_{2^n}(k)) = 1, \tag{9}$$
$$f_{2^n}(k)p_{i,j}^{2^n} + \cdots + f_1(k)p_{i,j} + f_0(k) = 0. \tag{10}$$

Proof
Let symmetric solution P of ARE (1) be

$$P = \begin{bmatrix} p_{1,1} & p_{1,2} & \cdots & p_{1,n} \\ p_{1,2} & p_{2,2} & \cdots & p_{2,n} \\ \vdots & \vdots & \vdots & \vdots \\ p_{1,n} & p_{2,n} & \cdots & p_{n,n} \end{bmatrix}.$$

Since matrix $PA + A^T P - PBR^{-1}B^T P + Q$ is symmetric, ARE (1) gives $\frac{n(n+1)}{2}$ algebraic equations with $\frac{n(n+1)}{2}$ unknowns $p_{1,1}, p_{1,2}, \cdots, p_{n,n}$ and parameter k. Computing Groebner basis of the equations with the lexicographic term ordering $p_{i,j} \succ p_{\tilde{i},\tilde{j}}$ $(i \ne \tilde{i}, j \ne \tilde{j}) \succ k$, we obtain a polynomial in $p_{i,j}$ and k in the form of

$$f_m(k)p_{i,j}^m + \cdots + f_1(k)p_{i,j} + f_0(k). \tag{11}$$

We can assume without loss of generality that the condition (9) is satisfied, since if it is not, we can divide (11) by $GCD_k(f_0(k), \cdots, f_{2^n}(k))$, not changing its roots with respect to $p_{i,j}$. From Lemma 1, the degree of the polynomial with respect to $p_{i,j}$ is 2^n, which proves the theorem.

From Theorem 2, we see that the defining polynomial of symmetric solutions of ARE is given as a factor of left-hand side of (10). Hereafter, we discuss the computation of polynomial (10).

3.2   Algorithm with Polynomial Interpolations

Suppose that we have a polynomial $\phi(k) \in \mathbf{Z}[k]$ that contains $f_{2^n}(k)$ in (10) as its factor, i.e.

$$\phi(k) = f_{2^n}(k)\bar{f}(k) \quad (\bar{f}(k), f_{2^n}(k) \in \mathbf{Z}[k]). \tag{12}$$

Then the polynomial in (10) can be computed as follows: Let k_r be an integer and let $\alpha_l(k_r)$ $(l = 1, \cdots, 2^n)$ denote 2^n roots of (10) with $k = k_r$. Since $\alpha_l(k_r)$ $(l = 1, \cdots, 2^n)$ are (i,j)th entries of 2^n symmetric solutions of (1), they can be computed numerically with a conventional numerical method after substitution $k = k_r$. Thus, a polynomial

$$\phi(k_r)\left\{(p_{i,j} - \alpha_1(k_r)) \cdots (p_{i,j} - \alpha_{2^n}(k_r))\right\},$$

in $p_{i,j}$ can be computed (we have $\phi(k)$ in (12)), which can be written as

$$\begin{aligned}
\phi(k_r) &\left\{(p_{i,j} - \alpha_1(k_r)) \cdots (p_{i,j} - \alpha_{2^n}(k_r))\right\} \\
&= f_{2^n}(k_r)\bar{f}(k_r)\left\{(p_{i,j} - \alpha_1(k_r)) \cdots (p_{i,j} - \alpha_{2^n}(k_r))\right\} \\
&= \bar{f}(k_r)\left\{f_{2^n}(k_r)p_{i,j}^{2^n} + \cdots + f_1(k_r)p_{i,j} + f_0(k_r)\right\} \\
&= \sum_{l=0}^{2^n} \bar{f}(k_r)f_l(k_r)p_{i,j}^l \quad (\in \mathbf{Z}[p_{i,j}]).
\end{aligned} \tag{13}$$

From the coefficients of (13), $\bar{f}(k_r)f_l(k_r) \in \mathbf{Z}$ can be calculated for any k_r, implying that a polynomial $\bar{f}(k)f_l(k) \in \mathbf{Z}[k]$ can be computed by polynomial interpolations. With $\bar{f}(k)f_l(k)$ obtained, $f_l(k)$ $(l = 0, \cdots, 2^n)$ can be computed by factoring $\sum_{l=0}^{2^n} \bar{f}(k)f_l(k)p_{i,j}^l$ as follows:

$$\sum_{l=0}^{2^n} \bar{f}(k)f_l(k)p_{i,j}^l = \bar{f}(k)\left\{f_{2^n}(k)p_{i,j}^{2^n} + \cdots + f_1(k)p_{i,j} + f_0(k)\right\}, \tag{14}$$

from which, polynomial (10) is obtained. Thus, the algorithm to compute (10) is as follows:

Algorithm 2 (Computation of (10))

⟨1⟩ Compute polynomial $\phi(k)$ in (12).

⟨2⟩ Let M be an initial guess of $\max_l \left(\deg\left(\bar{f}(k)f_l(k)\right)\right)$.

⟨3⟩ From $r = 1$ to M, perform the following computation.
Let $k_r \in \mathbf{Z}$ be adequate integer and compute 2^n roots of (10), which are symmetric solutions of (1). Then compute $\bar{f}(k_r)f_l(k_r) \in \mathbf{Z}$ in (13).

⟨4⟩ Compute $\bar{f}(k)f_l(k) \in \mathbf{Z}[k]$, using $\bar{f}(k_r)f_l(k_r) \in \mathbf{Z}$ $(r = 1, \cdots, M)$ in ⟨2⟩ and polynomial interpolations.

⟨5⟩ Validate the correctness of $\bar{f}(k)f_l(k)$, with the following two check points:
  – $\deg_k(\bar{f}(k)f_l(k)) < M$, which should be true for a large enough M.
  – Check $\bar{f}(k_0)f_l(k_0)$ is the same as the one obtained from (13), where k_0 is a randomly generated integer.
If $\bar{f}(k)f_l(k)$ is not validated, increase M and go to step ⟨3⟩.

⟨6⟩ Compute $f_l(k)$ $(l = 0, 1, \cdots, 2^n)$ in (14) by factoring $\sum_{l=0}^{2^n} \bar{f}(k)f_l(k)p_{i,j}^l$.

3.3   Computation of the Head Coefficient

Let a set of the eigenvalues of H in (3) be $\{\lambda_1, \cdots, \lambda_n, -\lambda_1, \cdots, -\lambda_n\}$. From Lemma 1, we see that 2^n symmetric solutions of ARE (1) are the solutions corresponding to eigenvalues $s_1\lambda_1, \cdots, s_n\lambda_n$ $(s_1, \cdots, s_n = \pm 1)$ of H. Thus, from Theorem 1, (i, j)th entry of the 2^n symmetric solutions are given by

$$\frac{|\bar{\Gamma}_{i,j}(s_1\lambda_1, \cdots, s_n\lambda_n)|}{|\Gamma_1(s_1\lambda_1, \cdots, s_n\lambda_n)|} \quad (s_l = \pm 1, \; l = 1, \cdots, n). \tag{15}$$

Since $|\Gamma_1(s_1\lambda_1, \cdots, s_n\lambda_n)|$ and $|\bar{\Gamma}_{i,j}(s_1\lambda_1, \cdots, s_n\lambda_n)|$ are both alternating polynomials in $s_1\lambda_1, \cdots, s_n\lambda_n$, they can be expressed in the form of

$$|\Gamma_1(s_1\lambda_1, \cdots, s_n\lambda_n)| = g(s_1\lambda_1, \cdots, s_n\lambda_n) \prod_{i<j}(s_i\lambda_i - s_j\lambda_j), \tag{16}$$

$$|\bar{\Gamma}_{i,j}(s_1\lambda_1, \cdots, s_n\lambda_n)| = h(s_1\lambda_1, \cdots, s_n\lambda_n) \prod_{i<j}(s_i\lambda_i - s_j\lambda_j), \tag{17}$$

where $g(s_1\lambda_1, \cdots, s_n\lambda_n)$ and $h(s_1\lambda_1, \cdots, s_n\lambda_n)$ are symmetric polynomials in $s_1\lambda_1, \cdots, s_n\lambda_n$.

Theorem 3. $\prod_{s_l=\pm 1} g(s_1\lambda_1, \cdots, s_n\lambda_n)$ *can be expressed as a polynomial in k, i.e. there exists $\bar{g}(k) \in \mathbf{Z}[k]$ such that*

$$\bar{g}(k) = \prod_{s_l=\pm 1} g(s_1\lambda_1, \cdots, s_n\lambda_n), \tag{18}$$

and $f_{2^n}(k)$ divides $\bar{g}(k)$.

Proof
First, we will prove former part of the theorem. Let $\{\lambda_1, \cdots, \lambda_n, -\lambda_1, \cdots, -\lambda_n\}$ be a set of the eigenvalues of H. The characteristic polynomial of H can be written as

$$\begin{aligned} H &= (x - \lambda_1) \cdots (x - \lambda_n)(x + \lambda_1) \cdots (x + \lambda_n) \\ &= (x^2 - \lambda_1^2) \cdots (x^2 - \lambda_n^2) \\ &= x^{2n} + g_{2(n-1)}(k)x^{2(n-1)} + \cdots + g_2(k)x^2 + g_0(k), \end{aligned}$$

where $g_{2i}(k)$ $(i = 0, \cdots, n-1)$ are polynomials in k. Thus, fundamental symmetric polynomials of $\lambda_1^2, \cdots, \lambda_n^2$ can be written as polynomials in k, since they are equal to $g_{2i}(k)$ $(i = 0, \cdots, n-1)$. Hence, $\prod_{s_l=\pm 1} g(s_1\lambda_1, \cdots, s_n\lambda_n)$ can be expressed as a polynomial in k, if we prove that

$$\prod_{s_l=\pm 1} g(s_1\lambda_1, \cdots, s_n\lambda_n) \text{ is a symmetric polynomial in } \lambda_1^2, \cdots, \lambda_n^2. \tag{19}$$

It is obvious that $\prod_{s_l=\pm 1} g(s_1\lambda_1, \cdots, s_n\lambda_n)$ is a symmetric polynomial in $\lambda_1, \cdots, \lambda_n$, since $g(s_1\lambda_1, \cdots, s_n\lambda_n)$ is symmetric polynomial in $s_1\lambda_1, \cdots, s_n\lambda_n$. Also note that $\prod_{s_l=\pm 1} g(s_1\lambda_1, \cdots, s_n\lambda_n)$ is an even function of λ_i, since its function value does not change even if we replace λ_i with $-\lambda_i$. Hence, the coefficients with odd power of λ_i are all zero, and $\prod_{s_l=\pm 1} g(s_1\lambda_1, \cdots, s_n\lambda_n)$ is a polynomial in $\lambda_1^2, \cdots, \lambda_n^2$. This proves (19), the former part of the theorem.

Next we will prove latter part of the theorem, i.e.

$$f_{2^n}(k) \text{ divides } \bar{g}(k). \tag{20}$$

From (7),(16) and (17), (i,j)th entry of 2^n symmetric solutions of ARE is given by

$$\frac{h(s_1\lambda_1, \cdots, s_n\lambda_n)}{g(s_1\lambda_1, \cdots, s_n\lambda_n)} \quad (s_1, \cdots, s_n = \pm 1).$$

Therefore, right-hand side of (10) can be written as

$$f_{2^n}(k) \prod_{s_l=\pm 1} \left(p_{i,j} - \frac{h(s_1\lambda_1, \cdots, s_n\lambda_n)}{g(s_1\lambda_1, \cdots, s_n\lambda_n)} \right).$$

Now, let us assume that

$$\begin{cases} \prod_{s_l=\pm 1} \{ g(s_1\lambda_1, \cdots, s_n\lambda_n)p_{i,j} - h(s_1\lambda_1, \cdots, s_n\lambda_n) \} \\ \text{can be express as a polynomial in } k \text{ and } p_{i,j}. \end{cases} \tag{21}$$

Then there exists $\Psi(p_{i,j}, k) \in \mathbf{Z}[p_{i,j}, k]$ such that

$$\begin{aligned} \Psi(p_{i,j}, k) &\overset{\text{def}}{=} \prod_{s_l=\pm 1} \{ g(s_1\lambda_1, \cdots, s_n\lambda_n)p_{i,j} - h(s_1\lambda_1, \cdots, s_n\lambda_n) \} \\ &= \bar{g}(k) \prod_{s_l=\pm 1} \left(p_{i,j} - \frac{h(s_1\lambda_1, \cdots, s_n\lambda_n)}{g(s_1\lambda_1, \cdots, s_n\lambda_n)} \right) \\ &= \left(\frac{\bar{g}(k)}{f_{2^n}(k)} \right) f_{2^n}(k) \prod_{s_l=\pm 1} \left(p_{i,j} - \frac{h(s_1\lambda_1, \cdots, s_n\lambda_n)}{g(s_1\lambda_1, \cdots, s_n\lambda_n)} \right) \\ &= \frac{\bar{g}(k)}{f_{2^n}(k)} \left\{ f_{2^n}(k)p_{i,j}^{2^n} + \cdots + f_1(k)p_{i,j} + f_0(k) \right\}. \end{aligned} \tag{22}$$

If $\frac{\bar{g}(k)}{f_{2^n}(k)}$ is not a polynomial, i.e. if $\frac{\bar{g}(k)}{f_{2^n}(k)}$ can be written as

$$\frac{r_1(k)}{r_2(k)} \quad (\text{GCD}(r_1(k), r_2(k)) = 1, \ \deg_k(r_2(k)) > 0),$$

then $r_2(k)$ is a common divisor of $f_i(k)$ $(i = 0, \cdots, 2^n)$ (recall that $\Psi(p_{i,j}, k)$ is a polynomial in $p_{i,j}$ and k), which contradicts (9). Therefore, $\frac{\bar{g}(k)}{f_{2^n}(k)}$ is a polynomial and $\bar{g}(k)$ divides $f_{2^n}(k)$. Thus, to prove (20), it is enough to prove (21). In a way similar to the former part of the proof, (21) is shown, if we prove that

$$\begin{cases} \prod_{s_l=\pm 1} \{ g(s_1\lambda_1, \cdots, s_n\lambda_n)p_{i,j} - h(s_1\lambda_1, \cdots, s_n\lambda_n) \} \\ \text{is a symmetric polynomial in } \lambda_1^2, \cdots, \lambda_n^2. \end{cases} \tag{23}$$

Since both of $g(s_1\lambda_1, \cdots, s_n\lambda_n)$ and $h(s_1\lambda_1, \cdots, s_n\lambda_n)$ are symmetric polynomials in $s_1\lambda_1, \cdots, s_n\lambda_n$, it is obvious that (23) is a symmetric polynomial in $\lambda_1, \cdots, \lambda_n$. Note also that (23) is an even function, since its function value does not change even if we replace λ_i with $-\lambda_i$. Thus, the coefficients with odd power of λ_i are all zero, and (23) is a function of $\lambda_1^2, \cdots, \lambda_n^2$. Thus, (23) is shown, and so is (21). This completes the proof.

From Theorem 3, we see that $\bar{g}(k)$ is a polynomial which has $f_{2^n}(k)$ as its factor. Thus, we can set $\phi(k) = \bar{g}(k)$ in $\langle 1 \rangle$ of Algorithm 2. Since $\bar{g}(k)$ is a polynomial in k, it can be computed with polynomial interpolations. More specifically, $\bar{g}(k)$ can be computed from $\bar{g}(k_l) \in \mathbf{Z}$, $k_l \in \mathbf{Z}$ $(l = 1, \cdots, L)$, where L is an integer satisfying $L > \deg_k(\bar{g}(k))$.

Algorithm 3 (Computation of $\bar{g}(k)$ $(= \phi(k))$)

$\langle 1 \rangle$ Let L be an initial guess of $\deg_k(\bar{g}(k))$.
$\langle 2 \rangle$ Let k_l $(l = 1, \cdots, L)$ be adequate integers.
$\langle 3 \rangle$ Let $l \leftarrow 1$.
$\langle 4 \rangle$ Compute a set of eigenvalues $\{\lambda_1, \cdots, \lambda_n, -\lambda_1, \cdots, -\lambda_n\}$ of $H|_{k=k_l}$.
$\langle 5 \rangle$ Compute $\bar{g}(k_l)$ by

$$\bar{g}(k_l) = \prod_{s_l=\pm 1} g(s_1\lambda_1, \cdots, s_n\lambda_n), \tag{24}$$

where

$$g(s_1\lambda_1, \cdots, s_n\lambda_n) = \frac{|\Gamma_1(s_1\bar{\lambda}_1, \cdots, s_n\bar{\lambda}_n)|}{\prod_{i<j}(s_i\lambda_i - s_j\lambda_j)}. \tag{25}$$

$\langle 6 \rangle$ If $l < L$, then let $l \leftarrow l + 1$ and go to $\langle 3 \rangle$. Otherwise compute $\bar{g}(k) \in \mathbf{Z}[k]$ from $\bar{g}(k_l)$ $(l = 1, \cdots, L)$ by polynomial interpolations.
$\langle 7 \rangle$ Validate the correctness of $\bar{g}(k)$ with the following two check points:
  – $\deg_k(\bar{g}(k)) < L$, which should be true for a large enough L.
  – Check $\bar{g}(k_0)$ is the same as the one obtained from (24), where k_0 is a randomly generated integer.
If $\bar{g}(k)$ is not validated, increase L and go to step $\langle 2 \rangle$.

Remark: As we mentioned before, $|\Gamma_1(s_1\bar{\lambda}_1, \cdots, s_n\bar{\lambda}_n)|$ is an alternating polynomial in $s_1\bar{\lambda}_1, \cdots, s_n\bar{\lambda}_n$, and contains factors $(s_i\lambda_i - s_j\lambda_j)$ $(i \neq j)$. It is possible to pull out the factors symbolically by column operations of matrix $\Gamma_1(s_1\bar{\lambda}_1, \cdots, s_n\bar{\lambda}_n)$, and we need not to compute denominator of (25) explicitly.

4   Numerical Experiments

Let $A \in \mathbf{Z}[k]^{n,n}$, $B \in \mathbf{Z}^{1,n}$ be the following matrices:

$$A = k\bar{E} + \Omega_{n,n}, \quad B = \Omega_{1,n}, \tag{26}$$

where matrix $\Omega_{n,n} \in \mathbf{Z}^{n,n}$ is a randomly generated $n \times n$ matrix whose entries are integers between -5 and 5, and $\bar{E}$ is a matrix whose (i,j)-th entry $\bar{e}_{i,j}$ is defined by

$$\bar{e}_{i,j} = \begin{cases} \tau, \text{ when } (i,j) = (1,1) \\ 0, \text{ otherwise} \end{cases}, \ \tau = \text{random integer } (\neq 0) \text{ between } -5 \text{ and } 5.$$

For example, when $n = 2$, an example of A, B in (26) is

$$A = \begin{bmatrix} 2k - 3 & 3 \\ -1 & -2 \end{bmatrix}, \ B = \begin{bmatrix} 0 \\ 1 \end{bmatrix}.$$

Numerical experiments are performed as follows: Matrices Q and R are set to the identity matrices, and we generate 5 set of A, B for each $n = 2, \cdots, 5$. Then the defining polynomial of symmetric solutions of ARE (1) is computed by the following two methods:

(M1) The method using Groebner basis.
(M2) The method using Algorithm 2, where $\phi(k)$ is computed by Algorithm 3.

The experiments are performed with Maple 10 on the machine equipped with Pentium M 2.0GHz and 1.5GByte memory. Table 1 shows the average of 5 computation times in seconds, where $\times$ denotes the failure of the computation due to either memory exhaustion or too much (> 24 hours) computation time.

Table 1. Computation time (in seconds)

n	2	3	4	5
M1	0.8844	$\times$	$\times$	$\times$
M2	2.044	16.71	766.6	$\times$

From the table, we see that (M2) is more efficient than (M1), computing the defining polynomial up to the system with $n = 4$. We also see that the method using Groebner basis is not practical for $n \geq 3$.

5 Conclusion

This paper presented the algorithm to compute the defining polynomial of symmetric solutions of ARE (1), where $A, W (= BR^{-1}B^T), Q$ contain an unknown parameter. The algorithm uses polynomial interpolations, where its head coefficient is computed from Theorem 3 in Section 3.3. Numerical experiments showed that for H_2 optimal problem, the algorithm is practical for the system with $n \leq 4$ (n denotes the size of matrices A, W, Q), where the method with Groebner basis is practical only for the system with $n \leq 2$. Since the algorithm is easily parallelizable, the computation time can be drastically reduced under multi-CPU environments.

References

1. Abdallah, C., Dorato, P., Yang, W., Liska, R., Steinberg, S.: Application of Quantifier Elimination Theory to Control System Design. In: Proc. of 4th IEEE Mediterranean Symposium of Control and Automation, Maleme, Crete, pp. 340–345 (1996)
2. Anai, H., Yanami, H.: SyNRAC: A maple-package for solving real algebraic constraints. In: Sloot, P.M.A., Abramson, D., Bogdanov, A.V., Gorbachev, Y.E., Dongarra, J.J., Zomaya, A.Y. (eds.) ICCS 2003. LNCS, vol. 2657, Springer, Heidelberg (2003)
3. Dorato, P., Yang, W., Abdallah, C.: Robust Multi-Objective Feedback Design by Quantifier Elimination. J. Symbolic Computation 24, 153–159 (1997)
4. Hong, H., Liska, R., Steinberg, S.: Testing Stability by Quantifier Elimination. J. Symbolic Computation 24, 161–187 (1997)
5. Kitamoto, T.: On the computation of H_∞ norm of a system with a parameter. The IEICE Trans. Funda. (Japanese Edition) J89-A(1), 25–39 (2006)
6. Kitamoto, T., Yamaguchi, T.: Parametric Computation of H_∞ Norm of a System. In: Proc. SICE-ICCAS2006, Busan, Korea (2006)
7. Kanno, M., Smith, M.C.: Validated numerical computation of the $\mathcal{L}_\infty$-norm for linear dynamical systems. J. of Symbolic Computation 41(6), 697–707 (2006)
8. Zhou, K., Doyle, J., Glover, K.: Robust and Optimal Control. Prentice-Hall. Inc, New Jersey (1996)
9. Nishimura, T., Kano, H.: Matrix Riccati Equations in Control Theory Asakurasyoten, Tokyo (1996) (in Japanese)

Symmetries and Dynamics of Discrete Systems

Vladimir V. Kornyak

Laboratory of Information Technologies
Joint Institute for Nuclear Research
141980 Dubna, Russia
kornyak@jinr.ru

Abstract. We consider discrete dynamical systems and lattice models in statistical mechanics from the point of view of their symmetry groups. We describe a C program for symmetry analysis of discrete systems. Among other features, the program constructs and investigates *phase portraits* of discrete dynamical systems *modulo groups* of their symmetries, searches dynamical systems possessing specific properties, e.g., *reversibility*, computes microcanonical *partition functions* and searches *phase transitions* in mesoscopic systems. Some computational results and observations are presented. In particular, we explain formation of moving soliton-like structures similar to *"spaceships"* in cellular automata.

1 Introduction

Symmetry analysis of continuous systems described by ordinary or partial differential equations is well developed and fruitful discipline. But there is a sense of incompleteness of the approach since the transformations used in the symmetry analysis of continuous systems — point and contact Lie, Bäcklund and Lie–Bäcklund, some sporadic instances of so-called *non-local* transformations — constitute negligible small part of all thinkable transformations. In this context finite discrete systems look more attractive since we can study *all possible* their symmetries.

Furthermore, there are many hints from quantum mechanics and quantum gravity that discreteness is more suitable for describing physics at small distances than continuity which arises only as a logical limit in considering large collections of discrete structures.

Both differential equations and cellular atomata are based on the idea of *locality* — behavior of a system as a whole is determined by interections of its closely situated parts. Recently [1,2] we showed that any collection of discrete points taking values in finite sets possesses some kind of locality. More specifically, let us consider collection of N "points", symbolically $\delta = \{x_1, \ldots, x_N\}$. We call δ *domain*. Each x_i takes value in its own set of values $Q_i = \left\{ s_i^1, \ldots, s_i^{q_i} \right\}$ or using the standard notation $Q_i = \{0, \ldots, q_i - 1\}$. Adopting Q^δ as symbolical notation for the Cartesian product $Q_1 \times \cdots \times Q_N$, we define *relation* on δ as an arbitrary subset $R^\delta \subseteq Q^\delta$. Then we define *consequence* of relation R^δ as an

V.G. Ganzha, E.W. Mayr, and E.V. Vorozhtsov (Eds.): CASC 2007, LNCS 4770, pp. 236–251, 2007.

arbitrary superset $S^\delta \supseteq R^\delta$ and *proper consequence* as a consequence which can be represented in the form $P^\alpha \times Q^{\delta \setminus \alpha}$, where P^α is *nontrivial* (i.e., $P^\alpha \neq Q^\alpha$) relation on the proper subset $\alpha \subset \delta$. We show that any relation R^δ allows a decomposition in terms of its proper consequences. This decomposition naturally imposes a structure of *abstract simplicial complex* — one of the mathematical abstractions of locality. Thus we call collections of discrete finite-valued points *discrete relations on abstract simplicial complexes*.

We demonstrated also that such relations in special cases correspond to *systems of polynomial equations* (if all points x_i take values in the same set Q and its cardinality is a power of a prime $|Q| = p^k$) and to *cellular automata* (if domain δ allows decomposition into congruent simplices with the same relation on the simplices and this relation is *functional*). The notion of discrete relations covers also discrete dynamical systems more general than cellular automata. The lattice models in statistical mechanics can also be included in this framework by considering *ensembles* of discrete relations on abstract simplicial complexes.

In this paper we study dependence of behavior of discrete dynamical systems on graphs — one-dimensional simplicial complexes — on symmetries of the graphs. We describe our C program for discrete symmetry analysis and results of its application to cellular automata and mesoscopic lattice models.

2   Symmetries of Lattices and Functions on Lattices

Lattices. A space of discrete dynamical system is a lattice L represented as a k-regular (k-valent) graph G_L. By a 'symmetry' of lattice L we mean the automorphism group $\mathrm{Aut}(G_L)$ of the graph of L. In applications one often assumes that the lattice L is embedded in some continuous space. In this case the notion of 'dimension' of lattice makes sense. Note that the same graph can be embedded regularly in different continuous spaces as it is clear from the example shown in Fig. 1. That is why the group $\mathrm{Aut}(G_L)$ is usually larger than the symmetry group of lattice placed in a space.

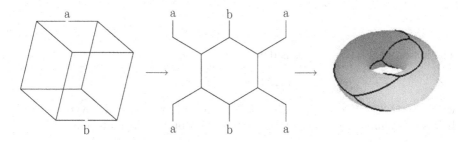

Fig. 1. The same graph forms 4-gonal (6 tetragons) lattice in sphere $\mathbb{S}^2$ and 6-gonal (4 hexagons) lattice in torus $\mathbb{T}^2$

The lattices we are concerned in this paper are shown in Fig. 2.

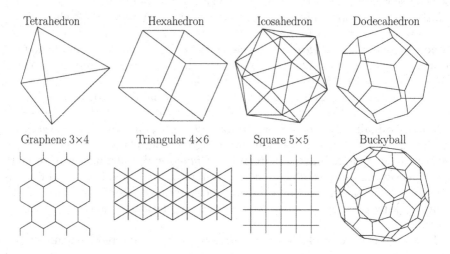

Fig. 2. Examples of lattices

Computing Automorphisms. The automorphism group of graph with n vertices may have up to $n!$ elements. However, McKay's algorithm [4], based on efficiently arranged search tree, determines the graph automorphisms by constructing compact set (not more than $n - 1$ elements, but usually much less) of generators of the group.

Functions on Lattices. To study the symmetry properties of a system on a lattice L we should consider action of the group $\mathrm{Aut}(G_L)$ on the space $\Sigma = Q^L$ of Q-valued functions on L, where $Q = \{0, \dots, q - 1\}$ is the set of values of lattice vertices. We shall call the elements of Σ *states* or (later in Sect. 5) *microstates*.

The group $\mathrm{Aut}(G_L)$ acts non-transitively on the space Σ splitting this space into the disjoint orbits of different sizes

$$\Sigma = \bigcup_{i=1}^{N_{orbits}} O_i.$$

The action of $\mathrm{Aut}(G_L)$ on Σ is defined by

$$(g\varphi)(x) = \varphi\left(g^{-1}x\right),$$

where $\varphi(x) \in \Sigma$, $g \in \mathrm{Aut}(G_L)$, x is a variable running over vertices of L. Burnside's lemma counts the total number of orbits in the state space Σ

$$N_{orbits} = \frac{1}{|\mathrm{Aut}(G_L)|} \sum_{g \in \mathrm{Aut}(G_L)} q^{N_{cycles}^g}.$$

Here N_{cycles}^g is the number of cycles (including unit ones) in the group element g.

Large symmetry group allows to represent dynamics on the lattice in more compact form. For example, the automorphism group of (graph of) icosahedron, dodecahedron and buckyball is S_5[1], and the information about behavior of any dynamical system on these lattices can be compressed nearly in proportion to $|S_5| = 120$.

Illustrative Data. In Table 1 we collect some quantitive information about the lattices from Fig.2 and their automorphism groups, namely, *number of vertices* $V(G_L)$, *size of automorphism group* $|\mathrm{Aut}(G_L)|$, total *number of states* $\Omega = |\Sigma| \equiv q^{V(G_L)}$ (assuming $q = 2$) and *number of group orbits* N_{orbits} in the space of states.

Table 1. Lattices, groups, orbits: numerical characteristics

Lattice	$V(G_L)$	$|\mathrm{Aut}(G_L)|$	$\Omega = q^{V(G_L)}$	N_{orbits}
Tetrahedron	4	24	16	5
Hexahedron	8	48	256	22
Icosahedron	12	120	4096	82
Dodecahedron	20	120	1048576	9436
Graphene 3×4 Torus	24	48	16777216	355353
Graphene 3×4 Klein bottle	24	16	16777216	1054756
Triangular 4×6	24	96	16777216	180070
Square 5×5	25	200	33554432	172112
Buckyball	60	120	1152921504606846976 $\approx 10^{18}$	9607679885269312 $\approx 10^{16}$

Note that the lattices marked in Fig. 2 as "Graphene 3×4", "Triangular 4×6" and "Square 5×5" can be closed by identifications of opposite sides of rectangles in several different ways. Most natural identifications form graphs embedded in the torus and in the Klein bottle. Computation shows that the Klein bottle arrangement (as well as others except for embeddings in the torus) leads to *nonhomogeneous* lattices. For example, the hexagonal lattice "Graphene 3×4" embedded in the Klein bottle has 16-element symmetry group and this group splits the set of vertices into two orbits of sizes 8 and 16. Since non-transitivity of points contradicts to our usual notion of space, we shall not consider further such lattices.

3   Computer Program and Its Functionality

We have written a C program to study different properties of deterministic and statistical lattice systems exploiting their symmetries. Input of the program consists of the following elements:

---

[1] Traditionally, the icosahedral group $I_h = A_5$ is adopted as a symmetry group for these polyhedra. A_5 is 60-element discrete subgroup of $SO(3)$. Adding reflections to A_5 we get twice larger (and hence more efficient for our purposes) group S_5.

- Graph of lattice $G_L = \{N_1, \ldots, N_n\}$. N_i is neighborhood of ith vertex, i.e., the set of k vertices adjacent to ith vertex.
- *Cellular automata branch:*
    Set of local rules $R = \{r_1, \ldots, r_m\}$. r_i is integer number representing bits of ith rule. The set R includes the rules we are interested in. In particular, this set may contain only one rule (for detailed study).
- *Statistical models branch:*
    Hamiltonian of the model.
- Some control parameters.

The program computes the automorphism group $\text{Aut}(G_L)$ and

- in the case of cellular automata the program constructs *phase portraits* of automata *modulo* $\text{Aut}(G_L)$ for all rules from R.
    Manipulating the above mentioned control parameters we can
    • select automata with specified properties, for example, *reversibility, conservation* of a given function on dynamical trajectories, etc.;
    • search automata whose phase portraits contain specific structures, for example, the limit cycles of a given length, *"gardens of Eden"* [5] or, more generally, isolated cycles, *"spaceships"*, etc.
- in the case of statistical lattice model the program computes the partition function and other characteristics of the system, searches phase transitions.

Example of timing
The full run of all 136 symmetric 3-valent binary cellular automata on the dodecahedron (number of vertices = 20, order of automorphism group = 120, number of states = 1048576, number of orbits = 9436) takes about 40 sec on a 1133MHz Pentium III personal computer.

4   Deterministic Dynamical Systems

In this section we point out a general principle of evolution of any causal dynamical system implied by its symmetry, explain formation of soliton-like structures, and consider some results of computing with symmetric 3-valent cellular automata.

Universal Property of Deterministic Evolution Induced by Symmetry. The splitting of the space Σ of functions on a lattice into the group orbits of different sizes imposes *universal restrictions* on behavior of a deterministic dynamical system for any law that governs evolution of the system. Namely, dynamical trajectories can obviously go only in the direction of *non-decreasing sizes of orbits*. In particular, *periodic trajectories* must lie *within the orbits of the same size*. Conceptually this restriction is an analog of the *second law of thermodynamics* — any isolated system may only lose information in its evolution.

Formation of Soliton-like Structures. After some lapse of time the dynamics of finite discrete system is governed by its symmetry group, that leads to appearance of *soliton-like* structures. Let us clarify the matter. Obviously phase portraits of the systems under consideration consist of attractors being limit cycles and/or isolated cycles (including limit and isolated fixed points regarded as cycles of period one). Now let us consider the behavior of the system which has come to a cycle, no matter whether the cycle is limit or isolated. The system runs periodically over some sequence of equal size orbits. The same orbit may occur in the cycle repeatedly. For example, the isolated cycle of period 6 in Fig. 5 — where a typical phase portrait *modulo* automorphisms is presented — passes through the sequence of orbits numbered[2] as 0, 2, 4, 0, 2, 4, i.e., each orbit appears twice in the cycle.

Suppose a state $\varphi(x)$ of the system running over a cycle belongs to ith orbit at some moment t_0: $\varphi(x) \in O_i$. At some other moment t the system appears again in the same orbit with the state $\varphi_t(x) = A_{t_0t}(\varphi(x)) \in O_i$. Clearly, the evolution operator A_{t_0t} can be replaced by the action of some group element $g_{t_0t} \in \mathrm{Aut}(G_L)$

$$\varphi_t(x) = A_{t_0t}(\varphi(x)) = \varphi\left(g_{t_0t}^{-1}x\right). \tag{1}$$

The element g_{t_0t} is determined uniquely *modulo* subgroup $\mathrm{Aut}(G_L; \varphi(x)) \subseteq \mathrm{Aut}(G_L)$ fixing the state $\varphi(x)$. Equation (1) means that the initial cofiguration (shape) $\varphi(x)$ is completely reproduced after some movement in the space L. Such soliton-like structures are typical for cellular automata. They are usually called "*spaceships*" in the cellular automata community.

Let us illustrate the group nature of such moving self-reproducing structures by the example of "*glider*" — one of the simplest spaceships of Conway's automaton "Life". This configuration moves along the diagonal of square lattice reproducing itself with one step diagonal shift after four steps in time. If one considers only translations as a symmetry group of the lattice, then, as it is clear from Fig. 3, the first configuration lying in the same orbit[3] with φ_1 is φ_5, i.e., for the translation group $\mathbf{T}^2$ glider is a cycle running over *four* orbits.

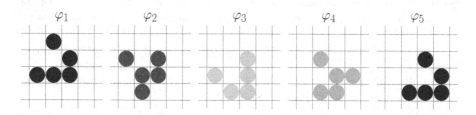

Fig. 3. Glider over translation group $\mathbf{T}^2$ is cycle in *four* group orbits

Our program constructs the maximum possible automorphism group for any lattice. For an $n \times n$ square toric lattice the maximal group is the *semidirect product* $\mathbf{T}^2 \rtimes \mathbb{D}_4$. Here $\mathbb{D}_4$ is the *dihedral* group, which in its turn is the semidirect product $\mathbb{D}_4 = \mathbb{Z}_4 \rtimes \mathbb{Z}_2$, where $\mathbb{Z}_4$ is generated by 90 degree rotations and $\mathbb{Z}_2$ are

---

[2] The program numbers orbits in the order of decreasing of their sizes and at equal sizes the lexicographic order of lexicograhically minimal orbit representatives is used.

[3] In Figs. 3 and 4 the configurations belonging to the same orbit have identical colors.

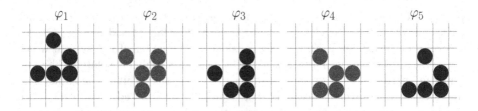

Fig. 4. Glider over maximal symmetry group $\mathbf{T}^2 \rtimes \mathbb{D}_4$ is cycle in *two* group orbits

reflections. The size of maximal group is $8n^2$, whereas the size of translation group[4] is only n^2. Now the glider is reproduced after two steps in time.

As one can see from Fig. 4, φ_3 is obtained from φ_1 and φ_4 from φ_2 by combinations of translations, 90 degree rotations and reflections. Thus, the glider in torus (and in the discrete plane obtained from the torus as $n \to \infty$) is a cycle located in two orbits of maximal automorphism group.

Note also that similar behavior is rather typical for continuous systems too. Many equations of mathematical physics have solutions in the form of running wave $\varphi(x - vt)$ $\left(= \varphi\left(g_t^{-1}x\right)\right.$ for Galilei group). One can see also an analogy between "*spaceships*" of cellular automata and *solitons* of KdV type equations. The solitons — like shape preserving moving structures in cellular automata — are often arise for rather arbitrary initial data.

Cellular Automata with Symmetric Local Rules. As a specific class of discrete dynamical systems, we consider 'one-time-step' cellular automata on k-valent lattices with local rules symmetric with respect to all permutations of k outer vertices of the neighborhood. This symmetry property is an immediate discrete analog of general local diffeomorphism invariance of fundamental physical theories based on continuum space. The diffeomorphism group $\mathrm{Diff}(M)$ of the *manifold* M is very special subgroup of the infinite symmetric group $\mathrm{Sym}(M)$ of the *set* M.

As we demonstrated in [3], in the binary case, i.e., if the number of vertex values $q = 2$, the automata with symmetric local rules are completely equivalent to generalized Conway's "Game of Life" automata [5] and, hence, their rules can be formulated in terms of "Birth"/"Survival" lists.

Adopting the convention that the outer points and the root point of the neighborhood are denoted $x_1, \ldots, x_k$ and x_{k+1}, respectively, we can write a *local rule* determining one-time-step evolution of the root in the form

$$x'_{k+1} = f(x_1, \ldots, x_k, x_{k+1}). \tag{2}$$

The total number of rules (2) symmetric with respect to permutations of points $x_1, \ldots, x_k$ is equal to $q^{\binom{k+q-1}{q-1}q}$. For the case of our interest ($k = 3$, $q = 2$) this number is 256.

---

[4] Translation group for $n \times n$ discrete torus is the *direct* product of two cyclic groups $\mathbf{T}^2 = \mathbb{Z}_n \times \mathbb{Z}_n$.

It should be noted that the rules obtained from each other by permutation of q elements in the set Q are equivalent since such permutation means nothing but renaming of values. Thus, we can reduce the number of rules to consider. The reduced number can be counted via *Burnside's lemma* as a number of orbits of rules (2) under the action of the group S_q. The concrete expression depends on the cyclic structure of elements of S_q. For the case $q = 2$ this gives the following number of non-equivalent rules

$$N_{rules} = 2^{2k+1} + 2^k.$$

Thus, studying 3-valent binary case, we have to consider 136 different rules.

Example of Phase Portrait. Cellular Automaton 86. As an example consider the rule 86 on hexahedron (cube). The number 86 is the "little endian" representation of the bit string 01101010 taken from the last column of the rule table with S_3-symmetric combinations of values for x_1, x_2, x_3

x_1	x_2	x_3	x_4	x_4'
0	0	0	0	0
0	0	0	1	1
1	0	0	0	1
1	0	0	1	0
1	1	0	0	1
1	1	0	1	0
1	1	1	0	1
1	1	1	1	0

The rule can also be represented in the "Birth"/"Survival" notation as B123/S0, or as polynomial over the Galois field $\mathbb{F}_2$ (see [3])

$$x_4' = x_4 + \sigma_3 + \sigma_2 + \sigma_1,$$

where $\sigma_1 = x_1 + x_2 + x_3$, $\sigma_2 = x_1x_2 + x_1x_3 + x_2x_3$, $\sigma_3 = x_1x_2x_3$ are *symmetric* functions. In Fig. 5 the group orbits are represented by circles. The ordinal numbers of orbits are placed within these circles. The numbers over orbits and within cycles are sizes of the orbits (recall that all orbits included in one cycle have the same size). The rational number p indicates the *weight* of the corresponding element of phase portrait. In other words, p is a probability to be in an isolated cycle or to be caught by an attractor at random choice of state: $p = (size$ $of\ basin)/(total\ number\ of\ states)$. Here *size of basin* is sum of sizes of orbits involved in the struture. The structures in Fig. 5 are placed in the decreasing order of their weights.

Note that most of cycles in Fig. 5 (36 of 45 or 80%) are *"spaceships"*. Other computed examples also confirm that soliton-like moving structures are typical for cellular automata.

Of course, in the case of large lattices it is impractical to output full phase portraits (the program easily computes tasks with up to hundreds thousands of different structures). But it is not difficult to extract structures of interest, e.g., *"spaceships"* or *"gardens of Eden"*.

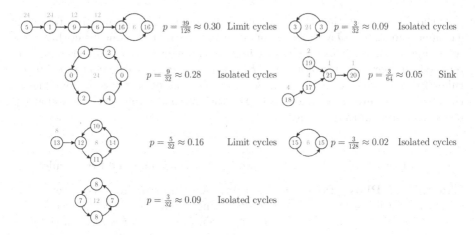

Fig. 5. Rule 86. Equivalence classes of trajectories on hexahedron. 36 of 45 cycles are *"spaceships"*.

Search for Reversibility. The program is able to select automata with properties specified at input. One of such important properties is *reversibility*.

In this connection we would like to mention recent works of G. 't Hooft. One of the difficulties of Quantum Gravity is a conflict between irreversibility of Gravity — information loss (dissipation) at the black hole horizon — with reversibility and unitarity of the standard Quantum Mechanics. In several papers of recent years (see, e.g., [6,7]) 't Hooft developed the approach aiming to reconcile both theories. The approach is based on the following assumptions

- physical systems have *discrete degrees of freedom* at tiny (Planck) distance scales;
- the states of these degrees of freedom form *primordial* basis of Hilbert space (with nonunitary evolution);
- primordial states form *equivalence classes*: two states are equivalent if they evolve into the same state after some lapse of time;
- the equivalence classes by construction form basis of Hilbert space with unitary evolution described by time-reversible Schrödinger equation.

In our terminology this corresponds to transition to limit cycles: in a finite time of evolution the limit cycle becomes physically indistinguishable from reversible isolated cycle — the system "forgets" its pre-cycle history. Fig. 6 illustrates construction of unitary Hilbert space from primordial.

This irreversibility hardly can be observed experimentally (assuming, of course, that considered models can be applied to physical reality). The system should probably spend time of order the Planck one ($\approx 10^{-44}$ sec) out of a cycle and potentially infinite time on the cycle. Nowadays, the shortest experimentally fixed time is about 10^{-18} sec or 10^{26} Planck units only.

Applying our program to all 136 symmetric 3-valent automata we have the following. There are two rules trivially reversible on all lattices

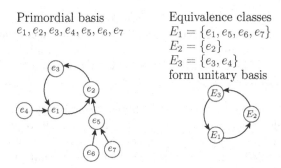

Primordial basis
$e_1, e_2, e_3, e_4, e_5, e_6, e_7$

Equivalence classes
$E_1 = \{e_1, e_5, e_6, e_7\}$
$E_2 = \{e_2\}$
$E_3 = \{e_3, e_4\}$
form unitary basis

Fig. 6. Transition from primordial to unitary basis

- 85 $\sim$ B0123/S $\sim$ $x_4' = x_4 + 1$,
- 170 $\sim$ B/S0123 $\sim$ $x_4' = x_4$.

Besides these uninteresting rules there are 6 reversible rules on *tetrahedron*

- 43 $\sim$ B0/S012 $\sim$ $x_4' = x_4(\sigma_2 + \sigma_1) + \sigma_3 + \sigma_2 + \sigma_1 + 1$,
- 51 $\sim$ B02/S02 $\sim$ $x_4' = \sigma_1 + 1$,
- 77 $\sim$ B013/S1 $\sim$ $x_4' = x_4(\sigma_2 + \sigma_1 + 1) + \sigma_3 + \sigma_2 + 1$,
- 178 $\sim$ B2/S023 $\sim$ $x_4' = x_4(\sigma_2 + \sigma_1 + 1) + \sigma_3 + \sigma_2$,
- 204 $\sim$ B13/S13 $\sim$ $x_4' = \sigma_1$,
- 212 $\sim$ B123/S3 $\sim$ $x_4' = x_4(\sigma_2 + \sigma_1) + \sigma_3 + \sigma_2 + \sigma_1$.

Note that all these reversible rules are symmetric with respect to permutation of values $Q = \{0, 1\}$. Two of the above rules, namely 51 and 204, are reversible on *hexahedron* too. There are no nontrivial reversible rules on all other lattices from Fig. 2. Thus we may suppose that 't Hooft's picture is typical for discrete dynamical systems.

5   Statistical Lattice Models and Mesoscopic Systems

Statistical Mechanics. The state of deterministic dynamical system at any point of time is determined uniquely by previous states of the system. A Markov chain — for which transition from any state to any other state is possible with some probability — is a typical example of *non-deterministic* dynamical system. In this section we apply symmetry approach to the lattice models in statistical mechanics. These models can be regarded as special instances of Markov chains. *Stationary distributions* of these Markov chains are studied by the methods of statistical mechanics.

The main tool of conventional statistical mechanics is the Gibbs *canonical ensemble* – imaginary collection of identical systems placed in a huge thermostat with temperature T. The statistical properties of canonical ensemble are encoded in the *canonical partition function*

$$Z = \sum_{\sigma \in \Sigma} e^{-E_\sigma/k_B T}. \tag{3}$$

Here Σ is the set of microstates, E_σ is energy of microstate σ, k_B is Boltzmann's constant. The canonical ensemble is essentially asymptotic concept: its formulation is based on approximation called "thermodynamic limit". For this reason, the canonical ensemble approach is applicable only to large (strictly speaking, infinite) homogeneous systems.

Mesoscopy. Nowadays much attention is paid to study systems which are too large for a detailed microscopic description but too small for essential features of their behavior to be expressed in terms of classical thermodynamics. This discipline, often called *mesoscopy*, covers wide range of applications from nuclei, atomic clusters, nanotechnological structures to multi-star systems [8,9,10]. To study *mesoscopic* systems one should use more fundamental *microcanonical ensemble* instead of canonical one. A microcanonical ensemble is a collection of identical isolated systems at fixed energy. Its definition does not include any approximating assumptions. In fact, the only key assumption of a microcanonical ensemble is that all its microstates are equally probable. This leads to the *entropy* formula

$$S_E = k_B \ln \Omega_E, \tag{4}$$

or, equivalently, to the *microcanonical partition function*

$$\Omega_E = e^{S_E/k_B}. \tag{5}$$

Here Ω_E is the number of microstates at fixed energy E. In what follows we will omit Boltzmann's constant assuming $k_B = 1$. Note that in the thermodynamic limit the microcanonical and canonical descriptions are equivalent and the link between them is provided by the Laplace transform. On the other hand, mesoscopic systems demonstrate observable experimentally and in computation peculiarities of behavior like heat flows from cold to hot, negative specific heat or "convex intruders" in the entropy versus energy diagram, etc. These anomalous – from the point of view canonical thermostatistics – features have natural explanation within microcanonical statistical mechanics [10].

Lattice Models. In this section we apply symmetry analysis to study mesoscopic lattice models. Our approach is based on exact enumeration of group orbits of microstates. Since statistical studies are based essentially on different simplifying assumptions, it is important to control these assumptions by exact computation, wherever possible. Moreover, we might hope to reveal with the help of exact computation subtle details of behavior of system under consideration.

As an example, let us consider the Ising model. The model consists of *spins* placed on a lattice. The set of vertex values is $Q = \{-1, 1\}$ and the interaction Hamiltonian is given by

$$H = -J \sum_{(i,j)} s_i s_j - B \sum_i s_i, \tag{6}$$

where $s_i, s_j \in Q$; J is a coupling constant ($J > 0$ and $J < 0$ correspond to *ferromagnetic* and *antiferromagnetic* cases, respectively); the first sum runs over all edges (i, j) of the lattice; B is an external "magnetic" field. The second sum $M = \sum_i s_i$ is called the *magnetization*. To avoid unnecessary technical details we will consider only the case $J > 0$ (assuming $J = 1$) and $B = 0$ in what follows.

Since Hamiltonian and magnetization are constants on the group orbits, we can count numbers of microstates corresponding to particular values of these functions – and hence compute all needed statistical characteristics – simply by summation of sizes of appropriate orbits.

Fig. 7 shows microcanonical partition function for the Ising model on dodecahedron. Here total number of microstates $\Omega = 1048576$, number of lattice vertices $V(G_L) = 20$, energy E is value of Hamiltonian.

Of course, other characteristics of the system can be computed easily via counting sizes of group orbits. For example, the magnetization is shown in Fig. 8.

Phase Transitions. Needs of nanotechnological science and nuclear physics attract special attention to phase transitions in finite systems. Unfortunately classical thermodynamics and the rigorous theory of critical phenomena in homogeneous infinite systems fails at the mesoscopic level. Several approaches have been proposed to identify phase transitions in mesoscopic systems. Most accepted of them is search of *"convex intruders"* [11] in the entropy versus energy diagram. In the standard thermodynamics there is a relation

$$\left.\frac{\partial^2 S}{\partial E^2}\right|_V = -\frac{1}{T^2}\frac{1}{C_V}, \tag{7}$$

where C_V is the specific heat at constant volume. It follows from (7) that $\partial^2 S/\partial E^2\big|_V < 0$ and hence the entropy versus energy diagram must be concave. Nevertheless, in mesoscopic systems there might be intervals of energy where $\partial^2 S/\partial E^2\big|_V > 0$. These intervals correspond to first-order phase transitions and are called *"convex intruders"*. From the point of view of standard thermodynamics one can say about phenomenon of *negative heat capacity*, of course, if one accepts that it makes sense to define the variables T and C_V as temperature and the specific heat at these circumstances. In [12] it was demonstrated via computation with exactly solvable lattice models that the convex intruders flatten and disappear in the models with local interactions as the lattice size grows, while in the case of long-range interaction these peculiarities survive even in the limit of an infinite system (both finite and long-range interacting infinite systems are typical cases of systems called *nonextensive* in statistical mechanics).

A convex intruder can be found easily by computer for the discrete systems we discuss here. Let us consider three adjacent values of energy E_{i-1}, E_i, E_{i+1} and corresponding numbers of microstates $\Omega_{E_{i-1}}, \Omega_{E_i}, \Omega_{E_{i+1}}$. In our discrete case the ratio $(E_{i+1} - E_i)/(E_i - E_{i-1})$ is always rational number p/q and we can write the convexity condition for entropy in terms of numbers of microstates as easily computed inequality

$$\Omega_{E_i}^{p+q} < \Omega_{E_{i-1}}^p \Omega_{E_{i+1}}^q. \tag{8}$$

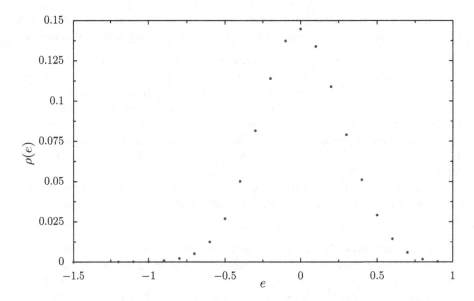

Fig. 7. Microcanonical density of states $\rho(e) = \Omega_E/\Omega$ versus energy per vertex $e = E/V(G_L)$ for the Ising model on dodecahedron

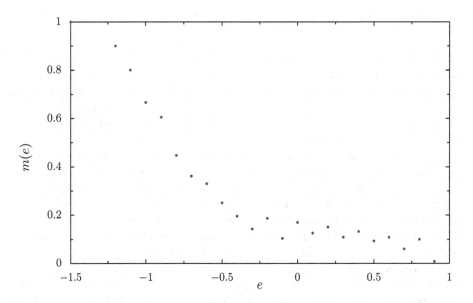

Fig. 8. Specific magnetization $m(e) = M(E)/V(G_L)$ vs. energy per vertex e for the Ising model on dodecahedron

As a rule $E_{i+1} - E_i = E_i - E_{i-1}$ and inequality (8) takes the form

$$\Omega_{E_i}^2 < \Omega_{E_{i-1}} \Omega_{E_{i+1}}.$$

This form means that within convex intruder the number of states with the energy E_i is less than *geometric mean* of numbers of states at the neighboring energy levels.

Fig. 9 shows the entropy vs. energy diagram for the Ising model on dodecahedron. The diagram has apparent convex intruder in the energy interval $[-24, -18]$. Exact computation reveals also a subtle convex intruder in the interval $[-16, -12]$. (In terms of specific energy, as in Fig. 9, these intervals are $[-1.2, -0.9]$ and $[-0.8, -0.6]$, respectively.) It is well known that one-dimensional Ising model has no phase transitions. To illustrate the difference between the diagrams for the cases with and without phase transitions, we place also in Fig. 9 the diagram for Ising model on the 1D circle lattice with 24 vertices.

In Fig. 10 we show the entropy-energy diagrams for lattices of different valences, namely, for 3-, 4- and 6-valent tori. These lattices are marked in Fig. 2 as "Graphene 3×4", "Square 5×5" and "Triangular 4×6", respectively. The diagram for 3-valent torus is symmetric with respect to change sign of energy and contains two pairs of adjacent convex intruders. One pair lies in the *e*-interval $[-1.25, -0.75]$ and another pair lies symmetrically in $[0.75, 1.25]$. The 4-valent torus diagram contains two intersecting convex intruders in the intervals $[-1.68, -1.36]$ and $[-1.36, -1.04]$. The 6-valent torus diagram contains a whole cascade of 5 intersecting or adjacent intruders. Their common interval is $[-2.5, -0.5]$.

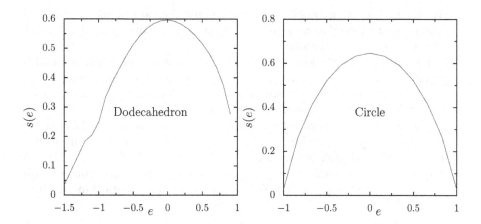

Fig. 9. Specific microcanonical entropy $s(e) = \ln(\Omega_E)/V(G_L)$ vs. energy per vertex e for the Ising model on dodecahedron (*left*) and on circle of length 24 (*right*). Left diagram contains distinct convex intruder in the interval $-1.2 \le e \le -0.9$ and subtle one in the interval $-0.8 \le e \le -0.6$. Right diagram is fully concave: one-dimensional Ising model has no phase transitions.

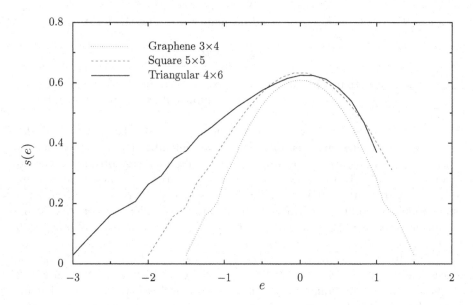

Fig. 10. Specific microcanonical entropy for the Ising model on 3-valent (*dot* line, 24 vertices), 4-valent (*dash* line, 25 vertices) and 6-valent (*solid* line, 24 vertices) tori

6 Summary

- A C program for symmetry analysis of finite discrete dynamical systems has been created.
- We pointed out that trajectories of any deterministic dynamical system go always in the direction of nondecreasing sizes of group orbits. Cyclic orbits run within orbits of the same size.
- After finite time evolution operators of dynamical system can be reduced to group actions. This lead to formation of moving soliton-like structures — "*spaceships*" in the case of cellular automata. Computer experiments show that "*spaceships*" are typical for cellular automata.
- Computational results for cellular automata with symmetric local rules allow to suppose that reversibility is rare property for discrete dynamical systems, and reversible systems are trivial.
- We demonstrated capability of exact computing based on symmetries in search of phase transitions for mesoscopic models in statistical mechanics.

Acknowledgments. I would like to thank Vladimir Gerdt whose comments improved the presentation significantly. This work was supported in part by the grants 07-01-00660 from the Russian Foundation for Basic Research and 5362.2006.2 from the Ministry of Education and Science of the Russian Federation.

References

1. Kornyak, V.V.: On Compatibility of Discrete Relations. In: Ganzha, V.G., Mayr, E.W., Vorozhtsov, E.V. (eds.) CASC 2005. LNCS, vol. 3718, pp. 272–284. Springer, Heidelberg (2005), http://arXiv.org/abs/math-ph/0504048
2. Kornyak, V.V.: Discrete Relations On Abstract Simplicial Complexes. Programming and Computer Software 32(2), 84–89 (2006)
3. Kornyak, V.V.: Cellular Automata with Symmetric Local Rules. In: Ganzha, V.G., Mayr, E.W., Vorozhtsov, E.V. (eds.) CASC 2006. LNCS, vol. 4194, pp. 240–250. Springer, Heidelberg (2006), http://arXiv.org/abs/math-ph/0605040
4. McKay, B.D.: Practical Graph Isomporphism. Congressus Numerantium 30, 45–87 (1981), http://cs.anu.edu.au/~bdm/nauty/PGI
5. Gardner, M.: On Cellular Automata Self-reproduction, the Garden of Eden and the Game of Life. Sci. Am. 224, 112–117 (1971)
6. Hooft, G.: Quantum Gravity as a Dissipative Deterministic System. SPIN-1999/07, gr-qc/9903084; Class. Quant. Grav. 16, 3263 (1999); Also published in: *Fundamental Interactions: from symmetries to black holes* (Conference held on the occasion of the "Eméritat" of François Englert, 24-27 March 1999, Frère, J.-M., et al. (ed.) by Univ. Libre de Bruxelles, Belgium, pp. 221–240 (1999)
7. 't Hooft, G.: The mathematical basis for deterministic quantum mechanics. ITP-UU-06/14, SPIN-06/12, quant-ph/0604008, pp. 1–17 (2006)
8. Imry, Y.: Introduction to Mesoscopic Physics (Mesoscopic Physics and Nanotechnology, 2), p. 256. Oxford University Press, USA (2002)
9. Gross, D.H.E.: Microcanonical thermodynamics: Phase transitions in "Small" Systems, p. 269. World Scientific, Singapore (2001)
10. Gross, D.H.E.: A New Thermodynamics from Nuclei to Stars. Entropy 6, 158–179 (2004)
11. Gross, D.H.E., Votyakov, E.V.: Phase Transitions in "Small" Systems. Eur. Phys. J. B 15, 115–126 (2000)
12. Ispolatov, I., Cohen, E.G.D.: On First-order Phase Transitions in Microcanonical and Canonical Non-extensive Systems. Physica A 295, 475–487 (2001)

Exact Solutions of Completely Integrable Systems and Linear ODE's Having Elliptic Function Coefficients

N.A. Kostov[1] and Z.T. Kostova[2]

[1] Institute of Electronics, Bulgarian Academy of Sciences,
Blvd. Tsarigradsko shosse 72, Sofia, 1784, Bulgaria
[2] National High School "Sofia", Blvd. Montevideo 21
Sofia, Bulgaria

Abstract. We present an algorithm for finding closed form solutions in elliptic functions of completely integrable systems. First we solve the linear differential equations in spectral parameter of Hermite-Halphen type. The integrability condition of the pair of equations of Hermite-Halphen type gives the large family of completely integrable systems of Lax-Novikov type. This algorithm is implemented on the basis of the computer algebra system MAPLE. Many examples, such as vector nonlinear Schödinger equation, optical cascaded equations and restricted three wave system are considered. New solutions for optical cascaded equations are presented. The algorithm for linear ODE's with elliptic functions coefficients is generalized to 2×2 matrix equations with elliptic coefficients.

1   Introduction

We consider a linear differential equation in spectral parameter λ

$$L\Psi = \left(p_0 \frac{d^m}{dx^m} + \sum_{j=1}^{m-1} p_{m-j}(x) \frac{d^{m-j}}{dx^{m-j}} \right) \Psi = \lambda \Psi, \qquad (1)$$

where p_j are expressed in terms of *doubly periodic* functions having the same periods. We will also require that our coefficients are in fact *elliptic functions*. New important development of algorithms for ODE's with elliptic coefficients is paper [BLH]. They are implemented in Maple 9 as DEsolve function. The geometry of ODE's with elliptic function is considered in [GK] using Hermite-Halphen algorithm. The aim of present paper is to present applications of algorithms for ODE's with elliptic coefficients to obtain periodic solutions of completely integrable and near to completely integrable systems.

Let $z(x)$ be a solution of

$$y'''(x) - 4r(x)y'(x) - 2r'(x)y(x) = 0, \qquad (2)$$

V.G. Ganzha, E.W. Mayr, and E.V. Vorozhtsov (Eds.): CASC 2007, LNCS 4770, pp. 252–264, 2007.
© Springer-Verlag Berlin Heidelberg 2007

and set

$$y_1(x) = \sqrt{z(x)} \exp\left(-\frac{C}{2}\int \frac{dx'}{z(x')}\right), \tag{3}$$

and

$$y_1(x) = \sqrt{z(x)} \exp\left(\frac{C}{2}\int \frac{dx'}{z(x')}\right), \tag{4}$$

where C is a constant given by

$$C^2 = z'(x)^2 - 2z(x)z''(x) + 4r(x)z(x)^2, \tag{5}$$

if $C \neq 0$, the $y_1(x), y_2(x)$ are linearly independent and form a basis of

$$y''(x) - r(x)y(x) = 0, \tag{6}$$

If $C = 0$, the the basis for solution space of (6) is given by $y_1(x)$ and $y_2(x) = \sqrt{z(x)} \int \frac{1}{z(x)} dx$.

For proof see for example [BLH].

Introduce Hermite polynomial as solution of the following nonlinear differential equation

$$\frac{1}{2}FF_{xx} - \frac{1}{4}F_x^2 - (u(x) + \lambda)F^2 + \frac{1}{4}R(\lambda) = 0, \tag{7}$$

where $r(x) = (u(x) + \lambda)$, $C^2 = R(\lambda)$, $F(x) = z(x)$. In modern literature (see for example references in [CEEK]) functions (3) and (4) are called Baker-Akhiezer (BA) functions.

1.1   Vector Nonlinear Schrödinger Equation

We consider the system of coupled nonlinear Schrödinger equations

$$i\frac{\partial}{\partial t}Q_j + s\frac{\partial^2}{\partial x^2}Q_j + \sigma\left(\sum_{k=1}^{n}|Q_k|^2\right)Q_j = 0, \quad j = 1,\ldots,n, \tag{8}$$

where $s = \pm 1$, $\sigma = \pm 1$. We seek solution of (8) in the following form [EEK])

$$Q_j = q_j(z) e^{i\Theta_j}, \quad j = 1,\ldots n, \tag{9}$$

where $z = x - ct$, $\Theta_j = \Theta_j(z,t)$, with q_j, Θ_j real. Substituting (9) into (8) and separating real and imaginary parts by supposing that the functions $\Theta_j, j = 1,\ldots n$ behave as

$$\Theta_j = \frac{1}{2}scx + (a_j - \frac{1}{4}sc^2)t - sC_j\int_0^z \frac{dz'}{q_j(z')^2} + \Theta_{j0},$$

we obtain the system ($\sigma = s = \pm 1$)

$$\frac{d^2}{dz^2}q_j + \left(\sum_{k=1}^{n}\frac{\sigma}{s}q_k^2 - \frac{a_j}{s}\right)q_j - \frac{C_j^2}{q_j^3} = 0, \quad k,j = 1,\ldots n, \tag{10}$$

where C_j, $j = 1,\ldots n$ are free parameters and Θ_{j0} are constants. These equations describe the integrable case of motion of a particle in a quartic potential perturbed with inverse squared potential, which is separable in ellipsoidal coordinates. The solutions of the system (10) are then given as

$$q_i^2(z) = 2\frac{\mathcal{F}(z, a_i - \Delta)}{\prod_{k\neq i}^{n}(a_i - a_k)}, \quad i = 1,\ldots,n, \tag{11}$$

where $\mathcal{F}(z,\lambda)$ is Hermite polynomial associated with Lamé potential and is defined as solution of (7). The final formula for the solutions of the system (8) then reads

$$Q_i(x,t) = \sqrt{2\frac{\mathcal{F}(z, a_i - \Delta)}{\prod_{k\neq i}^{n}(a_i - a_k)}}\exp(\Theta_i), \tag{12}$$

where

$$\Theta_j = \left\{\frac{1}{2}icx + i(a_j - \frac{1}{4}c^2)t - \frac{1}{2}\nu(a_j - \Delta)\int_0^z\frac{dz'}{\mathcal{F}(z', a_j - \Delta)}\right\},$$

and $i = 1,\ldots,n$ and we have made use of (11) and (9). To obtain the special class of periodic solution of (10) we introduce the following ansatses

$$q_i(\zeta) = \sqrt{A_i\wp(\zeta + \omega') + B_i}, \quad i = 1,2,3, \text{ or } i = 1,\ldots,4. \tag{13}$$

As a result we obtain:

$$\sum_{k=1}^{m}A_k = -2, \quad a_i = \sum_{k=1}^{m}B_k - \frac{B_i}{A_i}, \quad m = 3 \text{ or } 4, \tag{14}$$

$$-\frac{4C_i^2}{A_i^2} = (4\lambda^3 - \lambda g_2 - g_3)|_{\lambda = -B_i/A_i}, \quad i = 1,\ldots,3 \text{ or } 4 \tag{15}$$

and using the well known relations

$$\int_0^z\frac{dz'}{\wp(z') - \wp(\tilde{a}_j)} = \frac{1}{\wp'(\tilde{a}_j)}\left(2z\zeta(\tilde{a}_j) + \ln\frac{\sigma(z - \tilde{a}_j)}{\sigma(z + \tilde{a}_j)}\right), \tag{16}$$

and

$$\wp(z + \omega') - \wp(\tilde{a}_j) = -\frac{\sigma(z + \omega' + \tilde{a}_j)\sigma(z + \omega' - \tilde{a}_j)}{\sigma(z + \omega')^2\sigma(\tilde{a}_j)^2}. \tag{17}$$

We derive the following result

$$Q_j = \sqrt{-A_j} \frac{\sigma(z + \omega' + \tilde{a}_j)}{\sigma(z + \omega')\sigma(\tilde{a}_j)} \times$$
$$\exp\left(\frac{i}{2}cx + i(a_j - \frac{1}{4}c^2)t - (z + \omega')\zeta(\tilde{a}_j)\right), \tag{18}$$

where

$$\sum_{j=1}^{\epsilon_1} A_j = -2, \quad a_j = \sum_{k=1}^{\epsilon_1} B_k - \frac{B_j}{A_j},$$
$$\frac{C_j}{A_j} = \frac{i}{2}\sqrt{4\lambda^3 - \lambda g_2 - g_3}\Big|_{\lambda = -\frac{B_j}{A_j}}$$
$$\wp(\tilde{a}_j) = -\frac{B_j}{A_j} = \hat{a}_j, \quad j = 1 \ldots \epsilon_1, \epsilon_1 = 3, 4 \tag{19}$$

To obtain the class of periodic solutions of system (10) for $n = 3, 4$ we introduce the following two ansatses in terms of the Weierstrass function $\wp(\zeta + \omega')$

$$q_i(\zeta) = \sqrt{A_i \wp(\zeta + \omega')^3 + B_i \wp(\zeta + \omega')^2 + C_i \wp(\zeta + \omega') + D_i}, \tag{20}$$

where $i = 1, \ldots 3$. Next for conciseness we denote $\wp = \wp(\zeta + \omega')$, then the second ansatz have the form

$$q_i(\zeta) = \sqrt{A_i \wp^4 + B_i \wp^3 + C_i \wp^2 + D_i \wp + E_i},$$
$$i = 1, \ldots 4 \tag{21}$$

with the constants A_i, B_i, C_i, D_i, E_i defined from the compatibility condition of the ansatz with the equations of motion (10). Inserting (20) and (21) into Eqs. (10), using the basic equations for Weierstrass $\wp$ function [WW]

$$\left(\frac{d}{d\zeta}\wp(\zeta)\right)^2 = 4\wp(\zeta)^3 - g_2\wp(\zeta) - g_3, \quad \frac{d^2}{d\zeta^2}\wp(\zeta) = 6\wp(\zeta) - \frac{g_2}{2}, \tag{22}$$

and equating to zero the coefficients at different powers of $\wp$ we obtain the following algebraic equations for the parameters of the solutions $A_i, B_i, C_i, D_i, i = 1, 2, 3$ for $n = 3$

$$A_1 + A_2 + A_3 = 0, \quad B_1 + B_2 + B_3 = 0, \tag{23}$$
$$C_1 + C_2 + C_3 = -12, \quad C_i = \frac{2}{3}\frac{B_i^2}{A_i} - \frac{1}{4}A_i g_2, \tag{24}$$
$$a_i = \sum_{i=1}^{3} D_i - 5\frac{B_i}{A_i}, \quad D_i = \frac{5}{9}\frac{B_i^3}{A_i^2} - \frac{1}{3}B_i g_2 - \frac{1}{4}A_i g_3. \tag{25}$$

The analogical algebraic system for $n = 4$ is as follows

$$A_1 + A_2 + A_3 + A_4 = 0, \quad B_1 + B_2 + B_3 + B_4 = 0, \tag{26}$$

$$C_1 + C_2 + C_3 + C_4 = 0, \quad D_1 + D_2 + D_3 + D_4 = -20, \tag{27}$$

$$C_i = \frac{3}{5}\frac{B_i^2}{A_i} - \frac{3}{10}A_i g_2, \quad D_i = \frac{14}{45}\frac{B_i^3}{A_i^2} - \frac{53}{180}B_i g_2 - \frac{2}{9}A_1 g_3 \tag{28}$$

$$E_i = \frac{49}{225}\frac{B_i^4}{A_i^3} - \frac{113}{450}\frac{B_i^2}{A_i}g_2 - \frac{11}{36}B_i g_3 + \frac{9}{400}A_i g_2^2. \tag{29}$$

$$a_i = \sum_{i=1}^{4} E_i - 7\frac{B_i}{A_i},$$

Another result from the algebraic systems is the expression for constants C_i which parametrise our solutions. For them we obtain

$$C_i^2 = -\frac{\nu(a_i - \Delta)^2}{\prod_{k \neq i}(a_i - a_k)},$$

where $i, k = 3$ or 4 and parameters ν are defined by (for $n = 3$)

$$\nu^2 = \lambda^7 - \frac{63}{2}g_2\lambda^5 + \frac{297}{2}g_3\lambda^4 + \frac{4185}{16}g_2^2\lambda^3 -$$
$$\frac{18225}{8}g_2 g_3\lambda^2 + \frac{91125}{16}g_3^2\lambda - \frac{3375}{16}g_2^3\lambda, \tag{30}$$

and (for $n = 4$)

$$\nu^2 = \lambda^9 - \frac{231}{2}\lambda^7 g_2 + \frac{2145}{2}g_3\lambda^6 + \frac{63129}{16}\lambda^5 g_2^2 - \frac{518505}{8}g_2 g_3\lambda^4$$
$$+ \left(-\frac{563227}{16}g_2^3 + \frac{4549125}{16}g_3^2\right)\lambda^3 + \frac{991515}{2}g_3 g_2^2\lambda^2 +$$
$$\left(\frac{361179}{4}g_2^4 - \frac{5273625}{4}g_2 g_3^2\right)\lambda$$
$$-972405 g_3 g_2^3 - 1500625 g_3^3. \tag{31}$$

Using the general formulae, we will consider below the physically important cases of $n = 3, 4$ [EEK] which are associated with the three-gap $12\wp(\zeta + \omega')$, and four-gap elliptic potentials $20\wp(\zeta + \omega')$.

The Hermite polynomial $\mathcal{F}(\wp(x), \lambda)$ associated to the Lamé potential $12\wp(\zeta)$ has the form

$$\mathcal{F}(\wp(\zeta), \lambda) = \lambda^3 - 6\wp(\zeta + \omega')\lambda^2 - 3 \cdot 5(-3\wp(\zeta + \omega')^2 + g_2)\lambda$$
$$-\frac{3^2 \cdot 5^2}{4}(4\wp(\zeta + \omega')^3 - g_2\wp(\zeta + \omega') - g_3). \tag{32}$$

The solution is real under the choice of the arbitrary constants $a_i, i = 1, \ldots, n$ in such way, that the constants $a_i - \Delta, i = 1, \ldots, n$ lie in *different* lacunae. Comparing (20) and (32) and using (11) the solutions of polynomial equations (23),(24),(25) can be given by

$$A_i = \frac{2 \cdot 5^2 \cdot 3^2}{\prod_{k \neq i}^{3}(a_i - a_k)}, \tag{33}$$

$$B_i = -\frac{2 \cdot 3^2 \cdot 5(a_i - \Delta)}{\prod_{k \neq i}^{n}(a_i - a_k)},$$ (34)

$$\Delta = \frac{2}{5} \sum_{i=1}^{3} a_i.$$ (35)

The Hermite polynomial $\mathcal{F}(\wp(\zeta), \lambda)$ associated to the Lamé potential $20\wp(\zeta)$ can be written as

$$\mathcal{F}(\wp(\zeta), \lambda) = 11025\wp(\zeta + \omega')^4 - 1575\wp(\zeta + \omega')^3\lambda +$$
$$(135\lambda^2 - \frac{6615}{2}g_2)\wp(\zeta + \omega')^2 +$$
$$(-10\lambda^3 + \frac{1855}{4}\lambda g_2 - 2450g_3)\wp(\zeta + \omega') +$$
$$\lambda^4 - \frac{113}{2}\lambda^2 g_2 + \frac{3969}{16}g_2^2 + \frac{195}{4}\lambda g_3.$$ (36)

Comparing (21) and (36) and using (11) the solutions of polynomial equations (26-29) can be given by

$$A_i = \frac{11025 \cdot 2}{\prod_{k \neq i}(a_i - a_k)},$$

$$B_i = -\frac{1575 \cdot 2(a_i - \Delta)}{\prod_{k \neq i}(a_i - a_k)}$$ (37)

$$\Delta = \frac{2}{7} \sum_{i=1}^{4} a_i.$$

Next solution of system $(10, n = 3)$ we obtain using the following ansatz

$$q_i(\zeta) = \sqrt{A_i\wp(\zeta + \omega')^2 + B_i\wp(\zeta + \omega') + C_i}, \quad i = 1, 2, 3,$$ (38)

then we have

$$\sum_{i=1}^{3} A_i = 0, \quad \sum_{i=1}^{3} B_i = -6,$$ (39)

$$a_i = \sum_{k=1}^{3} C_k - 3\frac{B_i}{A_i}, \quad C_i = \frac{B_i^2}{A_i} - \frac{1}{4}A_i g_2,$$ (40)

$$\frac{C_i^2 \cdot 3^3 \cdot 4}{A_i^2} = (4\lambda^5 + 27\lambda^2 g_3 + 27\lambda g_2^2 - 21\lambda^3 g_2 - 81g_2 g_3),$$ (41)

where $\lambda = -3B_i/A_i$.

1.2 Optical Cascading Equations

Let us consider the system of two ordinary differential equations,

$$q_{1\xi\xi} + A_0 q_1 + B_0 q_1 q_2 = 0,$$ (42)
$$q_{2\xi\xi} + C_0 q_2 + D_0 q_2^2 = 0,$$ (43)

where we have A_0, B_0, C_0, D_0 are constants.

Introducing new variable

$$q_1^2 = \frac{4F}{B_0 D_0}, \quad F = \lambda^2 - 3\wp\lambda + 9\wp^2 - \frac{9}{4}g_2 \tag{44}$$

where F is Hermite polynomial [WW], g_2, g_3 are elliptic invariants defined in [WW]. $\wp = \wp(\xi + \omega')$ is Weierstrass function shifted by half period ω' is related to sn Jacobian elliptic function with modulus k

$$\wp(\xi + \omega'; g_2, g_3) = \alpha^2 k^2 \mathrm{sn}^2(\alpha\xi, k) - (1 + k^2), \tag{45}$$

where $\alpha = \sqrt{e_1 - e_3}$ and $e_i, i = 1, 2, 3, e_3 \le e_2 \le e_1$ are the real roots of the cubic equation

$$4\lambda^3 - g_2\lambda - g_3 = 0. \tag{46}$$

Using wave height α and modulus $k = \sqrt{(e_2 - e_3)/(e_1 - e_2)}$ we have the following relations

$$e_1 = \frac{1}{3}(2 - k^2)\alpha^2, \quad e_2 = \frac{1}{3}(2k^2 - 1)\alpha^2, \quad e_3 = -\frac{1}{3}(1 + k^2)\alpha^2,$$

$$g_2 = -4(e_1 e_2 + e_1 e_3 + e_2 e_3) = \frac{4}{3}\alpha^2(1 - k^2 + k^4),$$

$$g_3 = 4e_1 e_2 e_3 = \frac{4}{27}\alpha^6(k^2 + 1)(2 - k^2)(1 - 2k^2). \tag{47}$$

Inserting this expression in (42) we have the following nonlinear differential equation with spectral parameter $\lambda = -C_0/2$

$$\frac{1}{2}FF_{\xi\xi} - \frac{1}{4}F_\xi^2 - (u(\xi) + \lambda)F^2 + \frac{1}{4}R(\lambda) = 0, \tag{48}$$

with eigenvalue equations

$$R(\lambda) = 4\lambda^5 - 21\lambda^3 g_2 + 27\lambda g_2^2 + 27\lambda^2 g_3 - 81g_2 g_3 = 0,$$
$$u(\xi) = -(B_0 q_2 + \lambda + A_0) = 6\wp(\xi + \omega'), \tag{49}$$

or in factorized form

$$R(\lambda) = 4\prod(\lambda - \lambda_i) = 0, \quad \lambda_1 = -\sqrt{3g_2}, \quad \lambda_2 = 3e_3$$
$$\lambda_3 = 3e_2, \quad \lambda_4 = 3e_1, \quad \lambda_5 = \sqrt{3g_2}. \tag{50}$$

It is well known that equation (48) is reduced to linear periodic spectral problem of one dimensional Schrödinger equation with two gap potential $u(x) = 6\wp(\xi + \omega')$ and with five normalized eigenfunctions $q_1^{(i)}, (i) = 1, \dots 5$:

$$\frac{d^2 q_1^{(i)}}{d^2\xi^2} - u(\xi)q_1^{(i)} = \lambda_i q_1^{(i)}, \quad (i) = 1, \dots, 5. \tag{51}$$

Under these conditions the second equation (43) is automatically satisfied. Second equation can be considered as "self-consistent" equation for potential $u(\xi)$. Finally the five spectral families of periodic solutions can be written in the following Table 1

Table 1. Five spectral families of periodic solutions

(I).	$q_1 = \dfrac{6}{\sqrt{B_0 D_0}} \alpha^2 k^2\, \mathrm{E}_2^{(u-)}$	$q_2 = -\dfrac{1}{B_0}\left(u(\xi) + \dfrac{3g_2}{\lambda_1} - 2\lambda_1\right)$	(i)=1
(II)	$q_1 = \dfrac{6}{\sqrt{B_0 D_0}} \alpha^2 k\, \mathrm{E}_2^{(cd)}$	$q_2 = -\dfrac{1}{B_0}\left(u(\xi) + \dfrac{3g_2}{\lambda_2} - 2\lambda_2\right)$	(i)=2
(III)	$q_1 = \dfrac{6}{\sqrt{B_0 D_0}} \alpha^2 k\, \mathrm{E}_2^{(sd)}$	$q_2 = -\dfrac{1}{B_0}\left(u(\xi) + \dfrac{3g_2}{\lambda_3} - 2\lambda_3\right)$	(i)=3
(IV)	$q_1 = \dfrac{6}{\sqrt{B_0 D_0}} \alpha^2 k^2\, \mathrm{E}_2^{(sc)}$	$q_2 = -\dfrac{1}{B_0}\left(u(\xi) + \dfrac{3g_2}{\lambda_4} - 2\lambda_4\right)$	(i)=4
(V)	$q_1 = \dfrac{6}{\sqrt{B_0 D_0}} \alpha^2 k^2\, \mathrm{E}_2^{(u+)}$	$q_2 = -\dfrac{1}{B_0}\left(u(\xi) + \dfrac{3g_2}{\lambda_5} - 2\lambda_5\right)$	(i)=5

where

$$
\mathrm{E}_2^{(sc)} = \mathrm{sn}(\alpha\xi, k)\mathrm{cn}(\alpha\xi, k),
$$
$$
\mathrm{E}_2^{(sd)} = \mathrm{sn}(\alpha\xi, k)\mathrm{dn}(\alpha\xi, k),
$$
$$
\mathrm{E}_2^{(cd)} = \mathrm{cn}(\alpha\xi, k)\mathrm{dn}(\alpha\xi, k),
$$
$$
\mathrm{E}_2^{(u\pm)} = \mathrm{sn}^2(\alpha\xi, k) - \frac{1 + k^2 \pm \sqrt{1 - k^2 + k^4}}{3k^2},
$$

(52)

are normalized two-gap Lamé functions [WW], cn, dn are Jacobian elliptic functions and potential $u(\xi)$ have the form

$$
u(\xi) = 6\alpha^2 k^2 \mathrm{sn}^2(\alpha\xi, k) - 2(1 + k^2)\alpha^2.
$$

(53)

2 2×2 Matrix Spectral Problems and Integrable Systems

2.1   Baker-Akhiezer Function

Let us start with two linear systems

$$
\frac{d\Psi_{1j}}{dx} + F\Psi_{1j} + G\Psi_{2j} = 0, \qquad \frac{d\Psi_{1j}}{dt} + \tilde{A}\Psi_{1j} + \tilde{B}\Psi_{2j} = 0, \qquad (54)
$$

$$
\frac{d\Psi_{2j}}{dx} + H\Psi_{1j} - F\Psi_{2j} = 0, \qquad \frac{d\Psi_{2j}}{dt} + \tilde{C}\Psi_{1j} - \tilde{A}\Psi_{2j} = 0, \qquad (55)
$$

which constitute [AKNS] scheme in particular case $F = -i\lambda, G = iu(x,t), H = \pm iu(x,t)$, where coefficients depend on an arbitrary spectral parameter λ. The compatibility conditions $\Psi_{j,xt} = \Psi_{j,tx}, j = 1, 2$ yield to the following nonlinear system of equations:

$$
F_t - \tilde{A}_x + \tilde{C}G - \tilde{B}H = 0,
$$
$$
G_t - \tilde{B}_x + 2(\tilde{B}F - \tilde{A}G) = 0, \qquad (56)
$$
$$
H_t - \tilde{C}_x + 2(\tilde{A}H - \tilde{C}F) = 0.
$$

The general system (56) is equivalent also to zero curvature representation

$$U_t - V_x + [U, V] = 0, \tag{57}$$

where

$$U = \begin{pmatrix} F & G \\ H & -F \end{pmatrix}, \quad V = \begin{pmatrix} A & B \\ C & -A \end{pmatrix}. \tag{58}$$

The periodic solutions in elliptic functions are generated through special matrices L whose representations are polynomials in the spectral parameter λ and L obey the following set of equations:

$$L_x = [U, L], \qquad L_t = [V, L], \qquad L\Psi = \nu\Psi, \tag{59}$$

where

$$V = \sum_{k=0}^{N} \begin{pmatrix} V_k^{11} & V_k^{12} \\ V_k^{21} & V_k^{22} \end{pmatrix} \lambda^{N-k} = \begin{pmatrix} \tilde{V}^{11} & \tilde{V}^{12} \\ \tilde{V}^{21} & \tilde{V}^{22} \end{pmatrix}. \tag{60}$$

or in explicit form

$$\frac{dA}{dx} = -HB + GC, \quad A(x, \lambda) = \sum_{j=0}^{n+1} A_{n+1-j}(x, t)\lambda^j, \tag{61}$$

$$\frac{dB}{dx} = 2FB - 2GA, \quad B(x, \lambda) = \sum_{j=0}^{n} B_{n-j}(x, t)\lambda^j, \tag{62}$$

$$\frac{dC}{dx} = -2FC + 2HA, \quad C(x, \lambda) = \sum_{j=0}^{n} C_{n-j}(x, t)\lambda^j, \tag{63}$$

and for $N = 1$

$$\frac{dA}{dt} = \tilde{V}^{12}C - \tilde{V}^{12}B, \quad A(\xi, \lambda) = \sum_{j=0}^{n+1} A_{n+1-j}(x, t)\lambda^j, \tag{64}$$

$$\frac{dB}{dt} = 2\tilde{V}^{11}B - 2\tilde{V}^{12}A, \quad B(\xi, \lambda) = \sum_{j=0}^{n} B_{n-j}(x, t)\lambda^j, \tag{65}$$

$$\frac{dC}{dt} = -2\tilde{V}^{11}C + 2\tilde{V}^{21}A, \quad C(\xi, \lambda) = \sum_{j=0}^{n} C_{n-j}(x, t)\lambda^j. \tag{66}$$

The equations (61),(62),(63) yield that

$$(A(x, \lambda)^2 - B(x, \lambda)C(x, \lambda))_x = 0 \tag{67}$$

and hence

$$\nu^2 = A(x, \lambda)^2 - B(x, \lambda)C(x, \lambda) = R_{2n+2}(\lambda), \tag{68}$$

where the integration constant R_{2n+2} is a polynomial in λ of degree $2n+2$. After a chain of simple transformations we obtain

$$\Psi_{1,x} = \frac{1}{2B}\left(B_x - 2R_{2n+2}G\right)\Psi_1, \tag{69}$$

$$\Psi_{2,x} = \frac{1}{2C}\left(C_x - 2R_{2n+2}H\right)\Psi_2. \tag{70}$$

After integration as result we have

$$\Psi_1 = \sqrt{B}\exp\left(\pm\sqrt{R_{2n+2}}\int^x \frac{G}{B}dx\right), \tag{71}$$

$$\Psi_2 = \sqrt{-C}\exp\left(\pm\sqrt{R_{2n+2}}\int^x \frac{H}{C}dx\right). \tag{72}$$

2.2   Restricted Multiple Three Wave Interaction System

Let us consider coupled quadratic nonlinear oscillators

$$i\frac{db_j}{d\xi} + uc_j - \frac{1}{2}\epsilon_j b_j = 0, \tag{73}$$

$$i\frac{dc_j}{d\xi} + u^*b_j + \frac{1}{2}\epsilon_j c_j = 0, \tag{74}$$

$$i\frac{du}{d\xi} + \sum_{j=1}^n b_j c_j^* = 0, \tag{75}$$

where ξ is the evolution coordinate and ϵ_j are constants. The equations (73-75) can be written as Lax representation

$$\frac{dL}{d\xi} = [M, L], \tag{76}$$

of the following linear system:

$$\frac{d\psi}{d\xi} = M(\xi, \lambda)\psi(\xi, \lambda) \quad L(\xi, \lambda)\psi(\xi, \lambda) = 0, \tag{77}$$

where L, M are 2×2 matrices and have the form

$$L(\xi, \lambda) = \begin{pmatrix} A(\xi, \lambda) & B(\xi, \lambda) \\ C(\xi, \lambda) & D(\xi, \lambda) \end{pmatrix}, \tag{78}$$

$$M(\xi, \lambda) = \begin{pmatrix} -i\lambda/2 & iu \\ u^* & i\lambda/2 \end{pmatrix}. \tag{79}$$

where

$$A(\xi, \lambda) = a(\lambda)\left(-i\frac{\lambda}{2} + \frac{i}{2}\sum_{j=1}^n \frac{(c_j c_j^* - b_j b_j^*)}{\lambda - \epsilon_j}\right), \tag{80}$$

$$B(\xi, \lambda) = a(\lambda) \left(iu - i \sum_{j=1}^{n} \frac{b_j c_j^*}{\lambda - \epsilon_j} \right), \tag{81}$$

$$C(\xi, \lambda) = a(\lambda) \left(iu^* - i \sum_{j=1}^{n} \frac{c_j b_j^*}{\lambda - \epsilon_j} \right), \tag{82}$$

where $D(\xi, \lambda) = -A(\xi, \lambda)$ and $a(\lambda) = \prod_{i=1}^{n} (\lambda - \epsilon_i)$. The Lax representation yields the hyperelliptic curve $K = (\nu, \lambda)$

$$\det(L(\lambda) - \frac{1}{2}\nu \mathbf{1}_2) = 0, \tag{83}$$

where $\mathbf{1}_2$ is the 2×2 unit matrix. The curve (83) can be written in canonical form as

$$\nu^2 = 4 \prod_{j=1}^{2n+2} (\lambda - \lambda_j) = R(\lambda), \tag{84}$$

where $\lambda_j \neq \lambda_k$ are branching points. Next we develop a method which allows to construct periodic solutions of system (73-75). The method is based on the application of spectral theory for self-adjoint one dimensional Dirac equation with periodic finite gap potential $\mathcal{U} = -u$ cf. Eqs. (73,74)

$$i\frac{d\Psi_{1j}}{d\xi} - \mathcal{U}\Psi_{2j} - i\lambda_j\Psi_{1j} = 0, \tag{85}$$

$$i\frac{d\Psi_{2j}}{d\xi} - \mathcal{U}^*\Psi_{1j} + i\lambda_j\Psi_{1j} = 0, \tag{86}$$

with spectral parameter λ and eigenvalues $\lambda_j = i\epsilon_j/2$. The equation (76) is equivalently written as

$$\frac{dA}{d\xi} = iuC - iu^*B, \quad A(\xi, \lambda) = \sum_{j=0}^{n+1} A_{n+1-j}(\xi)\lambda^j, \tag{87}$$

$$\frac{dB}{d\xi} = -i\lambda B - 2iuA, \quad B(\xi, \lambda) = \sum_{j=0}^{n} B_{n-j}(\xi)\lambda^j, \tag{88}$$

$$\frac{dC}{d\xi} = i\lambda C + 2iu^*A, \quad C(\xi, \lambda) = \sum_{j=0}^{n} C_{n-j}(\xi)\lambda^j, \tag{89}$$

or in different form we have

$$A_{j+1,\xi} = iuC_j - iu^*B_j, \; A_0 = 1, A_1 = c_1, \tag{90}$$

$$iB_{j+1} = -B_{j,\xi} - 2iuA_{j+1}, \quad B_0 = -2u, \tag{91}$$

$$iC_{j+1} = C_{j,\xi} - 2iu^*A_{j+1} \quad C_0 = -2u^*, \tag{92}$$

where c_1 is the constant of integration. Differenciating Eq. (87) and using (83) we can obtain

$$BB_{\xi\xi} - \frac{u_\xi}{u}BB_\xi - \frac{1}{2}B_\xi^2 + \left(\frac{\lambda^2}{2} - i\lambda\frac{u_\xi}{u} + |u|^2\right)B^2 = 2u^2\nu. \qquad (93)$$

Using (69) the eigenfunction Ψ_1 for finite-gap potential $\mathcal{U}$ have the form

$$\Psi_1(\xi,\lambda) = \left[\frac{\mathcal{U}(\xi)}{\mathcal{U}(0)}\frac{B(\xi,\lambda)}{B(0,\lambda)}\right]^{1/2}\exp\left\{-i\int_0^\xi\frac{\sqrt{R(\lambda)}}{B(\xi',\lambda)}d\xi'\right\}.$$

Analogously we can write expression for $\Psi_2(\xi,\lambda)$ and finally elliptic solutions of initial system of restricted three interaction system take the form

$$b_j(\xi) = b_j^0\Psi_1(\xi,\lambda_j), \qquad c_j(\xi) = c_j^0\Psi_2(\xi,\lambda_j), \qquad j = 1\ldots n, \qquad (94)$$

where b_j^0, c_j^0 are constants fixed by initial conditions.

3 Implementation

In [GK] H-H (Hermite-Halphen) algorithm is presented and implemented in computer algebra REDUCE. Geometric interpretation of solutions found in [GK] is discussed in [EK]. Recently [BLH] Maple 9 [CGGMW] implementation of algorithm for solving linear ODE's having elliptic function coefficients is reported. New algorithm is found. This implementation is very important for deriving new solutions of integrable and nonintegrable dynamical systems with elliptic solutions [EK, GR, GH, B]. Important problem in deriving elliptic solutions is factorization of algebraic curves [CGHKW]. Algorithm for deriving elliptic solutions presented above is implemented on computer algebra Maple 10. The source code is available under request. Using this implementation new solutions of Manakov system in external potential are derived in [KEGKS].

References

[BLH]   Burger, R., Labahn, G., van Hoeji, M.: Closed form solutions of linear odes having elliptic function coefficients. In: Proceedings of ISSAC'04, pp. 58–64 (2004)

[GK]   Gerdt, V., Kostov, N.: Computer algebra in the theory of ordinary differential equations of Halphen type, Computers and mathematics. In: Proc.of Int. Conf., Cambridge, USA, pp. 178–188. Springer, New-York (1989)

[WW]   Whittaker, E., Watson, G.: A Course of Modern Analysis. Cambridge University Press, Cambridge (1986)

[EK]   Enolskii, V., Kostov, N.: On the geometry of elliptic solitons. Acta Applicandae Math. 36, 57–86 (1994)

[CEEK]   Christiansen, P., Eilbeck, J., Enolskii, V., Kostov, N.: Quasi-periodic and periodic solutions for Manakov type systems of coupled nonlinear Schrodinger equations. Proc. Royal Soc. London A 456, 2263–2281 (2000)

[EEK]   Eilbeck, J., Enolskii, V., Kostov, N.: Quasi-periodic and periodic solutions for vector nonlinear Schrodinger equations. Journ. Math. Phys. 41, 8236–8248 (2000)

[K]   Kostov, N.: Quasi-periodic and periodic solutions for dynamical systems related to Korteweg-de Vries equation. The European Physical Journal B. 29, 255–260 (2002)

[AKNS]   Ablowitz, M., Kaup, D., Newell, A., Segur, H.: The inverse scattering transform–Fourier analysis for nonlinear problems. Studies in Appl. Math. 53, 249–315 (1974)

[KEGKS]   Kostov, N., Enolskii, V., Gerdjikov, V., Konotop, V., Salerno, M.: Two-component Bose-Einstein condensates in periodic potential. Phys. Rev E. 70, 056617 (2004)

[S]   Smirnov, A.: Elliptic soliton and Heun's equation, The Kowalevski property. CRM Proc. Lecture Notes 32, 287–305 (2002)

[CGGMW]   Char, B.W., Geddes, K.O., Gonnet, G.H., Monagan, M.B., Watt, S.M.: Maple Reference Manual, Watcom Publications, Waterloo (1988)

[CGHKW]   Corless, R., Giesbrecht, M., van Hoeij, M., Kotsireas, I., Watt, S.: Towards factoring bivariate approximate polynomials, International Conference on Symbolic and Algebraic Computation. In: Proceedings of ISSAC'01, ACM Press, New York (2001)

[GH]   Gesztesy, F., Holden, H.: Soliton Equations and Their Algebro-Geometric Solutions. vol. I: $(1+1)$-Dimensional Continuous Models. Cambridge University Press, Cambridge (2003)

[GR]   Gesztesy, F., Ratneseelan, R.: An alternative approach to algebro-geometric solutions of AKNS hierarchy. Rev. Math. Phys. 10, 345–391 (1998)

[B]   Brezhnev, Y.: Elliptic solitons and Gröbner bases. Journal of Mathematical Physics 45, 696–712 (2004)

Dynamics of Nonlinear Parabolic Equations with Cosymmetry

Ekaterina S. Kovaleva[1], Vyacheslav G. Tsybulin[1], and Kurt Frischmuth[2]

[1] Department of Computational Mathematics, Southern Federal University,
Rostov-na-Donu, Russia
[2] Department of Mathematics, University of Rostock, Germany

Abstract. Dynamics of a cosymmetric system of nonlinear parabolic equations is studied to model of population kinetics. Computer algebra system Maple is applied to perform some stages of analytical investigation and develop a finite-difference scheme which respects the cosymmetry property. We present different scenarios of evolution for coexisted nonstationary regimes and families of equilibria branched off of the state of rest.

1 Population Kinetics Model

Cosymmetry [1,2] is an essentially nonlinear effect and a number of problems in mathematical physics (Darsy convection of an incompressible fluid saturating a porous medium, some models of anti-ferromagnetism, etc.) are cosymmetric ones. The systems with cosymmetry property give the emergence of continuous families of steady states and nontrivial dynamics. In contrast to symmetry problems, equilibria belonging to the family have variable spectrum of stability. The calculation of the cosymmetric families is a complicated problem due to degeneration in the vicinity of the family.

Mathematical modeling for biological problems with spatial distribution grows significantly last time [3]. Analysis of such systems has shown interesting scenario of transitions, nontrivial dynamics and coexistence of the regimes. We model kinetics of three populations which inhabit a common domain [4,5] and consider one-dimensional in space problem. This system of parabolic equations admits a cosymmetry and appearance of a family of stationary solutions with variable spectrum of stability.

We consider an initial boundary value problem for a system of nonlinear parabolic equations [5] for densities of species w_i

$$\dot{w} = Kw'' + Mw' + F(w, w') \equiv \Phi(w), \tag{1}$$
$$w(x, 0) = w^0(x), \quad x \in \Omega. \tag{2}$$
$$w(x, t) = 0, \quad x \in \partial\Omega, \tag{3}$$

here $w = (w_1, w_2, w_3)^\top$, the dot and prime mean differentiation with respect to time t and space coordinate $x \in \Omega = [0, a]$, respectively.

V.G. Ganzha, E.W. Mayr, and E.V. Vorozhtsov (Eds.): CASC 2007, LNCS 4770, pp. 265–274, 2007.

The right-hand side of (1) is composed of the diffusion term and transportation terms. The diffusion is given by second order derivatives and the diagonal matrix of diffusive coefficients $K = diag(k_1, k_2, k_3)$. Matrix M (linear transport coefficients) and the nonlinear interaction term $F = (f_1, f_2, f_3)^\tau$ are given as:

$$M = \begin{pmatrix} 0 & \nu & -\lambda \\ \nu & 0 & 0 \\ \lambda & 0 & 0 \end{pmatrix}, \quad F = \begin{pmatrix} \eta k_1(-3w'_1 w_1) \\ \eta k_2(w'_1 w_2 + 2w'_2 w_1) \\ \eta k_3(w'_1 w_3 + 2w'_3 w_1) \end{pmatrix}. \tag{4}$$

We try to construct a system with usual properties of biological space distributed systems. Diffusive and linear transport terms are traditional ones. We govern the nonlinear terms to provide an additional property namely cosymmetry. We consider the linear cosymmetry $\Psi(w) = BK^{-1}Mw$, where $B = diag(1, -1, -1)$. The choice of vector B was made by Maple to provide the cosymmetry identity which is given here as orthogonality of $\Phi(w)$ to $\Psi(w)$

$$(\Psi, \Phi)_{L_2} = \int_\Omega \Psi \cdot \Phi = 0. \tag{5}$$

To realize it we need in simple integration by parts and usual simplification rules as in [4]. The cosymmetry identity can be checked explicitly by multiplying $\Psi(w)$ on the right hand side of the system (1). Linear cosymmetry $\Psi(w)$ means that w^* is noncosymmetrical equilibrium ($\Psi(w^*) \neq 0$) and a family of steady states (equilibria) occurs to which given equilibrium w^* belongs [1,2].

The system of equations (1)–(4) is invariant with respect to the transformations:

$$R_x : \{\lambda, \nu, w_1, w_2, w_3\} \rightarrow \{\lambda, -\nu, w_1, -w_2, w_3\} \tag{6}$$
$$R_y : \{\lambda, \nu, w_1, w_2, w_3\} \rightarrow \{-\lambda, \nu, w_1, w_2, -w_3\}.$$

The main parameters of the problem are the transport parameters λ and ν, the diffusivity coefficients k_j and the growth parameter η.

Some properties of the nonlinear system (1)–(4) can be determine by analytically. Multipling each equation on the corresponding component of the concentration vector, adding and integrating by Ω, we obtain

$$\frac{d}{dt} \int_\Omega \frac{w_1^2 + w_2^2 + w_3^2}{2} dx = -\int_\Omega (k_1 w'_1 + k_2 w'_2 + k_3 w'_3)^2 dx + \tag{7}$$

$$\lambda \int_\Omega (w'_1 w_3 - w'_3 w_1) dx + \int_\Omega (f_1 w_1 + f_2 w_2 + f_3 w_3) dx.$$

The last integral is disappeared due to special nonlinear terms (4) and boundary conditions (3).

The system (1)–(4) subdivided into two systems when $\lambda = 0$: subsystem with respect to w_1, w_2 and equation for w_3. When $\nu = 0$, the system (1)–(4) is subdivided into subsystem with respect to w_1 and w_3 and equation for w_2.

Consider the case $\lambda = 0$, $w_3 = 0$, then equation (7) is given by

$$\frac{d}{dt} \int_\Omega \frac{w_1{}^2 + w_2{}^2}{2} dx = - \int_\Omega (k_1 w_1' + k_2 w_2')^2 dx. \tag{8}$$

Because of right hand side of the system (8) is nonnegative, any equilibrium of the system (1)–(4) is globally stable when $\lambda = 0$ and any ν.

In order to analyze stability of the zero equilibrium $w = 0$, we consider a system after linearization

$$\dot{w}_1 = k_1 w_1'' + \nu w_2' - \lambda w_3', \quad \dot{w}_2 = k_2 w_2'' + \nu w_1', \quad \dot{w}_3 = k_2 w_3'' + \lambda w_1', \tag{9}$$

with boundary conditions

$$w_k|_{x=0,a} = 0, \quad k = 1, 2, 3. \tag{10}$$

Substituting $w_k = \exp(pt) v_k$ to (9)–(10), we receive a spectral problem for parameter p (decrement). The cases $p = 0$ and $p = \pm iw$, $w > 0$ correspond respectively to monotonic and oscillatory instability. So, we can find the neutral curves on the plane (λ, ν) for these two cases of instability. Monotonic instability of $w = 0$ implies branching off of the zero equilibria new steady state $w^* \neq 0$. It means existence of the family of steady states to which the equilibrium w^* belongs.

Let analyze the case of monotonic instability. Take $\nu = 0$ and consider a subsystem with respect to v_1 and v_3. After the change of variables

$$v_1 = u_1 \sqrt{k_3 \lambda}, \quad v_3 = u_3 \sqrt{k_1 \lambda}, \quad \sigma = \frac{1}{2} \sqrt{\frac{\lambda^2}{k_1 k_3}}, \tag{11}$$

we obtain the system

$$0 = u_1'' + 2\sigma u_3', \quad 0 = u_3'' + 2\sigma u_1',$$

By introducing the complex function $U = u_1 + i u_3$, we deduce the problem for U:

$$0 = U'' - 2\sigma i U', \quad U|_{x=0,a} = 0. \tag{12}$$

Finally, we substitute $U = Z \exp(i\sigma x)$ and obtain the spectral problem

$$0 = Z'' + \sigma^2 Z, \quad Z|_{x=0,a} = 0, \tag{13}$$

with eigenvalues $\sigma_j = j\pi/a$, where $j \in \mathbf{Z}$. Thus, for all values of the parameter λ being greater than the critical value $\lambda_{crit} = 2\pi\sqrt{k_1 k_3}/a$ the zero equilibrium $w = 0$ is unstable.

2   Solution Scheme

To study the given system we use the finite-difference approach [7]. The uniform grid on $\Omega = [0, a]$ is considered

$$x_j = jh, \quad j = 0, \ldots, n+1, \quad h = a/(n+1). \tag{14}$$

The notation $w_{i,j}$ is used for discrete values of w_i, boundary conditions take the following form

$$w_{i0} = w_{in+1} = 0. \tag{15}$$

We apply method of lines and transform (1) to the following system of ordinary differential equations

$$\dot{w}_{ij} = K D_j^2 w_i + M D_j^1 w_i + \widetilde{F}_{ij} \equiv \widetilde{\Phi}_k, \quad k = n(i-1) + j, \tag{16}$$
$$j = 1, \ldots, n, \quad i = 1, 2, 3.$$

Here to keep cosymmetry of the system we use the centered difference operators for first and second order derivatives

$$D_j^1(u) = \frac{u_{j+1} - u_{j-1}}{2h}, \tag{17}$$

$$D_j^2(u) = \frac{u_{j+1} - 2u_j + u_{j-1}}{h^2}, \tag{18}$$

and special form of second order operator d_j [7] to approximate nonlinear terms

$$d_j(u, v) = \frac{2(u_{j+1} - u_{j-1})v_j - u_j(v_{j+1} - v_{j-1}) + u_{j+1}v_{j+1} - u_{j-1}v_{j-1}}{6h}. \tag{19}$$

Using d_j we derive for F (see (4))

$$\widetilde{F}_{1j} = -3\eta k_1 d_j(w_1, w_1),$$
$$\widetilde{F}_{2j} = \eta k_2 [d_j(w_1, w_2) + 2d_j(w_2, w_1)],$$
$$\widetilde{F}_{2j} = \eta k_3 [d_j(w_1, w_3) + 2d_j(w_3, w_1)].$$

The system (16) can be rewriten as

$$\dot{Y} = (P + \lambda Q + \nu S)Y + \widetilde{F}(Y) \equiv \widetilde{\Phi}, \tag{20}$$

here $Y = (w_{11}, \ldots, w_{1n}, w_{21}, \ldots, w_{2n}, w_{31}, \ldots, w_{3n})$, P is a positive-definite matrix, Q and S is a skew-symmetric matrix, and $\widetilde{F}(Y)$ denotes a nonlinear term.

The discrete version of cosymmetry $\widetilde{\Psi} = (\widetilde{\Psi}_1, \ldots, \widetilde{\Psi}_{3n})$ for the system (20) is given by:

$$\widetilde{\Psi}_j = \frac{\nu}{k_1} w_{2j} + \frac{\lambda}{k_1} w_{3j}, \quad \widetilde{\Psi}_{n+j} = \frac{\nu}{k_1} w_{1j}, \quad \widetilde{\Psi}_{2n+j} = -\frac{\lambda}{k_1} w_{1j}, \quad j = 1, \ldots, n. \tag{21}$$

Cosymmetry property (5) transforms to the sum

$$\sum_{k=1}^{3n} \widetilde{\Psi}_k \widetilde{\Phi}_k = 0. \qquad (22)$$

Using Maple we checked that this equality holds for the system (16) and cosymmetry $\widetilde{\Psi}$ (21).

To compute the family of steady states we apply the method [6,7] based on the Implicit Function Theorem [1]. When the zero equilibrium lost stability monotonically, one-parameter family of stationary states branches off from it. To calculate this family, we first find one equilibrium of the family by modified Newton method. Guess to the next equilibrium is obtained by Adams method. This procedure can be used for calculation both stable and unstable members of the family and repeats till we receive a whole family.

3   Numerical Results

Due to invariance (6) of the system it is sufficient to consider only positive values of transport parameters λ and ν. We fix $k_1 = k_3 = 1$, $\eta = 10$, $a = 1$ and find boundary of the stability region for the zero equilibrium – neutral curve. Figure 1 shows the neutral curves of the zero equilibrium for several values of diffusion coefficient k_2. While parameter $\lambda \leq \lambda_{crit}$ the zero equilibrium is stable. The

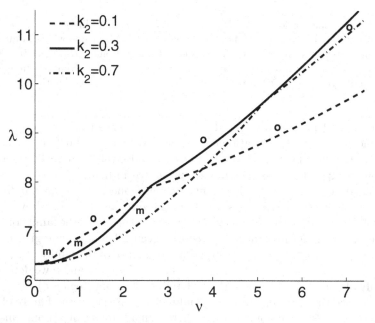

Fig. 1. Neutral curves for several value of diffusion coefficient k_2; $k_1 = k_3 = 1$

letters m and o correspond respectively the cases of monotonic and oscillatory instabilities. Let parameters λ and ν being such that corresponding point lies above the part of curve marked m and o. Then zero equilibrium becomes unstable and new regimes arise. In the case of monotonic instability (passing through the curve m) we have found the emergence of the family of steady states. In the case of oscillatory instability (parameters lie above curve marked o) we have detected nonstationary regimes. One can see that critical values of parameter λ grows when parameter ν increases. Boundaries of instability shift when diffusion coefficient k_2 becomes greater.

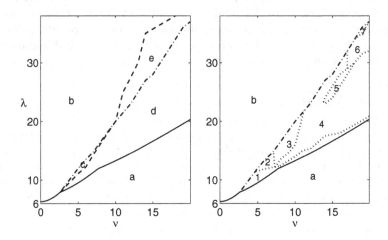

Fig. 2. Pattern diagram for $k_2 = 0.3$. Left: neutral curve (solid line), a region of stable zero equilibrium (a), a region with families of equilibria (b), a region where families and nonstationary regimes are coexist (c), a region of nonstationary regimes (d), a region where limit cycles is progressed to the nonzero equilibrium (e). Right: regions of the different limit cycles ($1,2,3,6$), a region of tori (4), a region of the chaotic regimes (5).

When control parameters λ and ν grow we observe transformation of the family and evolution of nonstationary regimes. We have found that coexistence of different nontrivial regimes is a typical situation for this system. The summary of computer experiments at $k_3 = 0.3$ are presented in figure 2.

The plane (λ, ν) may be divided on five regions (zones), see left part of figure 2. The region a corresponds to the stability of the zero equilibrium. Above neutral curve (solid line) we see the regions b and d where respectively the family of steady states (equilibria) and nonstationary regimes occur. The narrow region c corresponds the parameter values for whose the coexistence of the family of equilibria and nonstationary regimes takes place. For large values of λ and ν we detect that the family of equilibria shrinks to the isolated equilibrium (region e).

One can see that the region corresponding to existence of the family of equilibria (region b) is large comparing with the small arc of monotonic instability (arc m on the neutral curve). Let fix ν and will increase λ after oscillatory

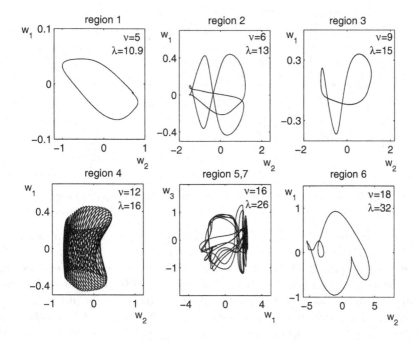

Fig. 3. Types of nonstationary regimes

instability arc on the neutral curve. We see that firstly the limit cycle is branched off of the zero equilibrium. After that coexistence of the limit cycle and the family of steady state takes place. Then this limit cycle lost its stability and only the family of equilibria exists.

More detailed description of nonstationary regimes (region d) is presented at the right part of the figure 2. Here we have the complicated picture of different transitions: transformation of limit cycles, torus appearance, chaos. The regions marked by $1,2,3$ and 6 correspond to different limit cycles. The region 4 gives the domain of parameters for whose the tori take place. Chaotic regimes are

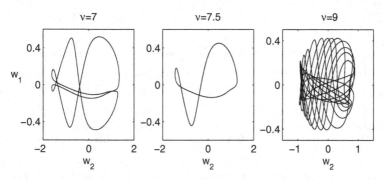

Fig. 4. Evolution of limit cycles for $\lambda = 14$

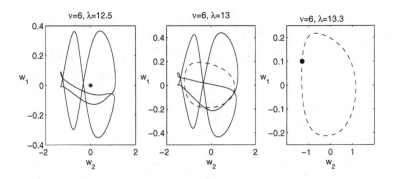

Fig. 5. Coexistence of the limit cycle and family of equilibrium

registered when the value λ and ν belong the region 5. Some typical examples of the regimes for each region are displayed in figure 3.

In figure 4 we present the transformation of the limit cycle. One can see the complex limit cycle (left part in figure 4) is decomposed to two limit cycles (one of them is given in middle picture) and subsequent toring (right picture).

Chaotic regimes were observed in rather small domain which contacts with the region 4 (see right part in figure 2), where tori exist. The region 6 is located near the region 4 and 7.

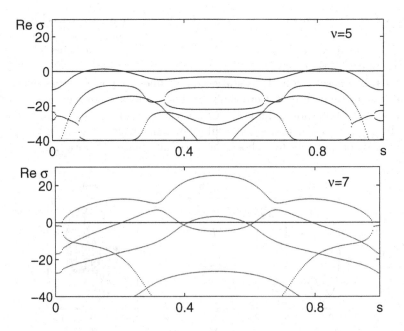

Fig. 6. Spectrum of steady states on the families for different ν; $\lambda = 15$

Figure 5 illustrates coexistence of the family of equilibria and the limit cycle (see the region c in figure 2) along the family. When $\nu = 6$ and $\lambda = 12.5$, the limit cycle is stable after braunching off of the zero equilibrium (marked by star). Then, at $\lambda = 13$ the limit cycle and the family of equilibrium (dashed line) coexists. Finally, at $\lambda = 13.3$ the limit cycle was shrunk to the equilibrium (disc) belonging to the family. One can see in figure 6 that stability spectrum of the members of the family changes.

4   Conclusion

In reality (biology, economics) we meet the situations (systems) characterized by a number of practically identical steady states. Symmetry may lead to such a behavior, another reason is cosymmetry. Dynamical cosymmetric systems may have a family consisting of infinitely many steady states with variable spectrum of stability. Even the cosymmetry property was destroyed under some perturbations the dynamics is being very close to the vanished states [8]. So, it is very important to investigate cosymmetric systems as some idealization that can help us to understand the dynamics of real system.

We consider dynamics of a cosymmetric system of nonlinear parabolic equations to model the population kinetics with many possible stationary regimes. Evolution of coexisted nonstationary regimes and families of equilibria was found. Computer algebra system Maple was used both to do the analytical work and to realize stability analysis for steady states. This technique may be useful for the problems where infinitely many steady states take place.

Acknowledgements

The authors wish to express their thanks to the unknown reviewers for careful reading and useful comments.This work was supported by the Russian Foundation for Basic Research (project 05-01-00567), programme 'Leading Scientific Schools' (project # 5457.2006.1) and by the development programme (grant #74) for Southern Federal University (former Rostov State University).

References

1. Yudovich, V.I.: Cosymmetry, degeneration of solutions of operator equations, and the onset of filtration convection. Mat. Zametki. 49, 142–148 (1991)
2. Yudovich, V.I.: Secondary cycle of equilibria in a system with cosymmetry, its creation by bifurcation and impossibility of symmetric treatment of it. Chaos. 5, 402–411 (1995)
3. Murray, J.D.: Mathematical biology, p. 766. Springer, New York (1993)
4. Frischmuth, K., Tsybulin, V.G.: Cosymmetry preservation and families of equilibria. In: Computer Algebra in Scientific Computing – CASC 2004, pp. 163–172 (2004)
5. Frischmuth, K., Tsybulin, V.G.: Families of equilibria and dynamics in a population kinetics model with cosymmetry. Physics Letters A 338, 51–59 (2005)

6. Govorukhin, V.N.: Calculation of one-parameter families of stationary regimes in a cosymmetric case and analysis of plane filtrational convection problem. Continuation methods in fluid dynamics, Notes Numer. Fluid Mech. Vieweg. Braunschweig 74, 133–144 (2000)
7. Frischmuth, K., Tsybulin, V.G.: Computation of a family of non-cosymmetrical equilibria in a system of two nonlinear parabolic equations. Computing 16, 67–82 (2002)
8. Yudovich, V.I.: Bifurcations under perturbations violating cosymmetry. Doklady Physics 49(9), 522–526 (2004)

Weak Integer Quantifier Elimination Beyond the Linear Case

Aless Lasaruk[1] and Thomas Sturm[2]

[1] FORWISS, Universität Passau, 94030 Passau, Germany
lasaruk@uni-passau.de
[2] FIM, Universität Passau, 94030 Passau, Germany
sturm@uni-passau.de

Abstract. We consider the integers using the language of ordered rings extended by ternary symbols for congruence and incongruence. On the logical side we extend first-order logic by bounded quantifiers. Within this framework we describe a weak quantifier elimination procedure for univariately nonlinear formulas. Weak quantifier elimination means that the results possibly contain bounded quantifiers. For fixed choices of parameters these bounded quantifiers can be expanded into finite disjunctions or conjunctions. In univariately nonlinear formulas all congruences and incongruences are linear and their modulus must not contain any quantified variable. All other atomic formulas are linear or contain only one quantified variable, which then may occur there with an arbitrary degree. Our methods are efficiently implemented and publicly available within the computer logic system REDLOG, which is part of REDUCE. Various application examples demonstrate the applicability of our new method and its implementation.

1   Introduction

After the fundamental work of Presburger [1] there has been considerable research on Presburger arithmetic, which is the additive theory of the integers with ordering and congruences. The largest part of this research was concerned with complexity issues and with decidability [2,3,4,5,6,7,8]. Weispfenning [9,10] was the first one who was explicitly interested in quantifier elimination as such in contrast to using it as a technique for decision. His quantifier elimination procedures are triply exponential, which is known to be optimal [3]. He managed, however, to optionally decrease that complexity by one exponential step to doubly exponential using the following technical trick: certain systematic disjunctions occurring during the elimination process are not written down explicitly. Instead one uses big $\bigvee$ (disjunction) and $\bigwedge$ (conjunction) operators with an index variable running over a finite range of integers. It is important to understand that at any time these big operators could be expanded such that one obtains a regular first-order formula at the price of considerably increasing the size of the representation. Independently, Weispfenning and others have developed virtual substitution techniques for quantifier elimination in various theories starting with the reals and including also valued fields and Boolean algebras [11,12,13,14,15].

V.G. Ganzha, E.W. Mayr, and E.V. Vorozhtsov (Eds.): CASC 2007, LNCS 4770, pp. 275–294, 2007.

In a recent publication [16] the authors of the present paper combined the two research areas by presenting integer quantifier elimination within the framework of virtual substitution. Furthermore, they extended that framework in order to cover a considerable generalization of Presburger arithmetic admitting as coefficients arbitrary polynomials in the parameters, i.e., the unquantified variables. This extension is called the *full linear theory of the integers*. It perfectly corresponds to what is referred to as linear quantifier elimination for the reals or for valued fields [12,14]. Recall that in regular Presburger arithmetic, in contrast, all coefficients must be numbers. The difference vanishes when considering decision problems. It is well-known that the full linear theory of the integers does not admit quantifier elimination in the traditional sense [10]. Instead one uses *weak quantifier elimination*. This does not necessarily deliver quantifier-free equivalents but formulas that possibly contain some *bounded quantifiers*. For this, one extends the language of logic by two additional quantifiers $\bigsqcup_{k:\,\beta}$ and $\bigsqcap_{k:\,\beta}$. Here k is a variable, and β is a formula not containing any quantifier. The semantics of the new quantifiers are defined as follows:

$$\bigsqcup_{k:\,\beta} \varphi \ \ \text{iff} \ \ \exists k(\beta \wedge \varphi), \qquad \bigsqcap_{k:\,\beta} \varphi \ \ \text{iff} \ \ \forall k(\beta \longrightarrow \varphi). \tag{1}$$

The quantifier $\bigsqcup_{k:\,\beta}$ is called an *existential bounded quantifier* if the solution set of β wrt. k is finite for all interpretations of all other variables. Under the same condition $\bigsqcap_{k:\,\beta}$ is called a *universal bounded quantifier*. Such formulas β are called k-bounds. Formulas containing no quantifiers at all are called *strictly quantifier-free*. Formulas containing exclusively bounded quantifiers are called *weakly quantifier-free*. The choice of notation obviously resembles Weispfenning's big disjunction and conjunction operators. In general, however, bounded quantifiers can be explicitly expanded only for fixed choices of all parameters occurring therein.

In this paper, we introduce weak quantifier elimination for a subset of first-order formulas, which considerably extends the full linear theory of the integers discussed in [16]: Our language is

$$L = \{0^{(0)}, 1^{(0)}, -^{(1)}, +^{(2)}, \cdot^{(2)}, \neq^{(2)}, \leq^{(2)}, >^{(2)}, \geq^{(2)}, <^{(2)}, \equiv^{(3)}, \not\equiv^{(3)}\}.$$

Consider a formula φ with parameters $a_1, \ldots, a_r$. Let φ contain quantifiers $Q_1, \ldots, Q_s$ with quantified variables $x_1, \ldots, x_s$, where each occurrence of any of our new quantifiers is in fact a bounded quantifier. Assume furthermore that all right hand sides of equations, inequalities, congruences, and incongruences in φ are 0, which can always be achieved by obvious equivalence transformations. Then we are able to eliminate from φ all the regular quantifiers provided that φ satisfies the following requirements:

(U$_1$) None of the quantified variables $x_1, \ldots, x_s$ occurs within moduli of congruences or incongruences. Note, however, that the moduli may be arbitrary polynomials in $a_1, \ldots, a_r$.

(U$_2$) Considering the left hand side terms of congruences and incongruences as polynomials in $x_1, \ldots, x_s$, over the coefficient ring $\mathbb{Z}[a_1, \ldots, a_r]$ each such term has a total degree less than or equal to 1.

(U$_3$) Considering the left hand side terms of equations and inequalities as polynomials in $x_1, \ldots, x_s$, over the coefficient ring $\mathbb{Z}[a_1, \ldots, a_r]$ each such term is either a nonlinear univariate polynomial or has a total degree less than or equal to 1.

We call formulas φ satisfying these three conditions *univariately nonlinear*. If especially in (U$_3$) every single left-hand side term matches the second case, then φ is a linear formula, and we are in the situation discussed in [16]. Thus note that according to our definition, every linear formula is also univariately nonlinear.

Accordingly, we refer to atomic subformulas of φ the left hand sides of which match (U$_2$) or the second case in (U$_3$) as *linear atomic formulas* (wrt. x_1, $\ldots$, x_s). Those matching the first case in (U$_3$) are called *superlinear univariate atomic formulas* (wrt. $x_1, \ldots, x_s$).

As an example, consider the following formula, which is univariately nonlinear:

$$\forall y \exists x (ax - y < 0 \land x^2 + x + a > 0). \tag{2}$$

The atomic formula $ax - y < 0$ is linear, and the atomic formula $x^2 + x + a > 0$ is superlinear univariate.

As within the framework of [16], the elimination of regular quantifiers possibly introduces several new bounded quantifiers. It is noteworthy that in contrast to similar elimination procedures for higher degrees over the reals [13], we can positively decide by inspection of the original input that we are able to eliminate *all* present regular quantifiers.

The plan of the paper is as follows: Section 2 recalls some basic definitions and results from [16] and generalizes these to our extended framework here. In Section 3 we formulate and prove our elimination theorem. Section 4 gives an overview of our implementation in REDLOG and discusses various computation examples in order to give an idea about possible applications as well as the practical efficiency and limitations of our method. In Section 5 we summarize and evaluate our results and mention some ideas for future research.

2   Extended Virtual Substitution Framework

Our quantifier elimination procedure for univariately nonlinear formulas is going to use distinct substitution procedures for test terms originating from superlinear univariate atomic formulas on the one hand and from linear atomic formulas on the other hand.

This gives rise to two extensions of the existing framework: First, with each test point there must be stored in addition the respective substitution procedure. Second, our new substitution for test terms from linear atomic formulas is going to considerably extend the existing concept of virtual substitution. It is going to be called *constrained virtual substitution*.

2.1   Parametric Elimination Sets

Let φ be a weakly quantifier-free formula. We recall some definitions and results from [16]. Originally, a parametric pre-elimination set for $\exists x \varphi$ had been defined there as a finite set

$$E = \big\{ (\gamma_i, t_i, B_i) \mid 1 \le i \le n \big\}, \quad \text{where} \quad B_i = \big((k_{ij}, \beta_{ij}) \mid 1 \le j \le m_i \big). \quad (3)$$

The *guards* γ_i are strictly quantifier-free formulas, the *test points* t_i are pseudo-terms possibly involving division, the k_{ij} are variables, and the β_{ij} are k_{ij}-bounds. Originally a parametric elimination set E for $\exists x \varphi$ is then a parametric pre-elimination set such that for some virtual substitution procedure ν the set E satisfies

$$\exists x \varphi \longleftrightarrow \bigvee_{(\gamma_i, t_i, B_i) \in E} \bigsqcup_{k_{i1}: \beta_{i1}} \cdots \bigsqcup_{k_{im_i}: \beta_{im_i}} \big(\gamma_i \wedge \nu(\varphi, t_i, x) \big). \quad (4)$$

With this definition, there is one single virtual substitution procedure used for all pseudo-terms in E. For the present paper we generalize this as follows: A *parametric pre-elimination set* for $\exists x \varphi$ is a finite set

$$E = \big\{ (\gamma_i, t_i, \sigma_i, B_i) \mid 1 \le i \le n \big\}, \quad (5)$$

where the definitions of γ_i, t_i and B_i are as before. Each σ_i is either the regular substitution $[\cdot / \cdot]$ of terms for variables or our new constrained virtual substitution $[\cdot /\!\!/ \cdot]$, which we are going to explain in detail in the next subsection. A *parametric elimination set* E for $\exists x \varphi$ is a parametric pre-elimination set such that

$$\exists x \varphi \longleftrightarrow \bigvee_{(\gamma_i, t_i, \sigma_i, B_i) \in E} \bigsqcup_{k_{i1}: \beta_{i1}} \cdots \bigsqcup_{k_{im_i}: \beta_{im_i}} \big(\gamma_i \wedge \sigma_i(\varphi, t_i, x) \big). \quad (6)$$

Assume that φ contains parameters $a_1, \ldots, a_r$. Let E be a parametric pre-elimination set for $\exists x \varphi$. For $z_1, \ldots, z_r \in \mathbb{Z}$, strictly quantifier-free formulas ψ, and pseudo-terms t we use for a moment the notational convention

$$\psi' = \psi[z_1/a_1, \ldots, z_r/a_r], \quad t' = t[z_1/a_1, \ldots, z_r/a_r]. \quad (7)$$

Furthermore, for formulas β' in at most one variable k we denote the solution set wrt. k by $S_{\beta'}^k = \{ z \in \mathbb{Z} \mid \beta'(z) \}$. The *projection* $\Pi(E, z_1, \ldots, z_r)$ of E is then defined as the finite set

$$\big\{ (\gamma'[y_1/k_1, \ldots, y_m/k_m], \, t'[y_1/k_1, \ldots, y_m/k_m], \, \sigma) \mid (\gamma, t, \sigma, B) \in E,$$
$$B = ((k_j, \beta_j) \mid 1 \le j \le m), \, y_1 \in S_{\beta_1'}^{k_1}, \, \ldots, \, y_m \in S_{\beta_m'[y_1/k_1, \ldots, y_{m-1}/k_{m-1}]}^{k_m} \big\}. \quad (8)$$

Lemma 1. *Let φ be a weakly quantifier-free formula with parameters $a_1, \ldots, a_r$. Let E be a parametric pre-elimination set for $\exists x \varphi$ with the following property: For each interpretation $z_1, \ldots, z_r \in \mathbb{Z}$ of the parameters $a_1, \ldots, a_r$, we have*

$$\exists x \varphi' \longleftrightarrow \bigvee_{(\gamma', t', \sigma) \in \Pi(E, z_1, \ldots, z_r)} \big(\gamma' \wedge \sigma(\varphi', t', x) \big),$$

where $\varphi' = \varphi[z_1/a_1, \ldots, z_r/a_r]$. Then E is a parametric elimination set for the formula $\exists x \varphi$. $\square$

2.2   Constrained Virtual Substitution

The essential idea of a virtual substitution ν is to be able to substitute for a variable x a parametric test point t that is formally not a term of the underlying language. For instance, in our language t might be a fraction. The virtual substitution maps atomic formulas to strictly quantifier-free formulas in such a way that whenever for some choice of values $z_1, \ldots, z_r \in \mathbb{Z}$ for the parameters $a_1, \ldots, a_r$ the test point evaluates to a number $t(z_1, \ldots, z_r) \in \mathbb{Z}$, then the following equivalence holds:

$$\varphi[t(z_1, \ldots, z_r)/x](z_1, \ldots, z_r) \longleftrightarrow \nu(\varphi, t, x)(z_1, \ldots, z_r). \tag{9}$$

Here, $[\cdot/\cdot]$ denotes regular substitution of terms for variables, where we allow ourselves to identify integers with corresponding sums of 1 or -1 representing them.

This notion of virtual substitution was sufficiently general for weak quantifier elimination from linear formulas as discussed in [16]. For our generalized setup here, however, we have to extend the concept of virtual substitution to *constrained virtual substitution*. Before giving a formal definition for this, let us return to our example in (2) in order to get a first idea about how our weak quantifier elimination would proceed for the elimination of $\exists x$:

$$\forall y \exists x (ax - y < 0 \wedge x^2 + x + a > 0). \tag{10}$$

From now on we allow ourselves to use absolute values within formulas as an abbreviated notation. Depending on the context they either stand for suitable case distinctions or for corresponding approximations by squares. The following suitable parametric elimination set for our example contains one entry originating from the first atomic formula and one entry from the second one:

$$E = \big\{ \big(a \neq 0 \wedge y + k \equiv_a 0,\ \tfrac{y+k}{a},\ [\cdot /\!/ \cdot],\ ((k, |k| \leq |a|))\big),$$
$$\big(\text{true},\ k,\ [\cdot/\cdot],\ ((k, |k| \leq |a| + 2))\big) \big\}. \tag{11}$$

The pseudo-term $\frac{y+k}{a}$ in the first entry describes a finite set of points around the solution of the equation $ax - y = 0$ corresponding to the first atomic formula. The guard $a \neq 0 \wedge y + k \equiv_a 0$ ensures that the pseudo-term evaluates to an integer. The k-bound $|k| \leq |a|$ describes the range of an existential bounded quantifier to be introduced for k. The substitution $(ax - y < 0)[\frac{y+k}{a} /\!/ x]$ is defined as regular substitution of terms for variables followed by multiplication with the square of the denominator that comes into existence.

Assume for a moment that we define the substitution $(x^2 + x + a > 0)[\frac{y+k}{a} /\!/ x]$ in the same fashion: This would yield $(y + k)^2 + a(y + k) + a^3 > 0$. This is neither linear nor superlinear univariate wrt. y and k. We thus make the following alternative definition:

$$(x^2 + x + a > 0)[\tfrac{y+k}{a} /\!/ x] := |ay + ak| > |a|^3 + 2a^2. \tag{12}$$

Notice that division of the right hand side of the definition by a^2 yields $|\frac{y+k}{a}| > |a| + 2$, where $|a| + 2$ is the Cauchy bound plus 1 of $x^2 + x + a$. So the right hand side of (12) formulates that $\frac{y+k}{a}$ satisfies $x^2 + x + a > 0$ due to the fact that it lies outside the Cauchy bounds of this parabola, which extends to $+\infty$. For the possible case that $\frac{y+k}{a}$ lies in contrast within the Cauchy bounds but still satisfies $x^2 + x + a > 0$ there is something left to do.

This turns us to the other entry in (11). Here $|k| \leq |a| + 2$ is the bound of a bounded quantifier that substituting k within its scope exactly covers every single point within the Cauchy bounds expanded by 1 of $x^2 + x + a$. Recall that the substitution $[\cdot / \cdot]$ is the regular substitution of terms for variables.

The overall elimination result for our example is the following weakly quantifier-free formula:

$$\bigsqcup_{k:\, |k| \leq |a|} \left(a \neq 0 \wedge y + k \equiv_a 0 \wedge k < 0 \wedge |ay + ak| > |a|^3 + 2a^2 \right) \vee$$

$$\bigsqcup_{k:\, |k| \leq |a|+2} \left(ak - y < 0 \wedge k^2 + k + a > 0 \right). \quad (13)$$

For understanding why it is important to consider the Cauchy bounds expanded by 1 in contrast to simply the Cauchy bounds themselves, consider the example $\exists x (x^2 - 1 > 0 \wedge x = 1)$.

We now turn to formal definitions for the virtual substitution $[\cdot /\!/ \cdot]$. Substitution into linear atomic formulas works exactly as usual [13,16]:

$$(ax = b)\left[\tfrac{b'}{a'} /\!/ x\right] := (ab' = a'b),$$

$$(ax \leq b)\left[\tfrac{b'}{a'} /\!/ x\right] := (aa'b' \leq a'^2 b),$$

$$(ax \equiv_m b)\left[\tfrac{b'}{a'} /\!/ x\right] := (ab' \equiv_{ma'} a'b). \quad (14)$$

One easily verifies that these substitutions satisfy Equivalence (9).

As already indicated by our example, we are going to use Cauchy bounds for substitution into superlinear univariate atomic formulas. Consider a parametric integer polynomial $p = c_n x^n + \cdots + c_0 \in \mathbb{Z}[a_1, \ldots, a_r][x]$. We define the *uniform Cauchy bound* of p as $|c_{n-1}| + \cdots + |c_0| + 1$.

Lemma 2. For $p = c_n x^n + \cdots + c_0 \in \mathbb{Z}[x]$ the following hold:

(i) For $c_n \neq 0$ the uniform Cauchy bound is always greater than or equal to the regular Cauchy bound:

$$|c_{n-1}| + \cdots + |c_0| + 1 \geq \max\left(1, \frac{|c_{n-1}| + \cdots + |c_0|}{|c_n|}\right).$$

(ii) Let $c_i \neq 0$ for at least one $i \in \{1, \ldots, n\}$. If $p(\xi) = 0$ for some $\xi \in \mathbb{R}$, then $|\xi| < |\xi| + 1 \leq |c_{n-1}| + \cdots + |c_0| + 1$.

Proof. To start with, note that $|c_n| \geq 1$. If in (i) the regular Cauchy bound equals 1, then our claim is obvious. Else our claim follows from the following observation by division by $|c_n|$:

$$|c_n| \cdot (|c_{n-1}| + \cdots + |c_0| + 1) \geq |c_{n-1}| + \cdots + |c_0| + 1 > |c_{n-1}| + \cdots + |c_0|.$$

In (ii) we have, of course, $|\xi|$ less than or equal to the regular Cauchy bound. If $|c_n| \leq |c_{n-1}| + \cdots + |c_0|$, then we obtain $|\xi| \leq |c_n| \cdot |\xi| \leq |c_{n-1}| + \cdots + |c_0|$, which implies $|\xi| + 1 \leq |c_{n-1}| + \cdots + |c_0| + 1$. If, in contrast, $|c_n| > |c_{n-1}| + \cdots + |c_0|$, then $|\xi| \leq 1$. In case $|c_{n-1}| + \cdots + |c_0| > 0$ we obtain $|\xi| + 1 \leq 2 \leq |c_{n-1}| + \cdots + |c_0| + 1$. In case $|c_{n-1}| + \cdots + |c_0| = 0$ we have $p = c_n x^n$ with $c_n \neq 0$, thus $\xi = 0$, and it follows that $|\xi| + 1 = 1 = |c_{n-1}| + \cdots + |c_0| + 1$. □

We adopt from [16] the definition of an interval boundary. For a subset $S \subseteq \mathbb{Z}$ a number $z \in S$ is an *interval boundary* if $z - 1 \notin S$ or $z + 1 \notin S$. In the former case, z is called a *lower* interval boundary. In the latter case, z is called an *upper* interval boundary. Let now α be an atomic formula in at most one variable x. The *characteristic points* of α are the interval boundaries of the solution set $S_\alpha^x = \{ z \in \mathbb{Z} \mid \alpha(z) \}$ wrt. x of α.

Lemma 3. *For $c_0, \ldots, c_n \in \mathbb{Z}$ consider an atomic formula $c_n x^n + \cdots + c_0 \varrho 0$ in at most one variable x, where $\varrho \in \{=, \neq, \leq, <, >, \geq\}$.*

(i) For all characteristic points k of $c_n x^n + \cdots + c_0 \varrho 0$ we have

$$|k| \leq |c_{n-1}| + \cdots + |c_0| + 1.$$

(ii) The atomic formula $c_n x^n + \cdots + c_0 \varrho 0$ has a constant truth value for choices l of x with $|c_{n-1}| + \cdots + |c_0| + 1 < l$. The same holds for choices l of x with $l < -(|c_{n-1}| + \cdots + |c_0| + 1)$.

Proof. If we have in part (i) that $c_1 = \cdots = c_n = 0$, then there are no characteristic points at all, and the statement is trivial. Else let $k \in \mathbb{Z}$ be a characteristic point of $c_n x^n + \cdots + c_0 \varrho 0$. Using the definition above and the intermediate value theorem it is easy to see that there is a real zero ξ of $c_n x^n + \cdots + c_0$ within the interval $[k - 1, k + 1]$. With Lemma 2(ii) it follows that $|k| \leq |\xi| + 1 \leq |c_{n-1}| + \cdots + |c_0| + 1$.

Part (ii) follows by induction from the observation that if an atomic formula has different truth values at l and $l + 1$, then either l or $l + 1$ is a characteristic point. □

Let now $p = c_n x^n + \cdots + c_0 \in \mathbb{Z}[a_1, \ldots, a_r][x]$, and let ϱ be any of the relations in our language or equality. We define

$$(p \varrho 0)\left[\tfrac{b'}{a'} /\!\!/ x\right] := \left(a'b' > a'^2(|c_{n-1}| + \cdots + |c_0| + 1) \wedge (p \varrho 0)[\infty /\!\!/ x]\right) \vee$$
$$\left(a'b' < -a'^2(|c_{n-1}| + \cdots + |c_0| + 1) \wedge (p \varrho 0)[-\infty /\!\!/ x]\right). \quad (15)$$

For substituting the nonstandard numbers $\pm\infty$ into atomic formulas we follow ideas by Weispfenning [13]. Substitution into equations and into negated equations is straightforward:

$$(p = 0)[\pm\infty /\!\!/ x] := \bigwedge_{i=0}^{n} c_i = 0, \quad (p \neq 0)[\pm\infty /\!\!/ x] := \bigvee_{i=0}^{n} c_i \neq 0. \quad (16)$$

For ordering inequalities ω the definition is recursive. Denote by $q = c_{n-1}x^{n-1} + \cdots + c_0$ the formal reductum of p, and let $\omega^s := (\omega \setminus =)$ be the strict part of ω. We first give the substitution for $+\infty$:

$$(p\ \omega\ 0)[\infty/\!/x] := c_n\ \omega^s\ 0 \vee \left(c_n = 0 \wedge (q\ \omega\ 0)[\infty/\!/x]\right) \quad \text{for} \quad n > 0,$$
$$(c_0\ \omega\ 0)[\infty/\!/x] := c_0\ \omega\ 0. \tag{17}$$

For the substitution of $-\infty$ one has to consider in addition the parities of the degrees during recursion:

$$(p\ \omega\ 0)[-\infty/\!/x] := (-1)^n c_n\ \omega^s\ 0 \vee \left(c_n = 0 \wedge (q\ \omega\ 0)[-\infty/\!/x]\right) \quad \text{for} \quad n > 0,$$
$$(c_0\ \omega\ 0)[-\infty/\!/x] := c_0\ \omega\ 0. \tag{18}$$

In contrast to our substitution (14) into linear atomic formulas our substitution (15) into superlinear univariate atomic formulas does not necessarily satisfy Equivalence (9). As a counterexample consider $((x - 1)^2(x - 2)^2 > 0)[0/\!/x]$. It satisfies, however, a weaker condition, which is made precise in the following lemma. It is one crucial technical observation of our paper that the weaker condition can still be exploited to establish an elimination theorem.

Lemma 4 (Constrained Virtual Substitution). *Consider a pseudo term* $t = b'/a'$ *and a superlinear univariate atomic formula* $p\ \varrho\ 0$, *where* $p = c_n x^n + \cdots + c_0$ *for* $n \geq 2$. *Set*

$$\lambda := |a'b'| > a'^2(|c_{n-1}| + \cdots + |c_0| + 1).$$

Whenever for some choice $\mathbf{z} = (z_1, \ldots, z_r) \in \mathbb{Z}^r$ *of the parameters* $a_1, \ldots, a_r$ *the test point* t *evaluates to a number* $t(\mathbf{z}) \in \mathbb{Z}$, *then the following holds:*

$$\lambda(\mathbf{z}) \longrightarrow \left((p\ \varrho\ 0)[t(\mathbf{z})/x](\mathbf{z}) \longleftrightarrow (p\ \varrho\ 0)[t/\!/x](\mathbf{z})\right).$$

Proof. To start with, it is noteworthy that the premise λ is equivalent to the disjunction of the two inequalities on the right hand side of the definition in (15). Furthermore, these inequalities exclude each other. Let $\mathbf{z} = (z_1, \ldots, z_r) \in \mathbb{Z}^r$ such that $t(\mathbf{z}) \in \mathbb{Z}$. Assume that $\lambda(\mathbf{z})$ holds. Then w.l.o.g. the first inequality $(a'b' > a'^2(|c_{n-1}| + \cdots + |c_0| + 1))(\mathbf{z})$ holds. This is equivalent to $t(\mathbf{z}) > (|c_{n-1}| + \cdots + |c_0| + 1)(\mathbf{z})$. By Lemma 3 we have

$$(p\ \varrho\ 0)[t(\mathbf{z})/x](\mathbf{z}) \longleftrightarrow (p\ \varrho\ 0)[l/x](\mathbf{z})$$

for all $l \geq t(\mathbf{z}) > (|c_{n-1}| + \cdots + |c_0| + 1)(\mathbf{z})$. The substitution of ∞ exactly simulates such points l:

$$(p\ \varrho\ 0)[t(\mathbf{z})/x](\mathbf{z}) \longleftrightarrow (p\ \varrho\ 0)[\infty/\!/x](\mathbf{z}).$$

Since we are already in a situation where $(a'b' > a'^2(|c_{n-1}| + \cdots + |c_0| + 1))(\mathbf{z})$ holds, we finally obtain

$$(p\ \varrho\ 0)[\infty/\!/x](\mathbf{z}) \longleftrightarrow (a'b' > a'^2(|c_{n-1}| + \cdots + |c_0| + 1))(\mathbf{z}) \wedge (p\ \varrho\ 0)[\infty/\!/x](\mathbf{z})$$
$$\longleftrightarrow (p\ \varrho\ 0)[t/\!/x](\mathbf{z}). \qquad \square$$

For clarity, we refer to virtual substitution procedures satisfying only the weaker condition described by the previous lemma as *constrained virtual substitution procedures*. The idea behind this notion is that λ serves as a constraint under which the virtual substitution behaves well. Note, however, that for substitution into linear formulas we still have Equivalence (9) without any constraints.

3 Univariate Quantifier Elimination

In this section we present a quantifier elimination procedure for the set of univariately nonlinear formulas.

3.1 Elimination of One Quantifier

The following representation lemma implies that characteristic points can generally be expressed by weakly quantifier-free formulas in terms of the coefficients of the input formula.

Lemma 5 (Representation Lemma). *Consider the superlinear univariate atomic formula $c_n x^n + \cdots + c_0 \varrho\, 0$ wrt. x where c_0, $\ldots$, $c_n \in \mathbb{Z}[a_1, \ldots, a_r]$. For a new variable k, we define the following strictly quantifier-free formula:*

$$\beta := |k| \le |c_{n-1}| + \cdots + |c_0| + 1.$$

Then β is linear in k. Furthermore β is a k-bound. Finally, for each interpretation z_1, $\ldots$, $z_r \in \mathbb{Z}$ of the parameters a_1, $\ldots$, a_r the solution set $S_\beta^k(z_1, \ldots, z_r)$ contains all characteristic points of $(c_n x^n + \cdots + c_0 \varrho\, 0)(z_1, \ldots, z_r)$.

Proof. The linearity of β and the finiteness of its solution set wrt. k are obvious. Choose interpretations z_1, $\ldots$, $z_r \in \mathbb{Z}$ of the parameters a_1, $\ldots$, a_r. If $c_i(z_1, \ldots, z_r) = 0$ for all $i \in \{1, \ldots, n\}$, then there are no characteristic points at all. Otherwise let $i \in \{1, \ldots, n\}$ be the largest index such that $c_i(z_1, \ldots, z_r) \ne 0$, and apply Lemma 3(i) to $(c_i x^i + \cdots + c_0 \varrho\, 0)(z_1, \ldots, z_r)$. □

Lemma 6. *Let σ be one of our substitutions $[\cdot/\cdot]$, $[\cdot/\!/\cdot]$. Let φ' be a weakly quantifier-free positive formula in one free variable x. Let t' be a variable-free pseudo-term that possibly contains division but describes an integer $t^* \in \mathbb{Z}$. Assume that for all atomic subformulas α of φ' and all interpretations y_1, $\ldots$, $y_n \in \mathbb{Z}$ of bound variables k_1, $\ldots$, k_n occurring in α the following holds:*

$$\sigma(\alpha, t', x)(y_1, \ldots, y_n) \longrightarrow \alpha[t^*/x](y_1, \ldots, y_n).$$

Then $\sigma(\varphi', t', x) \longrightarrow \varphi'[t^*/x]$.

Proof. We proceed by induction on the word length of the formula φ'. If φ' is an atomic formula, then it follows from the requirements of the lemma that $\sigma(\varphi', t', x)(y_1, \ldots, y_n) \longrightarrow \varphi'[t^*/x](y_1, \ldots, y_n)$ for all possible interpretations y_1, $\ldots$, $y_n \in \mathbb{Z}$ of the bound variables k_1, $\ldots$, k_n occurring in φ'. Since both

$\sigma(\varphi', t', x)$ and $\varphi'[t^*/x]$ do not contain any other variables besides $k_1, \ldots, k_n$, it follows that $\sigma(\varphi', t', x) \longrightarrow \varphi'[t^*/x]$. Consider now the case that φ' not atomic. Since φ' is positive, it suffices to consider formulas of the form $\varphi' = \varphi'_1 \vee \varphi'_2$, $\varphi' = \varphi'_1 \wedge \varphi'_2$, $\varphi' = \bigsqcup_{k:\,\beta} \varphi'_1$, and $\varphi' = \bigsqcap_{k:\,\beta} \varphi'_1$.

Consider the case $\varphi' = \varphi'_1 \vee \varphi'_2$. Assume that $\sigma(\varphi', t', x) = \sigma(\varphi'_1 \vee \varphi'_2, t', x)$ holds. By our induction hypothesis we have both

$$\sigma(\varphi'_1, t', x) \longrightarrow \varphi'_1[t^*/x] \quad \text{and} \quad \sigma(\varphi'_2, t', x) \longrightarrow \varphi'_2[t^*/x].$$

Since both our substitutions are defined in terms of substitutions for atomic formulas it follows that $\sigma(\varphi'_1 \vee \varphi'_2, t', x) = \sigma(\varphi'_1, t', x) \vee \sigma(\varphi'_2, t', x)$. Thus at least one of $\sigma(\varphi'_1, t', x)$, $\sigma(\varphi'_2, t', x)$ holds and, accordingly, at least one of $\varphi'_1[t^*/x]$, $\varphi'_2[t^*/x]$ holds. Hence $\varphi'_1[t^*/x] \vee \varphi'_2[t^*/x]$ holds. The case $\varphi' = \varphi'_1 \wedge \varphi'_2$ is similar.

Next, consider the case $\varphi' = \bigsqcup_{k:\,\beta} \varphi'_1$. Assume that the premise of our desired implication holds:

$$\sigma(\varphi', t', x) = \sigma\left(\bigsqcup_{k:\,\beta} \varphi'_1, t', x\right) = \bigsqcup_{k:\,\beta} \sigma(\varphi'_1, t', x).$$

Then there is $y \in S^k_\beta$ such that $\sigma(\varphi'_1, t', x)[y/k] = \sigma(\varphi'_1[y/k], t', x)$. By the induction hypothesis it follows that $\varphi'_1[y/k][t^*/x] = \varphi'_1[t^*/x][y/k]$ holds. Hence by our choice of y we obtain that the conclusion of our implication holds:

$$\bigsqcup_{k:\,\beta} (\varphi'_1[t^*/x]) = \left(\bigsqcup_{k:\,\beta} \varphi'_1\right)[t^*/x].$$

The case of a bounded universal quantifier is similar. Notice that then the induction hypothesis has to be applied several but finitely many times.   □

It is not hard to see that the previous lemma does not hold for non-positive formulas φ'.

In [16] we have explicitly given a parametric elimination set for the subset of linear formulas in the uniform Presburger arithmetic. We are going to use that very set as a subset of our elimination set for the more general case discussed here. Note that in the following lemma the elimination set E does not depend on the logical structure of φ but only on the bounded quantifiers and the set of atomic formulas contained therein.

Lemma 7 (Elimination of One Quantifier, Linear Case). *Consider a linear formula $\exists x \varphi$ with parameters $a_1, \ldots, a_r$, where φ is weakly quantifier-free, positive, and in prenex normal form:*

$$\varphi = \mathop{Q_1}_{k_1:\beta_1} \ldots \mathop{Q_n}_{k_n:\beta_n} \psi.$$

Let the set of all atomic formulas of ψ that contain x be

$$\{\, n_i x \varrho_i s_i + r_i \mid i \in I_1 \,\dot{\cup}\, I_2 \,\}.$$

Here, the n_i and r_i are polynomials in the parameters $a_1, \ldots, a_r$. The s_i are polynomials in both the parameters $a_1, \ldots, a_r$ and the bound variables $k_1, \ldots, k_n$. For $i \in I_1$, we have $\varrho_i \in \{=, \neq, <, \leq, \geq, >\}$. For $i \in I_2$, we have that ϱ_i is either $\equiv_{m_i}$ or $\not\equiv_{m_i}$, where m_i is a polynomial in $a_1, \ldots, a_r$. Let $k, k_1^*, \ldots, k_n^*$ denote new variables. Define

$$\beta_1^* = \beta_1[k_1^*/k_1, \ldots, k_n^*/k_n], \quad \ldots, \quad \beta_n^* = \beta_n[k_1^*/k_1, \ldots, k_n^*/k_n].$$

Define $m = \mathrm{lcm}\{ m_i^2 + 1 \mid i \in I_2 \}$. For $i \in I_1 \cup I_2$ define

$$s_i^* = s_i[k_1^*/k_1, \ldots, k_n^*/k_n] \quad and \quad \delta_i = -|n_i|m \leq k - s_i^* \leq |n_i|m.$$

Then $E = \{ (\gamma_i, t_i, B_i) \mid i \in I_1 \cup I_2 \} \cup \{(\mathrm{true}, 0, \varnothing)\}$, where

$$\gamma_i = (n_i \neq 0 \wedge r_i + k \equiv_{n_i} 0), \ t_i = \frac{r_i + k}{n_i}, \ B_i = \big((k_1^*, \beta_1^*), \ldots, (k_n^*, \beta_n^*), (k, \delta_i)\big),$$

is a parametric elimination set for $\exists x \varphi$.   □

Note that the definition of γ_i is such that whenever γ_i holds, then the corresponding t_i is defined and evaluates to an integer.

Lemma 8 (Elimination of One Quantifier). *Consider a univariately non-linear formula $\exists x \varphi$ with parameters $a_1, \ldots, a_r$, where*

$$\varphi = \underset{k_1:\beta_1}{Q_1} \ldots \underset{k_n:\beta_n}{Q_n} \psi.$$

is weakly quantifier-free, positive, and in prenex normal form. Let E_0 be the (regular) parametric elimination set according to Lemma 7 for the subset of linear atomic formulas in ψ and the bounded quantifiers occurring in φ. Let $\{ p_i \varrho_i 0 \mid i \in I \}$ be the subset of superlinear univariate atomic formulas of ψ. Let $\{ t_j \varrho_j 0 \mid j \in J \}$ be the set of all congruences and incongruences occurring in φ, i.e., ϱ_j is either $\equiv_{m_j}$ or $\not\equiv_{m_j}$. Let $m = \mathrm{lcm}\{ m_j^2 + 1 \mid j \in J \}$. For $i \in I$, denote by u_i the uniform Cauchy bound of p_i, and define for a new variable k the following strictly quantifier-free formula:

$$\delta := \bigvee_{i \in I} |k| \leq u_i + m.$$

Then the following is a parametric elimination set for $\exists x \varphi$:

$$E = \{ (\gamma, t, [\cdot /\!\!/ \cdot], B) \mid (\gamma, t, B) \in E_0 \} \cup \{(\mathrm{true}, k, [\cdot / \cdot], ((k, \delta)))\}.$$

Proof. Fix an interpretation $z_1, \ldots, z_r$ of the parameters $a_1, \ldots, a_r$. According to Lemma 1 it is sufficient to show that the projection $\Pi(E, z_1, \ldots, z_r)$ satisfies the following equivalence for $\varphi' := \varphi[z_1/a_1, \ldots, z_r/a_r]$:

$$\exists x \varphi' \longleftrightarrow \bigvee_{(\gamma', t', \sigma) \in \Pi(E, z_1, \ldots, z_r)} \big(\gamma' \wedge \sigma(\varphi', t', x)\big).$$

We first prove the implication from the right to the left. In contrast to the linear case and due to our constrained virtual substitution, this is not trivial. Suppose that the right hand side holds. Then for at least one $(\gamma'_j, t'_j, \sigma_j) \in \Pi(E, z_1, \ldots, z_r)$ the corresponding $\gamma'_j \wedge \sigma_j(\varphi', t'_j, x)$ holds. Recall that t'_j is a pseudo-term possibly containing division. On the other hand, the validity of γ'_j guarantees that t'_j corresponds to an integer. Denote that integer by t^*_j. We are now going to prove the following, which by Lemma 6 implies that our $t^*_j \in \mathbb{Z}$ is one possible choice for x, such that $\exists x \varphi'$ holds: Let α be any atomic sub-formula of φ' with bound variables $k_1, \ldots, k_n$. Let $y_1, \ldots, y_n \in \mathbb{Z}$ be an interpretation of $k_1, \ldots, k_n$. Then

$$\sigma_j(\alpha, t'_j, x)(y_1, \ldots, y_n) \longrightarrow \alpha[t^*_j/x](y_1, \ldots, y_n).$$

If σ_j is the regular substitution $[\cdot/\cdot]$, then the implication is trivial. Else σ_i is our constrained virtual substitution $[\cdot/\!/\cdot]$. If α is linear, then $[\cdot/\!/\cdot]$ satisfies Equivalence 9, and our implication is just the direction from the right to the left of that equivalence. If, in contrast, α is a superlinear univariate atomic formula $p_i \varrho_i 0$, where $i \in I$, then we make a case distinction on $t^*_j \in \mathbb{Z}$. If $|t^*_j| > u_i(z_1, \ldots, z_r)$, i.e., it lies outside the uniform Cauchy bound of α, then our implication follows from Lemma 4. Otherwise, one verifies by inspection of Definition (15) that $\sigma_i(\alpha, t'_j, x)(k_1, \ldots, k_n) \longleftrightarrow$ false such that the implication holds trivially.

Assume vice versa that $\exists x \varphi'$ holds. Consider first the degenerate case that $S^x_{\varphi'} = \mathbb{Z}$. If $I \neq \varnothing$, then we have $(\text{true}, 0, [\cdot/\cdot]) \in \Pi(E, z_1, \ldots, z_r)$. Otherwise, we have inherited from Lemma 7 $(\text{true}, 0, [\cdot/\!/\cdot]) \in \Pi(E, z_1, \ldots, z_r)$, and in the absence of superlinear univariate formulas $[0/\!/x] = [0/x]$. Let now $\varnothing \subsetneq S^x_{\varphi'} \subsetneq \mathbb{Z}$.

If $S^x_{\varphi'} \cap S^k_\delta \neq \varnothing$, say, $z \in S^x_{\varphi'} \cap S^k_\delta$, then there is $(\text{true}, z, [\cdot/\cdot]) \in \Pi(E, z_1, \ldots, z_r)$ originating from the test point $(\text{true}, k, [\cdot/\cdot], ((k, \delta)))$.

Assume now that, in contrast, $S^x_{\varphi'} \cap S^k_\delta = \varnothing$. Then we are in a situation, where we can consider instead φ the formula $\bar{\varphi}$ as follows: We replace in φ each superlinear univariate atomic formula $p_i \varrho_i 0$ by the following strictly quantifier-free formula:

$$\bigl(x < -u_i \wedge (p_i \varrho_i 0)[-\infty/\!/x]\bigr) \vee \bigl(x > u_i \wedge (p_i \varrho_i 0)[\infty/\!/x]\bigr).$$

Defining $\bar{\varphi}' := \bar{\varphi}[z_1/a_1, \ldots, z_r/a_r]$ and on our assumption that $S^x_{\varphi'} \cap S^k_\delta = \varnothing$, we have $\bar{\varphi}' \longleftrightarrow \varphi'$, from which it follows that $\exists x \bar{\varphi}'$ holds. Since $\exists x \bar{\varphi}$ obtained this way is a linear formula we know by Lemma 7 a regular parametric elimination set for this:

$$E_0 \cup \bigl\{ (\text{true}, -u_i + k, ((k, |k| \leq m))), (\text{true}, u_i + k, ((k, |k| \leq m))) \mid i \in I \bigr\}.$$

We adapt this set to our constrained virtual substitution framework by adding to each test point the constrained virtual substitution:

$$\bar{E} := \{ (\gamma, t, [\cdot/\!/\cdot], B) \mid (\gamma, t, B) \in E_0 \} \cup$$
$$\bigl\{ (\text{true}, \pm u_i + k, [\cdot/\!/\cdot], ((k, |k| \leq m))) \mid i \in I \bigr\}.$$

Recall from the definition of our constrained virtual substitution $[\cdot /\!/\cdot]$ that for linear formulas it equals the virtual substitution used in [16]. It thus follows that $\Pi(\bar{E}, z_1, \ldots, z_r)$ is an elimination set for $\exists x \bar{\varphi}'$. Consequently from the validity of $\exists x \bar{\varphi}'$ it follows that the following formula holds:

$$\bigvee_{(\gamma', t', \sigma) \in \Pi(\bar{E}, z_1, \ldots, z_r)} \left(\gamma' \wedge \sigma(\bar{\varphi}', t', x) \right).$$

Let $(\gamma', t', \sigma) \in \Pi(\bar{E}, z_1, \ldots, z_r)$ such that $\gamma' \wedge \sigma(\bar{\varphi}', t', x)$. We make a case distinction on the origin of (γ', t', σ). If

$$(\gamma', t', \sigma) \in \Pi(\{ (\gamma, t, [\cdot /\!/\cdot], B) \mid (\gamma, t, B) \in E_0 \}, z_1, \ldots, z_r),$$

then it follows from $\{ (\gamma, t, [\cdot /\!/\cdot], B) \mid (\gamma, t, B) \in E_0 \} \subseteq E$ that $(\gamma', t', \sigma) \in \Pi(E, z_1, \ldots, z_r)$. In the other case, where

$$(\gamma', t', \sigma) \in \Pi(\{ (\mathrm{true}, \pm u_i + k, [\cdot /\!/\cdot], ((k, |k| \leq m))) \mid i \in I \}, z_1, \ldots, z_r),$$

recall from the formulation of the present lemma the definition of m, and observe that the following relation holds for all y with $|y| \leq m(z_1, \ldots, z_r)$:

$$| \pm u_i(z_1, \ldots, z_r) + y| \leq (\pm u_i + m)(z_1, \ldots, z_r).$$

So there is a test point $(\gamma', t', [\cdot / \cdot]) \in \Pi(E, z_1, \ldots, z_r)$, which differs from our considered point only by the substitution procedure. In both cases, we have found a test point $(\gamma', t', \sigma^*) \in \Pi(E, z_1, \ldots, z_r)$, which differs from our considered point at most by the substitution procedure. We are now going to show that

$$\gamma' \wedge \sigma^*(\varphi', t', x).$$

Note that this is not trivial even in the first case where $\sigma = \sigma^* = [\cdot /\!/\cdot]$, because in φ' there possibly occur superlinear univariate formulas. Recall that we are in a situation where in particular γ' holds, which implies that $t' \in \mathbb{Z}$. This allows to apply Equivalence (9), and it follows that $|t'| > u_i(z_1, \ldots, z_r)$ for all $i \in I$. Hence, using Lemma 4 and Equivalence (9), $\sigma^*(\varphi', t', x')$ equivalently replaces every single atomic formula in φ' such that we obtain our desired observation $\gamma' \wedge \sigma^*(\varphi', t', x)$. Hence

$$\bigvee_{(\gamma', t', \sigma) \in \Pi(E, z_1, \ldots, z_r)} \left(\gamma' \wedge \sigma(\varphi', t', x) \right),$$

which is what had to be shown.   □

3.2   Elimination Theorem

In order to possibly iterate weak quantifier elimination we next have to make sure that the output of our elimination procedure is again univariately nonlinear; in other words, it satisfies the defining conditions (U_1)–(U_3) in the introduction. In contrast to the linear case, this observation is not trivial:

Lemma 9. *Let φ be weakly quantifier-free, positive and prenex. Assume that $\exists x \varphi$ occurs within a univariately nonlinear formula $\hat{\varphi}$. Then replacing $\exists x \varphi$ in $\hat{\varphi}$ with the result of the application of the parametric elimination set E from Lemma 8 is again univariately nonlinear.*

Proof. Let $x_1, \ldots, x_s$ be the quantified variables occurring in $\hat{\varphi}$. We have to show, that the formula

$$\varphi' = \bigvee_{(\gamma_i, t_i, \sigma_i, B_i) \in E} \bigsqcup_{k_{i1} : \beta_{i1}} \cdots \bigsqcup_{k_{im_i} : \beta_{im_i}} \left(\gamma_i \wedge \sigma_i(\varphi, t_i, x) \right). \tag{19}$$

satisfies our conditions (U_1)–(U_3) wrt. $x_1, \ldots, x_s$. The bounds of our newly created bounded quantifiers obtained according to Lemma 5 do not contain any of the variable $x_1, \ldots, x_s$. Since each nontrivial guard originates from a regular elimination set, all guards also satisfy the conditions (U_1)–(U_3). It is hence sufficient to consider formulas of the form $\sigma_i(\alpha, t_i, x)$ for each atomic formula α occurring in φ. If α is a linear formula the statement is trivial. For the case α is univariately nonlinear the statement is easily obtained by inspection of the definition in (15). □

Theorem 10 (Elimination Theorem). *The ordered ring of the integers with congruences admits weak quantifier elimination for univariately nonlinear formulas.*

Proof. Let $\hat{\varphi}$ be a univariately nonlinear formula. We proceed by induction on the number n of regular quantifiers in $\hat{\varphi}$. If $n = 0$, then $\hat{\varphi}$ is already weakly quantifier-free. So there is nothing to do. Consider now the case $n > 0$. There is then a subformula of $\hat{\varphi}$ of one of the forms $\exists x \varphi$ or $\forall x \varphi$, where φ is weakly quantifier-free. The latter case can be reduced to the former one by means of the equivalence $\forall x \varphi \longleftrightarrow \neg \exists x \neg \varphi$. We may w.l.o.g. assume that φ is in prenex normal form and positive. By Lemma 8, there exists a parametric elimination set E for $\exists x \varphi$. That is, $\exists x \varphi$ is equivalent to

$$\varphi' = \bigvee_{(\gamma_i, t_i, \sigma_i, B_i) \in E} \bigsqcup_{k_{i1} : \beta_{i1}} \cdots \bigsqcup_{k_{im_i} : \beta_{im_i}} \left(\gamma_i \wedge \sigma_i(\varphi, t_i, x) \right),$$

where $B_i = \left((k_{ij}, \beta_{ij}) \mid 1 \leq j \leq m_i \right)$. We obtain $\hat{\varphi}'$ from $\hat{\varphi}$ by equivalently replacing $\exists x \varphi$ with φ'. Lemma 9 states that $\hat{\varphi}'$ is again univariately nonlinear. Hence we can eliminate the remaining quantifiers from $\hat{\varphi}'$ by our induction hypothesis. □

Corollary 11 (Decidability of Sentences). *In the ordered ring of the integers with congruences univariately nonlinear sentences are decidable.*

Proof. Consider a univariately nonlinear sentence. Apply weak quantifier elimination. The result is an equivalent sentence containing only bounded quantifiers. In the absence of parameters these can be expanded into disjunctions and conjunctions. After this, all atomic formulas are variable-free such that we straightforwardly obtain either "true" or "false." □

4   Implementation and Computation Examples

The procedure described in this paper has been implemented in REDLOG, which stands for REDUCE *logic system* [17,18]. It provides an extension of the computer algebra system REDUCE to a computer logic system implementing symbolic algorithms on first-order formulas with respect to temporarily fixed first-order languages and theories. Such a choice of language and theory is called a *domain* or, alternatively, a *context*.

Before turning to the integer context relevant for our work here, we briefly summarize the other existing domains together with short names and alternative names, which are supported for backward compatibility:

BOOLEAN, B, IBALP. The class of Boolean algebras with two elements. These algebras are uniquely determined up to isomorphisms. BOOLEAN comprises quantified propositional calculus [15].

COMPLEX, C, ACFSF. The class of algebraically closed fields such as the complex numbers over the language of rings.

DIFFERENTIAL, DCFSF. A domain for computing over differentially closed fields. There is no natural example for such a field, but in special cases the methods can be used for obtaining relevant and interpretable results also for reasonable differential fields [19].

PADICS, DVFSF. One prominent example for discretely valued fields are the p-adic numbers for some prime p with abstract divisibility relations encoding order between values. All PADICS algorithms are optionally uniform in p [14].

QUEUES, QQE. A (two-sided) queue is a finite sequence of elements of some basic type. There are two sorts of variables, one for the basic type and one for the queue type. Accordingly, there is first-order quantification possible for both sorts. So far, the implementation is restricted to the reals as basic type [20].

REALS, R, OFSF. The class of real closed fields such as the real numbers with ordering. This context was the original motivation for REDLOG. It is still the most important and most comprehensive one [21].

TERMS, TALP. Free Malcev-type term algebras. The available function symbols and their arity can be freely chosen. [22].

The work discussed here has been integrated into another such domain:

INTEGERS, Z, PASF. The full linear theory of the integers.

This domain had been originally introduced for the methods described in [16]. It now naturally extends to univariately nonlinear formulas without loosing any of its previous features.

The idea of REDLOG is to combine methods from computer algebra with first-order logic thus extending the computer algebra system REDUCE to a computer logic system. In this extended system both the algebraic side and the logic side greatly benefit from each other in numerous ways. The current release REDLOG 3.0 is an integral part of the computer algebra system REDUCE 3.8. The implementation of our methods described here is part of the current development version of REDLOG. It is going to be distributed with REDUCE 3.9. Until then it is freely available on the REDLOG homepage.[1]

We are now going to discuss various computations with our implementation. The idea is to illustrate the possible application range but also the limits of our method and of the current implementation. All our computations have been performed on a 1.66 GHz Intel Core 2 Duo processor T5500 using only one core and 128 MB RAM.

4.1   Optimization

We define a *parametric linear optimization problem with univariately nonlinear constraints* as follows: Minimize a cost function $\gamma_1 x_1 + \cdots + \gamma_n x_n$ subject to

$$A\mathbf{x} \geq \mathbf{b}, \quad p_1 \varrho_1 0, \quad \ldots, \quad p_r \varrho_r 0.$$

As usual, $A = (\alpha_{ij})$ is an $m \times n$-matrix, and $\mathbf{b} = (\beta_1, \ldots, \beta_m)$ is an m-vector. For $i \in \{1, \ldots, m\}$ and $j \in \{1, \ldots, n\}$ we have $\alpha_{ij}, \beta_i, \gamma_j \in \mathbb{Z}[a_1, \ldots, a_k]$, i.e., all these coefficients are possibly parametric. For each $s \in \{1, \ldots, r\}$ we have $p_s \in \mathbb{Z}[a_1, \ldots, a_k][x_j]$ for some $j \in \{1, \ldots, n\}$, i.e., the $p_1, \ldots, p_r$ are parametric univariate polynomials. Each corresponding ϱ_s is one of $=$, $\neq$, $\leq$, $>$, $\geq$, or $<$.

Using a new variable z for the minimum such a problem can be straightforwardly translated to our framework as follows:

$$\exists x_1 \ldots \exists x_n \Big(\sum_{j=1}^{n} \gamma_j x_j \leq z \wedge \bigwedge_{i=1}^{m} \sum_{j=1}^{n} \alpha_{ij} x_j \geq \beta_i \wedge \bigwedge_{s=1}^{r} p_s \varrho_s 0 \Big).$$

Example 12. Minimize $x + y$ subject to the following constraints:

$$x \geq 0, \quad y \geq 0, \quad x + y \geq 0, \quad \text{and} \quad x^2 + a < 0.$$

The formulation as a quantifier elimination problem reads as follows:

$$\exists x \exists y \big(x + y \leq z \wedge x \geq 0 \wedge y \geq 0 \wedge x + y \geq 0 \wedge x^2 + a < 0 \big).$$

For this REDLOG computes within 20 ms a weakly quantifier-free equivalent containing 103 atomic formulas. Setting then $a = 10$ and automatically simplifying yields within 2190 ms the result $z > 3$, i.e., the minimum for $x + y$ is 4. This final simplification step includes in particular expansion of all present bounded quantifiers. If we plug in $a = 10$ before the elimination, then we directly obtain $z > 3$ in only 330 ms. This amazing difference in time, we had already observed for the full linear theory of the integers [16]. It can be explained as follows: In both Lemma 7 and Lemma 8, we compute the least common multiple of the squares of all moduli. For non-parametric moduli we optimize this by using instead the absolute values of the moduli.

Generalizing our method discussed in the present paper to *extended quantifier elimination* [23,13,24,15] would admit to obtain in addition a sample point for the computed optimum. The optimization addressed above with the absolute value instead of squares could be applied in the parametric case as well by adding to the language a symbol for the absolute value.

---

[1] www.redlog.eu

4.2   Software Security

Information flow control is one important issue in software security [25,26]. The question is whether it is possible to manipulate parameters in such a way that sensitive information can become accessible outside of special code segments. We are going to discuss a modification of an example from [16].

Example 13. For the following piece of code there is a security risk if there are choices for a and b such that y is assigned the value of some A[n^2].

```
if (a < b) then
    if (a+b mod 2 = 0) then
        n := (a+b)/2
    else
        n := (a+b+1)/2
    fi
    A[n^2] := get_sensitive_data(x)
    send_sensitive_data(trusted_receiver,A[n^2])
fi
y := A[abs(b-a)].
```

An attacker would be interested in a description of all values of a and b such that this happens. This can be formulated as follows:

$$\exists n \big((a < b \land a + b \equiv_2 0 \land 2n = a + b \land$$
$$((a < b \land b - a = n^2) \lor (a \geq b \land a - b = n^2))) \lor$$
$$(a < b \land a + b \not\equiv_2 0 \land 2n = a + b + 1 \land$$
$$((a < b \land b - a = n^2) \lor (a \geq b \land a - b = n^2)))\big).$$

Our implementation computes in less than 10 ms the following weakly quantifier-free description:

$$\bigsqcup_{k:\,|k|\leq(a-b)^2+2} (a - b < 0 \land a - b + k^2 = 0 \land a + b \not\equiv_2 0 \land a + b - 2k + 1 = 0) \lor$$

$$\bigsqcup_{k:\,|k|\leq(a-b)^2+2} (a - b < 0 \land a - b + k^2 = 0 \land a + b \equiv_2 0 \land a + b - 2k = 0).$$

4.3   Integer Roots

Example 14. Consider the generic polynomial $p = ax^2 + bx + c$. The question whether p has an integer root can be expressed by a univariately nonlinear formula as follows:

$$\exists x(ax^2 + bx + c = 0).$$

Our elimination procedure yields after less than 10 ms the following weakly quantifier-free equivalent:

$$\bigsqcup_{k:\,|k|<|b|+|c|+2} ak^2 + bk + c = 0.$$

This result exactly substitutes all integers inside the uniform Cauchy bounds of p expanded by 1. This expansion is the least common multiple of the (non-existing) moduli in the input.

This result obviously does not provide much mathematical insight. A helpful though imprecise intuition about our method is the following: *Its intelligence works mostly outside of the relevant Cauchy bounds.* Anyway, a slight modification of our previous example yields useful information:

Example 15. Given suitable n_1, $n_2 \in \mathbb{Z}$ and d_1, $d_2 \in \mathbb{N} \setminus \{0\}$ we look for integer zeros of $p(x) = \alpha x^2 + \beta x + \gamma$ within the interval $[n_1/d_1, n_2/d_2]$. This can be formulated as follows:

$$\exists x \big(p = 0 \wedge d_1 x \geq n_1 \wedge d_2 x \leq n_2 \big).$$

Let us consider the polynomial $p = x^5 - 3x^2 + 1$. We want to know whether there is a zero of p in $[1/3, 3]$. This yields the following input:

$$\exists x (x^5 - 3x^2 + 1 = 0 \wedge 3x \geq 1 \wedge x \leq 3).$$

For this, our implementation computes "false" in less than 10 ms. In fact, our chosen p has no integer zeros at all.

This last example illustrates the fact that our method combined with automatic simplification yields a *decision procedure* for univariately nonlinear sentences. So for sentences, we are able to obtain as a result either "true" or "false," which both do not contain any bounded quantifiers. Hence, concerning the decision of sentences, we provide a considerable extension of the original Presburger framework, where the user need not accept any additional syntactic constructs.

5   Conclusions

We have considered the integers using the language of ordered rings extended by ternary symbols for congruence and incongruence. On this basis we have given a weak quantifier elimination procedure for the set of univariately nonlinear formulas. The notion of weak quantifier elimination refers to the fact that the result possibly contains bounded quantifiers. For fixed choices of parameters these bounded quantifiers can be expanded into disjunctions or conjunctions. For decision problems they can be completely avoided. Our methods are efficiently implemented and publicly available within the computer logic system REDLOG, which is part of REDUCE. The applicability of our new method and its implementation has been demonstrated by means of various application examples. For the future it is planned to provide also an extended quantifier elimination procedure within the framework considered here. Furthermore, it appears to be a promising idea to extend the language by a symbol for the absolute value. This would allow to considerably reduce the ranges of the bounded quantifiers coming into existence.

References

1. Presburger, M.: Über die Vollständigkeit eines gewissen Systems der Arithmetik ganzer Zahlen, in welchem die Addition als einzige Operation hervortritt. In: Comptes Rendus du premier congres de Mathematiciens des Pays Slaves, Warsaw, Poland, pp. 92–101 (1929)
2. Cooper, D.C.: Theorem proving in arithmetic without multiplication. Machine Intelligence 7, 91–99 (1972)
3. Fischer, M., Rabin, M.: Super-exponential complexity of Presburger arithmetic. SIAM-AMS Proceedings 7, 27–41 (1974)
4. Ferrante, J., Rackoff, C.W.: A decision procedure for the first-order theory of real addition with order. SIAM Journal on Computing 4, 69–77 (1975)
5. Ferrante, J., Rackoff, C.W.: The Computational Complexity of Logical Theories. Lecture Notes in Mathematics, vol. 718. Springer, Berlin (1979)
6. von zur Gathen, J., Sieveking, M.: A bound on solutions of linear integer equalities and inequalities. Proceedings of the AMS 72, 155–158 (1978)
7. Berman, L.: Precise bounds for Presburger arithmetic and the reals with addition. In: FOCS 1977. 18th Annual Symposium on Foundations of Computer Science, Providence, RI, USA, October 3–November 2, pp. 95–99. IEEE Press, Los Alamitos (1977)
8. Berman, L.: The complexity of logical theories. Theoretical Computer Science 11, 71–77 (1980)
9. Weispfenning, V.: The complexity of almost linear diophantine problems. Journal of Symbolic Computation 10(5), 395–403 (1990)
10. Weispfenning, V.: Complexity and uniformity of elimination in Presburger Arithmetic. In: Küchlin, W.W. (ed.) ISSAC 97. Proceedings of the 1997 International Symposium on Symbolic and Algebraic Computation, Maui, HI, pp. 48–53. ACM Press, New York, NY (1997)
11. Weispfenning, V.: The complexity of linear problems in fields. Journal of Symbolic Computation 5(1&2), 3–27 (1988)
12. Loos, R., Weispfenning, V.: Applying linear quantifier elimination. The Computer Journal 36(5), 450–462 (1993) (special issue on computational quantifier elimination)
13. Weispfenning, V.: Quantifier elimination for real algebra—the quadratic case and beyond. Applicable Algebra in Engineering Communication and Computing 8(2), 85–101 (1997)
14. Sturm, T.: Linear problems in valued fields. Journal of Symbolic Computation 30(2), 207–219 (2000)
15. Seidl, A.M., Sturm, T.: Boolean quantification in a first-order context. In: Ganzha, V.G., Mayr, E.W., Vorozhtsov, E.V. (eds.) Computer Algebra in Scientific Computing. Proceedings of the CASC 2003, Institut für Informatik, Technische Universität München, München, Germany, pp. 329–345 (2003)
16. Lasaruk, A., Sturm, T.: Weak quantifier elimination for the full linear theory of the integers. a uniform generalization of presburger arithmetic. Technical Report MIP-0604, FMI, Universität Passau, D-94030 Passau, Germany (2006) (to appear in the journal AAECC)
17. Dolzmann, A., Sturm, T.: Redlog: Computer algebra meets computer logic. ACM SIGSAM Bulletin 31(2), 2–9 (1997)
18. Dolzmann, A., Sturm, T.: Redlog user manual. Technical Report MIP-9905, FMI, Universität Passau, D-94030 Passau, Germany, Edition 2.0 for Version 2.0 (1999)

19. Dolzmann, A., Sturm, T.: Generalized constraint solving over differential algebras. In: Ganzha, V.G., Mayr, E.W., Vorozhtsov, E.V. (eds.) Computer Algebra in Scientific Computing. Proceedings of the CASC 2004, Institut für Informatik, Technische Universität München, München, Germany, pp. 111–125 (2004)
20. Straßer, C.: Quantifier elimination for queues. In: Draisma, J., Kraft, H. (eds.) Rhine Workshop on Computer Algebra. Proceedings of the RWCA 2006, pp. 239–248. Universität Basel, Basel (2006)
21. Dolzmann, A., Sturm, T., Weispfenning, V.: Real quantifier elimination in practice. In: Matzat, B.H., Greuel, G.M., Hiss, G. (eds.) Algorithmic Algebra and Number Theory, pp. 221–247. Springer, Berlin (1998)
22. Sturm, T., Weispfenning, V.: Quantifier elimination in term algebras. The case of finite languages. In: Ganzha, V.G., Mayr, E.W., Vorozhtsov, E.V. (eds.) Computer Algebra in Scientific Computing. Proceedings of the CASC 2002, Institut für Informatik, Technische Universität München, München, Germany, pp. 285–300 (2002)
23. Weispfenning, V.: Simulation and optimization by quantifier elimination. Journal of Symbolic Computation 24(2), 189–208 (1997) (special issue on applications of quantifier elimination)
24. Dolzmann, A., Sturm, T.: P-adic constraint solving. In: Dooley, S. (ed.) ISSAC 99. Proceedings of the 1999 International Symposium on Symbolic and Algebraic Computation, Vancouver, BC, pp. 151–158. ACM Press, New York, NY (1999)
25. Snelting, G.: Quantifier elimination and information flow control for software security. In: Dolzmann, A., Seidl, A., Sturm, T. (eds.) Algorithmic Algebra and Logic. Proceedings of the A3L 2005, BoD, Germany, Norderstedt, pp. 237–242 (2005)
26. Snelting, G., Robschink, T., Krinke, J.: Efficient path conditions in dependence graphs for software safety analysis. ACM Transactions on Software Engineering and Methodology 15(4), 410–457 (2006)

Polynomial Division Using Dynamic Arrays, Heaps, and Packed Exponent Vectors*

Michael Monagan and Roman Pearce

Department of Mathematics, Simon Fraser University
Burnaby, B.C. V5A 1S6, Canada
mmonagan@cecm.sfu.ca, rpearcea@cecm.sfu.ca

Abstract. A common way of implementing multivariate polynomial multiplication and division is to represent polynomials as linked lists of terms sorted in a term ordering and to use repeated merging. This results in poor performance on large sparse polynomials.

In this paper we use an auxiliary heap of pointers to reduce the number of monomial comparisons in the worst case while keeping the overall storage linear. We give two variations. In the first, the size of the heap is bounded by the number of terms in the quotient(s). In the second, which is new, the size is bounded by the number of terms in the divisor(s).

We use dynamic arrays of terms rather than linked lists to reduce storage allocations and indirect memory references. We pack monomials in the array to reduce storage and to speed up monomial comparisons. We give a new packing for the graded reverse lexicographical ordering.

We have implemented the heap algorithms in C with an interface to Maple. For comparison we have also implemented Yan's "geobuckets" data structure. Our timings demonstrate that heaps of pointers are comparable in speed with geobuckets but use significantly less storage.

1   Introduction

In this paper we present and compare algorithms and data structures for polynomial division in the ring $P = F[x_1, x_2, ..., x_n]$ where F is a field. We are interested in (i) exact division of $f \in P$ by a single polynomial $g \in P$, that is testing whether $g|f$ and if so, computing the quotient $q = f/g$, (ii) exact division of $f \in P$ by a polynomial $g \in P$ modulo a triangular set of polynomials in $F[x_i, x_{i+1}, ..., x_n]$, and (iii) computing the remainder of $f \in P$ divided by a set of polynomials $\{g_1, g_2, ..., g_s\} \in P$. Since many algorithms in computer algebra use modular methods for efficiency, that is, they compute modulo primes, we will want to divide over characteristic p as well as characteristic 0.

We consider distributed polynomial representations that sort the terms of the polynomial with respect to a monomial ordering. See [3] or [4] for background material on monomial orderings. The orderings that we are most interested in are the pure lexicographical ordering (lex), the graded lexicographical ordering

---

* This work was supported by NSERC of Canada and the MITACS NCE of Canada.

V.G. Ganzha, E.W. Mayr, and E.V. Vorozhtsov (Eds.): CASC 2007, LNCS 4770, pp. 295–315, 2007.
© Springer-Verlag Berlin Heidelberg 2007

(grlex), and the graded reverse lexicographical ordering (grevlex). In the grlex ordering one first sorts the terms by total degree and then by lexicographical order. For example, the polynomial

$$-9x^4 - 7x^3yz + 6x^2y^3z + 8y^2z^3$$

when written with terms in descending grlex order with $x > y > z$ is

$$6x^2y^3z - 7x^3yz + 8y^2z^3 - 9x^4.$$

The data structure used to represent polynomials will have a direct impact on the efficiency of the division algorithm. The data structure used by the Axiom system [7] for Gröbner basis computations is the SDMP (Sparse Distributed Multivariate Polynomial) data structure. This is a linked list of terms where each term is a pair (c, e), where c is a (pointer to) a coefficient and e is a pointer to the exponent vector, which is an array of machine integers. Using $\langle a, b, c, ... \rangle$ to denote an array, $[a, b, c, ...]$ to denote a linked list, and (c, e) to denote a pair of pointers, the polynomial above would be represented as

$$[\, (6, \langle 2, 3, 1 \rangle), \, (-7, \langle 3, 1, 1 \rangle), \, (8, \langle 0, 2, 3 \rangle), \, (-9, \langle 4, 0, 0 \rangle) \,].$$

Recall the division algorithm. Following the notation of Cox, Little, and O'Shea [3], we let $LT(f)$, $LM(f)$, and $LC(f)$ denote the leading term, the leading monomial, and the leading coefficient of a polynomial f, respectively. These depend on the term ordering but satisfy $LT(f) = LC(f)LM(f)$.

The Division Algorithm
Input: $f, g_1, g_2, ..., g_s \in F[x_1, ..., x_n]$, F a field.
Output: $q_1, q_2, ..., q_s, r \in F[x_1, ...x_n]$ satisfying $f = q_1g_1 + q_2g_2 + ... + q_sg_s + r$.
1: Set $(q_1, q_2, ..., q_s) := (0, 0, ..., 0)$.
2: Set $p := f$.
3: While $p \neq 0$ do
    4: Find the first g_i s.t. $LM(g_i)|LM(p)$.
    5: If no such g_i exists then set $r := r + LT(p)$ and $p := p - LT(p)$
    6: else set $(q_i, p) := (q_i + t, p - t \times g_i)$ where $t = LT(p)/LT(g_i)$.
7: Output $(q_1, q_2, ..., q_s, r)$.

Remark: If one wishes to test if $(g_1, ..., g_s)|f$ with 0 remainder then Step 5 should be modified to stop execution and output false.

If polynomials are represented as linked lists of terms sorted in descending order in the term ordering then accessing the leading term $LT(f)$ takes constant time, the operation $p - LT(p)$ (link to the remaining terms of p) is constant time and $r + LT(p)$ can be done in constant time by maintaining a pointer to the last term of r. The most expensive step is the subtraction $p - t \times g_i$. This requires a "merge" – one simultaneously walks down the linked list of terms in p and the linked list of terms in g_i comparing monomials. In the worst case the merge must walk to the end of both p and g_i.

1.1 Storage Management and Non-local Memory References

We identify two sources of inefficiency in the division algorithm when the SDMP data structure is used. The first is the many intermediate pieces of storage that need to be allocated when we multiply $t\,g_i$, for example, storage for new exponent vectors in $t\,g_i$. The second is the memory references that occur during the merge when we walk down the linked lists and, for each term, link to the exponent vectors to compare monomials. These memory references cause a loss in efficiency when the polynomials are too large to fit inside the computer's cache. On a 2.4 GHz AMD Opteron 150 with 400 MHz RAM we measured the loss of speed at a factor of 6.

These two problems can be eliminated by representing polynomials as arrays with the coefficients and exponents stored in place. For example, $6x^2y^3z - 7x^3yz + 8y^2z^3 - 9x^4$ could be stored as

$$\langle\, 6, 2, 3, 1, -7, 3, 1, 1, 8, 0, 2, 3, -9, 4, 0, 0 \,\rangle.$$

The difference $p - tg_i$ can be computed efficiently by merging with two arrays: one, p, that we are copying terms out of, and another, p', that we are forming the difference $p - t \times g_i$ inside. When the merge is complete we interchange the roles of p and p' for the next iteration of the division algorithm. If p' is too small to store all of the terms of p and $-t \times g_i$ we allocate a new p' with 50% more terms than are needed to reduce the chance of another allocation in the future.

But, there is a loss of efficiency; instead of copying pointers (one word) we must now copy exponent vectors (n words). This loss can be reduced by packing multiple exponents into each word. For example, Macaulay [5] uses dynamic arrays and packed exponent vectors. Macaulay identifies the monomials $1, z, y, x, z^2, zy, y^2, zx, yx, x^2, \dots$ with non-negative integers $0, 1, 2, 3, \dots$ to encode each monomial as an integer. The polynomial $6x^2y^3z - 7x^3yz + 8y^2z^3 - 9x^4$ would be represented as an array of 8 words

$$\langle\, +6, 63, -7, 49, +8, 36, -9, 33 \,\rangle.$$

This encoding gives a very compact representation with fast monomial comparisons, but monomial multiplication and division are slow. In [1], Bachmann and Schönemann compare different monomial packings including the Macaulay encoding. They show that packing exponent vectors produces a modest speedup (a factor of 1.5 to 2) for Gröbner basis computations modulo a machine prime with the SDMP data structure. They also show that simpler packing schemes are more efficient overall than the Macaulay encoding.

1.2 The Problem of Too Many Monomial Comparisons

When using merging to subtract $p - tg_i$, a serious inefficiency may occur when $\#p$, the number of terms in p, is much larger than $\#g_i$, the number of terms in a divisor g_i. Consider $g = (x + 1)$, $q = y^n + \dots + y^2 + y$ and let $p = gq = xy^n + \dots + x + y^n + \dots + y$. If we compute p by adding $x\,q$ to q using merging, the merge does n comparisons which is efficient. In dividing f by g the first quotient is y^n and we subtract $y^n g = xy^n + y^n$ from $p = xy^n + ... + xy + y^n + ... + y$. The merge does n comparisons to find y^n in p. The full division does n such merges so the total number of comparisons is $O(n^2)$, much worse than multiplication.

One solution is to represent the polynomial p as a binary search tree. Then $\mathrm{LT}(p)$ can be computed with $O(\log \#p)$ monomial comparisons and the difference $p - tg_i$ can be computed with $O(\#g_i \log \#p)$ comparisons. However binary trees suffer from the same cache performance problems as linked lists.

A very nice solution is the "geobucket" data structure of Yan [12], which is used by the Singular [6] computer algebra system and others. Geobuckets are described in detail in Section 2. In the geobucket data structure a polynomial p with $\#p$ terms is represented by an array of $O(\log \#p)$ "buckets" where the i'th bucket p_i is a linked list of at most 2^i terms. To subtract $t \times g_i$ from p one subtracts $t \times g_i$ from the i'th bucket of p where $2^{i-1} < \#g_i \leq 2^i$. Subtraction is done by merging two linked lists. The idea is that asymptotic efficiency is not lost when we merge linked lists with a similar number of terms, e.g., their length differs by at most a factor of two.

In this paper we use an auxiliary "heap of pointers" instead. When dividing p by $\{g_1, g_2, ..., g_s\}$ we maintain a heap of pairs with quotient terms and pointers back into the divisors $\{g_1, g_2, ..., g_s\}$. The pointers indicate which terms have yet to be multiplied and subtracted from p.

Suppose we are dividing f by g. Let $f = gq + r$ where q is the quotient and r the remainder. With geobuckets, division does $O(\#g \, \#q(\log \#g + \log \#q))$ comparisons [12]. If we use a heap, division does $O(\#g \, \#q \log \#q)$ comparisons. A second key advantage of using a heap is that it requires only $O(\#q)$ space, and, if we need to compute the remainder, $O(\#r)$ space to write down the remainder. By comparison, the simple merge and geobucket algorithms may require $O(\#g \, \#q + \#r)$ space. The main disadvantage of using a heap is that for dense polynomials the merge and geobucket algorithms are better; they do only $O(\#g \, \#q)$ comparisons. A third advantage of using a heap is that we delay all coefficient arithmetic until we need to do it. This can result in significant speedups when we want to test if g divides f but g does not divide f.

The idea of using a heap for sparse polynomial arithmetic was first investigated by Johnson in 1974 [8]. Heaps were used in Altran [2], one of the earliest computer algebra systems. We are not aware of any other computer algebra system that has used heaps for polynomial arithmetic despite their good asymptotic performance. Heaps were not considered by Stoutemyer in [11] which, as far as we are aware, is the only systematic experiment ever done comparing different polynomial data structures on a computer algebra system's test suite.

1.3   Organization of the Paper

In Section 2 we describe how we encode and pack monomials for different term orderings. Our packing for graded reverse lexicographical order is new. In Section 3 we give the main algorithms that use heaps of pointers. Two algorithms are presented. The first algorithm bounds the size of the heap by the number of terms in the quotients $\{q_1, q_2, ..., q_s\}$. In the second algorithm, the size of the heap is bounded by the number of terms in the divisors $\{g_1, g_2, ..., g_s\}$. This algorithm is new, and it is particularly useful for polynomial GCD computations because the gcd G of two polynomials A and B typically has fewer terms, often much fewer, than the quotients A/G and B/G.

We have implemented the division algorithms in the C programming language. We create polynomials in Maple and call our C code from Maple using Maple's foreign function interface (see Ch. 8 of [10]). For comparison we have also implemented Yan's geobucket data structure using dynamic arrays with packed exponent vectors. Details of our geobucket implementation are given in Section 2. In Section 4 we give some benchmarks comparing the simple merging algorithm with Yan's geobucket representation and our heap algorithms, using packed and unpacked exponent vectors.

Our conclusions may be summarized as follows. Simple merging is not competitive with either heaps or geobuckets on sparse problems. The heap algorithms are as fast as geobuckets but use far less memory. Geobuckets do the fewest monomial comparisons, but heaps tend to be faster on large problems because they use cache more efficiently. For all algorithms, packing exponents significantly improves performance, especially on 64-bit machines.

2   Dynamic Array Implementation

Consider the minimum amount of work that a sparse algorithm must do. As noted by Johnson [8], a multiplication fg must construct all $\#f\#g$ products of terms because the monomials generated may be distinct. These terms are merged to form the result. Similarly, to divide f by g we construct the quotient q incrementally while subtracting qg from f, merging $\#f + \#q(\#g - 1)$ terms to do the division. Note, it is $\#g - 1$ and not $\#g$ because $-q \times LT(g)$ cancels terms so only $-q \times (g - LT(g))$ needs to be merged. Let $r = f - qg$. The number of monomial divisions attempted is $\#q + \#r$. To divide f by $\{g_1, ..., g_s\}$ with quotients $\{q_1, ..., q_s\}$ we merge $\#f + \sum_{i=1}^{s} \#q_i(\#g_i - 1)$ terms and attempt $\sum_{i=1}^{s}(\#q_i)i + (\#r)s$ monomial divisions if for each term we loop through the divisors in order.

Sorting the result imposes an additional cost in monomial comparisons if a function is called to compare terms with respect to an ordering. The nm terms of a product can be naively sorted using $O(nm \log(nm))$ comparisons, but if the polynomials are sorted we can exploit that fact to do only $O(nm \log(\min(n, m)))$ comparisons. In either case the logarithmic factor is significant – it means that monomial comparisons dominate sparse polynomial computations when the cost of coefficient arithmetic is low.

2.1   Packed Monomial Representations

After an initial experiment we decided to base our monomial representations on Bachmann and Schönemann's scheme [1], which is used in Singular. The

defining feature of this scheme is that a monomial stores two components: a
(possibly weighted) total degree and a vector of exponents. An inline function
compares the degree and the exponent vector in lexicographic order, and two
global variables invert the results of these comparisons separately. To compare
in reverse lexicographic order we reverse the variables and invert all the comparisons. Figure 1 shows the unpacked representations of $x^2y^3z^4$ with respect
to four common orders with $x > y > z$. Shading is used to indicate where the
results of comparisons are inverted.

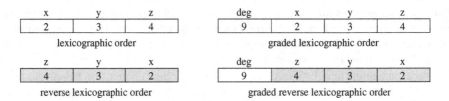

Fig. 1. Unpacked $x^2y^3z^4$ with $x > y > z$

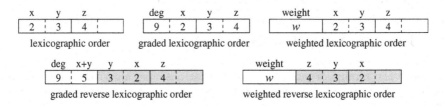

Fig. 2. Packed $x^2y^3z^4$ with $x > y > z$

To pack monomials we use bitwise or and shift operations on machine words so
that byte order is automatically taken into account. Our diagrams use big-endian
format. We reserve the most significant bit of each exponent as a guard bit for
monomial division. This operation subtracts machine words and uses a bit-mask
to detect if an exponent is negative. The mask also stores the length of the
monomials which is needed by every routine. Weighted orders use the entire first
word for the weighted degree since this can be large. We restrict the weights to
non-negative integers so that the weighted degree is also a non-negative integer.

For graded orders we use the same number of bits for the total degree as for
each exponent so that all monomials up to the maximum degree are encoded
efficiently. Note that it is especially easy to determine an optimal packing for
these orders using bounds on the total degree. If the polynomials are already
sorted with respect to the order then we can examine their leading terms and
repack the polynomials in linear time.

Figure 2 shows the packed representations of $x^2y^3z^4$ for five monomial orders with two exponents per machine word. Notice how monomial comparisons
are reduced to lexicographic and reverse lexicographic comparisons of machine

words. The encodings should all be straightforward except for graded reverse lexicographic order. In that case recall that the total degree only requires as many bits as a single packed exponent. The first word of the monomial, which must be compared lexicographically unlike the rest, would contain relatively little information if it only stored the total degree.

Our first idea was to pack more information into the first word to decide monomial comparisons. Observe that the matrices A and B in Figure 3 both describe graded reverse lexicographic order in four variables. Let V be an exponent vector. Then AV is encoded in the first $|V|$ words of the unpacked representation. The matrix B is obtained from A by adding the previous rows of A to each row of A, eliminating all negative entries. Thus BV contains only non-negative integers that are compared lexicographically. We pack as much of BV as possible into the first word of the monomial.

$$A = \begin{bmatrix} 1 & 1 & 1 & 1 \\ 0 & 0 & 0 & -1 \\ 0 & 0 & -1 & 0 \\ 0 & -1 & 0 & 0 \end{bmatrix} \qquad B = \begin{bmatrix} 1 & 1 & 1 & 1 \\ 1 & 1 & 1 & 0 \\ 1 & 1 & 0 & 0 \\ 1 & 0 & 0 & 0 \end{bmatrix}$$

Fig. 3. Matrix representations of graded reverse lexicographic (grevlex) order

deg	x..v	x..u	x..t	w	v	u	t	z	y	x
28	21	15	10	7	6	5	4	3	2	1

deg	x..v	x..u	x..t	t	z	y	x	w	v	u
28	21	15	10	4	3	2	1	7	6	5

Fig. 4. Packed representations of $x^1 y^2 z^3 t^4 u^5 v^6 w^7$ in grevlex order with 4 exponents per word. *The exponents for w, v, u, and x are redundant. In the second representation, all monomial comparisons can be decided on the basis of the first seven exponents, after looking at only two words.*

However, this does not actually fix the problem since now the second word of the monomial contains information that can be derived from the first. Refer to the top of Figure 4, where $w = 7$, $v = 6$ and $u = 5$ are known from $28 - 21$, $21 - 15$, and $15 - 10$. Thus the second word now provides only one exponent with new information, but we can easily fix this by moving all but the last exponent of the second word to the end of the monomial, as in the bottom of Figure 4. Then for n variables the first n exponents encode all of the information necessary to decide monomial comparisons in grevlex order.

One might wonder why we do not simply encode the vector BV. The reason is that for monomial division one must unpack and decode quotients to check that they are valid. An example is given below. In fact, we tried this representation initially and found that while it was quite compact for grevlex order, weighted orders were inefficient and reverse lexicographic order could not be implemented.

Eventually we decided to store all of the exponents explicitly, and Bachmann and Schönemann's scheme was the obvious choice.

Example 1. Consider x^2 and y^3 in graded reverse lexicographic order with $x > y > z$. The exponent vectors are $U = [3,0,0]$ and $V = [0,2,0]$ respectively, and the matrix B is shown below. The difference $BU - BV$ is non-negative even though $U - V = [3,-2,0]$.

$$B = \begin{bmatrix} 1 & 1 & 1 \\ 1 & 1 & 0 \\ 1 & 0 & 0 \end{bmatrix} \qquad BU = \begin{bmatrix} 3 \\ 3 \\ 3 \end{bmatrix} \qquad BV = \begin{bmatrix} 2 \\ 2 \\ 0 \end{bmatrix} \qquad BU - BV = \begin{bmatrix} 1 \\ 1 \\ 3 \end{bmatrix}$$

2.2 Iterated Merging with Dynamic Arrays

The classical approach to polynomial arithmetic is an iterated merge. To multiply f by g we compute $\sum_{i=1}^{\#f} f_i g$ by adding each $f_i g$ to the previous partial sum using a merge. Similarly, to divide f by g we compute terms of the quotient q while subtracting each $q_i g$ from an intermediate polynomial p, which is initially f.

Our first goal was to implement these algorithms while avoiding memory allocation. We use two global arrays or "merge buffers" p and p' which grow dynamically, and all merging takes place from p into p'. If p' does not have sufficient storage to hold the objects being merged then it is enlarged. To amortize this cost we allocate a new p' with 50% more storage than required. To further amortize the cost of memory allocation we reuse p and p' in the next call to an algorithm rather than free them each time.

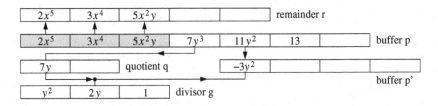

Fig. 5. Division using two dynamic arrays. *The fourth term of p produced the quotient term y, and we are beginning to merge the rest of p (two terms) with $-y$ times the rest of g (two terms). The buffer p' is large enough to store the result. Otherwise we would enlarge it to six terms.*

We describe our implementation of the division algorithm. To divide f by g, we copy f into p and increment along the terms of p until we reach the end or we find a term p_i that is divisible by $LT(g)$. We copy the previous terms of p to the remainder and if a reducible term was found, say $p_i = q_j LT(g)$, we merge the rest of p with $-q_j(g - LT(g))$ into p', as shown in Figure 5. The terms of $-q_j(g - LT(g))$ are constructed during the merge. Finally we interchange p and p' by swapping pointers so that p' becomes p for the next iteration of the algorithm and the storage for p is recycled.

The complexity of this approach was analyzed by Johnson [8]. He observed that for a multiplication fg where f has n terms, g has m terms, and fg has nm terms, adding each f_ig to the partial sum can require up to $im-1$ monomial comparisons, making the total number of comparisons $\sum_{i=2}^{n} im - n + 1 \in O(n^2m)$. A similar result holds for division when the quotient has n terms, the divisor has m terms, and the dividend has nm terms. Thus iterated merging can be very bad when the quotient is large.

It is interesting to note that $O(n^2m)$ comparisons may be required even if the product or dividend does not have $O(nm)$ terms, if terms introduced by the first $n/2$ summands are canceled by the last $n/2$ summands. We call this an *intermediate blowup in the number of terms*. One unfortunate feature of algorithms that add each f_ig or q_ig to a partial sum is that they allocate storage for all of these terms even when the end result is zero, as it will be for exact division. In Section 3 we will see that the heap algorithms avoid this problem by merging all of the partial products simultaneously.

For sparse polynomials an iterated merge uses about $2nm$ terms of storage where nm is the size of the largest intermediate sum. If we always enlarge the buffers by 50% then we will use storage for about $3nm$ terms on average. Quotient(s) and the remainder require additional storage if they are needed.

2.3   Divide and Conquer Merging – Geobuckets

A well-known alternative to iterated merging is divide-and-conquer merging, which is often used for polynomial multiplication. Let f have n terms and let g have m terms. If we compute $\sum_{i=1}^{n} f_ig$ by summing the first $n/2$ and the last $n/2$ summands recursively and adding their sums, then at most $C(n) \leq 2C(n/2) + nm - 1 \in O(nm\log n)$ monomial comparisons are required. The method is efficient because it merges polynomials of similar size.

But how much memory is required? If each recursive call allocates memory for its own result then we can solve the same recurrence to find that $O(nm\log n)$ memory is needed. This is an order of magnitude larger than any possible result. Instead we could reuse a set of geometrically increasing buckets with $\{2m, 4m, \ldots, nm/2\}$ terms for polynomials that we are waiting to merge, plus two arrays with nm and $nm/2$ terms for polynomials that we are currently merging. This simple "geobucket" algorithm is described below.

Geobucket Multiplication
Input: $f = f_1 + \cdots + f_n$, $g = g_1 + \cdots + g_m$.
Output: fg.
1: Allocate buckets with $\{2m, 4m, \ldots, 2^{\lceil \log_2(n)\rceil - 1}m\}$ terms.
2: Allocate dynamic arrays p and p'.
3: For $i := 1$ while $i \leq n$ do
    4: Compute f_ig and store it in p.
    5: If $i < n$ merge p and $f_{i+1}g$ into p' and swap p and p'.
    6: Set $i := i + 2$.
7: For $j := 1$ while $bucket[j] \neq 0$ do

8: Merge p and $bucket[j]$ into p' and swap p and p'.
9: Set $bucket[j] := 0$ and $j := j + 1$.
10: If $i \leq n$ set $bucket[j] := p$ and $p := 0$.
11: For $j := 1$ to $2^{\lceil \log_2(n) \rceil - 1}$ do
12: If $bucket[j] \neq 0$ merge p and $bucket[j]$ into p' and swap p and p'.
13: Output p.

Thus $f_1 g$ and $f_2 g$ are merged and their sum is stored in bucket 1, then $f_3 g$ and $f_4 g$ are merged and their sum is merged with $f_1 g + f_2 g$ and stored in bucket 2, then $f_5 g$ and $f_6 g$ are merged and their sum is stored in bucket 1, and so on, continuing in the manner of a depth-first search. If $n = 2^k$ it is easy to see that $O(nm)$ storage is used. The buckets contain $(n-2)m$ terms, the array that stores the result will need nm terms, but the other array can have $nm/2$ terms. The total amount of storage required is $2.5nm$ terms – only 50% more than for an iterated merge. If we always grow the arrays by an extra 50% then we can expect to allocate storage for about $3.25nm$ terms in total.

Geobuckets were proposed by Yan [12] with three significant improvements. First, Yan's buckets have a small base and ratio that are independent of any problem to ensure good performance when objects of varying sizes are added to the geobucket. In the algorithm above the base is $2m$ and the ratio is 2, so objects with fewer than m terms could be added more efficiently with a smaller bucket. Second, Yan always tries to store $p + bucket[j]$ in $bucket[j]$ if possible to avoid creating $bucket[j+1]$. This decreases the amount of memory and increases the likelihood of combining terms on dense problems, resulting in fewer monomial comparisons. Finally, Yan describes a reasonably efficient scheme for coalescing the leading terms of the buckets to compute the leading term of the polynomial. This allows us to run the division algorithm with the intermediate polynomial p stored as a geobucket. We state Yan's algorithm below for completeness.

Geobucket Leading Term
Input: polynomial f stored in $bucket[1 \ldots k]$.
Output: $LT(f)$ or FAIL when $f = 0$, set $bucket[1 \ldots k] := f - LT(f)$.

1: Set $j := 0$, the bucket containing the leading term.
2: For $i := 1$ while $i \leq k$ do
   3: If $bucket[i] \neq 0$ and ($j = 0$ or $LM(bucket[i]) > LM(bucket[j])$)
      4: Set $j := i$
   5: else if $bucket[i] \neq 0$ and $LM(bucket[i]) = LM(bucket[j])$
      6: Set $LC(bucket[j]) := LC(bucket[j]) + LC(bucket[i])$.
      7: Remove $LT(bucket[i])$ from $bucket[i]$.
   8: Set $i := i + 1$.
9: If $j = 0$ then $f = 0$ so output FAIL.
10: If $LC(bucket[j]) = 0$ remove this term from $bucket[j]$ and goto step 1.
11: Set $t := LT(bucket[j])$.
12: Remove $LT(bucket[j])$ from $bucket[j]$.
13: Output t.

We implemented Yan's geobuckets using a single dynamic array so that its storage could be reused in subsequent calls. We chose a ratio of two because that is optimal for merging and our smallest bucket (the base) has four terms. We found that geobuckets performed very well, often using fewer monomial comparisons than expected.

For a sparse multiplication producing nm terms geobuckets do $O(nm \log n)$ comparisons and store about $3.6nm$ terms. This number can be derived as follows. The arrays (merge buffers) require nm and $nm/2$ terms, but we will allocate an extra 50% for each. The buckets have nm terms, but the base (two) is independent of m so we expect each bucket to be 75% full. The total is $4nm/3 + (3/2)(nm + nm/2) = (43/12)nm$ terms.

We can make a similar estimate for exact division when the dividend has nm terms, however the complexity is $O(nm \log(nm))$ because of how leading terms are computed. The dividend is placed into the largest bucket, which we expect to be 75% full, so the storage for buckets is $2(4nm/3) = 8nm/3$. Nothing is merged with the largest bucket since $\sum_{i=1}^{\#q} q_i g$ fits entirely in the smaller buckets, so the largest merge that we expect to do is to construct $\sum_{i=1}^{\#q/2} q_i g$ which has $nm/2$ terms. This requires arrays with $nm/2$ and $nm/4$ terms, plus the extra 50% that we allocate, bringing the total number of terms to $8nm/3 + (3/2)(3nm/4) = (91/24)nm$.

The actual amount of memory that geobuckets need for exact division tends to vary. It can be lower if the leading term computations frequently cancel terms in the buckets, reducing the size of the polynomials that are merged. For random sparse divisions we found that approximately $3.6nm$ terms were used – about the same as for multiplication. The dynamic arrays were often the same size, about $3nm/5$ terms each.

3   Heap Algorithms for Polynomial Arithmetic

The heap algorithms are based on the following idea: rather than merge polynomials one by one into an intermediate object, we do a simultaneous n-ary merge using a heap. Consider the multiplication fg where we merge $f_i g$ for $1 \leq i \leq \#f$. If we maintain a heap of $\#f$ pointers into g, sorted by the monomial of $f_i g_j$, we can repeatedly extract the largest $f_i g_j$ from the heap, merge it onto the end of the result, and insert its successor $f_i g_{j+1}$ into the heap if $j < \#g$. We illustrate this process in Figure 6 below.

The monomial of $f_i g_j$ is computed and stored in the heap when the term is inserted. It is used to determine the maximum element of the heap. This storage is reused for $f_i g_{j+1}$ so only $O(\#f)$ storage is required, in addition to storage for the result.

To divide f by g we merge the dividend f with $-q_i g$ for each term q_i of the quotient. The heap maintains a pointer into f and we add a pointer into $-q_i g$ when q_i is constructed. The algorithm extracts the largest term from the heap and continues to extract terms with an equal monomial, adding their coefficients to produce the next term of $f - \sum_{j=1}^{i} q_j g$. If this term is not zero we divide it by

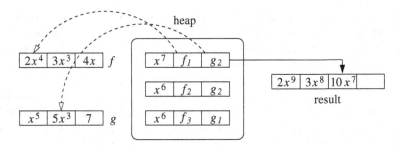

Fig. 6. Multiplication of $f = 2x^4 + 3x^3 + 4x$ and $g = x^5 + 5x^3 + 7$ using a heap. *The products $f_1 g_1$ and $f_2 g_1$ have been extracted and replaced by $f_1 g_2$ and $f_2 g_2$. We are now extracting $f_1 g_2 = 10x^7$ and writing it to the result. Its successor $f_1 g_3 = 14x^4$ will be inserted into the heap, and we will extract $f_2 g_2$ and $f_3 g_1$ to obtain $15x^6 + 4x^6 = 19x^6$, the fourth term of the result.*

$LT(g)$ to obtain either a new term of the quotient q_{i+1}, or the next term of the remainder. When a quotient term is found we insert the *second* term of $-q_{i+1}g$ into the heap, increasing the size of the heap by one, along with the successors of the other terms that were extracted. There is no intermediate blowup in the number of terms that are stored – the maximum number of terms in the heap is $\#q + 1$. We call this a "quotient heap" division.

The heap algorithms above were analyzed by Johnson [8] and used in Altran, one of the first computer algebra systems. For a binary heap of size n, inserting and extracting each term does $O(\log n)$ monomial comparisons. A multiplication that passes nm terms through a heap of size n does $O(nm \log n)$ comparisons – the same as divide-and-conquer. Exact division $f \div g$ with $\#f = nm$, $\#g = m$, and the quotient $\#q = n$, passes $2nm - n$ terms through a heap of size $n + 1$, which is also $O(nm \log n)$ comparisons.

One problem with the heap algorithms is that they do $O(nm \log n)$ comparisons even when the polynomials are dense, whereas the simple merge and the divide-and-conquer algorithms do only $O(nm)$ comparisons. In Section 3.2 we show how to modify the heap to make the heap algorithms efficient in the dense case as well.

Our main contribution is to modify the heap division algorithm to increment along the quotient(s) instead of the divisor(s). The resulting "divisor heap" algorithm does $O(nm \log m)$ comparisons and uses $O(m)$ storage, where m is the size of the divisor(s). Our incentive comes from the gcd problem, where we compute $G = \gcd(A, B)$ and divide A/G and B/G to recover the cofactors. The divisor G is typically small and the quotients (cofactors) are often big. The algorithm is also useful for computing over small towers of algebraic extensions, where the number of reductions usually exceeds the size of the extensions.

The modification is easy to do. The algorithm merges f with $-g_i q$ for $2 \le i \le \#g$ using a heap of size $\#g$, however we may merge $g_i q_{j-1}$ before q_j is computed, in which case we can not insert the next term $g_i q_j$ into the heap because we can not compute its monomial. However, since $LT(g)q_j > g_i q_j$ for all $i > 1$, we can safely wait for q_j to be computed to insert the terms $g_i q_j$ with $i > 1$ into the heap. We exploit the fact that the term $g_i q_j$ is greater than the term $g_{i+k} q_j$ for $k > 0$, so if $\#q = j - 1$ we encounter the strictly descending sequence $g_2 q_j > g_3 q_j > g_4 q_j > \dots$ in order. For each divisor g we store an index s of the largest $g_s q_j$ that is missing from the heap because q_j is unknown. When a new term of the quotient is computed ($\#q = j$) we compute all of the missing terms $\{g_2 q_j, \dots g_s q_j\}$ and insert them into the heap. Here we give the algorithm for one divisor.

Divisor Heap Division

Input: $f, g \in F[x_1, \dots, x_n], F$ a field, $g \neq 0$.
Output: $q, r \in F[x_1, \dots x_n]$ with $f = qg + r$.
1: If $f = 0$ then output $(0, f)$.
2: Initialize $(q, r, s) := (0, 0, \#g)$.
3: Create an empty heap H of size $\#g$ and insert $(-1)f_1$ into H.
4: While the heap H is not empty do
   6: Set $t := 0$.
   7: Repeat
      8: Extract $x := H_{max}$ from the heap and set $t := t - x$.
      9: Case $x = (-1)f_i$ and $i < \#f$: Insert $(-1)f_{i+1}$ into H.
      10: Case $x = g_i q_j$ and $j < \#q$: Insert $g_i q_{j+1}$ into H.
      11: Case $x = g_i q_j$ and $j = \#q$: Set $s := s + 1$ $(s = i)$.
   12: Until H is empty or $LM(t) \neq LM(H_{max})$.
   13: If $t \neq 0$ and $LT(g) | t$ then
      14: Copy $t/LT(g)$ onto the end of q.
      15: For $i = 2, 3, \dots, s$ compute $g_i \times (t/LT(g))$ and insert it into H.
      16: Set $s := 1$.
   17: Else if $t \neq 0$ copy t onto the end of r.
18: Output (q, r).

Theorem 1. *The divisor heap algorithm divides f by g producing the quotient q and remainder r using $O((\#f + \#q \#g) \log \#g)$ monomial comparisons and using storage for $O(\#g + \#q + \#r)$ terms.*

Proof. We show that at Step 4, $|H| + s - 1 = \#g$ if some $(-1)f_i \in H$ or $|H| + s = \#g$ otherwise. The first time Step 4 is executed, $|H| = 1$, $s = \#g$, and $(-1)f_1$ is in the heap, so the loop invariant holds. Steps 7-11 extract a term from H and either replace it or increment s, unless it was the last term of f. Step 15 inserts $s - 1$ terms into H and sets $s := 1$, maintaining the invariant.

Then $|H| \leq \#g$ since $s \geq 1$. Therefore the storage required is at most $\#g$ terms in the heap plus the terms of q and r. It should be clear that the algorithm adds terms of f, subtracts terms of each $g_i q$, and uses $LT(g)$ to cancel terms if possible, otherwise moving them to r, so that $f = qg + r$. Since we pass $\#f + \#q(\#g - 1)$ terms through a heap of size $|H| \leq \#g$, the number of monomial comparisons is $O((\#f + \#q \#g) \log \#g)$.

3.1   Heap Optimizations

We present two optimizations that are necessary to reproduce our results. The first is to implement the heap carefully. Many people are only aware of a bad algorithm for extracting the largest element from a heap, so we present a classical algorithm that is roughly twice as fast on average. As LaMarca and Ladner [9] observe, about 90% of the time is spent extracting elements from the heap so the resulting speedup is almost a factor of two.

We store the heap in a global dynamic array H, with the convention that $H[0]$ is the largest element and the children of $H[i]$ are $H[2i+1]$ and $H[2i+2]$.

```
inline heap_elem heap_extract_max(heap_elem *H, int *n)
{   int i, j, s = --(*n);
    heap_elem x = H[0];
    /* H[0] now empty - promote largest child */
    for (i=0, j=1; j < s; i=j, j=2*j+1) {
        j = (H[j] > H[j+1]) ? j : j+1;
        H[i] = H[j];
    }
    /* H[i] now empty - insert last element into H[i] */
    for (j=(i-1)/2; i>0 && H[s]>H[j]; H[i]=H[j], i=j, j=(j-1)/2);
    H[i] = H[s];
    return x;
}
```

The extraction algorithm promotes the largest child into the empty space at a cost of one comparison per level of the heap H. Then it inserts the last element of the heap into the empty slot on the lowest level. However, since the last element was already a leaf, we do not expect it to travel very far up the heap. The number of comparisons required is $\log_2(n) + O(1)$ on average.

Compare this with the more commonly known algorithm for shrinking a heap, which moves the last element to the top and, at a cost of two comparisons per level (to find the maximum child and compare with it), sifts it down the heap. Since the last element was already a leaf it is likely to go all the way back down to the bottom, requiring $2\log_2(n)$ comparisons on average.

Our second optimization improves performance when multiple terms are extracted from the heap. It is also necessary to obtain $O(nm)$ comparisons in the totally dense case. We insert and extract batches of terms instead of extracting a term and immediately inserting its successor. This requires a queue to store the extracted terms, however we can partition the heap to store this queue in place, as in heapsort. At the end of each iteration, we insert the successors of all of the extracted terms at once. As LaMarca notes [9], this strategy also produces favorable caching effects.

3.2   Chaining Terms with Equal Monomials

Our next improvement chains heap elements with equal monomials to reduce the number of comparisons. Johnson [8] also experimented with this idea, however our scheme is simpler and we will show that multiplication and division of dense polynomials does $O(nm)$ comparisons.

We chain elements only as they are inserted into the heap, using an additional pointer in the structure that points to f_i and g_j. In our implementation the pointers to f_i and g_j are not stored in the heap, but in a secondary structure that is accessed only when terms are inserted or extracted. Heap elements store a pointer to this structure and a pointer to the monomial product used for comparisons. The overhead of chaining elements in this way is negligible. The algorithms must be modified to check for chains and to extract all the elements of a chain without doing any monomial comparisons.

One final optimization is needed for multiplication. When multiplying fg, we must start with $f_1 g_1$ in the heap and insert each $f_i g_1$ only after $f_{i-1} g_1$ has been extracted from the heap. This leads to the following results.

Lemma 1. *Let f and g be dense univariate polynomials with n and m terms, respectively. A heap multiplication fg with chaining does $nm - n - m + 1$ comparisons.*

Proof. We prove a loop invariant: at the beginning of each iteration the heap contains exactly one element or chain. This is true initially since the only element is $f_1 g_1$. Each iteration removes the chain without doing a comparison, producing an empty heap. When we insert the successor terms into the heap all of the monomials are equal because the problem is dense, so all of the terms are chained together at the top of the heap. There are nm terms and $n + m - 1$ unique monomials. The first term with each monomial is inserted for free while the rest use one comparison each to chain. The total number of comparisons is thus $nm - n - m + 1$.

Lemma 2. *Let q and g be dense univariate polynomials with n and m terms and let $f = qg$. Then a quotient heap division $f \div g$ with chaining does $nm - n$ comparisons.*

Proof. We use the same loop invariant: the heap contains exactly one element or chain, which is initially f_1. Each iteration extracts the terms of this chain, adding their coefficients without a comparison, producing an empty heap. If the term is not zero, a new term of the quotient q_i is computed and the monomial of $q_i g_1$ equal to the monomial of the extracted terms. When we insert its successor $q_i g_2$ and the successors of all the other terms their monomials are all equal because the problem is dense, and all of the terms are chained together at the top of the heap. If each of the $n + m - 1$ monomials of f is inserted first without any comparisons, the remaining $n(m - 1)$ terms of $-q(g - LT(g))$ will be chained using one comparison each.

Remark: The divisor heap algorithm can also be modified to do nm comparisons in the dense univariate case. Each term q_j of the quotient should insert only $g_2 q_j$ if it is not in the heap, and each $g_{i+1} q_j$ should be inserted only after $g_i q_j$ is extracted from the heap. We have not yet implemented this modification.

4   Benchmarks

4.1   The Number of Monomial Comparisons

Our first benchmark (see Table 1 and Table 2) is due to Johnson [8]. We multiply and divide sparse univariate polynomials and report the number of comparisons divided by the total number of terms that are merged. Recall that for a sparse multiplication fg this is $(\#f)(\#g)$ and for a sparse division $f = qg$ this is $\#f + \#q(\#g - 1)$. A "structure parameter" S is used to randomly generate polynomials $f = a_0 + a_1 x^{e_1} + a_2 x^{e_2} + \cdots + a_k x^{e_k}$ with the difference between the exponents satisfying $1 \leq e_{i+1} - e_i \leq S$.

For each problem we generate f and g with n and m terms respectively, multiply $p = fg$ and divide p/g. For multiplication we test both chained and unchained heaps, and for division we test the "quotient heap" and the "divisor heap" algorithms.

Table 1. Multiplication fg and the number of comparisons divided by $(\#f)(\#g)$

S	$\#(fg)$	$\#f, \#g$	unchained heap	chained heap	geobuckets	direct merge
1	199	100	6.138	.980	1.114	1.475
	1999	1000	9.329	.998	1.027	1.497
10	1025	100	8.339	5.970	2.905	7.239
	10747	1000	11.717	8.478	3.065	8.025
100	5728	100	8.671	8.282	4.690	32.893
	97051	1000	11.879	11.334	5.798	69.191
1000	9364	100	8.805	8.748	5.274	48.073
	566984	1000	11.925	11.852	7.511	324.135

Table 2. Division $fg \div g$ and the number of comparisons divided by $\#(fg) + \#f(\#g - 1)$

S	$\#(fg)$	$\#f$	$\#g$	quotient heap	divisor heap	geobuckets	direct merge
1	199	100	100	.980	2.627	.980	.980
	1099	100	1000	.989	7.622	.989	.989
	1099	1000	100	.989	1.155	.989	.999
	1999	1000	1000	.998	4.170	.998	.998
10	1025	100	100	5.692	6.480	2.647	4.300
	5856	100	1000	6.493	8.244	2.738	4.872
	5949	1000	100	6.503	7.825	2.748	4.934
	11162	1000	1000	8.646	9.124	2.916	5.473
100	5725	100	100	7.106	7.580	3.945	14.502
	44725	100	1000	7.884	10.594	3.954	19.381
	45358	1000	100	7.696	7.938	4.405	18.231
	96443	1000	1000	10.898	11.438	5.471	42.262
1000	9403	100	100	7.116	7.522	3.992	17.307
	90884	100	1000	7.682	10.608	4.253	23.978
	91141	1000	100	7.658	7.747	4.596	22.736
	571388	1000	1000	10.563	11.056	6.574	142.095

We make a couple of remarks concerning tables 1 and 2. First it should be clear that our implementation of the divisor heap algorithm is not fully optimized. As discussed at the end of Section 3 we should delay inserting products $g_i q_j$ into the heap until after the previous product $g_{i-1}q_j$ is extracted from the heap. This is needed to obtain $O(nm)$ comparisons in the dense case ($S = 1$).

Second, it is interesting to see that geobuckets do roughly half the number of comparisons as the heap algorithms in the sparse case, and this ratio improves as the problems become more dense. We tried some improvements to the heap algorithms such as chaining elements while shrinking the heap, however these changes tended to decrease the real world performance of the algorithms.

4.2   7 Variable Cofactor Problem

Our next benchmark (see Table 3) simulates a GCD problem. A large sparse polynomial is divided by one of its factors (the GCD) to compute the cofactor. To generate this example we constructed four polynomials $\{f_1, f_2, f_3, f_4\}$ and divided their product $p = f_1f_2f_3f_4$ by f_1, f_1f_2, and $f_1f_2f_3$ over $\mathbb{Z}_{32003}$ using graded lexicographic order. The polynomials have $\#f_i = 50$ and $\deg(f_i) = 10$.

Table 3. Sparse multiplications and divisions in 7 variables over $\mathbb{Z}_{32003}$ using graded lex order with $\{1, 2, 4, 8\}$ exponents packed into each 64-bit word. $\#f_i = 50$, $\deg(f_i) = 10$, $\#(f_1f_2) = 2492$, $\#(f_3f_4) = 2491$, $\#(f_1f_2f_3) = 121903$, $\#(f_1f_2f_3f_4) = 4523085$.

$(f_1f_2) \times (f_3f_4)$

expon/wd	size of result	chained heap		geobuckets		direct merge	
1	310.57 MB	2.630 s	(0.38 MB)	7.720 s	(994 MB)	332.230 s	(371 MB)
2	172.54 MB	1.860 s	(0.31 MB)	4.230 s	(552 MB)	185.780 s	(206 MB)
4	103.52 MB	1.450 s	(0.27 MB)	2.550 s	(331 MB)	111.960 s	(124 MB)
8	69.01 MB	1.240 s	(0.25 MB)	1.760 s	(221 MB)	75.560 s	(83 MB)

$f_1 \times (f_2f_3f_4)$

expon/wd	size of result	chained heap		geobuckets		direct merge	
1	310.57 MB	1.700 s	(0.07 MB)	4.770 s	(1143 MB)	8.070 s	(483 MB)
2	172.54 MB	1.240 s	(0.06 MB)	2.660 s	(635 MB)	4.500 s	(216 MB)
4	103.52 MB	0.980 s	(0.06 MB)	1.690 s	(381 MB)	2.800 s	(161 MB)
8	69.01 MB	0.880 s	(0.06 MB)	1.230 s	(254 MB)	1.910 s	(107 MB)

$(f_1f_2f_3f_4)/(f_1f_2f_3)$

x	quotient heap		divisor heap		geobuckets		direct merge	
1	2.000 s	(0.13 MB)	8.820 s	(18.6 MB)	5.190 s	(1793 MB)	7.530 s	(944 MB)
2	1.450 s	(0.13 MB)	6.570 s	(14.9 MB)	2.960 s	(996 MB)	4.250 s	(524 MB)
4	1.250 s	(0.10 MB)	5.270 s	(13.0 MB)	1.950 s	(598 MB)	2.610 s	(315 MB)
8	1.060 s	(0.10 MB)	4.530 s	(12.1 MB)	1.500 s	(398 MB)	1.770 s	(210 MB)

$(f_1f_2f_3f_4)/(f_1f_2)$

x	quotient heap		divisor heap		geobuckets		direct merge	
1	3.270 s	(0.72 MB)	3.380 s	(0.30 MB)	8.020 s	(1461 MB)	330.730 s	(932 MB)
2	2.290 s	(0.65 MB)	2.430 s	(0.31 MB)	4.460 s	(812 MB)	183.060 s	(518 MB)
4	1.840 s	(0.62 MB)	1.930 s	(0.27 MB)	2.760 s	(487 MB)	110.290 s	(311 MB)
8	1.520 s	(0.60 MB)	1.620 s	(0.25 MB)	2.040 s	(321 MB)	74.540 s	(207 MB)

$(f_1f_2f_3f_4)/f_1$

x	quotient heap		divisor heap		geobuckets		direct merge	
1	8.010 s	(28.46 MB)	1.990 s	(0.07 MB)	8.320 s	(1371 MB)	–	
2	5.900 s	(25.69 MB)	1.480 s	(0.06 MB)	4.640 s	(762 MB)	–	
4	4.750 s	(24.29 MB)	1.240 s	(0.06 MB)	2.890 s	(457 MB)	–	
8	3.970 s	(23.60 MB)	1.080 s	(0.06 MB)	2.210 s	(305 MB)	3526.750 s	(207 MB)

The computations were performed on an AMD Opteron 254 2.8 GHz with 8GB of 400MHz RAM and 1 MB of L2 cache running 64-bit Red Hat Enterprise Linux 5 with a 2.6.18 kernel.

We report times and memory for $\{1, 2, 4, 8\}$ exponents per 64-bit word. For multiplications we subtracted the size of the product from the memory totals for the geobucket and merge algorithms, and for divisions we did not include memory for the quotient. For heap algorithms we report the size of the heap and products. Thus we report the memory overhead of the algorithms, not the total memory used. For divisions the largest quotient $(f_2 f_3 f_4)$ is at most 8.3 MB.

The heap algorithms performed very well on this example despite their higher cost in monomial comparisons. We attribute this to the fact that their working memory (the heap of pointers and the monomial products) fits in the L2 cache, whereas geobuckets and direct merging work mostly in RAM, which is 7 times slower than the processor.

Also note the effect of packing exponents. The performance of merging and geobuckets is practically linear in the size of the terms, which is 9, 5, 3, or 2 words with the coefficient. The heap algorithms do not benefit as much, but the improvement is worthwhile. Going from 64-bit (1 exponent per word) to 16-bit (4 exponents per word) exponents places only modest restrictions on the total degree and improves performance by 40%.

4.3   The Effect of Faster RAM and a Larger L2 Cache

In the previous benchmark the performance of geobuckets was constrained by the speed of the RAM and the size of the L2 cache. We thought that geobuckets should outperform the heap algorithms under different conditions, because they typically do fewer monomial comparisons.

Table 4. Sparse multiplications and divisions in 4 variables over $\mathbb{Z}_{32003}$. Lexicographic order was used with 32-bit words. Each f_i has degree 30 in each variable. $\#f_1 = 96$, $\#f_2 = 93$, $\#f_3 = 93$, $\#(f_1 f_2) = 8922$, $\#(f_2 f_3) = 8639$, $\#(f_1 f_2 f_3) = 795357$.

$f_1 \times (f_2 f_3)$

expon/word	size of result	chained heap		geobuckets		direct merge	
1	15.17 MB	0.200 s	(0.03 MB)	0.210 s	(55.74 MB)	0.650 s	(23.21 MB)
2	9.10 MB	0.150 s	(0.03 MB)	0.140 s	(33.44 MB)	0.470 s	(13.92 MB)
4	6.07 MB	0.120 s	(0.03 MB)	0.110 s	(22.30 MB)	0.360 s	(9.28 MB)

$(f_1 f_2 f_3)/(f_1 f_2)$

x/w	quotient heap		divisor heap		geobuckets		direct merge	
1	0.260 s	(0.06 MB)	0.460 s	(0.55 MB)	0.280 s	(70.91 MB)	0.600 s	(38.38 MB)
2	0.210 s	(0.05 MB)	0.370 s	(0.48 MB)	0.220 s	(37.38 MB)	0.440 s	(27.46 MB)
4	0.170 s	(0.05 MB)	0.300 s	(0.45 MB)	0.180 s	(22.36 MB)	0.350 s	(18.30 MB)

$(f_1 f_2 f_3)/f_1$

x/w	quotient heap		divisor heap		geobuckets		direct merge	
1	0.430 s	(0.53 MB)	0.280 s	(0.03 MB)	0.390 s	(55.90 MB)	44.000 s	(45.52 MB)
2	0.350 s	(0.47 MB)	0.230 s	(0.03 MB)	0.300 s	(33.54 MB)	28.790 s	(27.30 MB)
4	0.280 s	(0.43 MB)	0.190 s	(0.03 MB)	0.260 s	(22.36 MB)	22.150 s	(18.20 MB)

Our third benchmark (see Table 4) is a smaller problem similar to the previous one. We created three random polynomials $\{f_1, f_2, f_3\}$ and divided their product by f_1 and $f_2 f_3$. This test was run on a 2.4 GHz Intel E6600 Core 2 Duo with 2 GB of 666 MHz RAM, 4 MB of L2 cache, and 32-bit words running Fedora Core 6. Thus RAM is now only 3.6 times slower than the CPU and the number of words in the L2 cache has increased by a factor of eight.

Table 4 shows that geobuckets are competitive with heap algorithms if they work in the L2 cache. The times include memory allocation, so in practice if the geobucket is reused it may be faster than a quotient heap on sparse problems, with an additional advantage on dense problems (see tables 1 and 2). However when the quotient is large, the divisor heap's lower complexity easily wins.

4.4 Algebraic Extensions

Our final benchmark (see Table 5) is a large division with algebraic extensions. We constructed four random polynomials $\{f_1, f_2, f_3, f_4\}$ in $\mathbb{Z}_{32003}[x, y, z, \alpha, \beta, s, t]$ with $\deg(f_i) = 10$ and $LT(f_i) = x^{10}$. We used lexicographic order with $x > y > z > \alpha > \beta > s > t$ with the extensions $\alpha^2 - 3 = 0$ and $\beta^2 + st - 1 = 0$. Thus we are effectively computing with polynomials in $\{x, y, z\}$ with coefficients in $\mathbb{Z}_{32003}[\alpha, \beta, s, t]/\langle \alpha^2 - 3, \beta^2 + st - 1 \rangle$.

We report the times to multiply $(f_1 f_2) \times (f_3 f_4)$ and $f_4 \times (f_1 f_2 f_3)$ and reduce the product mod $\{\alpha^2 - 3, \beta^2 + st - 1\}$. Next we divide the reduced product by f_1, $(f_1 f_2)$, and $(f_1 f_2 f_3)$ mod $\{\alpha^2 - 3, \beta^2 + st - 1\}$ and reduce the quotients mod $\{\alpha^2 - 3, \beta^2 + st - 1\}$. The divisors in each case are already reduced mod $\{\alpha^2 - 3, \beta^2 + st - 1\}$.

We performed the test on a 3 GHz Intel Xeon 5160 with 16 GB of 666 MHz RAM and 4 MB of L2 cache running 64-bit Red Hat Enterprise Linux 5. Memory numbers are reported differently since the heap algorithms must store the quotients of $\{\alpha^2 - 3, \beta^2 + st - 1\}$ which are large, whereas geobuckets discards

Table 5. Sparse multiplications and divisions with algebraic extensions. Lexicographic order was used with 7 exponents per 64-bit word. We include the times, the number of monomial comparisons (upper right), and the total memory allocated. $\#f_1 = 106$, $\#f_2 = 96$, $\#f_3 = 105$, $\#f_4 = 98$, $\#(f_1 f_2) = 8934$, $\#(f_3 f_4) = 8982$, $\#(f_1 f_2 f_3) = 256685$, $\#(f_1 f_2 f_3 f_4) = 1663235$.

	quotient heap		divisor heap		geobuckets	
$p = (f_1 f_2)(f_3 f_4)$	11.080 s	9.713×10^8	11.100 s	9.267×10^8	8.510 s	4.218×10^8
reduce product	0.700 s	458.75 MB	0.300 s	166.73 MB	0.610 s	646.54 MB
$p = f_4(f_1 f_2 f_3)$	1.690 s	1.966×10^8	1.680 s	1.546×10^8	2.130 s	8.184×10^7
reduce product	0.670 s	446.07 MB	0.300 s	163.12 MB	0.560 s	642.30 MB
$p/(f_1 f_2 f_3)$	3.060 s	2.862×10^8	11.910 s	6.949×10^8	3.360 s	1.218×10^8
reduce quotient	0.000 s	208.02 MB	0.000 s	64.34 MB	0.000 s	479.98 MB
$p/(f_1 f_2)$	51.430 s	4.097×10^9	35.040 s	2.860×10^9	35.520 s	1.732×10^9
reduce quotient	0.010 s	733.72 MB	0.010 s	81.45 MB	0.010 s	1205.19 MB
p/f_1	49.790 s	2.005×10^9	5.980 s	4.616×10^8	13.140 s	9.100×10^8
reduce quotient	0.190 s	752.61 MB	0.080 s	113.25 MB	0.180 s	1038.96 MB

them. We report the total memory allocated by each routine, including reallocations to enlarge the geobucket and speculative allocations of quotients by the heap algorithms. We pack all seven exponents into one 64-bit word. The results with less packing are consistent with our previous benchmarks.

The divisor heap algorithm performs well on this example (and the quotient heap algorithm poorly) because $\{\alpha^2 - 3, \beta^2 + st - 1\}$ are small divisors with large quotients, i.e., they are frequently used to reduce terms during the division. The time and space requirements of the divisor heap algorithm scale linearly with the total number of reduction steps, so we expect it to be especially useful for divisions in the presence of algebraic extensions.

Geobuckets also perform well on this benchmark. Their overall memory usage is low because they do not need to store all of the quotients and the number of monomial comparisons they do is very competitive. However, performance is not dictated entirely by monomial comparisons. Consider the fourth benchmark $p/(f_1 f_2)$, where geobuckets do half the number of monomial comparisons as a divisor heap only to finish in the same amount of time.

The performance of geobuckets suffers because they access a large amount of memory randomly, and this decreases the effectiveness of the cache. Imagine what happens when $\beta^2 + st - 1$ is used to reduce one million terms in a row. Geobuckets will merge multiples of this polynomial into the smallest bucket 10^6 times, interspersed with 500,000 merges into the second bucket, 250,000 merges into the third, and so on. When a large bucket is merged the smaller buckets are evicted from the cache, producing cache misses the next time those buckets are accessed. If the problem is sufficiently large or the L2 cache is small, this will happen frequently.

By contrast, the divisor heap algorithm will do two simultaneous passes over the quotient of $\beta^2 + st - 1$ while randomly accessing a heap with three elements, two monomial products, and the terms of the divisor. This is a tiny amount of memory, so almost all of the cache is used to load terms from the quotient, and very few cache misses will occur.

5   Conclusions and Future Work

We have shown how a heap of pointers can be very efficient for sparse polynomial division and multiplication. This performance is primarily due to the very low memory requirements of the algorithms and their cache-friendly design. We have also presented a new division algorithm that scales linearly with the size of the quotient(s) by using a heap the size of the divisor(s). This algorithm should have many applications for polynomial computations with algebraic extensions.

In the future we plan to combine the quotient and divisor heap algorithms to produce a division algorithm which is $O(nm \log(\min(n, m)))$, which we believe is optimal. We also plan to implement versions of the heap algorithms that use GMP for large integer arithmetic, and we are experimentally trying to parallelize the heap algorithms as well.

References

[1] Bachmann, O., Schönemann, H.: Monomial representations for Gröbner bases computations. In: Proceedings of ISSAC 1998, pp. 309–316. ACM Press, New York (1998)

[2] Brown, W.S.: Altran Users Manual, 4th edn. Murray Hill, N.J. (1977)

[3] Cox, D., Little, J., O'Shea, D.: Ideals, Varieties and Algorithms: An Introduction to Computational Algebraic Geometry and Commutative Algebra. Springer, Heidelberg (1992)

[4] Geddes, K.O., Czapor, S.R., Labahn, G.: Algorithms for Computer Algebra. Kluwer Academic, Dordrecht (1992)

[5] Grayson, D.R., Stillman, M.E.: Macaulay 2, a software system for research in algebraic geometry. Available at http://www.math.uiuc.edu/Macaulay2/

[6] Greuel, G.-M., Pfister, G., Schönemann, H.: Singular 3.0. A Computer Algebra System for Polynomial Computations. Centre for Computer Algebra, University of Kaiserslautern (2005), http://www.singular.uni-kl.de

[7] Jenks, R., Sutor, R., Morrison, S.: AXIOM: The Scientific Computation System. Springer, Heidelberg (1992)

[8] Johnson, S.C.: Sparse polynomial arithmetic. ACM SIGSAM Bulletin 8(3), 63–71 (1974)

[9] LaMarca, A., Ladner, R.: The Influence of Caches on the Performance of Heaps. J. Experimental Algorithms 1, Article 4 (1996)

[10] Monagan, M., Geddes, K., Heal, K., Labahn, G., Vorkoetter, S., McCarron, J., DeMarco, P.: Maple 10 Introductory Programming Guide Maplesoft (2005) ISBN 1-894511-76

[11] Stoutemyer, D.: Which Polynomial Representation is Best? In: Proceedings of the 1984 Macsyma Users Conference, Schenectedy, NY, pp. 221–244 (1984)

[12] Yan, T.: The Geobucket Data Structure for Polynomials. J. Symb. Comput. 25, 285–293 (1998)

Ruppert Matrix as Subresultant Mapping

Kosaku Nagasaka

Kobe University, Japan
nagasaka@main.h.kobe-u.ac.jp

Abstract. Ruppert and Sylvester matrices are very common for computing irreducible factors of bivariate polynomials and computing polynomial greatest common divisors, respectively. Since Ruppert matrix comes from Ruppert criterion for bivariate polynomial irreducibility testing and Sylvester matrix comes from the usual subresultant mapping, they are used for different purposes and their relations have not been focused yet. In this paper, we show some relations between Ruppert and Sylvester matrices as the usual subresultant mapping for computing (exact/approximate) polynomial GCDs, using Ruppert matrices.

1 Introduction

Computing irreducible factors and greatest common divisors is the most popular arithmetic for symbolic algebraic computations. In fact, there are lots of studies for exact factorization ([1],[2],[3],[4] and more), approximate factorization ([5], [6] and more), polynomial GCD ([7], [8], [9] and more) and approximate GCD ([10], [11], [12] and more). For computing GCDs, the Sylvester matrix or its variants play important roles in most of the algorithms. The structure, properties and useful lemmas related to Sylvester matrix are widely known and well published. For computing irreducible factors, there are several approaches but their basic ideas have the common idea: converting the problem to linear equations. Such linear systems form Berlekamp, Niederreiter and Ruppert matrices for example. Hence, such structured matrices are very important for symbolic computations and studying those matrices is one of interesting topics: Lee and Vanstone [13] show Berlekamp and Niederreiter subspaces and their relation, the structure of Ruppert matrix is given by Nagasaka [14] and the displacement structure of Sylvester matrix for computing approximate GCD is studied by Zhi [15].

In this paper, we show some relations between Ruppert and Sylvester matrices as the usual subresultant mapping for computing (exact/approximate) polynomial GCDs via Ruppert matrix.

1.1 Notations and Sylvester Matrix

In this paper, $P(f)$ denotes the Newton polytope of the support of polynomial f. $\mathcal{P}_k$ denotes the set of polynomials of degree k. Φ_{k_1,k_2} $(k_1 \leq k_2)$ is the natural injection from $\mathbb{C}^{k_1 \times 1}$ to $\mathbb{C}^{k_2 \times 1}$ such that $\Phi_{k_1,k_2}(\boldsymbol{a}) = {}^t(b_1 \cdots b_{k_2-k_1} a_1 \cdots a_{k_1})$ where $\boldsymbol{b} = {}^t(b_i)$ is the $(k_2 - k_1)$-dimensional zero vector and $\boldsymbol{a} = {}^t(a_i)$. For

V.G. Ganzha, E.W. Mayr, and E.V. Vorozhtsov (Eds.): CASC 2007, LNCS 4770, pp. 316–327, 2007.

polynomial $f(x, y_1, \ldots, y_m)$, we abbreviate it to $f(x, \boldsymbol{y})$. The range of matrix $A = (\boldsymbol{a}_1 \cdots \boldsymbol{a}_m)$ where $\boldsymbol{a}_i$s are k-dimensional column vectors, is defined as range(A) $= \{A\boldsymbol{b} \mid \boldsymbol{b} \in \mathbb{C}^{k \times 1}\}$. We consider about polynomial GCDs of the following polynomials $f_0(x), f_1(x), \cdots, f_k(x)$.

$$
\begin{aligned}
f_0(x) &= f_{0,n_0} x^{n_0} + \cdots + f_{0,1} x + f_{0,0}, \\
f_1(x) &= f_{1,n_1} x^{n_1} + \cdots + f_{1,1} x + f_{1,0}, \\
&\vdots \\
f_k(x) &= f_{k,n_k} x^{n_k} + \cdots + f_{k,1} x + f_{k,0}.
\end{aligned} \tag{1.1}
$$

We assume that $n_i \geq n_{i+1}$ and $f_{i,n_i} \neq 0$.

$C_k(p)$ denotes the following convolution matrix of polynomial $p(x)$, of size $(n + k) \times k$.

$$
C_k(p) = \begin{pmatrix}
p_n & 0 & \cdots & 0 & 0 \\
p_{n-1} & p_n & \ddots & \vdots & \vdots \\
\vdots & p_{n-1} & \ddots & 0 & \vdots \\
p_0 & \vdots & \ddots & p_n & 0 \\
0 & p_0 & \ddots & p_{n-1} & p_n \\
\vdots & 0 & \ddots & \vdots & p_{n-1} \\
\vdots & \vdots & \ddots & p_0 & \vdots \\
0 & 0 & \cdots & 0 & p_0
\end{pmatrix},
$$

where $p(x) = p_n x^n + \cdots + p_1 x + p_0$.

Let S_r be the following subresultant mapping.

$$
S_r : \begin{cases} \mathcal{P}_{n_1-r-1} \times \mathcal{P}_{n_0-r-1} \to \mathcal{P}_{n_0+n_1-r-1}, \\ (u_0, u_1) \mapsto u_1 f_0 + u_0 f_1, \end{cases} \tag{1.2}
$$

where $\mathcal{P}_k$ denotes the set of univariate polynomials of degree k. This mapping can be expressed by the following Sylvester subresultant matrix $S_r(f_0, f_1)$.

$$
S_r(f_0, f_1) = \begin{pmatrix} C_{n_0-r}(f_1) & C_{n_1-r}(f_0) \end{pmatrix}.
$$

We note a well known fact: if r is the largest integer that S_r is not injective, we can compute the greatest common divisor of $f_0(x)$ and $f_1(x)$ from the right null vector of $S_r(f_0, f_1)$ (see the proof in Rupprecht [16] and so on). Moreover, the greatest common divisor also can be computed by QR-decomposition of $S_0(f_0, f_1)$ (see the proof in [8,9] and so on): the last non-zero row vector of the upper triangular matrix is the coefficient vector of the polynomial GCD of f_0 and f_1. $S_0(f_0, f_1)$ also has another useful property that the dimension of the null space is the degree of the polynomial GCD.

1.2 Ruppert Matrix

Ruppert matrix is the coefficient matrix of the corresponding linear equation of the following absolute irreducibility criterion due to Ruppert [17] (Gao and Rodrigues [18] studied the sparse polynomial version of this criterion).

$$f\frac{\partial g}{\partial y} - g\frac{\partial f}{\partial y} + h\frac{\partial f}{\partial x} - f\frac{\partial h}{\partial x} = 0, \quad g, h \in \mathbb{C}[x, y], \tag{1.3}$$

$$\deg_x g \leq \deg_x f - 1, \ \deg_y g \leq \deg_y f,$$
$$\deg_x h \leq \deg_x f, \qquad \deg_y h \leq \deg_y f - 2.$$

The criterion is that $f(x, y)$ is absolutely irreducible if and only if this differential equation does not have any non-trivial solutions. The matrix is useful for computing irreducible factors [6,1] and the irreducibility radius [19,14,20]. Since Ruppert matrix is the set of coefficient vectors w.r.t. unknowns of g and h, matrices by different term orders are not the same. For the Ruppert matrix of f, we use the lexicographic order of x, y and $x, y_1, \ldots, y_m$, as in Nagasaka [14,20], and by $R(f)$ we denote the Ruppert matrix of polynomial f.

For multivariate polynomials, May [21] studied the generalized version of the Ruppert criterion, with the following differential equation and degree constraints.

$$f\frac{\partial g}{\partial y_i} - g\frac{\partial f}{\partial y_i} + h_i\frac{\partial f}{\partial x} - f\frac{\partial h_i}{\partial x} = 0, \quad g, h \in \mathbb{C}[x, y_1, \cdots, y_m], \tag{1.4}$$

$$\deg_x g \leq \deg_x f - 2, \quad \deg_{y_i} g \leq \deg_{y_i} f,$$
$$\deg_x h_i \leq \deg_x f, \quad \deg_{y_j} h_i \leq \begin{cases} \deg_{y_j} f & i \neq j \\ \deg_{y_j} f - 1 & i = j \end{cases}$$

May [21] also studied the generalized Ruppert criterion with degree bounds via Newton polytopes as follows.

$$P(xg) \subseteq P(f) \text{ and } P(y_i h_i) \subseteq P(f). \tag{1.5}$$

The generalized two criteria have the same argument that the given polynomial $f(x, \boldsymbol{y})$ is absolutely irreducible if and only if this differential equation does not have any non-trivial solutions. For these criteria, we can also construct the coefficient matrix of the corresponding linear system, with the lexicographic order of $x, y_1, \ldots, y_m$.

2   GCD of Two Polynomials

In this section, we consider the subresultant mapping of two polynomials via Ruppert matrix. We define the following polynomial $f(x, y)$.

$$f(x, y) = f_0(x) + f_1(x)y. \tag{2.1}$$

It is obvious that $f(x, y)$ is reducible if and only if $f_0(x)$ and $f_1(x)$ have a non-trivial GCD. This means that we can check whether $f_0(x)$ and $f_1(x)$ have a non-trivial GCD or not via the differential equation (1.3) of the Ruppert criterion, with $f(x, y) = f_0(x) + f_1(x)y$. We note that $f_1(x) + f_0(x)y$ can be used instead of $f_0(x) + f_1(x)y$ for our purpose, since the degree constraints of Ruppert criterion are given by each variables separately.

2.1   Case 1-1: Simple Result

Substituting degrees of $f(x, y)$ for that of f in (1.3), we have

$$f\frac{\partial g}{\partial y} - g\frac{\partial f}{\partial y} = 0, \quad g \in \mathbb{C}[x, y], \quad \deg_x g \le n_0 - 1, \quad \deg_y g \le 1. \qquad (2.2)$$

Let $g(x, y)$ be the following polynomial satisfying (2.2).

$$g(x, y) = g_0(x) + g_1(x)y.$$

Substituting $g(x, y)$ for g in (2.2), we have

$$(f_0(x) + f_1(x)y)g_1(x) - (g_0(x) + g_1(x)y)f_1(x) = 0.$$

Collecting terms with respect to y, we have

$$g_1(x)f_0(x) - g_0(x)f_1(x) = 0. \qquad (2.3)$$

This equations can be represented as a linear equation w.r.t. coefficients of polynomials $g(x, y)$. The coefficient matrix is the Ruppert matrix of $f(x, y)$ and its structure is given by Nagasaka [14]. Moreover, the structure of this matrix is the Sylvester matrix $S_0(f_0, -f_1)$ since the degree constraints of $u_i(x)$ and $g_i(x)$ are the same if $n_0 = n_1$. For $n_0 > n_1$, the Ruppert matrix has extra column vectors that are not included in the Sylvester matrix, hence we have $\Phi_{n_0+n_1, 2n_0}(\mathrm{range}(S_0(f_0, f_1))) \subset \mathrm{range}(R(f))$. By comparing between the both sides of (2.3), degrees of $\deg(g_1 f_0)$ and $\deg(g_0 f_1)$ must be the same. Therefore, we have the following lemma.

Lemma 1. *For any polynomials $f_0(x)$ and $f_1(x)$, the Sylvester matrix and the Ruppert matrix of $f_0(x)$ and $f_1(x)$ have the same information for computing their GCD, with the Ruppert's original differential equation and constraints.* ◁

2.2   Case 1-2: Alternative Result

The degree bounds of the differential equation (1.3) are not the same as the following general version of the Ruppert criterion by John May [21] for bivariate polynomials , though the difference is only the roles of variables and not essential.

$$f\frac{\partial g}{\partial y} - g\frac{\partial f}{\partial y} + h\frac{\partial f}{\partial x} - f\frac{\partial h}{\partial x} = 0, \quad g, h \in \mathbb{C}[x, y], \qquad (2.4)$$

$$\deg_x g \le \deg_x f - 2, \quad \deg_y g \le \deg_y f, \\ \deg_x h \le \deg_x f, \qquad \deg_y h \le \deg_y f - 1. \qquad (2.5)$$

We have the following corollary (see [21] or [6]).

Corollary 1. *For a given $f(x, y) \in \mathbb{C}[x, y]$ that is square-free over $\mathbb{C}(y)$, the dimension (over $\mathbb{C}$) of the null space of $R(f)$ is equal to " (the number of absolutely irreducible factors of f over $\mathbb{C}$) $- 1$ ".*

Substituting degrees of $f(x, y)$ for that of f in (2.4), we have

$$f\frac{\partial g}{\partial y} - g\frac{\partial f}{\partial y} + h\frac{\partial f}{\partial x} - f\frac{\partial h}{\partial x} = 0, \quad g, h \in \mathbb{C}[x, y], \tag{2.6}$$

$\deg_x g \leq n_0 - 2$, $\deg_y g \leq 1$, $\deg_x h \leq n_0$, $\deg_y h \leq 0$.

Let $g(x, y)$ and $h(x, y)$ be the following polynomials satisfying (2.6).

$$g(x, y) = g_0(x) + g_1(x)y, \quad h(x, y) = h_0(x).$$

Substituting $g(x, y)$ and $h(x, y)$ for g and h, respectively, in (2.6), we have

$$(f_0(x) + f_1(x)y)g_1(x) - (g_0(x) + g_1(x)y)f_1(x)$$
$$+h_0(x)(\tfrac{\partial f_0(x)}{\partial x} + \tfrac{\partial f_1(x)}{\partial x}y) - (f_0(x) + f_1(x)y)\tfrac{\partial h_0(x)}{\partial x} = 0.$$

Collecting terms with respect to y, we have

$$g_1(x)f_0(x) - g_0(x)f_1(x) + h_0(x)\frac{\partial f_0(x)}{\partial x} - f_0(x)\frac{\partial h_0(x)}{\partial x} = 0, \tag{2.7}$$

$$h_0(x)\frac{\partial f_1(x)}{\partial x} - f_1(x)\frac{\partial h_0(x)}{\partial x} = 0. \tag{2.8}$$

This is not as same as the subresultant mapping in the previous subsection and is not reduces to the usual subresultant mapping (2.3).

Lemma 2. *For any polynomials $u_1(x) \in \mathcal{P}_{n_1-1}$ and $u_0(x) \in \mathcal{P}_{n_0-1}$ satisfying $\deg(u_1 f_0 + u_0 f_1) < n_0 + n_1 - 1$, there exist polynomials $g_0(x)$, $g_1(x)$ and $h_0(x)$ satisfying their degree constraints, the equation (2.8) and $g_1 f_0 - g_0 f_1 - f_0\frac{\partial h_0}{\partial x} + h_0\frac{\partial f_0}{\partial x} = u_1 f_0 + u_0 f_1$.* ◁

Proof. If $\deg(u_1) \leq n_1 - 2 \leq n_0 - 2$ and $\deg(u_0) \leq n_0 - 2$, the lemma follows from (2.7) and (2.8), with $g_0(x) = -u_0(x)$, $g_1(x) = u_1(x)$ and $h_0(x) = 0$. We suppose that $\deg(u_1) = n_1 - 1$ and $\deg(u_0) = n_0 - 1$ since the leading coefficients of $u_1 f_0$ and $u_0 f_1$ must be canceled. We put $u_0(x) = \sum_{i=0}^{n_0-1} u_{0,i} x^i$ and $u_1(x) = \sum_{i=0}^{n_1-1} u_{1,i} x^i$, and transform $u_1 f_0 + u_0 f_1$ as follows.

$$u_1 f_0 + u_0 f_1 = (u_1 - u_{1,n_0-1}x^{n_0-1})f_0 - (-u_0 + u_{0,n_0-1}x^{n_0-1})f_1$$
$$+u_{1,n_0-1}x^{n_0-1}f_0 + u_{0,n_0-1}x^{n_0-1}f_1$$
$$= (u_1 - u_{1,n_0-1}x^{n_0-1} + \tfrac{u_{0,n_0-1}}{n_0 f_{0,n_0}}(f_1 - f_{1,n_0}x^{n_0})')f_0$$
$$-(-u_0 + u_{0,n_0-1}x^{n_0-1} + \tfrac{u_{0,n_0-1}}{n_0 f_{0,n_0}}(f_0 - f_{0,n_0}x^{n_0})')f_1$$
$$+u_{1,n_0-1}x^{n_0-1}f_0 + u_{0,n_0-1}x^{n_0-1}f_1$$
$$-\tfrac{u_{0,n_0-1}}{n_0 f_{0,n_0}}(f_1 - f_{1,n_0}x^{n_0})'f_0 + \tfrac{u_{0,n_0-1}}{n_0 f_{0,n_0}}(f_0 - f_{0,n_0}x^{n_0})'f_1$$
$$= (u_1 - u_{1,n_0-1}x^{n_0-1} + \tfrac{u_{0,n_0-1}}{n_0 f_{0,n_0}}(f_1 - f_{1,n_0}x^{n_0})')f_0$$
$$-(-u_0 + u_{0,n_0-1}x^{n_0-1} + \tfrac{u_{0,n_0-1}}{n_0 f_{0,n_0}}(f_0 - f_{0,n_0}x^{n_0})')f_1$$
$$-\tfrac{u_{0,n_0-1}}{n_0 f_{0,n_0}}f_1' f_0 + \tfrac{u_{0,n_0-1}}{n_0 f_{0,n_0}}f_0' f_1 + (u_{1,n_0-1} + u_{0,n_0-1}\tfrac{f_{1,n_0}}{f_{0,n_0}})x^{n_0-1}f_0,$$

where $f' = \frac{\partial f}{\partial x}$. If $n_0 = n_1$, we have $u_{1,n_0-1} + u_{0,n_0-1}\frac{f_{1,n_0}}{f_{0,n_0}} = 0$ since the leading coefficients of $u_1 f_0 + u_0 f_1$ must be canceled. For $n_0 > n_1$, we also have $u_{1,n_0-1} + u_{0,n_0-1}\frac{f_{1,n_0}}{f_{0,n_0}} = 0$ since $u_{1,n_0-1} = f_{1,n_0} = 0$. Therefore, the following $g_0(x)$, $g_1(x)$ and $h_0(x)$ prove the lemma.

$$g_0(x) = -(u_0(x) - u_{0,n_0-1}x^{n_0-1}) + \frac{u_{0,n_0-1}}{n_0 f_{0,n_0}}(f_0 - f_{0,n_0}x^{n_0})'$$
$$g_1(x) = (u_1(x) - u_{1,n_0-1}x^{n_0-1}) + \frac{u_{0,n_0-1}}{n_0 f_{0,n_0}}(f_1 - f_{1,n_0}x^{n_0})'$$
$$h_0(x) = \frac{u_{0,n_0-1}}{n_0 f_{0,n_0}}f_1.$$

$\square$

Lemma 3. *For any polynomials $g_0(x)$, $g_1(x)$ and $h_0(x)$ satisfying their degree constraints and the equations (2.7) and (2.8), there exist polynomials $u_1(x) \in \mathcal{P}_{n_1-1}$ and $u_0(x) \in \mathcal{P}_{n_0-1}$ satisfying $u_1 f_0 + u_0 f_1 = g_1 f_0 - g_0 f_1 - f_0\frac{\partial h_0}{\partial x} + h_0\frac{\partial f_0}{\partial x}$, if $f(x,y) = f_0(x) + f_1(x)y$ is square-free over $\mathbb{C}(y)$.* ◁

Proof. Let $g_0(x)$, $g_1(x)$ and $h_0(x)$ be a solution of (2.7) and (2.8) with $f(x,y) = f_0(x) + f_1(x)y$. By the lemma 3.1 in John May [21] (or see [6]), we have

$$h_0(x) = \lambda f_1(x) \text{ with } \lambda \in \mathbb{C}.$$

If $n_1 \le \deg(g_1) \le n_0 - 2$, we have $\deg(g_1 f_0) \le n_0 + n_1 - 1$ since $\max\{\deg(g_0 f_1), \deg(f_0 f_1'), \deg(f_1 f_0')\} = n_0 + n_1 - 1$. However, $\deg(g_1 f_0) \le n_0 + n_1 - 1$ contradicts $n_1 \le \deg(g_1)$. Hence, we have $\deg(g_1) \le n_1 - 1$ and the following polynomials $u_0(x)$ and $u_1(x)$ prove the lemma.

$$u_0(x) = -g_0(x) + \lambda\frac{\partial f_0(x)}{\partial x}, \quad u_1(x) = g_1(x) - \lambda\frac{\partial f_1(x)}{\partial x}.$$

$\square$

The following theorem follows from the above lemmas, directly.

Theorem 1. *The polynomial GCD of $f_0(x)$ and $f_1(x)$ can be computed by Singular Value Decomposition (SVD) of Ruppert matrix of $f(x,y) = f_0(x) + f_1(x)y$ in (2.4), if $f(x,y)$ is square-free over $\mathbb{C}(y)$.* ◁

For computing polynomial GCDs, one of well known methods is computing the QR decomposition of the Sylvester matrix of $f_0(x)$ and $f_1(x)$ as in [8], [9] and so on. In the below, we show that we can compute polynomial GCDs by the QR decomposition of the Ruppert matrix of $f(x,y) = f_0(x) + f_1(x)y$. The figure 1 illustrates the structure of the Ruppert matrix of $f(x,y) = f_0(x) + f_1(x)y$, as in Nagasaka [14]. The size of this matrix is $(4n_0) \times (3n_0 - 1)$.

Lemma 4. *The range (and the span of column vectors) of the Ruppert matrix of $f(x,y)$ includes the descending coefficient vector (its constant term is the last element) of $n_0(f_{0,n_0}f_1(x) - f_{1,n_0}f_0(x))x^{n_0-1}$.* ◁

$$R(f) = \begin{pmatrix} R_{1,1} & \mathbf{0} \\ R_{2,1} & R_{2,2} \end{pmatrix}, \quad R_{2,2} = \begin{pmatrix} -f_{0,n_0} & 0 & f_{1,n_0} & 0 \\ -f_{0,n_0-1} & \ddots & f_{1,n_0-1} & \ddots \\ \vdots & \ddots & -f_{0,n_0} & \vdots & \ddots & f_{1,n_0} \\ \vdots & \ddots & -f_{0,n_0-1} & \vdots & \ddots & f_{1,n_0-1} \\ \vdots & \ddots & \vdots & \vdots & \ddots & \vdots \\ -f_{0,1} & \ddots & \vdots & f_{1,1} & \ddots & \vdots \\ -f_{0,0} & \ddots & \vdots & f_{1,0} & \ddots & \vdots \\ & \ddots & -f_{0,1} & & \ddots & f_{1,1} \\ 0 & & -f_{0,0} & 0 & & f_{1,0} \end{pmatrix},$$

$$R_{1,1} = \begin{pmatrix} 0 & 0 & \cdots & 0 & 0 & 0 \\ f_{1,n_0-1} & -f_{1,n_0} & \ddots & \vdots & \vdots & 0 \\ 2f_{1,n_0-2} & 0 & \ddots & 0 & \vdots & \vdots \\ \vdots & f_{1,n_0-2} & \ddots & (2-n_0)\,f_{1,n_0} & 0 & \vdots \\ \vdots & \vdots & \ddots & \vdots & (1-n_0)\,f_{1,n_0} & 0 \\ n_0\,f_{1,0} & \vdots & \ddots & \vdots & \vdots & -n_0\,f_{1,n_0} \\ 0 & (n_0-1)\,f_{1,0} & \ddots & 0 & \vdots & \vdots \\ \vdots & 0 & \ddots & f_{1,1} & -f_{1,2} & \vdots \\ 0 & \vdots & \ddots & 2f_{1,0} & 0 & -2f_{1,2} \\ 0 & 0 & \cdots & 0 & f_{1,0} & -f_{1,1} \\ 0 & 0 & \cdots & 0 & 0 & 0 \end{pmatrix},$$

$$R_{2,1} = \begin{pmatrix} f_{0,n_0-1} & -f_{0,n_0} & \ddots & \vdots & \vdots & 0 \\ 2f_{0,n_0-2} & 0 & \ddots & 0 & \vdots & \vdots \\ \vdots & f_{0,n_0-2} & \ddots & (2-n_0)\,f_{0,n_0} & 0 & \vdots \\ \vdots & \vdots & \ddots & \vdots & (1-n_0)\,f_{0,n_0} & 0 \\ n_0\,f_{0,0} & \vdots & \ddots & \vdots & \vdots & -n_0\,f_{0,n_0} \\ 0 & (n_0-1)\,f_{0,0} & \ddots & 0 & \vdots & \vdots \\ \vdots & 0 & \ddots & f_{0,1} & -f_{0,2} & \vdots \\ 0 & \vdots & \ddots & 2f_{0,0} & 0 & -2f_{0,2} \\ 0 & 0 & \cdots & 0 & f_{0,0} & -f_{0,1} \end{pmatrix}$$

Fig. 1. Ruppert matrix $R(f) = R(f_0(x) + f_1(x)y)$

Proof. Put $R(f) = (r_1, \cdots, r_{3n_0-1})$ where r_i is $4n_0$ dimensional column vector. We note that lower $2n_0 - 1$ rows of $(r_{n_0+2}, \cdots, r_{3n_0-1})$ is the usual Sylvester subresultant matrix $S_1(f_1, -f_0)$. If we apply fraction-free column reductions to the first column r_1 by $r_2, \cdots, r_{n_0+1}$, then the first column becomes

$$\bar{r}_1 = \sum_{i=0}^{n_0} f_{1,n_0-i} r_{i+1}.$$

Let $\hat{r}$ be the following column vector.

$$\hat{r} = \bar{r}_1 + \sum_{i=0}^{n_0-2} (n_0 - 1 - i) f_{1,n_0-1-i} r_{n_0+2+i} + \sum_{i=0}^{n_0-2} (n_0 - 1 - i) f_{0,n_0-1-i} r_{2n_0+1}.$$

$\hat{r}$ is the descending coefficient vector (its constant term is the last element) of $n_0(f_{0,n_0} f_1(x) - f_{1,n_0} f_0(x)) x^{n_0-1}$. □

Theorem 2. *The polynomial GCD of $f_0(x)$ and $f_1(x)$ can be computed by applying the QR decomposition to the transpose of the last $3n_0$ rows of their Ruppert matrix $R(f) = R(f_0(x) + f_1(x)y)$. The last non-zero row vector of the triangular matrix is the coefficient vector of their polynomial GCD.* ◁

Proof. Let $\bar{R}$ be the transpose of the last $3n_0$ rows of their Ruppert matrix $R(f)$. As in the proof [8], the last non-zero row vector of the triangular matrix of the QR decomposition is the coefficient vector of the lowest degree non-constant polynomial of linear combinations of polynomials whose coefficient vectors are row vectors of $\bar{R}$. Hence, we show that the lowest degree non-constant polynomial is the polynomial GCD of $f_0(x)$ and $f_1(x)$.

The rank of the upper $(n_0 + 1) \times 3n_0$ submatrix is n_0 at least since its upper left $n_0 \times n_0$ submatrix is a triangular matrix and its diagonal elements are non-zero elements: $n_0 f_{1,0}, \ldots, 2f_{1,0}, f_{1,0}$. If the linear combination includes some of the first n_0 row vectors, the degree of the combination is larger than $2n_0$. Since $\bar{R}$ has row vectors whose corresponding degrees are less than or equal to $2n_0$, the lowest degree non-constant polynomial does not include the first n_0 rows.

However, as in the proof of the lemma 4, another row vector generated from the first n_0 rows, can be included in the linear combination for the lowest degree non-constant polynomial. Hence, we only have to prove that the lowest degree non-constant polynomial of linear combinations of polynomials whose coefficient vectors are row vectors of the following matrix $\hat{R}$ is the polynomial GCD of $f_0(x)$ and $f_1(x)$.

$$\hat{R} = \begin{pmatrix} 0 & \text{the coefficient vector of } n_0(f_{0,n_0} f_1(x) - f_{1,n_0} f_0(x)) x^{n_0-1} \\ 0 & -f_{0,n_0} & -f_{0,n_0-1} & \cdots & \cdots & \cdots & -f_{0,1} & -f_{0,0} & 0 \\ \vdots & \ddots & \ddots & \ddots & \ddots & \ddots & \ddots & \ddots & \ddots \\ 0 & \cdots & 0 & -f_{0,n_0} & -f_{0,n_0-1} & \cdots & \cdots & \cdots & -f_{0,1} & -f_{0,0} \\ 0 & f_{1,n_0} & f_{1,n_0-1} & \cdots & \cdots & \cdots & f_{1,1} & f_{1,0} & 0 \\ \vdots & \ddots & \ddots & \ddots & \ddots & \ddots & \ddots & \ddots & \ddots \\ 0 & \cdots & 0 & f_{1,n_0} & f_{1,n_0-1} & \cdots & \cdots & \cdots & f_{1,1} & f_{1,0} \end{pmatrix}.$$

The last $2n_0 - 2$ rows form the usual Sylvester subresultant matrix $S_1(f_1, -f_0)$ whose range is the set of coefficient vectors of $u_1 f_0 + u_0 f_1$ where u_0 and u_1 are polynomials of degree $n_0 - 2$ at most, and this is enough to compute non-trivial GCDs of f_0 and f_1. Moreover, the first row of $\hat{R}$ is the $(n_1 + 1)$-th row reduced by the first row, of $S_0(f_1, -f_0)$. Therefore, the last non-zero row vector of the triangular matrix of the QR decomposition of $\hat{R}$ is that of $S_0(f_1, -f_0)$.   □

We note that for practical computations of polynomial GCDs, we do not have to use the Ruppert or Sylvester matrices because the usual Sylvester subresultant matrix which is smaller, is enough for GCDs, especially for approximate GCDs.

3   GCD of Several Polynomials

In this section, we show brief overview of relations between Sylvester matrix and Ruppert matrix for several polynomials $f_0(x), \ldots, f_k(x)$. Basically, the relations are natural extensions of the results in the previous section.

3.1   Generalized Sylvester Matrix for Several Polynomials

Let $\mathcal{S}_r$ be the following generalized subresultant mapping.

$$
\mathcal{S}_r : \begin{cases} \prod_{i=0}^{k} \mathcal{P}_{n_i - r - 1} \to \prod_{i=1}^{k} \mathcal{P}_{n_0 - n_i - r - 1}, \\ \begin{pmatrix} u_0 \\ \vdots \\ u_k \end{pmatrix} \mapsto \begin{pmatrix} u_1 f_0 + u_0 f_1 \\ \vdots \\ u_k f_0 + u_0 f_k \end{pmatrix}. \end{cases}
\tag{3.1}
$$

This mapping can be expressed by the following Sylvester subresultant matrix $\mathcal{S}_r(f_0, \ldots, f_k)$.

$$
\mathcal{S}_r(f_0, \ldots, f_k) = \begin{pmatrix} C_{n_0 - r}(f_1) & C_{n_1 - r}(f_0) & 0 & \cdots & 0 \\ C_{n_0 - r}(f_2) & 0 & C_{n_2 - r}(f_0) & \cdots & 0 \\ \vdots & & & \ddots & \vdots \\ C_{n_0 - r}(f_k) & 0 & \cdots & 0 & C_{n_k - r}(f_0) \end{pmatrix}.
$$

We note a well known fact: if r is the largest integer that $\mathcal{S}_r$ is not injective, we can compute the greatest common divisor of $f_0(x), \ldots, f_k(x)$ from the right null vector of $\mathcal{S}_r(f_0, \ldots, f_k)$ (see [16]).

3.2   Extension for Several Polynomials

Let $f(x, y)$ be the following polynomial.

$$
f(x, y) = f_0(x) + f_1(x) y_1 + \cdots + f_k(x) y_k.
\tag{3.2}
$$

This polynomial is irreducible if and only if the polynomials $f_0(x)$, $f_1(x)$, $\ldots$, $f_{k-1}(x)$ and $f_k(x)$ do not have any non-trivial GCD. As in the previous section, we can check whether $f_i(x)$ have a non-trivial GCD or not by the differential equation (1.4) of the generalized Ruppert criterion, with $f(x, y) = f_0(x) + \sum_{j=1}^{k} f_j(x)y_j$. The degree constraints of (1.4) with $f(x, y)$ of (3.2), becomes

$$\deg_x g \le n_0 - 2, \quad \deg_{y_i} g \le 1,$$

$$\deg_x h^{(i)} \le n_0, \quad \deg_{y_j} h^{(i)} \le \begin{cases} 1 \ i \ne j \\ 0 \ i = j \end{cases}$$

In the previous section, we define $g(x, y)$ and $h^{(k)}(x, y)$ satisfying the following differential equation. However, the degree constraints are not by total-degrees so the number of possible terms increases exponentially.

$$f \frac{\partial g}{\partial y_i} - g \frac{\partial f}{\partial y_i} + h_i \frac{\partial f}{\partial x} - f \frac{\partial h_i}{\partial x} = 0. \tag{3.3}$$

Hence, we limit the solution polynomials $g(x, y)$ and $h^{(i)}(x, y)$ as follows.

$$g(x, y) = g_0(x) + \sum_{j=1}^{k} g_j(x)y_j, \ h^{(1)}(x, y) = h_0^{(1)}(x), \ \ldots, \ h^{(k)}(x, y) = h_0^{(k)}(x).$$

We note that this limitation may be harmless since by the lemma 3.1 in John May [21] (or see [6]), we have

$$h_0^{(i)}(x) = \lambda_i f_i(x) \text{ with } \lambda_i \in \mathbb{C}.$$

Substituting the above $g(x, y)$ and $h^{(i)}(x, y)$ for g and h_i, respectively, in (3.3), we have

$$f(x, y)g_i(x) - g(x, y)f_i(x) + h_0^{(i)}(x)\frac{\partial f(x, y)}{\partial x} - f(x, y)\frac{\partial h_0^{(i)}(x)}{\partial x} = 0.$$

Collecting terms with respect to y and substituting $\lambda_i f_i(x)$, we have

$$\begin{cases} f_0 g_i - g_0 f_i + \lambda_i(f_i f_0{'} - f_0 f_i{'}) = 0, \\ f_1 g_i - g_1 f_i + \lambda_i(f_i f_1{'} - f_1 f_i{'}) = 0, \\ \quad \vdots \\ f_k g_i - g_k f_i + \lambda_i(f_i f_k{'} - f_k f_i{'}) = 0. \end{cases}$$

This system of equations is the system of equation (2.7) for all the combinations of $f_0, \ldots, f_k$ since the equation (2.8) with $f_0, \ldots, f_k$ is always satisfied by $h_0^{(i)}(x) = \lambda_i f_i(x)$. As in the proof of lemma 3, for the solution of the above system, there exist polynomials $u_i(x) \in \mathcal{P}_{n_i-1}$ $(i = 0, \ldots, k)$:

$$\begin{cases} u_0(x) = g_0(x) - \lambda_0 f_0{'}, \\ u_i(x) = -g_i(x) + \lambda_0 f_i{'} \ (i = 1, \ldots, k). \end{cases}$$

For the other lemma and theorem for two polynomials, the author thinks that the same relations are hold for several polynomials since the ranks of null spaces of Ruppert matrix and generalized Sylvester subresultant matrix are the same. However, these problems are postponed as a future work.

4 Conclusion

In this paper, we show some relations on Ruppert matrix and Sylvester matrix from the point of computing the greatest common divisors of two polynomials. Though no algorithm is present in this paper and does not compete with the finest recent algorithms for computing approximate GCDs, the author hopes that factoring polynomials and computing polynomial GCDs are the basics of symbolic computations, and revealing their relations will make some progress in the future.

References

1. Gao, S.: Factoring multivariate polynomials via partial differential equations. Math. Comp. 72, 801–822 (2003) (electronic)
2. Salem, F.A., Gao, S., Lauder, A.G.B.: Factoring polynomials via polytopes. In: ISSAC 2004, pp. 4–11. ACM, New York (2004)
3. van Hoeij, M.: Factoring polynomials and the knapsack problem. J. Number Theory 95, 167–189 (2002)
4. Chèze, G.: Absolute polynomial factorization in two variables and the knapsack problem. In: ISSAC 2004, pp. 87–94. ACM, New York (2004)
5. Sasaki, T.: Approximate multivariate polynomial factorization based on zero-sum relations. In: ISSAC 2001. Proceedings of the 2001 International Symposium on Symbolic and Algebraic Computation, pp. 284–291 (2001)
6. Gao, S., Kaltofen, E., May, J., Yang, Z., Zhi, L.: Approximate factorization of multivariate polynomials via differential equations. In: ISSAC 2004. Proceedings of the 2004 International Symposium on Symbolic and Algebraic Computation, pp. 167–174 (2004)
7. Cheng, H., Labahn, G.: On computing polynomial gcds in alternate bases. In: ISSAC '06. Proceedings of the 2006 international symposium on Symbolic and algebraic computation, pp. 47–54. ACM Press, New York, NY, USA (2006)
8. Laidacker, M.A.: Another theorem relating Sylvester's matrix and the greatest common divisor. Math. Mag. 42, 126–128 (1969)
9. Corless, R.M., Watt, S.M., Zhi, L.: QR factoring to compute the GCD of univariate approximate polynomials. IEEE Trans. Signal Process 52, 3394–3402 (2004)
10. Zeng, Z., Dayton, B.H.: The approximate GCD of inexact polynomials. II. A multivariate algorithm. In: ISSAC 2004, pp. 320–327. ACM, New York (2004)
11. Kaltofen, E., Yang, Z., Zhi, L.: Approximate greatest common divisors of several polynomials with linearly constrained coefficients and singular polynomials. In: ISSAC '06. Proceedings of the 2006 international symposium on Symbolic and algebraic computation, pp. 169–176. ACM Press, New York, NY, USA (2006)
12. Pan, V.Y.: Computation of approximate polynomial GCDs and an extension. Inform. and Comput. 167, 71–85 (2001)
13. Lee, T.C.Y., Vanstone, S.A.: Subspaces and polynomial factorizations over finite fields. Appl. Algebra Engrg. Comm. Comput. 6, 147–157 (1995)
14. Nagasaka, K.: Towards more accurate separation bounds of empirical polynomials. SIGSAM/CCA 38, 119–129 (2004)
15. Zhi, L.: Displacement structure in computing approximate GCD of univariate polynomials. In: Computer mathematics. Lecture Notes Ser. Comput., vol. 10, pp. 288–298. World Sci. Publ., River Edge, NJ (2003)

16. Rupprecht, D.: An algorithm for computing certified approximate GCD of n univariate polynomials. J. Pure Appl. Algebra 139, 255–284 (1999)
17. Ruppert, W.M.: Reducibility of polynomials $f(x, y)$ modulo p. J. Number Theory 77, 62–70 (1999)
18. Gao, S., Rodrigues, V.M.: Irreducibility of polynomials modulo p via newton polytopes. J. Number Theory 101, 32–47 (2003)
19. Kaltofen, E., May, J.: On approximate irreducibility of polynomials in several variables. In: ISSAC 2003. Proceedings of the 2003 International Symposium on Symbolic and Algebraic Computation, pp. 161–168 (2003)
20. Nagasaka, K.: Towards more accurate separation bounds of empirical polynomials II. In: Ganzha, V.G., Mayr, E.W., Vorozhtsov, E.V. (eds.) CASC 2005. LNCS, vol. 3718, pp. 318–329. Springer, Heidelberg (2005)
21. May, J.P.: Approximate Factorization of Polynomials in Many Variables and Other Problems in Approximate Algebra via Singular Value Decomposition Methods. PhD thesis, North Carolina State Univ., Raleigh, North Carolina (2005)

Construction of Computer System for Microobjects Recognition Based on Neural Networks

Ulugbek Kh. Narzullaev, Akmal R. Akhatov, and Olim I. Jumanov

Samarkand Branch of Tashkent University of Information Technologies,
Samarkand State University Named after Alisher Navoyi,
Samarkand, Uzbekistan
ulug1956@hotmail.com, akmalar@rambler.ru

Abstract. We propose a new and efficient approach for solving the tasks of the microobjects recognition based on using the neural network (NN) and work out a computer system of image visualization, recognition, and classification of the microobjects on the samples of the pollen grains. The technology is developed for a preliminary processing of images of the microobjects on the basis of the "Snake" model. The principle of teaching of formal neuron and mathematical model of teaching multilayer perceptron for recognition of the microobjects is proposed. An algorithm is developed for teaching the NN of the returning distribution, subject domain, and methods of classes of computer system.

1 Introduction

Solution of problems of the computer processing of visual information, in particular, tasks of computer tomography, recognition of the finger prints, texts of natural languages as well as recognition of the microobjects, for example, pollen grains is related to the elaboration and using of algorithms of teaching of neural networks (NN). Urgency of researches related to the recognition of microobjects consists of the fact that the microobjects differ from each other by their external structure, and while processing their pictures it is required to identify the variety of microobjects quickly and exactly, which belong to some class on the basis of their geometrical forms and other specific characteristics. Besides, it is required to supply the large efficiency of account of the images, classification, and recognition of the objects for formation and ordering of data of the research processes. It will, in turn, free the experts from tiresome counting of the objects and reconsideration of huge encyclopedias classifying the microobjects. Therefore, realization of researches and practical elaboration with the purpose of creation of the software system of visuality of images, recognition and classification as well as the systematization of microobjects are of considerable practical interest, but despite a big topicality and practical importance the solution of this scientific problem is still at its initial stages.

V.G. Ganzha, E.W. Mayr, and E.V. Vorozhtsov (Eds.): CASC 2007, LNCS 4770, pp. 328–338, 2007.

2   Creation of the Technology of Microobjects Classification

The construction of the technology of classification of the microobjects, in particular, pollen is carried out in four stages. The first stage is the preparation of samples. Various ways can be applied here: preparation of slides, coloring, etc. The second stage is the transitional stage. The pollen sample is counted under microscope. Then the scanning by the computer is performed. Particles are getting localized and then getting split into sub-samples consisting of pollen and not pollen. Besides, for the realization of the high-grade analysis, every pollen sub-sample is represented in a three-dimensional form, for which the reconstruction of pollen grains is made. The third stage is the identification stage. Identification of the microobjects is carried out on the basis of comparison with the reference database. Estimation of oriented invariant features of microobject is essential here. For the estimation of invariant features of pollen, the schemes of optimal and textural features are applied, frequency characteristics, pollen exit area, disputes, reticule, etc. are used for textural analysis [1].

Two kinds of extraction of textural characteristics are used: Haralick's measurement and measurement of Law Mask [2]. These elaborations of valuables reflect the degree of display features of images such as contrast, entropy, waves, etc. The fourth stage is the final stage. Pollen is subdivided into groups on various textures. Two different classification techniques are applied, which can correct each other subsequently. The first method uses the standard static qualifier constructed on the basis of Fisher Liner's discriminant function [3], and the second is based on Rumelhart's neural network [4].

The allocation of these stages serves important making part of the construction of computer system for visuality of images, recognition, and classification.

3   Initial Processing of Pollen Image on the Basis of "Snake" Model

In a general view, computer system for visualization of images, recognition, classification, and stock-taking of the microobjects represents a difficult software-technological complex, which is carrying out the functions of the initial processing of images, analog and discrete transformation at input of the microobjects, coding and decoding, compression, visualization, recognition, classification, calculation, formation of the database, and information output. The initial processing of images consists of filtration of the pollen from noise and in splitting of image into many sets of components of the original image of the slide. At the solution of this task, the principal part is searching "for double borders" in type of pollen, for identification of which the "Snake" model [5] is required. Functioning of the "Snake" model is based on many kinds of pollen that have thin endings or double borders (exines) which can be noticed under the microscope, and a number of types of these borders are the main tools in separating the pollen grains from the dust.

The "Snake" model reveals "internal reality" of the sub-sample and makes such actions as "splayn fine plate" or membrane. "Splayn fine plate" serves for identification of parts of the pollen grains. Grain must be clear with the first or the second order of breakup. In order to connect parts of the pollen grains and mark the necessary point the "external links" of the model are used. The model allows one to calculate the passing of the object through any double borders on the image by filtration way. The identification of the double borders is determined by using, basically, the round nature of pollen. Therefore, it is necessary to accomplish simple transformation from Cartesian space into the (r, Q)-space. As a result, borders of pollen will be approximated by a straight line of dimensionality Q. The transformation is realized by Saborov's determinant, which connects the width of the borders and orientation in the Q-section. The oriented determinant is used with double borders and can be analyzed as follows. The filter $f(l, g)$ is used

$$f(l, g) = \begin{bmatrix} 1\,1\,1\ldots 1 \\ 0\,0\,0\ldots 0 \\ \vdots\,\vdots\,\vdots\quad\vdots\quad\vdots \\ 0\,0\,0\ldots 0 \\ 1\,1\,1\ldots 1 \end{bmatrix}$$

The filter defines double lines, where lines are divided by g-pixels, and l is the pixel length. Using the filter takes out the hindrances from isolated pairs of g-pixels and brings to the change of the width of the double borders. Therefore, the result of the transformation of the pollen and not pollen object in (r, Q)-space is the recognizer of the double borders.

4 Principles of the Training of the Formal Neuron

Important forming part of the software system of the recognition and classification of the microobjects is using algorithms for NN training, which are recognizing perceptron model, and which allow one to classify objects for their similarity with given specification and specific characteristics keeping as standard. It is necessary to notice that the methods of the recognition and classification of the pollen based on NN create fixed classes of the data for training. If there is a need to enter the new type of pollen, these methods require that a full set of data both as old images and new ones were for retraining of the classifier. Moreover, the adaptation of the processes of the retraining is executed by the active participation of experts in the processing of the images of the microobjects.

Let us consider the mathematical principles underlying the training of the neural network. The main element of artificial neural network is a formal neuron, which represents a multi-output nonlinear converter with adaptive weighting of input signal. All neurons of the network are connected with each other by synapse links in definite architecture. We show in Fig. 1 the McCalloh–Pitts' model of the formal neuron. The following notation is used in the model: $x_1, x_2, ..., x_n$ are the input signals; $W_{j1}, W_{j2}, ..., W_{jn}$ are the adjusted synapse weights of jth neuron of the network; θ_j is a displacing signal; $\delta(\cdot)$ is a sigmoid function of activations; γ_j is a parameter assigning "steepness" of function of activation; Γ_j is the coefficient of consolidation defining the maximum and minimum values of outgoing signal.

The outgoing signal of such a neuron may be written as follows:

$$y_j = \Gamma_j \sigma(\gamma_j W_j^T x) \tag{1}$$

where $w_j = (\vartheta_j, w_{j1}, w_{j2}, \ldots, w_{jn})^T$.

It was proved in [6] that the NN with one hidden layer formed with such neuron can approximate any function with any accuracy given in advance. Therefore, the outgoing signal of the network consisting of N neurons has the form:

$$y = \sum_{j=1}^{n} \Gamma_j \sigma(\gamma_j W_j^T x)$$

and approximates the continuous function $f(x)$ providing the condition $|y - f(x)| < \varepsilon$ for all possible entries of h belonging to some hypercube.

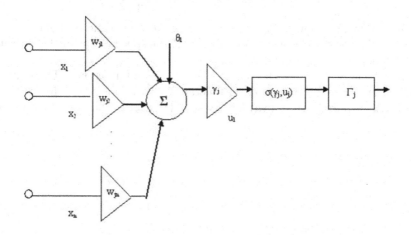

Fig. 1. The model of the formal neuron

Note that approximating properties of the specific neuron heavily depend on the choice of the form of the function $\delta(\cdot)$. As such a function we can propose the sigmoid function

$$0 < \sigma(\gamma y) = (1 + e^{-\gamma u})^{-1} < 1, \tag{2}$$

determined on the set of all real numbers and taking only positive values. Curved hyperbolic tangent turned out to be the most suitable

$$-1 < \tanh(\gamma u) = \frac{1 - e^{-2\gamma u}}{1 + e^{-2\gamma u}} < 1,$$

which is connected with the sigmoid function by the relation

$$\sigma(\gamma u) = \frac{1}{2}\left(\tanh\left(\frac{\gamma u}{2}\right) + 1\right).$$

Assigning the restrictions on the square $-1 \le u_j \le 1$, $-1 < \gamma_j < 1$ as possible functions of activation of neuron (1) the following is mostly used:

$$\sigma^1(\gamma u) = \tanh(\gamma u) = \frac{1 - e^{-2\gamma u}}{1 + e^{-2\gamma u}}, \quad \Gamma < \frac{1}{\tanh\gamma} \tag{3}$$

$$\sigma^2(\gamma u) = \frac{\gamma u}{\sqrt{1 + \gamma^2 u^2}}, \quad \Gamma^2 < \frac{\sqrt{1 + \gamma^2}}{\gamma} \tag{4}$$

$$\sigma^3(\gamma u) = \sin\left(\frac{\pi}{2}\gamma u\right), \quad \Gamma^3 < \frac{1}{\sin\left(\frac{\pi}{2}\gamma\right)} \tag{5}$$

$$\sigma^4(\gamma u) = \frac{2}{\pi}\arctan(\gamma u), \quad \Gamma^4 < \frac{\pi}{2\arctan\gamma} \tag{6}$$

$$\sigma^5(\gamma u) = \gamma u - \frac{y^3}{3}u^3, \quad \Gamma^5 < \frac{3}{3\gamma - \gamma^3} \tag{7}$$

As an example, we show in Fig. 2 the graphs of the dependencies of data for functions (3)–(7). As can be seen from the graphs, the specific type of the activation function depends on parameter γ_j.

Their identity can be achieved by the corresponding choice. The curves in Fig. 2 correspond to the following functions: $\sigma^1(\gamma u)$ for $\gamma = 3.5$; $\sigma^2(\gamma u)$ for $\gamma = 4$; $\sigma^3(\gamma u)$ for $\gamma = 1$; $\sigma^4(\gamma u)$ for $\gamma = 6$; $\sigma^5(\gamma u)$ for $\gamma = 1$.

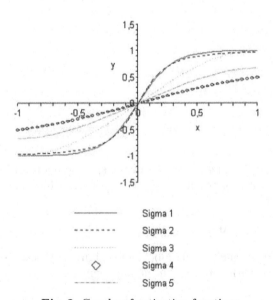

Fig. 2. Graphs of activation functions

It is determined that presented activation functions allow one not only to simplify the process of training but also to enter unified scheme of the training of the formal neuron.

5   Mathematical Model and Algorithms of Training of Multilayer NN for the Recognition of Microobjects

Software system of recognition and classification can be built on the basis of different algorithms of training NN, for example, Hopfield's and Hemming's algorithms, training of RpRop with teacher and without teacher, on the basis of bidirectional associative memory, etc. [7].

As a rule, for training of multilayer perceptron (MLP) problems of recognition of microimages, for example, five different incoming classes (by types of objects) it is necessary to compose a network with five neurons at the outgoing layer. In such a network, every outgoing neuron corresponds to some class of objects and neurons with the greatest outgoing value, it is used for determination for what class the network carries the given output. Quantity of neurons in incoming layer is usually equal to the quantity of properties allocated from the image at early stages of processing. During training the NN, the studying example is represented to the input of the network, and the resulting output is calculated. It is compared with the desirable answer, and errors are calculated. These errors are then used for changing the weights of neurons so that when this sample will be given next time, the output would be closer to the desirable result. The training set is presented to the network repeatedly so long till the network will learn to identify examples effectively. The various obtained weights of the multilayer perceptron must be settled. Any attempt to train the network to the new data will lead to the result that it will forget the old data. As an example, let us consider the algorithm of training the return distribution, which concerns the algorithms of training with teacher. Perceptron consists of several layers of neurons, in which every neuron of the layer i is connected with every neuron of layer $i + 1$. In general, such problem, with the limited set of input data, has an infinite set of solutions. For the restriction of space of search at training, the task of minimizing the target function of the NN error is defined, which is decided by the method of the least squares:

$$E(w) = \frac{1}{2} \sum_{j=1}^{p} (y_i - d_j)^2, \tag{8}$$

where y_j is the value of the jth output of neural network; d_j is a target value of the jth output; p is the number of neurons at the outgoing layer. Training of the neural network is made by the gradient descent method, in particular, at every iteration changing the weight is represented by the formula:

$$\Delta w_{ij} = -\eta \cdot \frac{\partial E}{\partial w_{ij}}, \tag{9}$$

where η is a parameter determining the speed of training.

Using the known relations of gradient descent method formula (9) may be rewritten in the open form

$$\Delta w_{ij} = -\eta \cdot \delta_j^{(n)} \cdot x_i^n, \tag{10}$$

where x_i^n is the value of the ith input of neuron of the nth layer; $\delta_i^{(n)}$ is the auxiliary variable of the nth layer for finding the targeted vector, in particular, vector of those values, which must give out NN in such a given set of input values.

General algorithm of training the NN may be presented as follows:

1. Submit to NN's input one of the required images and identify the values of output of the neurons of neuron network.
2. Calculate $\delta^{(n)}$ for the output layer of NN and calculate the changes of the weights $\Delta w_{ij}^{(N)}$ of output layer of N by formula (10).
3. Calculate (N) and $\Delta w_{ij}^{(N)}$ for the rest of layers of NN, $n = N - 1...1$.
4. Correct all weights of NN by formula

$$w_{ij}^{(n)}(t) = w_{ij}^{(n)}(t-1) + \Delta w_{ij}^{(n)}(t) \tag{11}$$

5. If the error is too large then move to step 1.

It is important to note that the simplest method of gradient descent considered above is inefficient in case when the derivatives with respect to various weights strongly differ. It corresponds to the situation when the value of the function S for several neurons is close to 1 in its absolute value or when some weights are much higher than 1 [8] in their absolute values.

Therefore, we will propose the simplest method of improvement of the gradient descent by introduction of the moment μ, when influence of the gradient on the changes of weights varies with the time. Then formula (11) takes the following form:

$$\Delta w_{ij}^{(n)}(t) = -\eta \cdot \delta_j^{(n)} \cdot x_i^n + \mu \cdot \Delta w_{ij}^{(n)}(t-1).$$

An additional advantage from the introduction of the moment is the ability of the algorithm to overcome fine local minima.

6   The Model of Functioning of the Software System for Recognition and Classification of Microobjects

Let us state the conceptual model of functioning of the software system for the recognition and classification constructed on the basis of the developed models and algorithms of NN. The hierarchy of classes has the following structure: neural network (object of the CNetH class) aggregates the set of layers of neurons (the file objects of the base class of CBaseLayer); every layer aggregates the set of neurons (the file objects of the base class CBaseNeuron). After beginning of work, the programs automatically load the previous adjustments of network (the size of the images to be processed by network; the number of the images-originals, etc.), and the NN is created after which it is training by itself, which takes place on the basis of the images-originals stored in database (the weights of synapses of all neuron networks are established). Recognizing images with microsamples, the network sends the array of points into the first layer (to the entrance of neurons of the first layer), which, after a certain processing of these signals, sends the array of signals-results of the processing to the second layer, etc. As a result, at the output of the network, the number of images-originals are composed, which are situated in the base of the originals, which is most similar to the image elaborated by the network, in particular, classification of the initial object takes place. Therefore, because of every microsample-original, which "is remembered" by the network, can be put in correspondence to any line (for example, the name of microobject) or number, then the index of the sample-original received at the output of a network can be used for transformation of any image into the textual or digital form in machine representation, in particular, to carry out recognition of the initial of arbitrary object. In case of changing the parameters of network, for instance, when changing the number of images-originals on which the network is training (changing of the base of images-originals) or changing the sizes of samples, which network works with, the old network is destroyed, and instead of it a new network with new characteristics is composed. The diagram of NN class will be represented in accordance with the described hierarchy of classes. The details of the names of classes, used variables, and their types are given in the diagram, the methods of objects in the class are specified.

CBaseLayer
pNeurons: CBaseNeuron*
nInputs: int
nNeurons: int
pInputs: float*
pSynapses: float*
pAxons: float*
+ CBaseLayer ()
+   CBaseLayer ()
+ GetInputs(): float*
+ GetAxons(): float*
+ Create(int nInps,int nNrns,float* pInpts,float* pSyns): float*
+ CalcLayer(): CHRESULT

↓

CBaseNeuron   State: float
pAxon: float*
nInputs: int
pInputs: float*
pSynapses: float*
SigmoidType: uns
signed char

SigmoidAlfa: float
+ CBaseNeuron()
+ CBaseNeuron()
♯ Sigmoid(): float
+ Create(int nInps,float* pInps, float* pSyns, float* pAxn,
unsigned char ST = ST_NONE, float SA = 0., float fRes = 0., int iRes = 0):
int
+ Randomize(float MaxRange=1.): void
+ CalcState(): float
+ Enter(): float

↓

CFirstLayerH + CFirstLayerH()
+ CFirstLayerH()
+ Create(int nInps,int nNrns,float* pInpts,float* pSyns): float*

↓

CSecondLayerH
+ CSecondLayerH ()
+ CSecondLayerH ()
+ Create(int nInps,int nNrns,float* pInpts,float* pSyns): float*
- IsConverged(): int
+ CalcLayer(): CHRESULT

↓

CFirstLayerHNeuron
+ CFirstLayerHNeuron ()
+ CFirstLayerHNeuron ()
+ Create(int nInps,float* pInps, float* pSyns, float* pAxn,
unsigned char $ST = ST\_NONE$, float SA = 0., float fRes = 0., int iRes = 0):
int

↓

CSecondLayerHNeuron
+ CSecondLayerHNeuron ()
+ CSecondLayerHNeuron ()
+ Create(int nInps,float* pInps, float* pSyns, float* pAxn,
unsigned char ST = ST_NONE, float SA = 0., float fRes = 0., int iRes = 0):
int

For simplicity of presentation of the working the methods of classes, let us submit the names and functions for some of them.

Methods of Class CBaseLayer

GetInputs (): float* returns pointer on array input layer.

GetAxons (): float* returns pointer on array axon layer.

Create (int nInps, int nNrns, float* pInpts, float* pSyns): float* is initialization of a layer. It dynamically creates an array neuron and the array axons (outputs) of neuron for the given layer; and also initializes statistical variables, for example, number of neurons or number of inputs in the given layer.

CalcLayer (): CHRESULT is the method expecting the given layer, i.e. starts all neurons of layer for processing of input signal of layer.

Methods of Class CBaseNeuron

Sigmoid (): float forms output values of neuron (value of axon) on the basis of the current condition of neuron with the help of the activation function used at the given moment of time.

Create (int nInps, float* pInps, float* pSyns, float* pAxn, unsigned char ST = ST_NONE, float SA = 0., float fRes = 0., int iRes = 0): int is initialization of neuron.

Randomize (float MaxRange = 1.): void establishes any values on synapse of neuron. The method is intended for debugging.

CalcState (): float forms a condition of neuron on the basis of signals on inputs of neuron and synapse weights.

Enter (): float makes complete account of neuron and returns result-value of its axon.

Thus, in the present work, an effective and new solving approach is proposed for recognition of the microobjects by the construction of computer system on the basis of models of preliminary processing of the images and using of NN, common principles and models of creating the system of recognition of microobjects is elaborated, methods of training and calculation of weight coefficients of synapses of the stage of initialization of the network as well as the algorithms and software-realizing models of functioning of the network are proposed.

References

1. Tomczak, R., Rouquet, C., Bonton, P.: Colour image segmentation in microscopy: application to the automation of pollen rates measurement. In: CGIP'2000. Proc. Int. Conf. on Color in Graphics and Image Processing, Saint-Etienne (France), pp. 87–96 (2000)
2. Haralick, R.M., Shanmagan, K.: Textural features for image classification. IEEE Trans. Syst., Man and Cybern. SMC-3, 610–621 (1973)
3. France, I., Duller, A.W.G, Lamb, H.F, Duller, G.A.T.: Comparison of the approaches for automatic recognition of pollen. Sch. of Electronic Engineering and Computer Systems, Inst. of Earth Studies, University of Wales, pp. 56–61 (2000)
4. Fukushima, K.: Neocognitron: A self organizing neural network model for a mechanism of pattern recognition unaffected by shift in pattern. Biological Cybernetics, 193–202 (1980)

5. Kass, M., Witkin, A., Terzopoulos, D.: Snakes: Active contour models. In: Proc. First Int. Conf. on Computer Vision, pp. 259–268 (1987)
6. Cichocli, A., Unbehauen, R.: Neural networks for optimization and signal processing. Teubner, Stuttgart (1993)
7. Narzullayev, U.Kh., Djumanov, O.I., Akhatov, A.R.: Information technologies for processing of microobjects images on the basis of neural networks with associative memory. In: The 5th Int. Conf. on ALPIT, 2006. - TUIT, UZBEKISTAN, Korea, pp. 114–118. Information Processing Society, Korea (2006)
8. Korotkiy, S.: Networks of neuron: algorithm of back propagation. Computer Press 7, 48–54 (1998) (in Russian)

Analytical Solution for Transient Flow of a Generalized Bingham Fluid with Memory in a Movable Tube Using Computer Algebra*

Juan Ospina and Mario Velez

Department of Physical Engineering
Logic and Computation Group
EAFIT University
Medellin, Colombia
{jospina,mvelez}@eafit.edu.co

Abstract. A rheological linear model for a certain generalized Bingham fluid with rheological memory, which flows in a movable tube is proposed and analytically solved. The model is a system of two linear and coupled partial differential equations with integral memory. We apply the Laplace transform method making the inverse transform by means of the Bromwich integral and the theorem of residues and the analytical solution are obtained using computer algebra. We deduce the explicit forms of the velocity and stress profiles for the generalized Bingham fluid in terms of Bessel and Struve functions. Various limit cases are obtained and the standard Hagen-Poiseuille and Buckingham-Reiner equations are recovered from more general equations. This works shows the powerful of Maple to solve complex rheological problems in an analytical form as it is presented here by the first time.

1 Introduction

Mathematical Rheology is a permanent source of very interesting computational problems. In consequence it is possible to speak about the Computational Rheology [1],[2],[3] as a separate and well defined discipline within the domain of Rheology. Inside the Computational Rheology it is possible to realize the existence of two different but linked trends. The first trend is named here Numeric Computational Rheology (NCR) [1,2,3] and the second trend is called here Symbolic Computational Rheology (SCR) [4]. As its name indicates, the NCR is concerned with the numerical solution of non-linear rheological equations, using software for numerical computation, such as Matlab. From the other side, the SCR is dedicated to calculation of analytical solutions for linear or linearized rheological equations, using computer algebra software [5], like Maple [6], which is able to make symbolic or algebraic computations. Given that Rheology is a highly non-linear science is natural that the dominant trend be actually justly the NCR [1,2,3]. But the SCR is a very interesting and practically unexplored

---

* This work is supported by EAFIT University.

V.G. Ganzha, E.W. Mayr, and E.V. Vorozhtsov (Eds.): CASC 2007, LNCS 4770, pp. 339–349, 2007.

land [4]. The present authors think that is worthwhile to make an exploration of SCR and this work intends to bring the reader some of the flavor of SCR. A source of inspiration for this work is the wonderful paper [7]. The example that was chosen here, was the Bingham fluid given such fluid has a linear rheological equation [8]. Specifically we consider here the case of transient flow for a Bingham fluid in a movable tube, when the fluid starts from the rest completely relaxed. The transport equation for such transient flow is linear and it can be solved analytically. More over in this work we introduced the extra complications that derive of the introduction of memory effects in the rheological equation of the Bingham fluid. In this case we are concerned with a generalized Bingham fluid with rheological memory [4]. The resultant transport equation for the transient flow of such generalized Bingham fluid in a movable tube, is again a linear equation and it can be solved analytically as it will be showed here. Our method of solution is the Laplace transform technique with Bromwich integral and residue theorem [4], [9]. The final analytical solutions will be given in terms of the special functions of Mathematical Physics such as Bessel functions [10] and Struve functions [11]. Given that the linear transport equations for our generalized Binghman fluid, are very large and the mathematical procedure of solution is very hard as for to be implemented by hand using only pen and paper, we use Computer Algebra Software (CAS) to make all symbolic computations. We find that CAS is a very valuable tool for the engineer that works with non-newtonian fluids when analytical solutions are demanded.

2   Mathematical Problem

We consider here the transient flow in a movable tube for a certain generalized Bingham fluid. Within such fluid we find two different kinds of fields, to know: the velocity field and the stresses field. The rheological relation for a generalized Bingham fluid with memory, which links the velocity field and the stresses field is the constitutive equation which reads

$$\sigma_{r,z}(r,t) + \mu \frac{\partial}{\partial r} v(r,t) - \tau_0 + \chi \int_0^t e^{-\epsilon(t-\tau)} \sigma_{r,z}(r,\tau) d\tau +$$

$$\mu_1 \int_0^t e^{-\epsilon_1(t-\tau)} \frac{\partial}{\partial r} v(r,\tau) d\tau = 0, \tag{1}$$

where $v(r,t)$ is the velocity field within the fluid, $\sigma_{r,z}(r,t)$ is the stresses field, μ denotes the Bingham plasticity, χ is the amplitude of the memory effect on the stresses field, being ϵ the attenuation factor for such memory; μ_1 is the amplitude of the memory effect on the velocity field, being ϵ_1 the corresponding attenuation factor; and τ_0 is the yield stress. The reader can note that when $\chi = 0$, $\epsilon \to \infty$, $\mu_1 = 0$ and $\epsilon_1 \to \infty$, the equation (1) is reduced to the standard Bingham rheological equation [8]. The equation (1) is for flow in a circular and infinitely long tube, being r the distance from the axis of the tube with radius a. For the transient flow in a tube we have at general two unknowns: $v(r,t)$ and

$\sigma_{r,z}(r,t)$. We need other equation which complements to (1). Such equation is the general equation of the movement of continuum media in a circular and very long tube, when a fully development laminar flow is established, namely [8]

$$\rho \frac{\partial}{\partial t} v(r,t) + \frac{d}{dz} P(z) + \frac{\sigma_{r,z}(r,t) + r \frac{\partial}{\partial r} \sigma_{r,z}(r,t)}{r} = 0 \qquad (2)$$

where r is the density of the fluid assumed incompressible and $dP(z)/dz$ is the pressure gradient. The initial condition that we use here, is that the fluid at $t = 0$ starts from the rest completely relaxed, it is to say

$$v(r,0) = 0, \qquad (3)$$

$$\sigma_{r,z}(r,0) = 0. \qquad (4)$$

From the other side, the boundary condition that we consider here is the corresponding to a movable tube. At particular we use the following boundary condition

$$v(a,t) = \nu e^{-\delta t}, \qquad (5)$$

where ν is the initial velocity and δ is the attenuation factor for such velocity.

Then, the mathematical problem that is proposed in this section consists in to obtain the analytical solution of the system (1)-(2) with the initial conditions (3)-(4) and with boundary condition (5) jointly with a natural finitude condition for solutions. As we can observe the problem (1)-(5) is a linear problem but it is a formidable linear problem given that the analytical solution exists by general mathematical theorems but the explicit computation of the solutions is very difficult.

3   Method of Solution

The mathematical problem given by (1)-(5) is a linear problem but it can not be solved directly using the standard method of separation of variables due to the inhomogeneous boundary condition (5). The most effective method of solution is the Laplace transform technique enriched with the Bromwich integral and residue theorem [4],[9]. Such method can be implemented by CAS and the details are as follows [4]: Taken the Laplace transform of (1) we have

$$\Sigma(r) + \mu \frac{d}{dr} V(r) - \frac{\tau_0}{s} + \frac{\chi \Sigma(r)}{s + \epsilon} + \frac{\mu_1 \frac{d}{dr} V(r)}{s + \epsilon_1} = 0, \qquad (6)$$

where $V(r)$ and $\Sigma(r)$ are the Laplace transforms of $v(r,t)$ and $\sigma_{r,z}(r,t)$ respectively. Solving (6) with respect to $\Sigma(r)$ gives

$$\Sigma(r) = A(s) \frac{d}{dr} V(r) + B(s), \qquad (7)$$

where

$$A\left(s\right) = -\frac{\left(\mu\,s + \mu\,\epsilon_1 + \mu_1\right)\left(s + \epsilon\right)}{\left(s + \epsilon_1\right)\left(s + \epsilon + \chi\right)}, \tag{8}$$

$$B\left(s\right) = \tau_0 s^{-1}\left(1 + \frac{\chi}{s + \epsilon}\right)^{-1}. \tag{9}$$

Now, taking the Laplace transform of (2) and using (3) and (7) the following differential equation is obtained

$$\rho\,sV\left(r\right) + \frac{A\left(s\right)\frac{d}{dr}V\left(r\right)}{r} + A\left(s\right)\frac{d^2}{dr^2}V\left(r\right) + \frac{\left(\frac{d}{dz}P\left(z\right)\right)r + sB\left(s\right)}{sr} = 0. \tag{10}$$

The general solution of (10) such as computed by Maple [6] is

$$V\left(r\right) = J_0\left(\sqrt{\frac{\rho\,s}{A\left(s\right)}}r\right)C_2 + Y_0\left(\sqrt{\frac{\rho\,s}{A\left(s\right)}}r\right)C_1 + $$
$$\frac{1}{2\rho s^2}\left(-2\frac{d}{dz}P\left(z\right) - \pi\,sB\left(s\right)\sqrt{\frac{\rho\,s}{A\left(s\right)}}H_0\left(\sqrt{\frac{\rho\,s}{A\left(s\right)}}r\right)\right), \tag{11}$$

where $J_0(x)$ and $Y_0(x)$ are the Bessel functions of zero order of the first class [10] and $H_0(x)$ is the Struve function of zero order [11]. Given that $Y_0(x)$ is singular at $x = 0$ [10], we chose that $C_1 = 0$ to avoid the singularities in the velocity profile. For hence $V(r)$ turns:

$$V\left(r\right) = J_0\left(\sqrt{\frac{\rho\,s}{A\left(s\right)}}r\right)C_2 + $$
$$\frac{1}{2\rho s^2}\left(-2\frac{d}{dz}P\left(z\right) - \pi\,sB\left(s\right)\sqrt{\frac{\rho\,s}{A\left(s\right)}}H_0\left(\sqrt{\frac{\rho\,s}{A\left(s\right)}}r\right)\right). \tag{12}$$

To determine the constant C_2, we use the boundary condition (5) which is Laplace-transformed to:

$$V\left(a\right) = \frac{\nu}{s + \delta}, \tag{13}$$

and the constant C_2 is computed with Maple [6] and the result is

$$C_2 = \frac{\left(2\,s + 2\,\delta\right)\frac{d}{dz}P\left(z\right) + s\pi\,B\left(s\right)\sqrt{\frac{\rho\,s}{A\left(s\right)}}\left(s + \delta\right)H_0\left(\sqrt{\frac{\rho\,s}{A\left(s\right)}}a\right) + 2\,\nu\,\rho\,s^2}{2\,s^2 J_0\left(\sqrt{\frac{\rho\,s}{A\left(s\right)}}a\right)\rho\left(s + \delta\right)}. \tag{14}$$

Now, with the substitution of (14), (8) and (9) in (12) we obtain the Laplace-transformed velocity profile which is given by

$$V\left(r\right) = \frac{NV(r)}{DV(r)}, \tag{15}$$

where

$$NV(r) = 2 J_0\left(\sqrt{\frac{\rho s}{A(s)}}r\right)\left(\frac{d}{dz}P(z)\right)s^2 + 2 J_0\left(\sqrt{\frac{\rho s}{A(s)}}r\right)\left(\frac{d}{dz}P(z)\right)s\epsilon +$$

$$2 J_0\left(\sqrt{\frac{\rho s}{A(s)}}r\right)\left(\frac{d}{dz}P(z)\right)s\chi + 2 J_0\left(\sqrt{\frac{\rho s}{A(s)}}r\right)\left(\frac{d}{dz}P(z)\right)\delta s +$$

$$2 J_0\left(\sqrt{\frac{\rho s}{A(s)}}r\right)\left(\frac{d}{dz}P(z)\right)\delta\epsilon + 2 J_0\left(\sqrt{\frac{\rho s}{A(s)}}r\right)\left(\frac{d}{dz}P(z)\right)\delta\chi +$$

$$J_0\left(\sqrt{\frac{\rho s}{A(s)}}r\right)\pi s^2\tau_0\sqrt{\frac{\rho s}{A(s)}}H_0\left(\sqrt{\frac{\rho s}{A(s)}}a\right) +$$

$$J_0\left(\sqrt{\frac{\rho s}{A(s)}}r\right)\pi s\tau_0\sqrt{\frac{\rho s}{A(s)}}H_0\left(\sqrt{\frac{\rho s}{A(s)}}a\right)\epsilon +$$

$$J_0\left(\sqrt{\frac{\rho s}{A(s)}}r\right)\pi \tau_0\sqrt{\frac{\rho s}{A(s)}}H_0\left(\sqrt{\frac{\rho s}{A(s)}}a\right)\delta s +$$

$$J_0\left(\sqrt{\frac{\rho s}{A(s)}}r\right)\pi\tau_0\sqrt{\frac{\rho s}{A(s)}}H_0\left(\sqrt{\frac{\rho s}{A(s)}}a\right)\delta\epsilon + 2 J_0\left(\sqrt{\frac{\rho s}{A(s)}}r\right)\nu\rho s^3 +$$

$$2 J_0\left(\sqrt{\frac{\rho s}{A(s)}}r\right)\nu\rho s^2\epsilon + 2 J_0\left(\sqrt{\frac{\rho s}{A(s)}}r\right)\nu\rho s^2\chi -$$

$$2 J_0\left(\sqrt{\frac{\rho s}{A(s)}}a\right)\left(\frac{d}{dz}P(z)\right)s^2 - 2 J_0\left(\sqrt{\frac{\rho s}{A(s)}}a\right)\left(\frac{d}{dz}P(z)\right)\delta s -$$

$$2 J_0\left(\sqrt{\frac{\rho s}{A(s)}}a\right)\left(\frac{d}{dz}P(z)\right)s\epsilon - 2 J_0\left(\sqrt{\frac{\rho s}{A(s)}}a\right)\left(\frac{d}{dz}P(z)\right)\delta\epsilon -$$

$$2 J_0\left(\sqrt{\frac{\rho s}{A(s)}}a\right)\left(\frac{d}{dz}P(z)\right)s\chi - 2 J_0\left(\sqrt{\frac{\rho s}{A(s)}}a\right)\left(\frac{d}{dz}P(z)\right)\delta\chi -$$

$$J_0\left(\sqrt{\frac{\rho s}{A(s)}}a\right)\pi\tau_0\sqrt{\frac{\rho s}{A(s)}}H_0\left(\sqrt{\frac{\rho s}{A(s)}}r\right)s^2 -$$

$$J_0\left(\sqrt{\frac{\rho s}{A(s)}}a\right)\pi\tau_0\sqrt{\frac{\rho s}{A(s)}}H_0\left(\sqrt{\frac{\rho s}{A(s)}}r\right)s\delta -$$

$$J_0\left(\sqrt{\frac{\rho s}{A(s)}}a\right)\pi\tau_0\sqrt{\frac{\rho s}{A(s)}}H_0\left(\sqrt{\frac{\rho s}{A(s)}}r\right)\epsilon s -$$

$$J_0\left(\sqrt{\frac{\rho s}{A(s)}}a\right)\pi\tau_0\sqrt{\frac{\rho s}{A(s)}}H_0\left(\sqrt{\frac{\rho s}{A(s)}}r\right)\epsilon\delta \qquad (16)$$

and

$$DV(r) = 2(s+\epsilon+\chi)s^2 J_0\left(\sqrt{\frac{\rho s}{A(s)}}a\right)\rho(s+\delta). \qquad (17)$$

As we can observe in (15), (16) and (17) the Laplace-transformed velocity profile is a super very large equation which is very difficult and maybe impossible to be obtained using only pen and paper with calculations by hand. Immediately we appreciate the invaluable useful of CAS and particularly Maple. Now the question is to obtain the inverse Laplace transform of the transformed profile

$V(r)$ which is given by (15), (16) and (17). Unfortunately the actual CAS are not able to make the transformation automatically and in consequence it is necessary to introduce within the Maple environment both the Bromwich integral and the residue theorem [4], [9]. The reader can note that the poles of $V(r)$ according with (15) and (17), are the roots of the following equations respect to the Laplace parameter s:

$$s^2 = 0, \tag{18}$$

$$s + \delta = 0, \tag{19}$$

$$J_0\left(\sqrt{-\frac{\rho s (s + \epsilon_1)(s + \epsilon + \chi)}{(\mu s + \mu \epsilon_1 + \mu_1)(s + \epsilon)}} a\right) = 0, \tag{20}$$

$$s + \epsilon + \chi = 0. \tag{21}$$

The equation (18) indicates the presence of a pole of second order at $s = 0$. Similarly (19) shows that a pole exists at $s = -\delta$. Now (20) gives a infinite set of poles as the roots of the infinite set of equations of the form

$$\sqrt{-\frac{\rho s (s + \epsilon_1)(s + \epsilon + \chi)}{(\mu s + \mu \epsilon_1 + \mu_1)(s + \epsilon)}} a = \alpha_n, \tag{22}$$

where α_n are the zeroes of J_0, it is to say the roots of the equation $J_0(x) = 0$, with $n = 1..\infty$ [10]. Finally the equation (21) says that there is a pole at $s = -\epsilon - \chi$. With all these poles, the inverse Laplace of $V(r)$ in the Figure 1 can be calculated as the summation of residues for all poles defined by (18)-(21). Such computations of residues can be implemented by Maple.

4   Results

The explicit formula for the velocity profile $v(r, t)$ which derives from $V(r)$ given in (15), (16) and (17), via the inverse Laplace transformation is:

$$v(r,t) = \frac{1}{4}\frac{\left(\epsilon r^2 - \epsilon a^2 + r^2\chi - \chi a^2\right)\epsilon_1 \frac{d}{dz}P(z)}{\epsilon(\mu_1 + \mu\epsilon_1)} + \frac{(-a+r)\epsilon_1\tau_0}{\mu_1 + \mu\epsilon_1} +$$
$$\frac{J_0\left(\sqrt{\frac{\rho\delta(\epsilon_1-\delta)(\epsilon+\chi-\delta)}{(-\mu\delta+\mu_1+\mu\epsilon_1)(-\delta+\epsilon)}}r\right)\nu e^{-\delta t}}{J_0\left(\sqrt{\frac{\rho\delta(\epsilon_1-\delta)(\epsilon+\chi-\delta)}{(-\mu\delta+\mu_1+\mu\epsilon_1)(-\delta+\epsilon)}}a\right)} +$$
$$\sum_{i=1}^{3}\sum_{n=1}^{\infty}\frac{J_0\left(\frac{\alpha_n r}{a}\right)F_{n,i}e^{S_{i,n}t}\alpha_n\left(\mu S_{i,n} + \mu\epsilon_1 + \mu_1\right)^2\left(S_{i,n} + \epsilon\right)^2}{a^3 S_{i,n}{}^2\rho^2\left(S_{i,n} + \epsilon + \chi\right)\left(S_{i,n} + \delta\right)J_1\left(\alpha_n\right)G_{n,i}}, \tag{23}$$

where

$$
F_{n,i} = 2\left(\frac{d}{dz}P(z)\right)S_{i,n}{}^2 a + 2\left(\frac{d}{dz}P(z)\right)S_{i,n}\epsilon\,a + 2\left(\frac{d}{dz}P(z)\right)S_{i,n}\chi\,a +
$$
$$
2\left(\frac{d}{dz}P(z)\right)\delta\,S_{i,n}a + 2\left(\frac{d}{dz}P(z)\right)\delta\epsilon\,a + 2\left(\frac{d}{dz}P(z)\right)\delta\chi\,a +
$$
$$
\pi\,S_{i,n}{}^2\tau_0\alpha_n H_0\left(\alpha_n\right) + \pi\,S_{i,n}\tau_0\alpha_n H_0\left(\alpha_n\right)\epsilon + \pi\,\tau_0\alpha_n H_0\left(\alpha_n\right)\delta\,S_{i,n} +
$$
$$
\pi\,\tau_0\alpha_n H_0\left(\alpha_n\right)\delta\,\epsilon + 2\,\nu\,S_{i,n}{}^3\rho\,a + 2\,\nu\,S_{i,n}{}^2\rho\epsilon\,a + 2\,\nu\,S_{i,n}{}^2\rho\chi\,a \qquad (24)
$$

and

$$
G_{n,i} = \epsilon_1\chi\,\epsilon\,\mu_1 + \epsilon_1{}^2\chi\epsilon\,\mu + 2\,S_{i,n}\epsilon^2\mu\epsilon_1 + 4\,S_{i,n}{}^2\mu\,\epsilon_1\epsilon + S_{i,n}{}^2\chi\,\mu\,\epsilon + 2\,S_{i,n}\chi\,\mu_1\epsilon +
$$
$$
2\,\epsilon_1{}^2 S_{i,n}\mu\,\epsilon + 2\,\epsilon_1 S_{i,n}\mu_1\epsilon + 2\,S_{i,n}{}^3\mu_1 + S_{i,n}{}^4\mu + 2\,S_{i,n}\chi\,\mu\,\epsilon_1\epsilon + \epsilon_1\epsilon^2\mu_1 +
$$
$$
\epsilon_1{}^2\epsilon^2\mu + 2\,S_{i,n}{}^3\mu\,\epsilon_1 + S_{i,n}{}^2\epsilon^2\mu + 2\,S_{i,n}\epsilon^2\mu_1 + S_{i,n}{}^2\chi\,\mu_1 + \epsilon_1{}^2 S_{i,n}{}^2\mu +
$$
$$
\epsilon_1 S_{i,n}{}^2\mu_1 + 2\,S_{i,n}{}^3\mu\,\epsilon + 4\,S_{i,n}{}^2\mu_1\epsilon, \qquad (25)
$$

and being $S_{i,n}$, with $i = 1..3$, $n = 1..\infty$ the roots of the equation (22) rewritten here as:

$$
\rho\,s^3 a^2 + \left(\rho\,a^2\chi + \rho\,a^2\epsilon + \rho\,a^2\epsilon_1 + \alpha_n{}^2\mu\right)s^2 +
$$
$$
\left(\alpha_n{}^2\mu\epsilon + \alpha_n{}^2\mu_1 + \rho\,a^2\chi\,\epsilon_1 + \rho\,a^2\epsilon\,\epsilon_1 + \alpha_n{}^2\mu\epsilon_1\right)s +
$$
$$
\alpha_n{}^2\mu\,\epsilon\,\epsilon_1 + \alpha_n{}^2\mu_1\epsilon = 0. \qquad (26)
$$

In (23), the transient velocity profile $v(r,t)$ is the summation of three residues. The first line in (23) derives from the residue at $s = 0$, and it is independent of time and corresponds to the stationary or permanent fully development laminar Bingham flow. The second line in (23) results from the residue at $s = -\delta$ and represents a transitory flow at consonance with the transitory movement of the tube itself. And the third line in (23) results of the summation of the infinitely many residues for the infinitely many poles at $s = S_{i,n}$ according with (26), and correspond with the transient flow determined by the rheological properties of the fluid. The corroboration that the third term is a transient term is reached using stability analysis according with the Routh-Hurwitz theorem (RHT) [12]. The application of RHT to (26) shows that the flow is always stable, it is to say all the $S_{i,n}$ have negative real parts and for hence we have an authentic transient flow that it turns to a permanent flow when a very large time had passed. The equation (23) gives the transient velocity profile for the fluid portion of the Bingham fluid, which corresponds to the region between $r_0(t)$ and a. The solid portion of the Bingham fluid corresponds to the region between $r = 0$ and $r = r_0(t)$, being $r_0(t)$ the variable radius of the solid portion. The initial condition for $r_0(t)$ is $r_0(0) = a$ and the asymptotic radius of the solid portion is denoted $r_0(\infty)$ which corresponds to the radius of solid portion of fluid for the case of a fully development laminar flow in a tube. Concretely the transient velocity profile for the solid portion is obtained as

$$
v(t) = v(r_0(t),t), \qquad (27)
$$

where $v(t)$ is the velocity of solid portion and $v(r,t)$ is the velocity profile for the fluid portion, given by (23). For hence the total transient discharge in the tube for this transient flow is given by:

$$Q(t) = \pi r_0^2 v(t) + 2\pi \int_{r_0(t)}^{a} v(r,t)\, r dr. \tag{28}$$

To determine the equation that gives $r_0(t)$ it is necessary to compute the explicit form of the stress field $\sigma_{r,z}(r,t)$. The substitution of $V(r)$ given by (15), (16) and (17), jointly with (8) and (9) on (7), and doing the inverse Laplace transform by residues, generates the stress field which is given by

$$\sigma_{r,z}(r,t) = -\frac{1}{2}\left(\frac{d}{dz}P(z)\right)r +$$

$$\frac{J_1\left(\sqrt{\frac{\rho\delta(\epsilon_1-\delta)(\epsilon+\chi-\delta)}{(-\mu\delta+\mu_1+\mu\epsilon_1)(-\delta+\epsilon)}}r\right)\sqrt{\frac{\rho\delta(\epsilon_1-\delta)(\epsilon+\chi-\delta)}{(-\mu\delta+\mu_1+\mu\epsilon_1)}}\,\nu\,(\mu\delta-\mu_1-\mu\epsilon_1)}{J_0\left(\sqrt{\frac{\rho\delta(\epsilon_1-\delta)(\epsilon+\chi-\delta)}{(-\mu\delta+\mu_1+\mu\epsilon_1)}}a\right)e^{\delta t}(-\epsilon_1+\delta)(\epsilon+\chi-\delta)(-\delta+\epsilon)^{-1}} +$$

$$\sum_{i=1}^{3}\sum_{n=1}^{\infty}\frac{-J_1\left(\frac{\alpha_n r}{a}\right)(S_{i,n}+\epsilon)^3 H_{n,i}e^{S_{i,n}t}\alpha_n\,(\mu S_{i,n}+\mu\epsilon_1+\mu_1)^2}{a^3 S_{i,n}^2 \rho^2 (S_{i,n}+\epsilon_1)(S_{i,n}+\epsilon+\chi)(S_{i,n}+\delta)J_1(\alpha_n)E_{i,n}}, \tag{29}$$

where

$$H_{n,i} = \rho S_{i,n}\pi \tau_0 H_0(\alpha_n)\delta\epsilon_1 a + \rho S_{i,n}^2\pi \tau_0 H_0(\alpha_n)\epsilon_1 a + \rho S_{i,n}^3\pi \tau_0 H_0(\alpha_n)a +$$

$$\rho S_{i,n}^2\pi \tau_0 H_0(\alpha_n)\delta a - 2\alpha_n\nu S_{i,n}^2\rho\mu_1 - 2\alpha_n\nu S_{i,n}^2\rho\mu\epsilon_1 -$$

$$2\alpha_n\nu S_{i,n}^3\rho\mu - 2\alpha_n\left(\frac{d}{dz}P(z)\right)S_{i,n}\mu\epsilon_1 - 2\alpha_n\left(\frac{d}{dz}P(z)\right)S_{i,n}\mu_1 -$$

$$2\alpha_n\left(\frac{d}{dz}P(z)\right)\delta\mu_1 - 2\alpha_n\left(\frac{d}{dz}P(z)\right)\delta\mu\epsilon_1 -$$

$$2\alpha_n\left(\frac{d}{dz}P(z)\right)\delta S_{i,n}\mu - 2\alpha_n\left(\frac{d}{dz}P(z)\right)S_{i,n}^2\mu \tag{30}$$

and

$$E_{n,i} = \epsilon_1\chi\epsilon\mu_1 + \epsilon_1^2\chi\epsilon\mu + 2S_{i,n}\epsilon^2\mu\epsilon_1 + 4S_{i,n}^2\mu\epsilon_1\epsilon + S_{i,n}^2\chi\mu\epsilon +$$

$$2S_{i,n}\chi\mu_1\epsilon + 2\epsilon_1^2 S_{i,n}\mu\epsilon + 2\epsilon_1 S_{i,n}\mu_1\epsilon + 2S_{i,n}^3\mu_1 + S_{i,n}^4\mu +$$

$$2S_{i,n}\chi\mu\epsilon_1\epsilon + \epsilon_1^2\mu_1 + \epsilon_1^2\epsilon^2\mu + 2S_{i,n}^3\mu\epsilon_1 + S_{i,n}^2\epsilon^2\mu +$$

$$2S_{i,n}\epsilon^2\mu_1 + S_{i,n}^2\chi\mu_1 + \epsilon_1^2 S_{i,n}^2\mu + \epsilon_1 S_{i,n}^2\mu_1 + 2S_{i,n}^3\mu\epsilon +$$

$$4S_{i,n}^2\mu_1\epsilon. \tag{31}$$

From the result that the equations (29), (30) and (31) display, the equation that determines $r_0(t)$ is

$$\sigma_{r,z}(r_0(t),t) = \tau_0. \tag{32}$$

In consequence the explicit form of the velocity for the solid portion is obtained when τ_0 given by (32) is substituted in (27). The resultant expression is very large and heavy as for to be presented here but it can be hosted and exploited in a computer with CAS. Also using (28) it is possible to compute the discharge $Q(t)$ but again the result is very long and it can not be presented here but it can to processed by computer with CAS.

5   Analysis of Results

From (29) we observe that $\sigma_{r,z}(0,t) = 0$ for all t, as must be; and for all r the stress grows with the time until the stationary value is reached, namely

$$\sigma_{r,z}(r,\infty) = -\frac{1}{2}\left(\frac{d}{dz}P(z)\right)r, \tag{33}$$

as must be too.

Now, from the equations (23) and (29) , many limit cases can be derived. For example when the tube is fixed, the boundary condition (5) is applied with $\nu = 0$ and then the resultant profiles are obtained (compare with [4]).

The particular case when the memory effects are ignored, it is to say with $\chi = 0$, $\mu_1 = 0$, $\epsilon \to \infty$ and $\epsilon_1 \to \infty$ leads to the following results:

$$v(r,t) = \frac{1}{4}\frac{(-4\,a + 4\,r)\,\tau_0}{\mu} + \frac{1}{4}\frac{(r^2 - a^2)\frac{d}{dz}P(z)}{\mu} +$$
$$\sum_{n=1}^{\infty}\frac{J_0\left(\frac{\alpha_n r}{a}\right)\left(2\left(\frac{d}{dz}P(z)\right)a + \pi\tau_0\alpha_n H_0\left(\alpha_n\right)\right)e^{-\frac{\alpha_n^2\mu t}{\rho a^2}}\,a}{\mu J_1\left(\alpha_n\right)\alpha_n^3}, \tag{34}$$

$$\sigma_{r,z}(r,t) = -\frac{1}{2}\left(\frac{d}{dz}P(z)\right)r +$$
$$\sum_{n=1}^{\infty}\frac{J_1\left(\frac{\alpha_n r}{a}\right)\left(2\left(\frac{d}{dz}P(z)\right)a + \pi\tau_0\alpha_n H_0\left(\alpha_n\right)\right)e^{-\frac{\alpha_n^2\mu t}{\rho a^2}}}{\alpha_n^2 J_1\left(\alpha_n\right)}. \tag{35}$$

From (34) and (35) we obtain immediately the asymptotic profiles which correspond to the fully developed laminar flow and using (27)-(33) with $t \to \infty$ we obtain the Buckingham-Reiner equation [8], being $Q(\infty) = Q$:

$$Q = -\frac{1}{8\mu}\left(\frac{d}{dz}P(z)\right)\pi a^4\left(1 - \frac{4}{3}\frac{\tau_0}{\tau_a} + \frac{1}{3}\frac{\tau_0^4}{\tau_a^4}\right), \tag{36}$$

where $\tau_a = \sigma_{r,z}(a,\infty)$, according with (33). We note that from (36) we recover the Hagen-Poiseuille equation for Newtonian flow in a tube, namely [8]

$$\lim_{\tau_0 \to 0}Q = -\frac{1}{8}\frac{\left(\frac{d}{dz}P(z)\right)\pi a^4}{\mu}. \tag{37}$$

Finally when the memory effects are considered, the asymptotic profiles which result from the equations (23) and (29) are:

$$v(r, \infty) = -\frac{1}{4} \frac{\left(-\epsilon r^2 + \epsilon a^2 - r^2 \chi + \chi a^2\right) \epsilon_1 \frac{d}{dz} P(z)}{\epsilon \left(\mu_1 + (\mu) \epsilon_1\right)} - \frac{(a - r) \epsilon_1 \tau_0}{\mu_1 + (\mu) \epsilon_1}, \quad (38)$$

$$v(\infty) = -\frac{1}{4} \frac{\epsilon_1 \left(\frac{d}{dz} P(z)\right) \left(\epsilon r_0^2 + \epsilon a^2 - r_0^2 \chi + \chi a^2 - 2 r_0 \epsilon a\right)}{\epsilon \left(\mu_1 + (\mu) \epsilon_1\right)}, \quad (39)$$

where (38) gives the velocity profile for the fluid portion of Bingham fluid, and (39) gives the velocity of the solid portion. Here the asymptotic radius of the solid portion, $r_0(\infty)$ is written as r_0. Now using (38) and (39) in (28) we obtain the formula of the total discharge in the tube:

$$Q = \frac{1}{192} \frac{\epsilon_1 \pi a^3 \left(16 \epsilon \tau_0^4 - 48 \chi \tau_0^4 + 48 \epsilon \tau_a^4 + 48 \chi \tau_a^4 - 64 \tau_0 \epsilon \tau_a^3\right)}{\tau_a^3 \epsilon \left(\mu_1 + \mu \epsilon_1\right)}. \quad (40)$$

We note that when the memory effects are neglected in (40), then we recover the Buckingham-Reiner equation (36). From other side, the equation (40) with $\tau_0 = 0$, provides a generalization of the Newtonian Hagen-Poiseuille equation, namely

$$Q = -\frac{1}{8} \frac{\epsilon_1 \pi a^4 \left(\frac{d}{dz} P(z)\right) (\epsilon + \chi)}{\epsilon \left(\mu_1 + \mu \epsilon_1\right)}. \quad (41)$$

The contrast between (41) and (37) shows that (41) has the same form of (37) but with an effective viscosity which incorporates the memory effects.

6   Conclusions

This work was an intent to convey the reader some notions about Symbolic Computational Rheology. The central problem was the transient laminar flow in a movable tube, for a certain generalized Bigham fluid with rheological memory. The object was to obtain the analytical solution for such linear rheological problem. The method that was used was the Laplace transform technique with Bromwich integral and residue theorem. All the mathematical procedure was implemented via Computer Algebra Software. The analytical solutions were given in terms of the Bessel functions and the Struve functions. The problem solved here is a linear problem with analytical solution but such solution is very difficult to obtain using only pen and paper. It is very remarkable that a system like Maple can be used to solve such problem in a very efficient way. We think that the present paper can be useful to engineers that work on rheological problems for which is possible and necessary to have an analytical solution. This work shows the powerful of Maple to solve a relatively complex rheological linear problem. From the results that were derived here, it is possible to design an experimental procedure that permits to observe the transient flow and then to measure some rheological magnitudes such as the Bingham plasticity or the memory parameters. Also, from our results is possible to establish contact with the Numerical Computational Rheology and to make animations that simulate the transient flow and the establishment of the fully development laminar flow. Another line for future research is the case of transient flow with Bingham plasticity with spatial variations. Finally we think that the Symbolic Computational Rheology is a very important branch of the Mathematical Rheology and deserves more investigation. We expect to have the opportunity both to continue the exploration of the wonderful land of the Symbolic Computational Rheology and to bring the reader the results.

Acknowledgments

This work was made in the course of Rheology which was teaching by the Emeritus Professor Jorge Alberto Naranjo, Head of School of Sciences and Humanities of EAFIT University. The authors are very indebted with Prof. Naranjo, by show us the wonderful land of Rheology and stimulate us to explore it. The author are also indebted with the Professor Felix Londoño, research director in EAFIT by permanent support and with the Professor Andres Sicard by permanent scientific and computational assistance.

References

1. Owens, R.G., Phillips, T.N.: Computational Rheology. World Scientific (2002)
2. Chin, W.C.: Computational Rheology for Pipeline and Annular Flow. Gulf Professional Publishing (2001)
3. Keunnings, R.: A survery of computational rheology (2001),
   www.inma.ucl.ac.be/~rk/K-Cam.pdf
4. Velez, M., Ospina, J.: Symbolic computation of generalized transient flow inside a movable tube using computer algebra. WSEAS Transactions on Fluid Mechanics 1(1), 3–8 (2006)
5. Wikipedia: Computer algebra (2005),
   en.wikipedia.org/wiki/Computer_algebra_system
6. Maplesoft: Maple (2007), www.maplesoft.com
7. Chatzimina, M., Georgiou, G.C., Argyropaidas, I., Mitsoulis, E., Huigol, R.R.: Cessation of couette and poiseuille flows of a bingham plastic and finite stopping times. J. Non-Newtonian Fluid Mech. 129, 117–127 (2005)
8. Bird, R.B., Stewart, W., Lightfoot, E.N.: Transport Phenomena. John Wiley, New York (2002)
9. Apostol, T.: Analisis Matematico. Ed. Reverte (1988)
10. Bowman, F.: Introduction to Bessel functions. Dover Publications Inc. (1958)
11. Mathworld: Struve functions (2006),
    http://mathworld.wolfram.com/StruveFunction.html
12. Mathworld, Wikipedia: Routh-hurwitz theorem (2006),
    mathworld.wolfram.com/Routh-HurwitzTheorem.html,
    en.wikipedia.org/wiki/Routh-Hurwitztheorem

Some Elimination Problems for Matrices

Wilhelm Plesken and Daniel Robertz

Lehrstuhl B für Mathematik, RWTH Aachen, Templergraben 64, D-52062 Aachen,
Germany
plesken@momo.math.rwth-aachen.de, daniel@momo.math.rwth-aachen.de

Abstract. New elimination methods are applied to compute polynomial
relations for the coefficients of the characteristic polynomial of certain
families of matrices such as tensor squares.

1 Introduction

The problems treated here concern polynomial relations between the coefficients
of characteristic polynomials of certain naturally defined varieties of matrices
such as tensor squares of matrices. These questions originated from the recognition problem of finite matrix groups, cf. [L-GO'B 97] and [PlR], where the
general question is addressed. The reason why we take up the problem here
with different examples is that these problems provide good challenges for testing elimination techniques. They arise in sequences parametrized by n, and two
quantities reflect their difficulty very well, the first being the Krull dimension
and the second the number of variables. The Krull dimension grows linearly with
n, the number of variables quadratically in n. We treat examples of Krull dimensions between 2 and 5 here and suggest that new elimination methods could
be evaluated using these series of problems.

The methods we apply are developed in [PlR]. In particular, elimination by
"degree steering" is the preferred strategy here. Janet bases of the same ideal are
computed repeatedly for different gradings of the polynomial ring. The degrees
of the variables to be eliminated are increased in each step until a Janet basis
for the elimination ideal can be read off (cf. Section 3.1). Other techniques which
come up in the course of [PlR] still wait for implementation.

All results were obtained by using implementations of the involutive basis
algorithm by V. P. Gerdt and Y. A. Blinkov [Ger 05], [GBY 01]. More precisely,
all Janet bases have been computed by the new open source software package
ginv [ginv] in connection with its Maple interface to the Involutive package
[BCG 03], which provides direct access to combinatorial tools like Hilbert series
[PlR 05].

The results of the examples treated here are either short and listed or are
lengthy and available on the web: http://wwwb.math.rwth-aachen.de/
elimination.

V.G. Ganzha, E.W. Mayr, and E.V. Vorozhtsov (Eds.): CASC 2007, LNCS 4770, pp. 350–359, 2007.
© Springer-Verlag Berlin Heidelberg 2007

2   The Problems

The problems posed in this section and treated in the next section can be summarized as follows.

Problem 1. Given a classical group G defined over a field K of characteristic zero and any finite dimensional representation ρ of G, find a generating set of the polynomial relations for the coefficients of the characteristic polynomial $\chi_{\rho(g)}(t)$ of $\rho(g)$, $g \in G$.

In [PlR] the case $G = \mathrm{GL}(n, K) \times \mathrm{GL}(m, K)$ for a field K of arbitrary characteristic is treated with $\rho : G \to \mathrm{GL}(nm, K) : (a, b) \mapsto a \otimes b$.

Remark 1. Problem 1 is an elimination problem for an ideal in a commutative polynomial ring and is therefore tackled by constructive methods from commutative algebra. Let $\nu : G \to \mathrm{GL}(n, K)$ be the natural representation of G. We define the polynomial ring $R := K[a_1, \ldots, a_n, c_1, \ldots, c_r]$ over K with the coefficients a_i of $\chi_{\nu(g)}(t)$ and c_j of $\chi_{\rho(g)}(t)$ as indeterminates. By expressing the coefficients c_j in terms of the coefficients a_i we obtain generators for an ideal I of R. The polynomial relations for the c_j as required by Problem 1 are given by the elimination ideal $J := I \cap K[c_1, \ldots, c_r]$. It is well known that the limits of computational feasibility are reached rather quickly for such elimination problems. New promising methods are developed in [PlR] and we apply here some of these methods to the more specific problems described below.

Remark 2. Very often there is a natural way to assign non-standard degrees to the variables a_i and c_j such that the polynomial relations in I become homogeneous with respect to this grading.

Remark 3. For the present problems the Krull dimension $\dim K[c_1, \ldots, c_r]/J$ and the number r of variables $c_1, \ldots, c_r$ are useful parameters which estimate how hard the given problem is. Although these parameters provide only a very rough description of the problem, we want to point out that problems up to Krull dimension 3 can be dealt with using our particular elimination methods quite well and sometimes cases with $\dim K[c_1, \ldots, c_r]/J = 4$ can be solved.

Remark 4. If a concrete instance of Problem 1 turns out to be too difficult to approach directly, the following strategy often allows to come much closer to the final elimination result if not even solving it. One factors the problem into two steps. First treat the problem for matrices of determinant 1. Depending on the kind of problem, one usually gets rid of the variable c_r and two or three variables a_i. Often this means a considerable reduction in complexity. The result for matrices of general determinant is then obtained from the result for matrices of determinant 1 by homogenizing the polynomials with the variable c_r.

In this paper we deal with the following more specific problems.

Problem 2.   1. Tensor square of $GL(n, K)$ and $SL(n, K)$

$G = GL(n, K)$ or $G = SL(n, K)$, $\rho : G \to GL(n^2, K) : g \mapsto g \otimes g$, i.e. each matrix g is mapped to the Kronecker product of g with itself.

In case $G = GL(n, K)$ we have $\dim K[c_1, \ldots, c_r]/J = n$, in case $G = SL(n, K)$ the Krull dimension as defined above is $n - 1$.

2. Compound representation of $GL(n, K)$ and $SL(n, K)$

$G = GL(n, K)$ or $G = SL(n, K)$, $\rho : G \to GL(\binom{n}{k}, K) : g \mapsto \Lambda^k g$, i.e. each matrix g is mapped to the matrix of its $(k \times k)$-minors, where $k = 2, \ldots, n-1$. These representations are obtained as the actions of G on the exterior powers $\Lambda^k V$ of the natural KG-module V.

In this case the resulting Krull dimension of $K[c_1, \ldots, c_r]/J$ is n resp. $n - 1$.

As already mentioned in Remark 4, for difficult cases of $G = GL(n, K)$ an intermediate step treating the compound representation of $SL(n, K)$ is often very helpful.

3. Exterior and symmetric square of $SO(n, K)$

$G = SO(n, K)$, $\rho : G \to GL(\binom{n}{2}, K)$ resp. $\rho : G \to GL(\binom{n+1}{2}, K)$ representing the action of G on the exterior square $\Lambda^2 V$ resp. the symmetric square $S^2 V$ of the natural KG-module V.

The Krull dimension of $K[c_1, \ldots, c_r]/J$ is $\lfloor \frac{n}{2} \rfloor$.

3   Results

3.1   Tensor Square of $GL(n, K)$

We start with Problem 2 1. Let $G = GL(n, K)$, where K is a field of characteristic zero, ν the natural representation of G, and $\rho : G \to GL(n^2, K) : g \mapsto g \otimes g$. We are going to compute the polynomial relations for the coefficients c_j of the characteristic polynomial

$$\chi_{\rho(g)}(t) = t^m - c_1 t^{m-1} + \ldots + (-1)^m c_m$$

of $\rho(g)$, $g \in G$, where $m = n^2$. Equating the c_j with the corresponding polynomials in the coefficients a_i of

$$\chi_{\nu(g)}(t) = t^n - a_1 t^{n-1} + \ldots + (-1)^n a_n$$

defines the ideal I of the polynomial ring $K[a_1, \ldots, a_n, c_1, \ldots, c_{n^2}]$. Hence, our ultimate goal is to compute $J := I \cap K[c_1, \ldots, c_{n^2}]$.

The ideal I consists of homogeneous polynomials w.r.t. the grading defined by assigning degree i to the variable a_i and degree $2j$ to the variable c_j. We refer to this grading as the natural grading. For the resulting ideal J the grading defined by $\deg c_j = j$ can be chosen afterwards. Since we assume K to be of characteristic zero, all computations will be done over the field of rational numbers.

n = 2: The ideal I is generated by:

$$[-a_1{}^2 + c_1, \ -2 a_2{}^2 + 2 a_1{}^2 a_2 - c_2, \ -a_1{}^2 a_2{}^2 + c_3, \ a_2{}^4 - c_4]$$

This case can still be dealt with simply by computing a Janet basis for I w.r.t. the block ordering on the monomials in $a_1, a_2, c_1, c_2, c_3, c_4$ which eliminates the variables a_1, a_2.

The Janet basis for $J = I \cap K[c_1, c_2, c_3, c_4]$ is obtained very quickly:

$$[\, c_1{}^2 c_4 - c_3{}^2,\ c_2{}^2 c_3 - 4\, c_1 c_3{}^2 + 4\, c_1 c_2 c_4 + 4\, c_3 c_4,\ c_1{}^2 c_2{}^2 - 4\, c_1{}^2 c_3 + 4\, c_2 c_3 + 4\, c_1 c_4,$$
$$c_1{}^2 c_2 c_4 - c_2 c_3{}^2,\ c_2{}^3 c_3 - 4\, c_1 c_2 c_3{}^2 + 16\, c_3{}^3 - 12\, c_2 c_3 c_4 - 16\, c_1 c_4{}^2,$$
$$c_2{}^4 - 16\, c_1{}^2 c_3{}^2 + 32\, c_2 c_3{}^2 - 8\, c_2{}^2 c_4 + 16\, c_4{}^2,\ c_1{}^2 c_2{}^2 - 4\, c_1{}^3 c_3 + 4\, c_1 c_2 c_3 + 4\, c_3{}^2 \,]$$

n = 3: The ideal I is generated by:

$$[\, -a_1{}^2 + c_1,\ -2\, a_2{}^2 + 2\, a_1{}^2 a_2 - c_2,\ -3\, a_3{}^2 + 6\, a_2 a_3 a_1 - 2\, a_1{}^3 a_3 - a_1{}^2 a_2{}^2 + c_3,$$
$$-4\, a_2{}^2 a_3 a_1 - a_1{}^2 a_3{}^2 + 2\, a_1{}^3 a_2 a_3 + a_2{}^4 - c_4,$$
$$a_2{}^2 a_3{}^2 + 4\, a_3{}^2 a_2 a_1{}^2 - 2\, a_3 a_2{}^3 a_1 - a_3{}^2 a_1{}^4 + c_5,$$
$$-6\, a_2 a_3{}^3 a_1 + a_3{}^2 a_2{}^2 a_1{}^2 + 2\, a_3{}^2 a_2{}^3 + 3\, a_3{}^4 - c_6,$$
$$2\, a_1{}^2 a_3{}^4 - 2\, a_3{}^3 a_2{}^2 a_1 + c_7,\ a_2{}^2 a_3{}^4 - c_8,\ -a_3{}^6 + c_9 \,]$$

As a first step we compute a Janet basis for I without intending to eliminate any variable. Of course, we chose the grading of $K[a_1, a_2, a_3, c_1, \ldots, c_9]$ which makes the relations homogeneous, i.e. the natural grading. In the computations which are described next, Janet bases are constructed with respect to the degree-reverse lexicographic ordering. In the present example we fix that ordering by choosing

$$c_9 > c_8 > \ldots > c_1 > a_3 > a_2 > a_1,$$

which turns out to be computationally beneficial.

Our strategy is to eliminate one variable at a time by computing a Janet basis of I repeatedly (using the result of the previous run as input) and increasing the degree of the variable to be eliminated in each step until we can extract a Janet basis of J from the Janet basis of I by intersecting it with $K[c_1, \ldots, c_9]$. This stage is reached as soon as the following criterion is satisfied.

Lemma 1 ([PlR]). *Let $J \subseteq K[X_1, \ldots, X_n, Y_1, \ldots, Y_m]$ be a Janet basis with respect to some term ordering. For any $0 \neq p \in K[X_1, \ldots, X_n, Y_1, \ldots Y_m]$ let $\lambda(p)$ be the leading monomial. If*

$$R := \{\, p \in J \mid p \in K[X_1, \ldots, X_n, Y_1, \ldots, Y_m] - K[Y_1, \ldots, Y_m]$$
$$and \quad \lambda(p) \in K[Y_1, \ldots, Y_m] \,\}$$

is empty, then $J \cap K[Y_1, \ldots, Y_m]$ generates $\langle J \rangle \cap K[Y_1, \ldots, Y_m]$.

Proof. The case $J \cap K[Y_1, \ldots, Y_m] = \{0\}$ is trivial. Let $q \in \langle J \rangle \cap K[Y_1, \ldots, Y_m]$ be non-zero. Since there is no X_i involved in q, it can be reduced by some element $p \in J$ with $\lambda(p)$ not divisible by any X_i. By hypothesis $p \in K[Y_1, \ldots, Y_m]$, so that the first step of involutive reduction replaces q by an element in $\langle J \rangle$ again without X_i's and with smaller leading monomial w.r.t. the given term ordering. So induction yields the result.

We refer to this elimination technique as "elimination by degree steering". For more information we refer to [PlR].

Coming back to the concrete problem, Lemma 1 shows that for eliminating a_1 it is enough to increase the degree of a_1 to 2. In the second step degree 3 for a_2 is already sufficient, and a_3 is eliminated by choosing degree 4 for a_3, again assured by Lemma 1. This yields a Janet basis of J consisting of 31 elements. The Hilbert series (with respect to the natural grading) of $K[c_1, \ldots, c_9]/J$ is

$$\frac{1 + t^4 + t^5 + t^6 + t^7 + 2\,t^8 + 2\,t^9 + t^{10} - 4\,t^{11} - 4\,t^{12} - 2\,t^{13} + 2\,t^{14} + 2\,t^{15}}{(1 - t)\,(1 - t^2)\,(1 - t^3)}.$$

The case $n = 4$ is much harder than $n = 3$. It is partially treated in the next section.

3.2   Tensor Square of $\mathbf{SL}(n, K)$

We give only one example of Problem 21. for $G = \mathrm{SL}(n, K)$. Since the Krull dimension and the number of variables is less in comparison to the previous case $G = \mathrm{GL}(n, K)$, these problems are easier to deal with.

n = 4: The generating set for the ideal I is obtained from the generating set for the case $G = \mathrm{GL}(4, K)$ by setting $a_4 = 1$, $c_{16} = 1$. Hence, I is an ideal of $K[a_1, a_2, a_3, c_1, \ldots, c_{15}]$. Starting with the natural grading and the degree-reverse lexicographic ordering defined by

$$c_{15} > c_{14} > \ldots > c_1 > a_3 > a_2 > a_1,$$

degree steering (cf. Lemma 1) produces a Janet basis of the elimination ideal consisting of 96 elements. The degrees for a_1, a_2, and a_3 had to be increased up to 10, 15 resp. 16. The elements of the resulting Janet basis have (natural) degrees between 12 and 22 and are all irreducible.

As explained in Remark 4, we could compute a generating set for J in case $G = \mathrm{GL}(4, K)$ by homogenizing the previous result with the variable c_{16}, but computing a Janet basis for J (of relations for general determinant) is actually a difficult task.

3.3   Compound Representation of $\mathbf{GL}(n, K)$

In this section we deal with Problem 22. Let $G = \mathrm{GL}(n, K)$, $k \in \{2, \ldots, n-1\}$, and $\rho : G \to \mathrm{GL}(\binom{n}{k}, K)$ be the representation mapping each $g \in G$ to the matrix of its $(k \times k)$-minors. We set $m = \binom{n}{k}$. As in the previous section, the polynomials in the ideal I of $K[a_1, \ldots, a_n, c_1, \ldots, c_m]$ are homogeneous with respect to the grading defined by assigning degree i to a_i and degree $2j$ to c_j.

We restrict to the case $k = 2$ here.

n = 4: The ideal I of $K[a_1, \ldots, a_4, c_1, \ldots, c_6]$ is generated by:

$$[\,-a_2 + c_1,\ a_1 a_3 - a_4 - c_2,\ 2\,a_2 a_4 - a_3{}^2 - a_1{}^2 a_4 + c_3,\ -a_4{}^2 + a_3 a_1 a_4 - c_4,$$
$$-a_2 a_4{}^2 + c_5,\ a_4{}^3 - c_6\,]$$

In this case a block ordering on the monomials of $K[a_1,\ldots,a_4,c_1,\ldots,c_6]$ which eliminates $a_1, \ldots, a_4$ gives a Janet basis of J very quickly. A second run of Janet's algorithm on the result which takes the natural grading into account yields the final answer:

$$[c_2{}^2c_5 - c_1c_4{}^2,\ c_4c_5 - c_1c_2c_6,\ c_2c_5{}^2 - c_1{}^2c_4c_6,\ c_4{}^3 - c_2{}^3c_6,\ c_5{}^2c_4 - c_6c_5c_2c_1,$$
$$c_5{}^3 - c_1{}^3c_6{}^2]$$

The Hilbert series (w.r.t. the natural grading) of $K[c_1,\ldots,c_6]/J$ is

$$\frac{1 + t^4 + t^5 + t^8 - t^9 + t^{10} - t^{12}}{(1-t)\,(1-t^2)\,(1-t^3)\,(1-t^6)}.$$

$\mathbf{n = 5}$: The ideal I of $K[a_1,\ldots,a_5,c_1,\ldots,c_{10}]$ is generated by:

$$[-a_2 + c_1,\ a_1a_3 - a_4 - c_2,\ 2\,a_2a_4 - a_3{}^2 - a_1{}^2a_4 + a_1a_5 + c_3,$$
$$a_5a_3 - a_4{}^2 - 3\,a_2a_1a_5 + a_3a_1a_4 + a_1{}^3a_5 - c_4,\ -2\,a_5{}^2 + 2\,a_5a_1a_4 + 2\,a_2a_5a_3$$
$$-a_3a_1{}^2a_5 - a_2a_4{}^2 + c_5,\ -3\,a_4a_5a_3 + a_2a_5a_1a_4 - a_5{}^2a_1{}^2 + a_4{}^3 + a_2a_5{}^2 - c_6,$$
$$-a_4{}^2a_5a_1 - a_2{}^2a_5{}^2 + 2\,a_3a_5{}^2a_1 + a_4a_5{}^2 + c_7,\ a_4a_5{}^2a_2 - a_1a_5{}^3 - c_8,$$
$$-a_5{}^3a_3 + c_9,\ a_5{}^4 - c_{10}]$$

We eliminate $a_1, \ldots, a_5$ (in this order) by degree steering (cf. Lemma 1) and start, of course, with the natural grading of $K[a_1,\ldots,a_5,c_1,\ldots,c_{10}]$. The degree-reverse lexicographic ordering which is defined by

$$c_{10} > c_9 > \ldots > c_1 > a_5 > \ldots > a_1$$

is used. In order to eliminate variable a_i, assigning degree $i+1$ to a_i is already sufficient. The main computational difficulty lies in the elimination of a_4 and a_5. By applying finally to the result Janet's algorithm which takes the grading defined by $\deg c_i = i$ into account, we obtain a Janet basis of J consisting of 254 irreducible polynomials of (natural) degrees between 24 and 38. The Hilbert series (w.r.t. the natural grading) of $K[a_1,\ldots,a_5,c_1,\ldots,c_{10}]/J$ is:

$$(1 + t^2 + t^4 + t^6 + t^7 + 2\,t^8 + 2\,t^9 + 3\,t^{10} + 2\,t^{11} + 3\,t^{12} + 2\,t^{13} + 4\,t^{14} + 3\,t^{15}$$
$$+ 6\,t^{16} + 5\,t^{17} + 8\,t^{18} + 6\,t^{19} + 9\,t^{20} + 7\,t^{21} + 10\,t^{22} + 9\,t^{23} + 12\,t^{24} + 8\,t^{25}$$
$$+ 3\,t^{26} - 26\,t^{27} - 54\,t^{28} - 70\,t^{29} - 65\,t^{30} - 31\,t^{31} + 2\,t^{32} + 65\,t^{33} + 79\,t^{34}$$
$$+ 85\,t^{35} + 47\,t^{36} + 14\,t^{37} - 33\,t^{38} - 45\,t^{39} - 38\,t^{40} - 27\,t^{41} - 2\,t^{42} + 5\,t^{43}$$
$$+ 16\,t^{44} - t^{45} - 2\,t^{46} - t^{47} + 2\,t^{48} + 2\,t^{49} + 2\,t^{50})\,/$$
$$((1-t)(1-t^3)(1-t^4)(1-t^5)(1-t^6)).$$

3.4   Exterior Square of SO(n, K)

We turn to Problem 24. Let $G = SO(n, K)$, ν the natural representation of G, and $\rho : G \to GL(\binom{n}{2}, K)$ the representation of G on the exterior square $\Lambda^2 V$ of the natural KG-module V. For each root ω of the characteristic polynomial $\chi_V(t)$ the negative root $-\omega$ is also a root. Moreover, all $g \in G$ have determinant 1. Therefore $\chi_V(t)$ has the form

$$\chi_V(t) = t^n - a_1 t^{n-1} + a_2 t^{n-2} - \ldots + (-1)^{n-2} a_2 t^2 + (-1)^{n-1} a_1 t + (-1)^n.$$

The exterior square of V has K-dimension $r = \binom{n}{2}$. The coefficients c_j of

$$\chi_\rho(t) = t^r - c_1 t^{r-1} + c_2 t^{r-2} - \ldots + (-1)^{r-2} c_2 t^2 + (-1)^{r-1} c_1 t + (-1)^r$$

are equated with the corresponding polynomials in $a_1, \ldots, a_m$ occurring as coefficients of the polynomial in (1), where $m = \lfloor \frac{n}{2} \rfloor$.

n = 5: The ideal I of $K[a_1, a_2, c_1, \ldots, c_5]$ is generated by:

$$[-a_2 + c_1,\ a_1 a_2 - a_1 - c_2,\ 2 a_1 a_2 - a_2^2 - a_1^3 + a_1 + c_3,$$
$$a_2 - a_1^2 - 3 a_1 a_2 + a_1^2 a_2 + a_1^3 - c_4,\ -2 + 2 a_1^2 + 2 a_2^2 - 2 a_1^2 a_2 + c_5]$$

Here a block ordering on the monomials of $K[a_1, a_2, c_1, \ldots, c_5]$ which eliminates a_1, a_2 gives a Janet basis of J immediately. We obtain a Janet basis consisting of 5 irreducible polynomials of (standard) degree up to 4. Exactly one of these polynomials is linear:

$$2 c_1 - 2 c_2 + 2 c_3 - 2 c_4 + c_5 - 2.$$

The Hilbert series (w.r.t. the standard grading) of $K[a_1, a_2, c_1, \ldots, c_5]/J$ is:

$$1 + 4t + 10 t^2 + 17 t^3 + t^4 \left(\frac{17}{(1-t)} + \frac{7}{(1-t)^2} \right).$$

n = 6: Now we deal with the ideal I in $K[a_1, a_2, a_3, c_1, \ldots, c_7]$ generated by the following polynomials:

$$[-a_2 + c_1,\ a_2 - a_1 a_3 + c_2,\ 2 a_2^2 - a_3^2 - a_1^2 a_2 - 1 + a_1^2 + c_3,$$
$$-2 a_2 + a_1^2 - a_1 a_3 + a_2^2 + 3 a_1^2 a_2 - a_3 a_2 a_1 - a_1^4 + c_4,$$
$$2 a_2 - 2 a_1^2 - 3 a_1 a_3 + 6 a_1^2 a_2 - 2 a_2^2 + 2 a_3 a_2 a_1 - a_3 a_1^3 - a_1^4 - a_2^3 + c_5,$$
$$-2 + 3 a_3^2 + 3 a_1^2 - 6 a_3 a_2 a_1 - a_1^4 + a_3 a_1^3 + a_2^2 a_1^2 + a_2^3 - c_6,$$
$$-a_1^2 - 3 a_1 a_3 - a_2^2 + a_3 a_2 a_1 + a_1^4 + 2 a_3 a_1^3 - 3 a_2^2 a_1^2 + 2 a_2^3 + c_7]$$

We eliminate a_1, a_2, a_3 (in this order) by degree steering (cf. Lemma 1). Although we cannot use a grading of the polynomial ring with respect to which the input is homogeneous, it turns out that assigning degree i to a_i and degree $2j$ to c_j is again very helpful. In the present case this only results in a filtration of $K[a_1, a_2, a_3, c_1, \ldots, c_7]$. We choose the degree reverse lexicographic ordering defined by

$$c_1 > c_2 > \ldots > c_7 > a_3 > a_2 > a_1.$$

The degrees of a_1, a_2, a_3 need to be increased up to 6, 3 resp. 6 in order to obtain a Janet basis of J consisting of 61 elements of (standard) degrees up to 10. There is exactly one linear relation:

$$c_7 - 3\,c_6 + 5\,c_5 - 7\,c_4 + 9\,c_3 - 11\,c_2 + 13\,c_1 - 15.$$

The Hilbert series is:

$$\frac{1 + t + t^2 + t^3 + t^4 + 2\,t^5 + 3\,t^6 + 3\,t^7 + t^8 - 3\,t^{10} - 2\,t^{11} - t^{12} + t^{13} + t^{14}}{(1 - t^2)\,(1 - t^3)\,(1 - t^4)}.$$

3.5 Reduced Symmetric Square of $SO(n, K)$

In this final section we deal with the symmetric square defined in Problem 2 4. Let $G = SO(n, K)$, ν the natural representation of G, and $\rho : G \to GL(\binom{n+1}{2}, K)$ the representation of G on the symmetric square $S^2 V$ of the natural KG-module V. As in the previous section, $\chi_\nu(t)$ has the form

$$\chi_\nu(t) = t^n - a_1 t^{n-1} + a_2 t^{n-2} - \ldots + (-1)^{n-2} a_2 t^2 + (-1)^{n-1} a_1 t + (-1)^n.$$

The symmetric square of V always has the trivial KG-module as a constituent. Therefore, we consider here the reduced symmetric square, which is the complement of the trivial KG-module in $S^2 V$. It has K-dimension $r = n^2 - \binom{n}{2} - 1$. From the characteristic polynomials $\chi_\nu^{\otimes 2}(t)$, $\chi_\nu^{S^2}(t)$, and $\chi_1(t)$ of the representation of G acting on the tensor square of V, the symmetric square resp. the trivial representation of degree 1 we obtain $\chi_\rho(t)$ as

$$\chi_\rho(t) = \frac{\chi_\nu^{\otimes 2}(t)}{\chi_\nu^{S^2}(t) \cdot \chi_1(t)}. \tag{1}$$

The coefficients c_j of

$$\chi_\rho(t) = t^r - c_1 t^{r-1} + c_2 t^{r-2} - \ldots + (-1)^{r-2} c_2 t^2 + (-1)^{r-1} c_1 t + (-1)^r$$

are equated with the corresponding polynomials in a_1, $\ldots$, a_m occurring as coefficients of the polynomial in (1), where $m = \lfloor \frac{n}{2} \rfloor$.

$n = 5$: Here the degrees of $\chi_\nu^{\otimes 2}(t)$, $\chi_\nu^{S^2}(t)$, and $\chi_\rho(t)$ in t are 25, 10 resp. 14. As generators for the ideal I of $K[a_1, a_2, c_1, \ldots, c_7]$ we obtain:

$$
\begin{aligned}
&[1 + a_2 - a_1{}^2 + c_1,\ a_2 a_1{}^2 - a_2{}^2 + a_1 + 1 - a_1 a_2 - a_1{}^2 + a_2 - c_2, \\
&\quad a_2 a_1{}^2 - a_1 a_2 - a_2 a_1{}^3 - a_1{}^2 + 1 + 4\,a_2{}^2 a_1 + a_2 - a_2{}^3 - 3\,a_2{}^2 + c_3, \\
&\quad 5\,a_2{}^2 a_1 - a_2{}^2 a_1{}^2 + 6\,a_2 a_1{}^2 - 5\,a_2 a_1{}^3 + a_2{}^3 a_1 - 3\,a_2{}^2 - 3\,a_1{}^2 + a_1{}^5 + 1 - 2\,a_2{}^3 \\
&\quad -c_4,\ -a_2{}^2 a_1{}^2 + a_2 a_1{}^4 - a_2{}^2 a_1{}^3 - 2\,a_2 a_1{}^2 - a_2{}^2 a_1 + 4\,a_2{}^3 a_1 + 3\,a_1{}^2 - a_1{}^4 \\
&\quad -2 + 3\,a_2{}^2 - 2\,a_2{}^3 - a_2{}^4 + c_5,\ -3\,a_2{}^3 a_1 - a_2{}^2 a_1{}^3 + 5\,a_2 a_1{}^3 - 4\,a_2{}^2 a_1 - a_2 a_1{}^4 \\
&\quad +2\,a_2{}^2 a_1{}^2 - 5\,a_2 a_1{}^2 + a_2{}^3 a_1{}^2 + 3\,a_2{}^2 - 2\,a_1{}^3 - 2 + a_1{}^4 + 2\,a_1{}^2 + a_1 a_2 + 3\,a_2{}^3 \\
&\quad -2\,a_2 - c_6,\ -4\,a_2{}^3 a_1 + 4\,a_2{}^2 a_1{}^3 + 2\,a_2 a_1{}^3 - 8\,a_2{}^2 a_1 - 2\,a_2 a_1{}^2 - 2\,a_2{}^3 a_1{}^2 \\
&\quad +2\,a_1 a_2 - 2\,a_1{}^5 - 2 + 2\,a_2{}^2 + 2\,a_2{}^4 + 2\,a_1{}^2 + 4\,a_1{}^3 - 2\,a_1 + 4\,a_2{}^3 - 2\,a_2 + c_7]
\end{aligned}
$$

We eliminate a_1, a_2 (in this order) by degree steering (cf. Lemma 1) and start with degree 1 for every variable. Note that the polynomial relations are not homogeneous with respect to this grading. In this case it turns out that the degree-reverse lexicographic ordering defined by

$$a_1 > a_2 > c_1 > \ldots > c_7$$

is superior to the reversed order. The first Janet basis of I has 83 elements. Then three steps of "degree steering" are sufficient to eliminate a_1. Two further steps are enough to produce the final result consisting of 95 polynomials. The Hilbert series (with respect to the standard grading) of $K[c_1, \ldots, c_7]/J$ is

$$1 + 6\,t + 21\,t^2 + 56\,t^3 + 96\,t^4 + 112\,t^5 + 134\,t^6 + 156\,t^7 + 179\,t^8 + 202\,t^9 + 225\,t^{10}$$
$$+ t^{11}\left(\frac{225}{(1-t)} + \frac{23}{(1-t)^2}\right).$$

There is exactly one linear relation in the Janet basis of J:

$$2\,c_1 - 2\,c_2 + 2\,c_3 - 2\,c_4 + 2\,c_5 - 2\,c_6 + c_7 - 2.$$

All other elements have (standard) degrees 4 to 11.

$n = 6$: Here we have an ideal I in the polynomial ring $K[a_1, a_2, a_3, c_1, \ldots, c_{10}]$. It again turns out to be advantageous to assign degree i to a_i and degree $2j$ to c_j, although the generating polynomials are not homogeneous. Hence, similar to Section 3.4 we only work with a filtration of the polynomial ring. Degree steering (cf. Lemma 1) accomplishes the elimination of a_1, a_2, a_3. It is necessary to increase the degree of a_1 up to 2, the degree of a_2 up to 15 and the degree of a_3 up to 7. We obtain a Janet basis of $J = I \cap K[c_1, \ldots, c_{10}]$ consisting of more than 1500 elements. The Hilbert series is:

$$(1 + t^4 + t^5 + t^6 + t^7 + 2\,t^8 + 2\,t^9 + 2\,t^{10} + 2\,t^{11} - t^{12} - 16\,t^{13} - 12\,t^{14} + 9\,t^{15}$$
$$+ 16\,t^{16} + 5\,t^{17} - 10\,t^{18})/((1-t)(1-t^2)(1-t^3)).$$

References

[BCG 03]   Blinkov, Y.A., Cid, C.F., Gerdt, V.P., Plesken, W., Robertz, D.: The MAPLE Package "Janet": I. Polynomial Systems. In: Ganzha, V.G., Mayr, E.W., Vorozhtsov, E.V. (eds.) Proc. of Computer Algebra in Scientific Computing CASC 2003, pp. 31–40. Institut für Informatik, TU München, Garching, Germany (2003), Also available together with the package from http://wwwb.math.rwth-aachen.de/Janet

[ginv]   ginv project, cf. http://invo.jinr.ru, http://wwwb.math.rwth-aachen.de/Janet

[Ger 05]   Gerdt, V.P.: Involutive Algorithms for Computing Gröbner Bases. In: Cojocaru, S., Pfister, G., Ufnarovski, V. (eds.) Computational Commutative and Non-Commutative Algebraic Geometry. In: NATO Science Series, pp. 199–225. IOS Press, Amsterdam (2005)

[GBY 01]   Gerdt, V.P., Blinkov, Y.A., Yanovich, D.A.: Construction of Janet Bases, II. In: Ganzha, V.G., Mayr, E.W., Vorozhtsov, E.V. (eds.) Polynomial Bases. In Computer Algebra in Scientific Computing CASC 2001, pp. 249–263. Springer, Heidelberg (2001)

[L-GO'B 97] Leedham-Green, C.R., O'Brien, E.: Recognising tensor products for matrix groups. Int. J. of Algebra and Computation 7, 541–559 (1997)

[PlR 05]   Plesken, W., Robertz, D.: Janet's approach to presentations and resolutions for polynomials and linear pdes. Arch. Math. 84(1), 22–37 (2005)

[PlR]   Plesken, W., Robertz, D.: Elimination for coefficients of special characteristic polynomials (submitted for publication)

A Full System of Invariants
for Third-Order Linear Partial Differential
Operators in General Form

Ekaterina Shemyakova and Franz Winkler

Research Institute for Symbolic Computation (RISC),
J.Kepler University,
Altenbergerstr. 69, A-4040 Linz, Austria
{kath,Franz.Winkler}@risc.uni-linz.ac.at
http://www.risc.uni-linz.ac.at

Abstract. We find a full system of invariants with respect to gauge
transformations $L \to g^{-1}Lg$ for third-order hyperbolic linear partial differential operators on the plane. The operators are considered in a normalized form, in which they have the symbol $\mathrm{Sym}_L = (pX + qY)XY$
for some non-zero bivariate functions p and q. For this normalized form,
explicit formulae are given. The paper generalizes a previous result for
the special, but important, case $p = q = 1$.

Keywords: Linear Partial Differential Operators, Invariants, Gauge
transformations.

1 Introduction

For a second-order hyperbolic Linear Partial Differential Operators (LPDOs) on
the plane in the normalized form

$$L = D_x \circ D_y + aD_x + bD_y + c, \tag{1}$$

where $a = a(x, y), b = b(x, y), c = c(x, y)$, it has been known for several centuries
that the quantities

$$h = c - a_x - ab, \quad k = c - b_y - ab \tag{2}$$

are its invariants with respect to the gauge transformations $L \to g^{-1}Lg$. These
two invariants were proved [2] to form together a full system of invariants for
operators of the form (1). Thus, if two operators of the form (1) are known to
have the same invariants h and k, then one may conclude that the operators
are equivalent with respect to such transformations. Any other invariant of the
operator, as well as all of its invariant properties, can be expressed in terms of
h and k.

The case of operators of order two has been actively investigated. For example, we can note the classical Laplace hyperbolic second-order LPDOs, scalar

V.G. Ganzha, E.W. Mayr, and E.V. Vorozhtsov (Eds.): CASC 2007, LNCS 4770, pp. 360–369, 2007.

hyperbolic non-linear LPDOs, and so on (sample references include [1,3,5]). For the case of hyperbolic operators of high orders, however, not much is known. A method for obtaining some invariants for a hyperbolic operator of arbitrary order was mentioned in [8]. In the paper [4] a method to compute some invariants for operators of order three was suggested.

Although the determination of some particular invariants is already important, there is an enormous area of applications for a full system of invariants. Whenever we have a full system of invariants for a certain class of LPDOs, we have an easy way to judge whether two operators of the class are equivalent, and it is possible to classify some of the corresponding partial differential equations in terms of their invariants. Thus, for example, classification has an immediate application to the integration of PDEs. Indeed, most integration methods work with operators given in some normalized form. Also a full system of invariants for a certain class of operators can be used for the description of all the invariant properties of the operators in terms of the invariants of the full system.

For third-order operators of the form

$$L = (D_x + D_y)D_xD_y + a_{20}D_x^2 + a_{11}D_{xy} + a_{02}D_y^2 + a_{10}D_x + a_{01}D_y + a_{00}, \quad (3)$$

where all the coefficients are functions in x and y, a full system of invariants was obtained in [6]. This is a special case — albeit an important one — of a general normalized form for a third-order hyperbolic bivariate LPDO. Indeed, the symbol of the normalized form of such operators has the form $(X + qY)XY$, where $q = q(x, y)$ is not zero.

Full systems of invariants have important applications, such as classification, integration algorithms, etc. So one needs them for as general a class of LPDOs as possible. In the present paper we establish a full system of invariants for operators of the form

$$L = (pD_x + qD_y)D_xD_y + a_{20}D_x^2 + a_{11}D_{xy} + a_{02}D_y^2 + a_{10}D_x + a_{01}D_y + a_{00}, \quad (4)$$

where $p = p(x, y)$ and $q = q(x, y)$ are not zero (Theorem 4).

2   Preliminaries

We consider a field K with a set $\Delta = \{\partial_1, \ldots, \partial_n\}$ of commuting derivations acting on it, and work with the ring of linear differential operators $K[D] = K[D_1, \ldots, D_n]$, where $D_1, \ldots, D_n$ correspond to the derivations $\partial_1, \ldots, \partial_n$, respectively.

Any operator $L \in K[D]$ is of the form

$$L = \sum_{|J| \le d} a_J D^J, \quad (5)$$

where $a_J \in K$, $J \in \mathbb{N}^n$ and $|J|$ is the sum of the components of J. Then we say that the polynomial

$$\mathrm{Sym}_L = \sum_{|J|=d} a_J X^J$$

is *the symbol* of L. Let K^* denotes the set of invertible elements in K. Then for $L \in K[D]$ and every $g \in K^*$ there is a gauge transformation

$$L \to g^{-1}Lg.$$

We also can say that this is the operation of conjugation. Then an algebraic differential expression I in the coefficients appearing in L is invariant under the gauge transformations if it is unaltered under these transformations. Trivial examples of an invariant are coefficients of the symbol of the operator.

An operator $L \in K[D]$ is said to be hyperbolic if its symbol is completely factorable (all factors are of first order) and each factor has multiplicity one.

3   Obstacles to Factorizations and Their Invariance

In this section we briefly recapitulate a few results from [7], because they are essential to the next sections.

Definition 1. *Let $L \in K[D]$ and suppose that its symbol has a decomposition* $\mathrm{Sym}_L = S_1 \ldots S_k$. *Then we say that the factorization*

$$L = F_1 \circ \ldots \circ F_k, \quad \text{where} \quad \mathrm{Sym}_{F_i} = S_i, \ \forall i \in \{1, \ldots, k\}, \tag{6}$$

is of the factorization type $(S_1)(S_2) \ldots (S_k)$.

Definition 2. *Let $L \in K[D]$, $\mathrm{Sym}_L = S_1 \ldots S_k$. An operator $R \in K[D]$ is called a* common obstacle *to factorization of the type $(S_1)(S_2) \ldots (S_k)$ if there exists a factorization of this type for the operator $L - R$ and R has minimal possible order.*

Remark 1. In general a common obstacle to factorizations of some factorization type is not unique.

Example 1. Consider a hyperbolic operator

$$L = D_{xy} - aD_x - bD_y - c,$$

where $a, b, c \in K$. An operator P_1 (in this particular case it is an operator of multiplication by a function) is a common obstacle to factorizations of the type $(X)(Y)$ if there exist $g_0, h_0 \in K$ such that

$$L - P_1 = (D_x - g_0) \circ (D_y - h_0).$$

Comparing the terms on the two sides of the equation, one gets $g_0 = b, h_0 = a$, and

$$P_1 = a_x - ab - c.$$

Analogously, we get a common obstacle to factorization of the type $(Y)(X)$:

$$P_2 = b_y - ab - c,$$

and the corresponding factorization for $(L - P_2)$: $L - P_2 = (D_x - a) \circ (D_y - b)$. Thus, the obtained common obstacles P_1 and P_2 are the Laplace invariants [2].

Theorem 1. *Consider a separable operator $L \in K[D_x, D_y]$ of order d, and the factorizations of L into first-order factors. Then*

1. *the order of common obstacles is less than or equal to $d - 2$;*
2. *a common obstacle is unique for each factorization type;*
3. *there are $d!$ common obstacles;*
4. *if $d = 2$, then the common obstacles of order 0 are the Laplace invariants;*
5. *the symbol of a common obstacle is an invariant.*

Corollary 1. *For an LPDO of the form*

$$L = (pD_x + qD_y)D_xD_y + a_{20}D_x^2 + a_{11}D_{xy} + a_{02}D_y^2 + a_{10}D_x + a_{01}D_y + a_{00}, \quad (7)$$

where all the coefficients belong to K, and p, q are not zero, consider its factorizations into first-order factors. Then

1. *the order of common obstacles is zero or one;*
2. *a common obstacle is unique for each factorization type, and therefore, the corresponding obstacles consist of just one element;*
3. *there are 6 common obstacles to factorizations into exactly three factors;*
4. *the symbol of a common obstacle is an invariant with respect to the gauge transformations $L \to g^{-1}Lg$.*

4   Computing of Invariants

Consider the operator (7). Since the symbol of an LPDO does not change under the gauge transformations $L \to g^{-1}Lg$, then the symbol, and therefore the coefficients of the symbol, are invariants with respect to these transformations. Thus, p and q are invariants.

Now we use Corollary 1 to compute a number of invariants for the operator L. Suppose for a while that

$$p = 1.$$

Denote the factors of the symbol $\mathrm{Sym}_L = (X + qY)XY$ of L by

$$S_1 = X, \;\; S_2 = Y, \;\; S_3 = X + qY.$$

Denote the common obstacle to factorizations of the type $(S_i)(S_j)(S_k)$ by Obst_{ijk}.

Then the coefficient of Y in the symbol of the common obstacle Obst_{123} is

$$(a_{01}q^2 + a_{02}^2 - (3q_x + a_{11}q)a_{02} + q_x qa_{11} - \partial_x(a_{11})q^2 + q\partial_x(a_{02}) + 2q_x^2 - q_{xx})/q^2.$$

By Theorem 1, this expression is invariant with respect to gauge transformations $L \to g^{-1}Lg$. Since the term $(2q_x^2 - q_{xx})/q^2$ and multiplication by q^2 does not influence the invariance property (because q is an invariant), the following expression is invariant also:

$$I_4 = a_{01}q^2 + a_{02}^2 - (3q_x + a_{11}q)a_{02} + q_x qa_{11} - \partial_x(a_{11})q^2 + q\partial_x(a_{02}).$$

The coefficient of Y in the symbol of the common obstacle Obst_{213} is

$$(I_4 - (\partial_x(a_{20})q^2 - \partial_y(a_{02})q + a_{02}q_y)q + a_{02}q_y)q - q_xq_yq + q_{xy}q^2 + 2q_x^2 - q_{xx}q)/q^2.$$

Again the expressions in q can be omitted, while I_4 is itself an invariant. Therefore,

$$I_2 = \partial_x(a_{20})q^2 - \partial_y(a_{02})q + a_{02}q_y$$

is an invariant.

Similarly, we obtain the invariants

$$I_1 = 2a_{20}q^2 - a_{11}q + 2a_{02},$$
$$I_3 = a_{10} + a_{20}(qa_{20} - a_{11}) + \partial_y(a_{20})q - \partial_y(a_{11}) + 2a_{20}q_y.$$

Generally speaking, by Corollary 1, there are six different obstacles to factorizations into exactly three factors. In fact, all the coefficients of the symbols of the common obstacles can be expressed in terms of four invariants

$$I_1, I_2, I_3, I_4.$$

Denote the symbol of the common obstacle Obst_{ijk} by Sym_{ijk}. Direct computations justify the following theorem:

Theorem 2

$$
\begin{aligned}
q^2\text{Sym}_{123} &= (q^2I_3 + I_2 - q_{xy}q + q_{yy}q^2 + q_xq_y)D_x + (I_4 + 2q_x^2 - q_{xx})D_y,\\
q^2\text{Sym}_{132} &= (i_2 + I_2)D_x & + (I_4 + 2q_x^2 - q_{xx})D_y,\\
q^2\text{Sym}_{213} &= (q^2I_3 + q^2q_{yy})D_x & + i_3D_y,\\
q^2\text{Sym}_{231} &= (q^2I_3 + q^2q_{yy})D_x & + i_1D_y,\\
q^2\text{Sym}_{312} &= (i_2 + I_2)D_x & + (i_1 + I_2q)D_y,\\
q^2\text{Sym}_{321} &= i_2D_x & + i_1D_y,
\end{aligned}
$$

where

$$
\begin{aligned}
i_1 &= I_4 - 2\partial_x(I_1)q + 4q_xI_1 - 2I_2q,\\
i_2 &= q^2I_3 - 2\partial_y(I_1)q + 2I_1q_y + I_2,\\
i_3 &= I_4 - I_2q - q_xq_yq + q_{xy}q^2 + 2q_x^2 - q_{xx}q.
\end{aligned}
$$

Note that neither of the obtained invariants I_1, I_2, I_3, I_4 depends on the "free" coefficient a_{00} of the operator L, and, therefore, we need at least one more invariant.

We guess the form of the fifth invariant by analyzing the structure of invariant

$$I_5 = a_{00} - a_{01}a_{20} - a_{10}a_{02} + a_{02}a_{20}a_{11} + (2a_{02} - a_{11} + 2a_{20})\partial_x(a_{20}) + \partial_{xy}(a_{20} - a_{11} + a_{02})$$

of the case $p = 1$, $q = 1$, considered in [6], and then perform some elimination. One of the difficulties here lies in the handling of large expressions, which appear during such manipulations. Naturally, a computer algebra system is needed, and we used MAPLE running our own package for linear partial differential operators with parametric coefficients. Thus, we get several candidates to be the fifth invariant. The most convenient of them has the form

$$I_5 = a_{00} - \frac{1}{2}\partial_{xy}(a_{11}) + q_x\partial_y(a_{20}) + q_{xy}a_{20} +$$
$$\left(2qa_{20} + \frac{2}{q}a_{02} - a_{11} + q_y\right)\partial_x(a_{20}) - \frac{1}{q}a_{02}a_{10} - a_{01}a_{20} + \frac{1}{q}a_{20}a_{11}a_{02}.$$

5   A Full System of Invariants for Third Order LPDOs

Here we prove that the obtained five invariants together form a full system of invariants for the case of operators with the symbol $(X + qY)XY$, and then, as the consequence, obtain a full system of invariants for operators with the symbol $(pX + qY)XY$.

One can prove that invariants I_1, I_2, I_3, I_4, I_5 form a full system in a similar way to that which was done for invariants of operators with the symbol $(X + Y)XY$ [6]. Below we suggest a simplification of such a way of proving, even though we consider a more general case.

Theorem 3. *For some non-zero $q \in K$, consider the operators of the form*

$$L = (D_x + qD_y)D_xD_y + a_{20}D_x^2 + a_{11}D_{xy} + a_{02}D_y^2 + a_{10}D_x + a_{01}D_y + a_{00}, \quad (8)$$

where the coefficients belong to K. Then the following is a full system of invariants of such an operator with respect to the gauge transformations $L \to g^{-1}Lg$:

$$I_1 = 2a_{20}q^2 - a_{11}q + 2a_{02},$$
$$I_2 = \partial_x(a_{20})q^2 - \partial_y(a_{02})q + a_{02}q_y,$$
$$I_3 = a_{10} + a_{20}(qa_{20} - a_{11}) + \partial_y(a_{20})q - \partial_y(a_{11}) + 2a_{20}q_y,$$
$$I_4 = a_{01}q^2 + a_{02}^2 - (3q_x + a_{11}q)a_{02} + q_xqa_{11} - \partial_x(a_{11})q^2 + q\partial_x(a_{02}),$$
$$I_5 = a_{00} - \frac{1}{2}\partial_{xy}(a_{11}) + q_x\partial_y(a_{20}) + q_{xy}a_{20} +$$
$$\left(2qa_{20} + \frac{2}{q}a_{02} - a_{11} + q_y\right)\partial_x(a_{20}) - \frac{1}{q}a_{02}a_{10} - a_{01}a_{20} + \frac{1}{q}a_{20}a_{11}a_{02}.$$

Thus, an operator $L' \in K[D]$

$$L' = (D_x + qD_y)D_xD_y + b_{20}D_x^2 + b_{11}D_xD_y + b_{02}D_y^2 + b_{10}D_x + b_{01}D_y + b_{00} \quad (9)$$

is equivalent to L (with respect to the gauge transformations $L \to g^{-1}Lg$) if and only if their corresponding invariants I_1, I_2, I_3, I_4, I_5 are equal.

Remark 2. Since the symbol of an LPDO L does not alter under the gauge transformations $L \to g^{-1}Lg$, we consider the operators with the same symbol.

Proof. 1. The direct computations show that the five expressions from the statement of the theorem are invariants with respect to the gauge transformations $L \to g^{-1}Lg$. One just has to check that these expressions do not depend on g, when calculate them for the operator $g^{-1}Lg$. Basically, we have to check the fifth expression I_5 only, since the others are invariants by construction.

2. Prove that these five invariants form a complete set of invariants, in other words, the operators L and L' are equivalent (with respect to the gauge transformations $L \to g^{-1}Lg$) if and only if their corresponding invariants are equal.

The direction "$\Rightarrow$" is implied from 1. Prove the direction "$\Leftarrow$". Let

$$I_1', I_2', I_3', I_4', I_5'$$

be the invariants computed from the coefficients of the operator L' by the formulas from the statement of the theorem, and

$$I_i = I_i', \quad i = 1, 2, 3, 4, 5. \tag{10}$$

Look for a function $g = e^f$, $f, g \in K$, such that

$$g^{-1}Lg = L'. \tag{11}$$

Equate the coefficients of D_{xx}, D_{yy} on both sides of (11), and get

$$\partial_y(f) = b_{20} - a_{20}, \tag{12}$$
$$\partial_x(f) = (b_{02} - a_{02})/q. \tag{13}$$

In addition, the assumption $I_2 = I_2'$ implies

$$(b_{20} - a_{20})_x = ((b_{02} - a_{02})/q)_y.$$

Therefore, there is only one (up to a multiplicative constant) function f, which satisfies the conditions (12) and (13).

Consider such a function f. Then substitute the expressions

$$b_{20} = a_{20} + f_y, \tag{14}$$
$$b_{02} = a_{02} + qf_x. \tag{15}$$

for b_{20}, b_{02} in (11), and prove that it holds for $g = e^f$.

Subtracting the coefficients of D_{xy} in $g^{-1}Lg$ from that in L' we get

$$b_{11} - a_{11} - 2f_x - 2qf_y,$$

which equals

$$2q(I_1 - I_1'),$$

which is zero by the assumption (10). Now we can substitute

$$b_{11} = a_{11} + 2f_x + 2qf_y.$$

Analogously, subtracting the coefficients of D_x, D_y in $g^{-1}Lg$ from those in L', correspondingly, we get

$$b_{10} - a_{10} - 2a_{20}f_x - a_{11}f_y - 2f_{xy} - 2f_xf_y - qf_{yy} - qf_y^2 =$$
$$I_3' - I_3 = 0,$$
$$b_{01} - a_{01} - 2a_{02}f_y - a_{11}f_x - 2qf_{xy} - 2qf_xf_y - f_{xx} - f_x^2 =$$
$$I_4' - I_4 = 0.$$

Now we can express b_{10} and b_{01}. Now, subtracting the "free" coefficient of $g^{-1}Lg$ from that of L', we get

$$b_{00} - a_{00} - a_{10}f_x - a_{01}f_y - a_{20}(f_{xx} + f_x^2) - a_{11}(f_{xy} + f_xf_y) - a_{02}(f_{yy} + f_y^2) -$$
$$f_{xxy} - 2f_{xy}f_x - f_yf_{xx} - f_yf_x^2 - qf_xf_{yy} - qf_xf_y^2 - qf_{xyy} - 2qf_yf_{xy} =$$
$$I_5' - I_5 = 0.$$

Thus, we proved that for the chosen function f, the equality (11) holds, and therefore, the operators L and L' are equivalent.

Remark 3. The Theorem 3 is a generalization of the result of [6], where the case $q = 1$ is considered.

Thus, a full system of invariants for the case $p = 1$ has been found. Now we give the formulae for the general case.

Theorem 4. *For some non-zero $p, q \in K$ consider the operators of the form*

$$L = (pD_x + qD_y)D_xD_y + a_{20}D_x^2 + a_{11}D_{xy} + a_{02}D_y^2 + a_{10}D_x + a_{01}D_y + a_{00}, \quad (16)$$

where the coefficients belong to K. Then the following is a full system of invariants of such an operator with respect to the gauge transformations $L \to g^{-1}Lg$:

$I_1 = 2a_{20}q^2 - a_{11}pq + 2a_{02}p^2,$

$I_2 = \partial_x(a_{20})pq^2 - \partial_y(a_{02})p^2q + a_{02}p^2q_y - a_{20}q^2p_x,$

$I_3 = a_{10}p^2 - a_{11}a_{20}p + 2a_{20}q_yp - 3a_{20}qp_y + a_{20}^2q - \partial_y(a_{11})p^2 + a_{11}p_yp + \partial_y(a_{20})pq,$

$I_4 = a_{01}q^2 - a_{11}a_{02}q + 2a_{02}qp_x - 3a_{02}pq_x + a_{02}^2p - \partial_x(a_{11})q^2 + a_{11}q_xq + \partial_x(a_{02})pq,$

$I_5 = a_{00}p^3q - p^3a_{02}a_{10} - p^2qa_{20}a_{01} +$
$\quad (pI_1 - pq^2p_y + qp^2q_y)a_{20x} + (qq_xp^2 - q^2p_xp)a_{20y}$
$\quad + (4q^2p_xp_y - 2qp_xq_yp + qq_{xy}p^2 - q^2p_{xy}p - 2qq_xpp_y)a_{20}$
$\quad + (\frac{1}{2}p_{xy}p^2q - p_xp_ypq)a_{11} - \frac{1}{2}p^3qa_{11xy} + \frac{1}{2}a_{11x}p_yp^2q + \frac{1}{2}a_{11y}p_xp^2q$
$\quad + p^2a_{02}a_{20}a_{11} + pqp_xa_{20}a_{11} - 2p_xq^2a_{20}^2 - 2p^2p_xa_{20}a_{02}.$

Proof. Since $p \neq 0$ we can multiply (16) by p^{-1} on the right, and get some new operator

$$L_1 = (D_x + \frac{q}{p}D_y)D_xD_y + \frac{a_{20}}{p}D_x^2 + \frac{a_{11}}{p}D_{xy} + \frac{a_{02}}{p}D_y^2 + \frac{a_{10}}{p}D_x + \frac{a_{01}}{p}D_y + \frac{a_{00}}{p}.$$

The invariants of the operator L and L_1 are the same. We compute the invariants of the operator L_1 by the formulae of Theorem 3, and get the invariants of the statement of the current theorem up to multiplication by integers and p, q.

Example 2. For some $p, q, c \in K$ consider the simple operator

$$L = (pD_x + qD_y)D_xD_y + c. \tag{17}$$

Compute the system of invariants of Theorem 4 for L:

$$0 = I_1 = I_2 = I_3 = I_4,$$
$$I_5 = p^3 qc.$$

Thus, every LPDO in $K[D_x, D_y]$ with the symbol $XY(pX + qY)$ that has the same set of invariants is equivalent to the simple operator (17). In fact, LPDOs that are equivalent to the operator (17) are not always trivial looking. Such operators have the form

$$L = (pD_x + qD_y)D_xD_y + pf_yD_x^2 + (2pf_x + 2qf_y)D_{xy} + qf_xD_y^2 +$$
$$(2pf_{xy} + 2pf_xf_y + qf_{yy} + qf_yf_y)D_x + (pf_{xx} + pf_xf_x + 2qf_{xy} + 2qf_xf_y)D_y +$$
$$c + pf_{xxy} + 2pf_{xy}f_x + pf_yf_{xx} + pf_yf_x^2 + qf_xf_{yy} + qf_xf_y^2 + qf_{xyy} + 2qf_yf_{xy},$$

for some $f \in K$.

6   Conclusion

For operators of the form

$$L = (pD_x + qD_y)D_xD_y + a_{20}D_x^2 + a_{11}D_{xy} + a_{02}D_y^2 + a_{10}D_x + a_{01}D_y + a_{00}, \tag{18}$$

where all the coefficients belong to K, we have found five invariants with respect to the gauge transformations $L \to g^{-1}Lg$ and proved that together they form a full system of operators.

In fact, Theorem 1 provides a way to find a number of invariants for hyperbolic bivariate LPDOs of arbitrary order, rather than just for those of order three. One of the difficulty lies in very large expressions, which appear already for third-order operators. Moreover, even if one manages to compute them, in general one gets a number of very large expressions. Then a challenge is to extract some nice looking invariants out of those large ones, so that these nice looking invariants generate the obtained ones. Thus, for the case of third-order LPDOs, we extracted four invariants out of twelve ones.

Another problem is that for applications one rather needs a full system of invariants. Thus, for the considered operators (18) we had to find a fifth invariant. However, even in this case it was not easy. Also for operators of high order, one needs to find more than one invariants so that they together with the obtained from obstacles ones form a full system.

Acknowledgments. This work was supported by Austrian Science Foundation (FWF) under the project SFB F013/F1304.

References

1. Anderson, I., Kamran, N.: The Variational Bicomplex for Hyperbolic Second-Order Scalar Partial Differential Equations in the Plane. Duke J. Math. 87, 265–319 (1997)
2. Darboux, G.: Leçons sur la théorie générale des surfaces et les applications géométriques du calcul infinitésimal, vol. 2. Gauthier-Villars (1889)
3. Ibragimov, N.: Invariants of hyperbolic equations: Solution of the Laplace problem. Prikladnaya Mekhanika i Tekhnicheskaya Fizika 45(2), 11–21 (2004), English Translation in Journal of Applied Mechanics and Technical Physics 45(2), 158166 (2004)
4. Kartashova, E.: Hierarchy of general invariants for bivariate LPDOs. J. Theoretical and Mathematical Physics, 1–8 (2006)
5. Morozov, O.: Contact Equivalence Problem for Linear Hyperbolic Equations, arxiv.orgpreprintmath-ph/0406004
6. Shemyakova, E.: A Full System of Invariants for Third-Order Linear Partial Differential Operators. In: Calmet, J., Ida, T., Wang, D. (eds.) AISC 2006. LNCS (LNAI), vol. 4120, pp. 978–973. Springer, Heidelberg (2006)
7. Shemyakova, E., Winkler, F.: Obstacles to the Factorization of Linear Partial Differential Operators into Several Factors. Programming and Computer Software 2 (accepted, 2007)
8. Tsarev, S.P.: Generalized Laplace Transformations and Integration of Hyperbolic Systems of Linear Partial Differential Equations. In: Proc. ISSAC'05 (2005)

Automatic Stability Analysis for a Diffusion Equation with Memories Using Maple

Daniel Esteban Sierra Sosa

Logic and Computer Group.
Deparment of Physics Engineering
School of Science and Humanities
Eafit University
dsierras@eafit.edu.co

Abstract. An efficient CAS helps the user to develop different Symbolic calculus problems, a clear example of this aid consist in the solution of the diffusion equation with and without memories, and its stability analysis working with Maple software package; the software gives the symbolic solution to this problem, but to do it, some basic definitions had to be implemented in the software, the stability analysis was not made automatically by the software, and when the problem was solved the necessity of an automatic solver were found.

1 Introduction

The CAS (computer algebra system) allows obtaining analytical solutions to mathematical problems with long expressions; proof of this is the analysis development of the diffusion equation, using Maple Software. An ideal solution algorithm to the diffusion equation will be presented, with an automatic Routh-Hurwitz theorem, it will be evaluated the advantages of the actually used CAS.

Also an stability analysis of the diffusion equation will be presented in this paper, with the stability analysis, the need of an intelligent CAS can be showed, not only for the results of the equation also with the mathematical theory that software should have in memory, and recommend to the user for the solution of the problem.

Despite this advantage the software interface demands some algorithmic knowledge to solve the problems; with the implementation of an intelligent CAS, the software user will be able to have an accurate interaction with the package, he does not have to be worried about his lack of mathematical knowledge, the software must be able to provide this aid. The ideal algorithm will be a proposal of new friendlier interfaces with user.

2 Model

The diffusion equation describes the behavior of a substance, in an autocatalytic reaction inside a circular reactor; it shows the relation between the creation, and

V.G. Ganzha, E.W. Mayr, and E.V. Vorozhtsov (Eds.): CASC 2007, LNCS 4770, pp. 370–376, 2007.

the diffusion of the substance that catalyze the reacting material [2]. With the stability analysis the critic radius will be estimated, employing some mathematical theorems and the Maple Software package.

The diffusion equation is a six terms equation, describes the concentration, reaction an diffusion behavior of the substance inside the reactor, it is also take in to account the memories in time for each considered variation [2]:

$$
\frac{\partial}{\partial t} u(r,t) + \int_0^t m_1(t-\tau) \frac{\partial}{\partial \tau} u(r,\tau) d\tau - \frac{\eta \left(\frac{\partial}{\partial r} u(r,t) + r \frac{\partial^2}{\partial r^2} u(r,t) \right)}{r} -
$$

$$
ku(r,t) - \int_0^t \frac{\chi \left(\frac{\partial}{\partial r} u(r,\tau) + r \frac{\partial^2}{\partial r^2} u(r,\tau) \right) m_2(t-\tau)}{r} d\tau -
$$

$$
q \int_0^t m_3(t-\tau) u(r,\tau) d\tau = 0, \tag{1}
$$

where the first couple of terms are the concentration and the concentration variation in time, the diffusion and reaction and each set of memories respectively; to solve this equation the initial condition take into account was U(r,0)=0, and a boundary condition $U(a,t) = \mu_b$, besides the function must be finite. Applying the Laplace Transform taking into account the initial condition, and renaming the transforms of corresponding memory functions of the equation; is a solving step to change the partial differential equation into an ordinary differential equation

Table 1. Symbol descriptions

Name	Symbol
Diffusivity	η
Deactivity	k
Amplitude coefficient of diffusion memory	χ
Amplitude coefficient of the reaction memory	q

The ordinary differential equation was solved taking into account the boundary and finitude conditions. Using the solver of differential equation in Maple software was obtained:

$$
U(r) = \frac{\mu_b J_0(f(s) r)}{J_0(f(s) a) s}. \tag{2}
$$

The inverse Laplace transform was obtained using the Bromwich integral which was solved using the residue theorem [4,1]

$$
U(r,t) = \frac{1}{2\pi i} \int \frac{\mu_b J_0(f(s) r) e^{st}}{J_0(f(s) a) s} ds, \tag{3}
$$

Table 2. Program code

```
> restart;
> with(VectorCalculus) :
> with(inttrans, laplace);
> eq := diff(u(r,t),t) + int(m[1](t − tau) * diff(u(r, tau), tau), tau = 0..t)
−eta * Laplacian(u(r,t), polar[r, theta]) − k * u(r,t)−
int(chi * Laplacian(u(r, tau), polar[r, theta]) * m[2](t − tau), tau = 0..t)−
q * int(m[3](t − tau) * u(r, tau), tau = 0..t) = 0;
> eq1 := laplace(eq, t, s);
> eq2 := subs([laplace(u(r,t), t, s) = U(r), u(r, 0) = 0, laplace(m[1](t), t, s) =
M[1](s), laplace(m[2](t), t, s) = M[2](s), laplace(m[3](t), t, s) =
M[3](s)], eq1);
> eq3 := dsolve(eq2, U(r));
> eq4 := subs(c2 = 0, eq3);
>c 1 := solve((subs(r = a, rhs(eq4)) = laplace(mu[b], t, s)),c 1);
> eq4;
> eq5A := expand(eq4);
> eq5B := subs([(−1/(chi * M[2](s) + eta) * s − 1/(chi * M[2](s) + eta)*
M[1](s) * s + 1/(chi * M[2](s) + eta) * q * M[3](s) + 1/(chi * M[2](s) + eta) * k)(1/2) =
f(s)], eq5A);
> eq5BP := subs([BesselJ(0, f(s) * r) = J[0](f(s) * r), BesselJ(0, f(s) * a)
= J[0](f(s) * a)], eq5B);
> eq5e := U(r,t) = (1/(2 * pi * i)) * Int(rhs(eq5BP) * exp(s * t), s);
> U(r,t) = residue(rhs(eq5B), s = 0) + Sum((exp(s * t) * (numer(rhs(eq5B))))/s)/
(diff((denom((rhs(eq5B)))/s), s)), n = 1..infinity);
> eq6 := C[r] = numer(simplify(((−1/(chi * M[2](s) + eta) * s − 1/(chi*
M[2](s) + eta) * M[1](s) * s + 1/(chi * M[2](s) + eta) * q * M[3](s)+
1/(chi * M[2](s) + eta) * k)(1/2))² − (alpha[n]/a)²));
> eq6A := subs(s = 0, rhs(eq6)) = 0;
> eq7 := U(r,t) = subs([s = s[ni], BesselJ(0, f(s) * r) =
J[0](alpha[n]/a * r), BesselJ(1, f(s) * a) = J[1](alpha[n]), BesselJ(0, f(0) * r) =
J[0](f(0) * r), BesselJ(0, f(0) * a) = J[0](f(0) * a)],
residue(rhs(eq5B), s = 0) + Sum(Sum((exp(s * t) * (numer(rhs(eq5B))))/s)/
(diff((denom((rhs(eq5B)))/s), s)), i = 1..degree * C[r]), n = 1..infinity));
> f(s)² − (alpha[n]/a)² = 0;
> eq8 := coeff(lhs(eq6A), s, 0) = 0;
> eq9 := (solve(eq6A, a));
> a[c, n] = eq9[1];
> a[c, 1] = (((chi * M[2](0) + eta)/(q * M[3](0) + k))(1/2)) * 2.405;
```

after the residue theorem was applied, the following expression was obtained [4]:

$$U(r,t) = \frac{\mu_b J_0 (f(0) r)}{J_0 (f(0) a)} +$$

$$\sum_{n=1}^{\infty} \left(\sum_{i=1}^{(deg) C_r} -e^{s_{ni}t} \mu_b J_0 \left(\frac{\alpha_n r}{a} \right) s_{ni}^{-1} (J_1 (\alpha_n))^{-1} \left(\frac{d}{ds_{ni}} f (s_{ni}) \right)^{-1} a^{-1} \right), \quad (4)$$

where C_r is the function defined as follows, and the maximum s variable degree it is the summation upper limit:

$$C_r = -sa^2 - M_1(s)\, sa^2 + qM_3(s)\, a^2 + ka^2 - \alpha_n{}^2 \chi\, M_2(s) - \alpha_n{}^2 \eta, \qquad (5)$$

and M_1, M_2 and M_3 are the Laplace transform of the function into diffusion equation memories. Then the Routh-Hurwitz theorem was applied to find the critical radius into reactor [3],

$$a_{c,n} = \sqrt{\frac{\chi M_2(0) + \eta}{qM_3(0) + k}}\, \alpha_n. \qquad (6)$$

2.1 Program Code

The code use in general case [Table 2] was implemented in Maple Software, where the first three lines are the used packages of the software, the next line is the input, in this case the diffusion equation; then the Laplace transform of the equation was obtained, and the initial, finitude and boundary condition were set. Solving the differential equation was obtained the U(r) function, then using the Bromwich Integral the inverse Laplace Transform was obtained, and at the end of the code lines the residue theorem was applied, and with the Routh-Hurwitz theorem the stability analysis was did.

3 Solution Method

To solve the diffusion equation according to the model, was employed the Maple Software package in order to obtain step by step each result that was presented in the model, but some of the procedures, as the right employment of the theorems were manual, the idea is develop an intelligent CAS to solve this kind of problems without the usage of any other reference.

3.1 Diffusion Equation

The diffusion equation describes the concentration change of a reacting substance in a reactor, besides the concentration change, the substance can move inside the reactor; and react creating another substance, the equation (1) also describes these conditions.

$$\frac{\partial}{\partial t} u(r,t) - \frac{\eta \left(\frac{\partial}{\partial r} u(r,t) + r \frac{\partial^2}{\partial r^2} u(r,t) \right)}{r} - ku(r,t) = 0. \qquad (7)$$

To solve this equation was applied the Bromwich integral

$$U(r,t) = \frac{1}{2\pi i} \int \frac{\mu_b J_0\left(f(s)\, r\right) e^{st}}{J_0\left(f(s)\, a\right) s} \, ds, \qquad (8)$$

in this case the residue theorem is define as:

$$U(r,t) = \frac{\mu_b J_0\left(f(0)\, r\right)}{J_0\left(f(0)\, a\right)} +$$

$$\sum_{n=1}^{\infty} -e^{s_n t} \mu_b J_0\left(\frac{\alpha_n r}{a}\right) s_n{}^{-1} \left(J_1(\alpha_n)\right)^{-1} \left(\frac{d}{ds_n} f(s_n)\right)^{-1} a^{-1}, \qquad (9)$$

where f(s) is the function:

$$f(s) = \sqrt{\frac{-s+k}{\eta}}. \tag{10}$$

Applying the Routh-Hurwitz theorem to the f(s) function and taking into account the definition of the critical radius was obtained:

$$s_n = \frac{ka^2 - \alpha_n{}^2 \eta}{a^2}, \tag{11}$$

$$a_{c,n} = \sqrt{\frac{\eta}{k}} \alpha_n, \tag{12}$$

$$a_{c,1} = 2.405 \sqrt{\frac{\eta}{k}}. \tag{13}$$

3.2   Diffusion Equation with Exponential Memory

Now the diffusion equation will be analyzed taking into account all the memory terms, and the functions $m_1(t-\tau)$, $m_2(t-\tau)$ and $m_3(t-\tau)$, displayed in the model will be exponential functions, as follows:

$$\frac{\partial}{\partial t}u(r,t) + \int_0^t e^{-\beta(t-\tau)}\frac{\partial}{\partial \tau}u(r,\tau)\,d\tau - \frac{\eta\left(\frac{\partial}{\partial r}u(r,t) + r\frac{\partial^2}{\partial r^2}u(r,t)\right)}{r} -$$

$$ku(r,t) - \int_0^t \frac{\chi\left(\frac{\partial}{\partial r}u(r,\tau) + r\frac{\partial^2}{\partial r^2}u(r,\tau)\right)e^{-\alpha(t-\tau)}}{r}\,d\tau -$$

$$q\int_0^t e^{-\delta(t-\tau)}u(r,\tau)\,d\tau = 0. \tag{14}$$

To solve this equation we applied the Bromwich integral

$$U(r,t) = \frac{1}{2\pi i}\int \frac{\mu_b J_0(f(s)r)e^{st}}{J_0(f(s)a)s}\,ds, \tag{15}$$

in this case the residue theorem is defined as:

$$U(r,t) = \frac{\mu_b J_0(f(0)r)}{J_0(f(0)a)} +$$

$$\sum_{n=1}^{\infty}\left(\sum_{i=1}^{4} -e^{s_{ni}t}\mu_b J_0\left(\frac{\alpha_n r}{a}\right)s_{ni}{}^{-1}(J_1(\alpha_n))^{-1}\left(\frac{d}{ds_{ni}}f(s_{ni})\right)^{-1}a^{-1}\right), \tag{16}$$

where f(s) is the function:

$$-a^2 s^4 + \left(-\alpha_n{}^2\eta - a^2\delta - a^2\beta - a^2\alpha - a^2 + a^2 k\right)s^3 + Bs^2 + Ws +$$
$$a^2 q\beta\,\alpha + a^2 k\beta\,\delta\,\alpha - \alpha_n{}^2\beta\,\delta\,\eta\,\alpha - \alpha_n{}^2\beta\,\delta\,\chi = f(s), \tag{17}$$

$$B = -a^2\alpha + a^2q - \alpha_n{}^2\delta\eta + a^2k\alpha - a^2\delta - \alpha_n{}^2\beta\eta - a^2\delta\alpha + a^2k\delta -$$
$$\alpha_n{}^2\eta\alpha - a^2\beta\alpha - a^2\beta\delta - \alpha_n{}^2\chi + a^2k\beta, \tag{18}$$
$$W = a^2q\beta - \alpha_n{}^2\beta\eta\alpha - a^2\beta\delta\alpha - \alpha_n{}^2\delta\eta\alpha + a^2k\beta\alpha - \alpha_n{}^2\beta\chi -$$
$$\alpha_n{}^2\beta\delta\eta + a^2k\delta\alpha - \alpha_n{}^2\delta\chi + a^2k\beta\delta + a^2q\alpha - a^2\delta\alpha. \tag{19}$$

Applying the Routh-Hurwitz theorem to the f(s) function and taking into account the definition of the critical radius was obtained:

$$a^2q\beta\alpha + a^2k\beta\delta\alpha - \alpha_n{}^2\beta\delta\eta\alpha - \alpha_n{}^2\beta\delta\chi = 0, \tag{20}$$

$$a_{c,n} = \sqrt{\frac{\delta\,(\chi + \eta\,\alpha)}{\alpha\,(k\delta + q)}}\,\alpha_n, \tag{21}$$

$$a_{c,1} = 2.405\sqrt{\frac{\delta\,(\chi + \eta\,\alpha)}{\alpha\,(k\delta + q)}}. \tag{22}$$

3.3   Diffusion Equation with a Power Function Memory

In this case the functions $m_1(t - \tau)$, $m_2(t - \tau)$ and $m_3(t - \tau)$ in the diffusion equation, displayed in the model will be power functions, as follows:

$$\frac{\partial}{\partial t}u\,(r,t) + \int_0^t (t - \tau)^n\,e^{-\beta\,(t-\tau)}\,\frac{\partial}{\partial\tau}u\,(r,\tau)\,d\tau - \frac{\eta\left(\frac{\partial}{\partial r}u\,(r,t) + r\frac{\partial^2}{\partial r^2}u\,(r,t)\right)}{r} -$$
$$ku\,(r,t) - \int_0^t \frac{(t - \tau)^n\,\chi\left(\frac{\partial}{\partial r}u\,(r,\tau) + r\frac{\partial^2}{\partial r^2}u\,(r,\tau)\right)\,e^{-\alpha\,(t-\tau)}}{r}\,d\tau -$$
$$q\int_0^t (t - \tau)^n\,e^{-\delta\,(t-\tau)}u\,(r,\tau)\,d\tau = 0. \tag{23}$$

To solve this equation ise applied the Bromwich integral

$$U\,(r,t) = \frac{1}{2\pi i}\int \frac{\mu_b J_0\left(\sqrt{\lambda\,(s)}r\right)e^{st}}{J_0\left(\sqrt{\lambda\,(s)}a\right)s}\,ds \tag{24}$$

where $\lambda(s)$ in this case is the function defined as follows, and one more time its s maximum degree is the summation upper limit:

$$\lambda\,(s) = -\frac{(s + \alpha)\,G\left(\eta\,(s + \alpha)^{2n}\,s + \Gamma\,(n)\,\chi\,n\,(s + \alpha)^n + \eta\,\alpha\,(s + \alpha)^{2n}\right)}{(s + \beta)\,(s + \delta)\,(\eta\,(s + \alpha)^n\,s + \eta\,(s + \alpha)^n\,\alpha + \chi\,\Gamma\,(n)\,n)^2}, \tag{25}$$
$$G = s^3 + s^2\Gamma\,(k + 1)\,(s + \beta)^{-n} - s^2k + s^2\delta + \beta\,s^2 - s\Gamma\,(k + 1)\,(s + \delta)^{-n}\,q -$$
$$sk\beta + \beta\,\delta\,s - sk\delta + s\Gamma\,(k + 1)\,(s + \beta)^{-n}\,\delta - k\beta\delta - \Gamma\,(k + 1)\,(s + \delta)^{-n}\,q\beta.(26)$$

Applying the Routh-Hurwitz theorem to the $\lambda(s)$ function and taking into account the definition of the critical radius was obtained:

$$-2\,\Gamma(n)\,\alpha_n{}^2\eta\,\alpha^n\alpha\,\chi\,n\beta\,\delta - \alpha_n{}^2\chi^2\,(\Gamma(n))^2\,n^2\beta\,\delta - \alpha_n{}^2\eta^2\alpha^{2\,n}\alpha^2\beta\,\delta +$$
$$a^2\alpha^2 k\eta\,\alpha^{2\,n}\beta\,\delta + \Gamma(n)\,a^2\alpha\,k\chi\,n\alpha^n\beta\,\delta +$$
$$\Gamma(k+1)\,a^2\alpha\,\delta^{-n}q\beta\,\Gamma(n)\,\chi\,n\alpha^n + \Gamma(k+1)\,a^2\delta^{-n}q\beta\,\eta\,\alpha^2\alpha^{2\,n} = 0, \quad (27)$$

$$a_{c,n} = \alpha_n\delta\,\sqrt{\frac{\eta\,\alpha^2 + \chi}{k\delta^2 + q}}\,\alpha^{-1}, \tag{28}$$

$$a_{c,1} = 2.405\,\sqrt{\frac{\eta\,\alpha^2 + \chi}{k\delta^2 + q}}\,\delta\alpha^{-1}. \tag{29}$$

4   Advantages of Intelligent CAS

The actual developed CAS are useful solving most of the symbolic calculus problems, but the user of the software has to interfere in some mechanical process that should be solved automatically by software; some theorems must be implemented by the user to solve a symbolic calculus problem, in order to obtain an answer. All this *aid* that user has to provide to software should be none.

In order to solve this problem in actual CAS is a real automatic solver, with a friendlier interface to the user, that provide help to the user and suggest solution ways.

5   Conclusions

The diffusion equation with memories is a good example to evaluate the advantages and disadvantages of the existing CAS, besides the solution of this problem proofs the utility that provides the software solving long term expressions. In order to find a solution to the problem the implemented algorithm was accurate to solve the problem but there was a lot theoretical interference that software should has.

The CAS is a powerful tool, but it has to be refined, in order to be friendlier with user, also more accurate, because it will be easier to the user to handle it, and in some cases this advantage could lead to automatic problem solution.

References

1. Arfken, G.: Mathematical Methods for Physicists. Academic Press, London (1985)
2. Lightfoot Bird, S.: Transport Phenomena. John Wiley and Sons, New York (2002)
3. Gradshteyn, I.S., Ryzhik, I.M.: Tables of Integrals, Series, and Products. Academic Press, London (2000)
4. Brown, R.C.J.: Complex Variables and Applications. McGraw-hill, New York (2003)

Bounds for Real Roots and Applications to Orthogonal Polynomials

Doru Ştefănescu

University of Bucharest, Romania
stef@rms.unibuc.ro

Abstract. We obtain new inequalities on the real roots of a univariate polynomial with real coefficients. Then we derive estimates for the largest positive root, which is a key step for real root isolation. We discuss the case of classic orthogonal polynomials. We also compute upper bounds for the roots of orthogonal polynomials using new inequalities derived from the differential equations satisfied by these polynomials. Our results are compared with those obtained by other methods.

1 Bounds for Real Polynomial Roots

The computation of the real roots of univariate polynomials with real coefficients is based on their isolation. To isolate the real positive roots, it is sufficient to estimate the smallest positive root (cf. [2] and [21]). This can be achieved if we are able to compute accurate estimates for the largest positive root.

1.1 Computation of the Largest Positive Root

Several bounds exist for the absolute values of the roots of a univariate polynomial with complex coefficients (see, for example, [15]). These bounds are expressed as functions of the degree and of the coefficients, and naturally they can be used also for the roots (real or complex) of polynomials with real coefficients. However, for the real roots of polynomials with real coefficients there also exist some specific bounds. In particular, some bounds for the positive roots are known, the first of which were obtained by Lagrange [11] and Cauchy [5]. We briefly survey here the most often used bounds for positive roots and discuss their efficiency in particular cases, emphasizing the classes of orthogonal polynomials. We then obtain extensions of a bound of Lagrange, and derive a result also valid for positive roots smaller than 1.

A Bound of Lagrange

Theorem 1 (Lagrange). *Let* $P(X) = a_0 X^d + \cdots + a_m X^{d-m} - a_{m+1} X^{d-m-1} \pm \cdots \pm a_d \in \mathbb{R}[X]$, *with all* $a_i \geq 0$, $a_0, a_{m+1} > 0$. *Let*

$$A = \max\left\{ a_i \, ; \, \operatorname{coeff}\left(X^{d-i}\right) < 0 \right\}.$$

V.G. Ganzha, E.W. Mayr, and E.V. Vorozhtsov (Eds.): CASC 2007, LNCS 4770, pp. 377–391, 2007.

The number

$$1 + \left(\frac{A}{a_0}\right)^{1/(m+1)}$$

is an upper bound for the positive roots of P.

The bound from Theorem 1 is one of the most popular (cf. H. Hong [8]), however it gives only bounds larger than one. For polynomials with subunitary real roots, it is recommended to use the bounds of Kioustelidis [9] or Ştefănescu [18]. A discussion on the efficiency of these results can be found in Akritas–Strzeboński–Vigklas [2] and Akritas–Vigklas [3].

Extensions of the Bound of Lagrange. We give a result that extends the bound $L_1(P)$ of Lagrange.

Theorem 2. *Let* $P(X) = a_0 X^d + \cdots + a_m X^{d-m} - a_{m+1} X^{d-m-1} \pm \cdots \pm a_d \in \mathbb{R}[X]$, *with all* $a_i \geq 0$, $a_0, a_{m+1} > 0$. *Let*

$$A = \max\left\{a_i \,; \operatorname{coeff}(X^{d-i}) < 0\right\}.$$

The number

$$\begin{cases} 1 + \max\left\{\left(\dfrac{pA}{a_0 + \cdots + a_s}\right)^{1/(m-s+1)}, \right. \\[2ex] \left. \left(\dfrac{qA}{sa_0 + \cdots + 2a_{s-2} + a_{s-1}}\right)^{1/(m-s+2)}, \right. \\[2ex] \left. \left(\dfrac{2rA}{s(s-1)a_0 + (s-1)(s-2)a_1 + \cdots + 2a_{s-2}}\right)^{1/(m-s+3)}. \right. \end{cases} \tag{1}$$

is an upper bound for the positive roots of P *for any* $s \in \{2, 3, \ldots, m\}$ *and* $p \geq 0, q \geq 0, r \geq 0$ *such that* $p + q + r = 1$.

The proof of Theorem 2 is similar to that of our Theorem 1 in [19].

Particular Cases of Theorem 2.

1. For $p = 1$, $q = r = 0$, we obtain the bound

$$1 + \left(\frac{A}{a_0 + \cdots + a_s}\right)^{1/(m-s+1)}.$$

This bound is also valid for $s = 0$ and $s = 1$. For $s = 0$, it reduces to the bound $L_1(P)$ of Lagrange.

2. For $p = q = r = 1/3$, we obtain Theorem 1 from [19].

3. For $p = q = 1/4$, $r = 1/2$, we obtain

$$1 + \max \left\{ \left(\frac{A}{4(a_0 + \cdots + a_s)} \right)^{1/(m-s+1)} \right. ,$$

$$\left(\frac{A}{4(sa_0 + \cdots + 2a_{s-2} + a_{s-1})} \right)^{1/(m-s+2)} ,$$

$$\left. \left(\frac{A}{s(s-1)a_0 + (s-1)(s-2)a_1 + \cdots + 2a_{s-2}} \right)^{1/(m-s+3)} \right\} .$$

4. For $p = q = \dfrac{1}{2}$, $r = 0$, we obtain

$$1 + \max \left\{ \left(\frac{A}{2(a_0 + \cdots + a_s)} \right)^{1/(m-s+1)} \right. ,$$

$$\left. \left(\frac{A}{2(sa_0 + \cdots + 2a_{s-2} + a_{s-1})} \right)^{1/(m-s+2)} \right\} ,$$

which is Theorem 3 from [18]. This bound is also valid for $s = 0$.

Example 1. Let

$$P_1(X) = X^{17} + X^{13} + X^{12} + X^9 + 3X^8 + 2X^7 + X^6 - 5X^4 + X^3 - 4X^2 - 6,$$

$$P_2(X) = X^{13} + X^{12} + X^9 + 3X^8 + 2X^7 + X^6 - 6X^4 + X^3 - 4X^2 - 7.$$

We denote by $B(P) = B(m, s, p, q, r)$ the bound given by Theorem 1, by $L_1(P)$ that of Lagrange (Theorem 1) and by LPR the largest positive root. For P_1 we have $A = 6$ and $m = 11$, and for P_2 we have $A = 7$ and $m = 6$. We obtain:

P	s	p	q	r	$B(P)$	$L_1(P)$	LPR
P_1	8	0.5	0.5	0	13.89	2.161	1.53
P_1	2	0.5	0.5	0	3.15	2.161	1.53
P_1	1	0.5	0.5	0	2.00	2.161	1.53
P_1	8	0.4	0.3	0.3	64.78	2.161	1.53
P_1	2	0.2	0.6	0.2	3.25	2.161	1.53
P_2	7	0.5	0.5	0	8.25	2.232	1.075
P_2	3	0.4	0.6	0	7.18	2.232	1.075
P_2	3	0.5	0.5	0	6.85	2.232	1.075
P_2	1	0.4	0.6	0	3.07	2.232	1.075
P_2	5	0.4	0.3	0.3	26.2	2.232	1.075
P_2	2	0.4	0.3	0.3	4.02	2.232	1.075
P_2	2	0.6	0.2	0.2	3.84	2.232	1.075

Comparison with the Bound of Lagrange. We compare the bound given by Theorem 2 with that of Lagrange

$$L_1(P) = 1 + \left(\frac{A}{a_0}\right)^{1/(m+1)}.$$

We consider $p = q = 0.25$, $r = 0.5$ and $s = 2$ in Theorem 2. With the previous notation we have

$$B(P) = 1 + \max\left\{\left(\frac{A}{4(a_0 + a_1 + a_2)}\right)^{1/(m-1)}, \left(\frac{A}{4(2a_0 + a_1)}\right)^{1/m}, \left(\frac{A}{2a_0}\right)^{1/(m+1)}\right\}.$$

We can see which of the bounds $B(P)$ and $L_1(P)$ is better by looking to the size of A with respect to a_0, a_1, a_2 and m. We obtain

$$B(P) < L_1(P) \quad \text{if} \quad A < \min\left\{\frac{\left(4(a_0 + a_1 + a_2)\right)^{(m+1)/2}}{a_0^{(m-1)/2}}, \frac{4^{m+1}(2a_0 + a_1)^{m+1}}{a_0^m}\right\}$$

and

$$B(P) > L_1(P) \quad \text{if} \quad A > \max\left\{\frac{\left(4(a_0 + a_1 + a_2)\right)^{(m+1)/2}}{a_0^{(m-1)/2}}, \frac{4^{m+1}(2a_0 + a_1)^{m+1}}{a_0^m}\right\}.$$

Example 2. We consider

$$\begin{aligned}
P(X) &= P(X, d, A) \\
&= X^d + 3X^{d-1} + X^{d-2} + 0.001\,X^{d-3} + 0.0003\,X^{d-4} \\
&\quad - AX^4 - AX^3 - AX - A + 1,
\end{aligned}$$

with $A > 0$.

We obtain the values in the following table:

d	A	$L_1(P)$	$B(P)$	LPR
10	3	2.201	2.069	1.146
11	3	2.201	2.069	1.126
8	4	2.256	2.122	1.287
9	4	2.256	2.122	1.230
10	4	2.256	2.122	1.193
10	20^6	20.999	43.294	19.687

1.2 Other Bounds for Positive Roots

Note that the bound $L_1(P)$ of Lagrange and its extensions give only numbers greater than one, so they cannot be used for some classes of polynomials. For example, the roots of Legendre orthogonal polynomials are subunitary.

J. B. Kioustelidis [9] gives the following upper bound for the positive real roots:

Theorem 3 (Kioustelidis). *Let* $P(X) = X^d - b_1 X^{d-m_1} - \cdots - b_k X^{d-m_k} + g(X)$, *with* $g(X)$ *having positive coefficients and* $b_1 > 0, \ldots, b_k > 0$. *The number*

$$K(P) = 2 \cdot \max\{b_1^{1/m_1}, \ldots, b_k^{1/m_k}\}$$

is an upper bound for the positive roots of P.

For polynomials with an even number of variations of sign, we proposed in [18] another bound. Our method can be applied also to polynomials having at least a sign variation. For this it is sufficient to make use of the following

Lemma 1. *Any polynomial* $P(X) \in \mathbb{R}[X]$ *having at least one sign variation can be represented as*

$$P(X) = c_1 X^{d_1} - b_1 X^{m_1} + c_2 X^{d_2} - b_2 X^{m_2} + \cdots + c_k X^{d_k} - b_k X^{m_k} + g(X),$$

with $g(X) \in \mathbb{R}_+[X]$, $c_i > 0$, $b_i > 0$, $d_i > m_i$ *for all* i.

Proof. If P has an even number of sign variations there is nothing to prove. Otherwise, suppose that

$$P(X) = c X^{d_1} + \sum_{i=2}^{k-t} c_i X^{d_i} - \sum_{i=1}^{k} b_i X^{m_i} + h(X),$$

with $h(X) \in \mathbb{R}_+[X]$, $c > 0$, $c_i > 0$, $b_i > 0$, $d_i > m_i$ for all $i \le k - t$ and $t \ge 1$. Then we put $c = c_1 + c_{k-t+1} + \ldots + c_k$ with $c_1 > 0$, $c_j > 0$ for any $j \in \{k - t + 1, \ldots, k\}$. □

Then we have the following estimation for positive roots:

Theorem 4. *Let* $P(X) \in \mathbb{R}[X]$ *and suppose that* P *has at least one sign variation. If*

$$P(X) = c_1 X^{d_1} - b_1 X^{m_1} + c_2 X^{d_2} - b_2 X^{m_2} + \cdots + c_k X^{d_k} - b_k X^{m_k} + g(X),$$

with $g(X) \in \mathbb{R}_+[X]$, $c_i > 0$, $b_i > 0$, $d_i > m_i$ *for all* i, *the number*

$$S(P) = \max\left\{ \left(\frac{b_1}{c_1}\right)^{1/(d_1 - m_1)}, \ldots, \left(\frac{b_k}{c_k}\right)^{1/(d_k - m_k)} \right\}$$

is an upper bound for the positive roots of P.

Proof. It is sufficient to observe that for $x > 0$, $x > S(P)$, we have

$$c_i\, x^{d_i} - b_i\, x^{m_i} > 0 \quad \text{for all } i\,,$$

so $P(x) > 0$. □

Remark 1. We obtained in [18], Theorem 2, another version of Theorem 4, under the additional assumption that the polynomial has an even number of sign variations and that $d_i > m_i > d_{i+1}$ for all i. Afterwards, Akritas *et al.* presented in [2] a result based on Theorem 2 from [18]. Their approach to adapt our theorem to any polynomial with sign variations uses a representation

$$P(X) = \sum_{i=1}^{m} (q_{2i-1}(X) - q_{2i}(X)) + g(X), \tag{2}$$

where all polynomials q_j and g have positive coefficients, and some supplementary inequalities among the degrees of the monomials of the successive polynomials q_{2i-1} and q_{2i} are satisfied. However, the proof of their result is a variation of ours from [18] and their upper bound is similar. There is no evidence that the consideration of the additional polynomials q_j is a gain from a computational point of view. Our statement of Theorem 4 is also more concise.

At the same time, the splitting of a positive coefficient in Lemma 1, respectively the decomposition (2) in [2], if needed, are not unique. We also note that Theorem 2 from [18] and the extensions of Akritas *et al.* were implemented in [2] and [3].

Remark 2. If a polynomial $P \in \mathbb{R}[X]$ has all real positive roots in the interval $(0, 1)$, using the transformation $x \to 1/x$ we obtain a polynomial — called the reciprocal polynomial — with positive roots greater than one. If we compute an bound ub for the positive roots of the reciprocal polynomial, the number $lb = 1/ub$ will be a lower bound for the positive roots of the initial polynomial P. This process can be applied to any real polynomial with positive roots, and is a key step in the Continued Fraction real root isolation algorithm (see [2] and [21]).

Note that in some special cases the following other bound of Lagrange can be useful:

Theorem 5. *Let F be a nonconstant monic polynomial of degree n over $\mathbb{R}$ and let $\{a_j\,;\, j \in J\}$ be the set of its negative coefficients. Then an upper bound for the positive real roots of F is given by the sum of the largest and the second largest numbers in the set*

$$\left\{ \sqrt[j]{|a_j|}\,;\, j \in J \right\}.$$

Theorem 5 can be extended to absolute values of polynomials with complex coefficients (see [14]).

Notation. The bounds of Lagrange from Theorems 1 and 5 will be denoted by $L_1(P)$, respectively $L_2(P)$.

Example 3. We consider $P(X) = 2X^7 - 3X^4 - X^3 - 2X + 1 \in \mathbb{R}[X]$. The polynomial P does not fulfill the assumption $d_i > m_i > d_{i+1}$ for all i from Theorem 2 in [18]. However, after the decomposition of the leading coefficient in a sum of positive numbers, as in Lemma 1, Theorem 4 can be applied. We use the following two representations:

$$P(X) = P_1(X) = (X^7 - 3X^4) + (0.5\, X^7 - X^3) + (0.5\, X^7 - 2X) + 1,$$

$$P(X) = P_2(X) = (1.1\, X^7 - 3X^4) + (0.4\, X^7 - X^3) + (0.5\, X^7 - 2X) + 1.$$

We denote $S_j(P) = S(P_j)$ for $j = 1, 2$, and obtain the bounds

$$S_1(P) = 1.442, \quad S_2(P) = 1.397.$$

On the other hand the largest positive root of P is 1.295. Other bounds give

$$K(P) = 2.289, \quad L_1(P) = 2.404, \quad L_2(P) = 2.214.$$

Both $S_1(P)$ and $S_2(P)$ are smaller than $L_1(P)$, $L_2(P)$ and $K(P)$.

2   Bounds for Roots of Orthogonal Polynomials

Classical orthogonal polynomials have real coefficients and all their zeros are real, distinct, simple and located in the interval of orthogonality.

We first evaluate the largest positive roots of classical orthogonal polynomials using the results in Section 1 and a bound considered by van der Sluis in [17]. We also obtain new bounds using properties of of the differential equations which they satisfy. These new bounds will be compared with known bounds.

Proposition 1. *Let P_n, L_n, T_n and U_n be the orthogonal polynomials of degree n of Legendre, respectively Laguerre and Chebyshev of first and second kind. We have*

i. *The number $S(P_n) = \sqrt{\dfrac{n(n-1)}{2(2n-1)}}$ is an upper bound for the roots of P_n.*

ii. *The number $S(L_n) = n^2$ is an upper bound for the roots of L_n.*

iii. *The number $S(T_n) = \dfrac{\sqrt{n}}{2}$ is an upper bound for the roots of T_n.*

iv. *The number $S(U_n) = \dfrac{\sqrt{n-1}}{2}$ is an upper bound for the roots of U_n.*

Proof. We use the representations

$$P_n(X) = \sum_{k=0}^{\lfloor n/2 \rfloor} (-1)^k \frac{(2n-2k)!}{k!(n-k)!(n-2k)!} X^{n-2k},$$

$$L_n(X) = \sum_{k=0}^{n} \binom{n}{n-k} \frac{(-1)^k}{k!} X^k,$$

$$T_n(X) = \frac{n}{2} \sum_{k=0}^{\lfloor n/2 \rfloor} \frac{(-1)^k 2^{n-2k}}{n-k} \binom{n-k}{k} X^{n-2k},$$

$$U_n(X) = \sum_{k=0}^{\lfloor n/2 \rfloor} (-1)^k 2^{n-2k} \binom{n-k}{k} X^{n-2k},$$

and Theorem 4 or Theorem 2 of [18].

For example, in the case of Legendre polynomials, Theorem 4 gives

$$\max \left\{ \frac{(n-2k+1)(n-2k+2)}{k(2n-2k+1)} ; 1 \le k \le \lfloor n/2 \rfloor \right\},$$

and we obtain the bound $S(P_n) = \sqrt{\dfrac{n(n-1)}{2(2n-1)}}.$ □

Because orthogonal polynomials are hyperbolic polynomials — i. e. all their roots are real numbers — for the estimation of their largest positive root we can also use the bounds given by van der Sluis [17]. He considers monic univariate polynomials

$$P(X) = X^n + a_1 X^{n-1} + a_2 X^{n-2} + \cdots + a_n \in \mathbb{R}[X]$$

and mentions the following upper bound for the roots in the hyperbolic case:

$$Nw(P) = \sqrt{a_1^2 - 2a_2}.$$

For orthogonal polynomials Newton's bound gives

Proposition 2. *Let P_n, L_n, T_n and U_n be the orthogonal polynomials of degree n of Legendre, respectively Laguerre and Chebyshev of first and second kind. We have*

i. The number $Nw(P_n) = \sqrt{\dfrac{2(2n-2)!}{(n-1)!(n-2)!}}$ is an upper bound for the roots of P_n.

ii. The number $Nw(L_n) = \sqrt{n^4 - n^2(n-1)^2}$ is an upper bound for the roots of L_n.

iii. The number $Nw(T_n) = 2^{(n-1)/2}$ is an upper bound for the roots of T_n.

iv. The number $Nw(U_n) = \sqrt{(n-1)2^{n-1}}$ is an upper bound for the roots of U_n.

Comparisons on Orthogonal Polynomials. In the following tables we denote by L_1 the bound of Lagrange from Theorem 1, by K the bound of Kioustelidis, by S our bound from [18], by Nw the bound of Newton and by LPR the largest positive root of the polynomial P. We used the `gp-pari` package for computing the entries in the tables.

I. Bounds for Zeros of Legendre Polynomials

n	$L_1(P)$	$K(P)$	$S(P)$	Nw	LPR
5	2.05	2.10	1.054	141.98	0.901
8	2.367	2.73	1.366	157822.9	0.960
15	2.95	3.80	1.902	2.08×10^{14}	0.987
50	47.043	7.035	3.517	1.96×10^{76}	0.9988
120	26868.98	10.931.97	5.465	1.091×10^{231}	0.9998

II. Bounds for Zeros of Laguerre Polynomials

n	$L_1(P)$	$K(P)$	$S(P)$	$Nw(P)$	LPR
5	600	25	25	15.0	12.61
8	376321.0	64	25	30.983	22.86
15	7.44×10^{13}	225	225	80.777	48.026
50	6.027×10^{68}	2500	2500	497.49	180.698
120	1.94×10^{206}	14400	14400	1855.15	487.696

III. Bounds for Zeros of Chebyshev Polynomials of First Kind

n	$L_1(P)$	$K(P)$	$S(P)$	Nw	LPR
5	2.118	2.236	1.118	4.0	0.951
8	2.41	2.83	1.41	11.313	0.994
15	3.072	3.872	1.936	128.0	0.994
50	48.822	7.416	3.708	2.37×10^7	0.9995
120	27917.33	10.00	5.00	8.15^{17}	0.99991

IV. Bounds for Zeros of Chebyshev Polynomials of Second Kind

n	$L_1(P)$	$K(P)$	$S(P)$	$Nw(P)$	LPR
5	2.00	2.00	1.00	8.0	0.87
8	2.322	2.83	1.41	29.933	0.994
15	2.87	3.74	1.87	478.932	0.98
50	45.348	9.96	4.98	1.66×10^8	0.9981
120	25864.44	9.96	4.98	8.89×10^{18}	0.9996

Note that for Legendre and Chebyshev polynomials we have $K(P) = 2\,S(P)$.

Other comparisons on roots of orthogonal polynomials were obtained by Akritas *et al.* in [3]. They consider the bounds of Cauchy and Lagrange, and also cite their result derived from our result in [18]. Obviously, in the case of classical orthogonal polynomials there exist an even number of sign variations, and thus Akritas *et al.* apply, in fact, our theorem.

We note that Newton bound gives the best results for Laguerre polynomials. Better estimates can be derived using the Hessian of Laguerre.

Bounds Derived Through the Hessian of Laguerre

Another approach for estimating the largest positive root of an orthogonal polynomial is the study of inequalities derived from the positivity of the Hessian associated to an orthogonal polynomial. They will allow us to obtain better bounds than known estimations.

If we consider

$$f(X) = \sum_{j=1}^{n} a_j\, X^j\,,$$

a univariate polynomial with real coefficients, its *Hessian* is

$$\mathrm{H}(f) = (n-1)^2\, f'^2 - n(n-1)\, ff' \geq 0\,.$$

The Hessian was introduced by Laguerre [12], who proved that $\mathrm{H}(f) \geq 0$.

Let now $f \in \mathbb{R}[X]$ be a polynomial of degree $n \geq 2$ that satisfies the second–order differential equation

$$p(x)\, y'' + q(x)\, y' + r(x)\, y = 0\,, \tag{3}$$

with p, q and r univariate polynomials with real coefficients, $p(x) \neq 0$. We recall the following

Theorem 6 (Laguerre). *If all the roots of f are simple and real, we have*

$$4(n-1)\Big(p(\alpha)r(\alpha) + p(\alpha)q'(\alpha) - p'(\alpha)q(\alpha)\Big) - (n+2)q(\alpha)^2 \geq 0 \tag{4}$$

for any root α of f.

The inequality (4) can be applied successfully for finding upper bounds for the roots of orthogonal polynomials.

Example 4. Consider the Legendre polynomial P_n, which satisfies the differential equation

$$(1 - x^2)y'' - 2xy' + n(n+1)y = 0\,.$$

From (3) it follows that $La(n) = (n-1)\sqrt{\dfrac{n+2}{n(n^2+2)}}$ is a bound for the roots of P_n. We have thus the following bounds for the largest zeros of Legendre polynomials:

n	La(P)	LPR
5	0.91084	0.90617
8	0.96334	0.96028
11	0.98021	0.97822
15	0.98922	0.98799
55	0.99917	0.99906
100	0.99975	0.99971

Example 5. Consider the Hermite polynomial H_n, which satisfies the differential equation

$$y'' - 2xy' + 2ny = 0.$$

From (3) it follows that $He(n) = (n-1)\sqrt{\dfrac{2}{n+2}}$ is a bound for the roots of H_n. We have the following bounds for the largest zeros of Hermite polynomials:

n	He(P)	LPR
3	1.264	1.224
8	3.130	2.930
12	4.156	3.889
20	5.728	5.387
50	9.609	9.182

For Hermite polynomials we obtain better estimates using the following

Theorem 7. *Let $f \in \mathbb{R}[X]$ be a polynomial of degree $n \geq 2$ that satisfies the second order differential equation*

$$p(x)\, y'' + q(x)\, y' + r(x)\, y = 0,\tag{5}$$

with p, q and r univariate polynomials with real coefficients, $p(x) \neq 0$.
If all the roots of f are simple and real we have

$$8(n-3)q_2(\alpha)^2 + 9(n-2)q(\alpha)q_3(\alpha) \geq 0,$$

where

$$q_2 = q^2 + p'q - pq' - pr,$$

$$q_3 = (2p' + q)\left(-q^2 - p'q + pq' - pr\right) - pq\left(p'' + 2q' + r\right) - p^2\left(q'' + 2r'\right).$$

for any root α of f.

Proof. Let α be a root of the polynomial $f(X)$ and consider the polynomial

$$g(X) = \frac{f(X)}{X - \alpha}.$$

We observe that

$$f^{(k)}(\alpha) = kg^{(k-1)}(\alpha) \quad \text{for all} \quad k \geq 1.$$

From the differential equation (5) we obtain

$$g'(\alpha) = -\frac{q(\alpha)}{2p(\alpha)} \cdot g(\alpha). \qquad (6)$$

We take derivatives in (5) twice and obtain

$$p(x)y''' + (p'(x) + q(x))\, y'' + (q'(x) + r(x))\, y' + r'(x)y = 0, \qquad (7)$$

respectively

$$p(x)y^{(iv)} + (2p'(x) + q(x))\, y'''$$
$$+ (p''(x) + 2q'(x) + r(x))\, y'' + (q''(x) + 2r'(x))\, y' + r''(x)y = 0. \qquad (8)$$

From the previous equations we get

$$3p(\alpha)g''(\alpha) + 2\, (p'(\alpha) + q(\alpha))\, g'(\alpha) + q'(\alpha)g(\alpha) = 0,$$

therefore

$$g''(\alpha) = \frac{q_2(\alpha)}{3p(\alpha)^2} \cdot g(\alpha) \qquad (9)$$

where

$$q_2 = q^2 + p'q - pq' - pr.$$

Similarly, from (8) we obtain

$$4p(\alpha)g'''(\alpha) + 3\, (2p'(\alpha) + q(\alpha))\, g''(\alpha) + 2\, (p''(\alpha) + 2q'(\alpha) + r(\alpha))\, g'(\alpha)$$

$$+ (q''(\alpha) + 2r'(\alpha))\, g(\alpha) = 0,$$

hence

$$g'''(\alpha) = \frac{q_3(\alpha)}{4p(\alpha)^3} \cdot g(\alpha), \qquad (10)$$

where

$$q_3 = -\, (2p' + q)\, q_2 + pq\, (p'' + 2q' + r) - p^2\, (q'' + 2r').$$

From the positivity of the Hessian

$$H\, (g') = (n - 3)^2 g''^2 - (n - 2)(n - 3)g'g'''$$

and the fact that $g(\alpha) \neq 0$, we obtain

$$8(n - 3)q_2(\alpha)^2 + 9(n - 2)q(\alpha)q_3(\alpha) \geq 0. \qquad (11)$$

□

Remark 3. If the polynomials p, q, and r satisfy the conditions $\deg(p) \leq 2$, $\deg(q) \leq 1$, $\deg(r) = 0$, we have

$$q_3 = (2p' + q)\left(-q^2 - p'q + pq' - pr\right) + pq\, (p'' + 2q' + r)$$

$$= -(2p' + q)q_2 + pq(p'' + 2q' + r).$$

These hypotheses are satisfied if the solutions of the differential equation (3) are classical orthogonal polynomials (v. G. Szegö [20]).

Other Upper Bounds for Zeros of Hermite Polynomials

Proposition 3. *The number*

$$\sqrt{\frac{2n^2 + n + 6 + \sqrt{(2n^2 + n + 6 + 32(n + 6)(n^3 - 5n^2 + 7n - 3)}}{4(n + 6)}}$$

is an upper bound for the positive roots of H_n.

Proof. We assume that $\alpha \in \mathbb{R}$ is a root of the nth Hermite polynomial H_n. By Theorem 7 it follows that $D(\alpha) \geq 0$, where

$$D(X) = (-2n - 12)X^4 + (2n^2 + n + 6)X^2 + 4n^3 - 20n^2 + 28n - 12.$$

The largest positive root of $D(X)$ is $\sqrt{\beta}$, with

$$\beta = \frac{2n^2 + n + 6 + \sqrt{(2n^2 + n + 6 + 32(n + 6)(n^3 - 5n^2 + 7n - 3)}}{4(n + 6)},$$

so

$$\alpha_0 = \sqrt{\frac{2n^2 + n + 6 + \sqrt{(2n^2 + n + 6 + 32(n + 6)(n^3 - 5n^2 + 7n - 3)}}{4(n + 6)}}$$

is an upper bound for the positive roots of H_n. $\qquad\square$

We consider

$$He(H_n) = (n - 1)\sqrt{\frac{2}{n + 2}},$$

$$Se(H_n) = \sqrt{\frac{2n^2 + n + 6 + \sqrt{(2n^2 + n + 6 + 32(n + 6)(n^3 - 5n^2 + 7n - 3)}}{4(n + 6)}}$$

and obtain

n	$He(H_n)$	$Se(H_n)$	LPR
3	1.264	1.224	1.224
8	3.130	2.995	2.930
12	4.156	4.005	3.889
16	4.999	4.844	4.688
20	5.728	5.574	5.387
25	6.531	6.382	6.164
50	9.609	9.484	9.182
60	10.596	10.478	10.159
100	13.862	13.765	13.406
120	15.236	15.146	14.776
150	17.091	17.009	16.629
200	19.801	19.729	19.339

Comparisons with Other Bounds. We remind some known bounds for the largest positive roots of Hermite polynomials:

$$Bott(H_n) = \sqrt{2n - 2\sqrt[3]{\frac{n}{3}}} \qquad\qquad \text{O. Bottema [4]}$$

$$Venn(H_n) = \sqrt{2(n+1) - 2(5/4)^{2/3}(n+1)^{1/3}} \qquad \text{S. C. Van Venn [22]}$$

$$Kras(H_n) = \sqrt{2n-2} \qquad\qquad\qquad \text{I. Krasikov [10]}$$

$$FoKr(H_n) = \sqrt{\frac{4n - 3n^{1/3} - 1}{2}} \qquad\qquad \text{W. H. Foster–I. Krasikov [7]}$$

Comparing them with our results we obtain

n	Bott	Venn	Kras	FoKr	He	Se	LPR
4	2.408	2.455	2.449	2.262	1.732	1.659	1.650
16	5.339	5.294	5.477	5.265	4.999	4.844	4.688
24	6.633	6.573	6.782	6.570	6.379	6.228	6.015
64	11.065	10.984	11.224	11.022	10.966	10.851	10.526
100	13.912	13.827	14.071	13.875	13.862	13.765	13.406
120	15.269	15.182	15.422	15.234	15.236	15.146	14.776

The bound $Se(H_n)$ gives the best estimates.

Acknowledgements. I would like to thanks the referees for their comments and suggestions and especially for the appropriate reference to the paper of A. van der Sluis [17].

References

1. Akritas, A.G., Strzeboński, A.W.: A comparative study of two real root isolation methods. Nonlin. Anal: Modell. Control 10, 297–304 (2005)
2. Akritas, A., Strzeboński, A., Vigklas, P.: Implementations of a new theorem for computing bounds for positive roots of polynomials. Computing 78, 355–367 (2006)
3. Akritas, A., Vigklas, P.: A Comparison of various methods for computing bounds for positive roots of polynomials. J. Univ. Comp. Sci. 13, 455–467 (2007)
4. Bottema, O.: Die Nullstellen der Hermitischen Polynome. Nederl. Akad. Wetensch. Proc. 33, 495–503 (1930)
5. Cauchy, A.-L.: Exercises de mathématiques, Paris (1829)
6. Derwidué, L.: Introduction à l'algèbre supérieure et au calcul numérique algébrique, Masson, Paris (1957)
7. Foster, W.H., Krasikov, I.: Bounds for the extreme zeros of orthogonal polynomials. Int. J. Math. Algorithms 2, 307–314 (2000)
8. Hong, H.: Bounds for Absolute Positiveness of Multivariate Polynomials. J. Symb. Comp. 25, 571–585 (1998)

9. Kioustelidis, J.B.: Bounds for positive roots of polynomials. J. Comput. Appl. Math. 16, 241–244 (1986)

10. Krasikov, I.: Nonnegative Quadratic Forms and Bounds on Orthogonal Polynomials. J. Approx. Theory 111, 31–49 (2001)

11. Lagrange, J.-L.: Traité de la résolution des équations numériques, Paris (1798), Reprinted in Œuvres, t. VIII, Gauthier–Villars, Paris (1879)

12. Laguerre, E.: Mémoire pour obtenir par approximation les racines d'une équation algébrique qui a toutes les racines réelles. Nouv. Ann. Math., 2ème série 19, 161–172, 193–202 (1880)

13. Laubenbacher, R., McGrath, G., Pengelley, D.: Lagrange and the solution of numerical equations. Hist. Math. 28(201), 220–231

14. Mignotte, M., Ştefănescu, D.: On an estimation of polynomial roots by Lagrange, IRMA Strasbourg 025/2002, pp. 1–17 (2002)

15. Mignotte, M., Ştefănescu, D.: Polynomials – An algorithmic approach, Springer, Heidelberg (1999)

16. Rouillier, F., Zimmermann, P.: Efficient isolation of polynomial's real roots. J. Comput. Appl. Math. 162, 33–50 (2004)

17. van der Sluis, A.: Upper bounds for the roots of polynomials. Numer. Math. 15, 250–262 (1970)

18. Ştefănescu, D.: New bounds for the positive roots of polynomials. J. Univ. Comp. Sc. 11, 2125–2131 (2005)

19. Ştefănescu, D.: Inequalities on Upper Bounds for Real Polynomial Roots. In: Ganzha, V.G., Mayr, E.W., Vorozhtsov, E.V. (eds.) CASC 2006. LNCS, vol. 4194, pp. 284–294. Springer, Heidelberg (2006)

20. Szegö, G.: Orthogonal Polynomials. Proc. Amer. Math. Soc. Colloq. Publ., Providence, RI 23 (2003)

21. Tsigaridas, E.P., Emiris, I.Z.: Univariate polynomial real root isolation: Continued fractions revisited. In: Azar, Y., Erlebach, T. (eds.) ESA 2006. LNCS, vol. 4168, pp. 817–828. Springer, Heidelberg (2006)

22. Van Veen, S.C.: Asymptotische Entwicklung un Nullstellenabschätzung der Hermitische Funktionen. Nederl. Akad. Wetensch. Proc. 34, 257–267 (1931)

23. Yap, C.K.: Fundamental problems of algorithmic algebra. Oxford University Press (2000)

Distance Computation from an Ellipsoid to a Linear or a Quadric Surface in $\mathbb{IR}^n$

Alexei Yu. Uteshev and Marina V. Yashina

Faculty of Applied Mathematics, St. Petersburg State University
Universitetskij pr. 35, Petrodvorets, St.Petersburg, Russia
Alexei.Uteshev@pobox.spbu.ru

Abstract. Given the equations of the surfaces, our goal is to construct a univariate polynomial one of the zeroes of which coincides with the square of the distance between these surfaces. To achieve this goal we employ the Elimination Theory methods.

1   Problem Statement

Find the distance d from the ellipsoid

$$X^T \mathbf{A}_1 X + 2B_1^T X - 1 = 0 \tag{1}$$

a) to linear surface given by the system of equations

$$C_1^T X = 0, \ldots, C_k^T X = 0 \tag{2}$$

b) to quadric

$$X^T \mathbf{A}_2 X + 2B_2^T X - 1 = 0. \tag{3}$$

Here $X = [x_1, \ldots, x_n]^T$ is the column of variables, $\{B_1, B_2, C_1, \ldots, C_k\} \subset \mathbb{IR}^n$ are the given columns, and $C_1, \ldots, C_k$ ($k \leq n$) are assumed to be linearly independent, $\mathbf{A}_1$ and $\mathbf{A}_2$ are the given symmetric matrices, and $\mathbf{A}_1$ is sign-definite.

Such problem arises in Computational Geometry [1,2], for instance, in the pattern recognition problem where one has to estimate the closeness of the objects given in n-dimensional parametric space.

The stated problem, being a problem of constrained optimization:

$$\min(X - Y)^T(X - Y) \text{ subject to } \begin{cases} X \in (1), Y \in (2) & \text{in the case of } a), \\ X \in (1), Y \in (3) & \text{in the case of } b), \end{cases}$$

can be reduced, via the conventional application of Lagrange multipliers method, to a problem of solving a system of algebraic equations. Thus, for instance, in the case of b)

$$\begin{cases} z - (X - Y)^T(X - Y) = 0 \\ X - Y - \lambda_1(\mathbf{A}_1 X + B_1) = \mathbf{O}, \quad -X + Y - \lambda_2(\mathbf{A}_2 Y + B_2) = \mathbf{O} \\ X^T \mathbf{A}_1 X + 2B_1^T X = 1, \quad Y^T \mathbf{A}_2 Y + 2B_2^T Y = 1. \end{cases} \tag{4}$$

V.G. Ganzha, E.W. Mayr, and E.V. Vorozhtsov (Eds.): CASC 2007, LNCS 4770, pp. 392–401, 2007.

The main objective of this paper is to eliminate all the variables from this system except for z, i.e., to construct an algebraic equation $\mathcal{F}(z) = 0$ one of the zeros of which coincides with the square of the distance [3]. On evaluation of the latter, one can generically express the coordinates of the nearest points on the given surfaces as rational functions of this value.

2   Elimination Theory

The constructive realization of the declared procedure can be performed either via the Gröbner basis construction or with the aid of the classical Elimination Theory toolkit. From the latter the most suitable tool for solving our problem turns out to be the **discriminant**. For the (uni- or multivariate) polynomial $g(X) \in \mathbb{R}[X]$ its discriminant is formally and up to a multiple defined as

$$\mathcal{D}_X(g) \overset{def}{=} \prod_{j=1}^{N} g(\Lambda_j),$$

where $\{\Lambda_1, \ldots, \Lambda_N\}$ is a set of zeros (counted in accordance with their multiplicities) of the system

$$\frac{\partial g}{\partial x_1} = 0, \ldots, \frac{\partial g}{\partial x_n} = 0.$$

Discriminant can be expressed as a rational function of the coefficients of $g(X)$ with the aid of several determinantal representations. For instance, by the Bézout method [4,5]

$$\mathcal{D}_X(g) = \det [b_{\ell j}]_{\ell,j=0}^{N-1}. \tag{5}$$

Here, for the univariate case and for $\deg g(X) = N + 1$, the element $b_{\ell j}$ stands for the coefficient of the remainder obtained on dividing $X^\ell g(X)$ by $g'(X)$:

$$X^\ell g(X) \equiv b_{\ell 0} + b_{\ell 1} X + \ldots + b_{\ell,N-1} X^{N-1} + q_\ell(X) g'(X), \ell \in \{0, \ldots, N - 1\}.$$

As for the bivariate case, the element $b_{\ell j}$ of the matrix (5) is the coefficient of the **reduction** of the polynomial $\mathcal{M}_\ell(X) g(X)$ **modulo** $\partial g / \partial x_1$ **and** $\partial g / \partial x_2$:

$$\mathcal{M}_\ell(X) g(X) \equiv b_{\ell 0} \mathcal{M}_0(X) + \ldots + b_{\ell,N-1} \mathcal{M}_{N-1}(X) + \\ + q_{\ell 1}(X) \partial g / \partial x_1 + q_{\ell 2}(X) \partial g / \partial x_2.$$

Here $\{q_{\ell 1}(X), q_{\ell 2}(X)\} \subset \mathbb{R}[X]$, while $\{\mathcal{M}_\ell(X)\}_{\ell=0}^{N-1} \subset \mathbb{R}[X]$ is a set of the appropriately chosen power products in X. For the particular case of the polynomial standing as an argument for the discriminant function in Theorem 3, one should take $N = (n + 1)^2$ and

$$\{\mathcal{M}_\ell(X)\}_{\ell=0}^{N-1} = \left\{x_1^{j_1} x_2^{j_2} \,\middle|\, 0 \leq j_1 < n + 1, 0 \leq j_2 \leq 2(n - j_1)\right\}. \tag{6}$$

The constructive reduction algorithm with respect to such a set was presented in [5].

If $\mathcal{D}_X(g) = 0$ then $g(X)$ possesses a multiple zero; if the latter is unique then it can be expressed rationally via the coefficients of the polynomial. This can be constructively performed with the aid of the minors of the determinant (5). For the univariate case this zero is given by

$$X = B_{N2}/B_{N1} \tag{7}$$

where B_{Nj} are the cofactors of the elements of the last row of the determinant (5). As for the bivariate polynomial from Theorem 3, let us reorder the power products of the set (6) in such a manner that $\mathcal{M}_0 = 1$, $\mathcal{M}_1 = x_1$, $\mathcal{M}_2 = x_2$ and denote by B_{Nj} the cofactors of the elements of the last row of the corresponding determinant (5). Then the components of the multiple zero can be expressed as

$$x_1 = B_{N2}/B_{N1}, \quad x_2 = B_{N3}/B_{N1}. \tag{8}$$

3   Distance to a Linear Surface

Theorem 1. *Construct the matrices* $\mathbf{C} \overset{def}{=} [C_1, \dots, C_k]$ *and* $\mathbf{G} \overset{def}{=} \mathbf{C}^T \mathbf{C}$ *(i.e.* $\mathbf{G}$ *is the Gram matrix for the columns* $C_1, \dots, C_k$*). The condition*

$$0 \le \begin{vmatrix} \mathbf{A}_1 & B_1 & \mathbf{C} \\ B_1^T & -1 & \mathbf{O} \\ \mathbf{C}^T & \mathbf{O} & \mathbf{O} \end{vmatrix} \times \begin{cases} (-1)^{k-1} \text{ if } \mathbf{A}_1 \text{ is positive definite} \\ (-1)^n \text{ if } \mathbf{A}_1 \text{ is negative definite} \end{cases} \tag{9}$$

is the necessary and sufficient one for the linear surface (2) to intersect the ellipsoid (1); in this case one has $d = 0$. *If this intersection condition does not satisfied then the value* d^2 *coincides with the minimal positive zero of the equation*

$$\mathcal{F}(z) \overset{def}{=} \mathcal{D}_\mu \left(\mu^k \begin{vmatrix} \mathbf{A}_1 & B_1 & \mathbf{C} \\ B_1^T & -1 + \mu z & \mathbf{O} \\ \mathbf{C}^T & \mathbf{O} & \frac{1}{\mu}\mathbf{G} \end{vmatrix} \right) = 0 \tag{10}$$

provided that this zero is not a multiple one.

Proof. **I. Finding the intersection condition.** Let us find first the critical value of[1] $V(X) = X^T \mathbf{A} X + 2B^T X - 1$ in the surface $\mathbf{C}^T X = \mathbf{O}$. The critical point of the Lagrange function

$$L = X^T \mathbf{A} X + 2B^T X - 1 - \nu_1 C_1^T X - \dots - \nu_k C_k^T X$$

satisfies the system of equations

$$2\mathbf{A}X + 2B - \mathbf{C}[\nu_1, \dots, \nu_k]^T = \mathbf{O}, \quad \mathbf{C}^T X = \mathbf{O}.$$

Therefrom

$$X = -\mathbf{A}^{-1}B + \frac{1}{2}\mathbf{A}^{-1}\mathbf{C}[\nu_1, \dots, \nu_k]^T \tag{11}$$

---

[1] To simplify the notation we will type matrices $\mathbf{A}$ and B without their subscript.

with

$$[\nu_1, \ldots, \nu_k]^T = 2 \left(\mathbf{C}^T \mathbf{A}^{-1} \mathbf{C}\right)^{-1} \mathbf{C}^T \mathbf{A}^{-1} B. \tag{12}$$

Substitution of (12) into (11) yields

$$X_e = -\mathbf{A}^{-1} B + \mathbf{A}^{-1} \mathbf{C} \left(\mathbf{C}^T \mathbf{A}^{-1} \mathbf{C}\right)^{-1} \mathbf{C}^T \mathbf{A}^{-1} B$$

and the corresponding critical value of $V(X)$ subject to $\mathbf{C}^T X = \mathbf{O}$ equals

$$V(X_e) = -(B^T \mathbf{A}^{-1} B + 1 - B^T \mathbf{A}^{-1} \mathbf{C} (\mathbf{C}^T \mathbf{A}^{-1} \mathbf{C})^{-1} \mathbf{C}^T \mathbf{A}^{-1} B).$$

With the aid of the Schur complement formula [6]:

$$\det \begin{pmatrix} \mathbf{U} \ \mathbf{V} \\ \mathbf{S} \ \mathbf{T} \end{pmatrix} = \det \mathbf{U} \det \left(\mathbf{T} - \mathbf{S} \mathbf{U}^{-1} \mathbf{V}\right) \tag{13}$$

(here $\mathbf{U}$ and $\mathbf{T}$ are square matrices and $\mathbf{U}$ is non-singular) one can transform the last expression into

$$V(X_e) = \frac{-\begin{vmatrix} \mathbf{C}^T \mathbf{A}^{-1} \mathbf{C} & \mathbf{C}^T \mathbf{A}^{-1} B \\ B^T \mathbf{A}^{-1} \mathbf{C} & B^T \mathbf{A}^{-1} B + 1 \end{vmatrix}}{\det(\mathbf{C}^T \mathbf{A}^{-1} \mathbf{C})} = \frac{(-1)^k \begin{vmatrix} \mathbf{A} & B & \mathbf{C} \\ B^T & -1 & \mathbf{O} \\ \mathbf{C}^T & \mathbf{O} & \mathbf{O} \end{vmatrix}}{\det(\mathbf{A}) \det(\mathbf{C}^T \mathbf{A}^{-1} \mathbf{C})}. \tag{14}$$

If $V(X_e) = 0$ then the linear surface (2) is tangent to the ellipsoid (1) at $X = X_e$. Otherwise let us compare the sign of $V(X_e)$ with the sign of $V(X)$ at infinity. These signs will be distinct iff the considered surfaces intersect. If $\mathbf{A}$ is positive definite then $V_\infty > 0$, $\det(\mathbf{A}) > 0$ and $\det(\mathbf{C}^T \mathbf{A}^{-1} \mathbf{C}) > 0$. Therefore, $V(X_e) < 0$ iff the numerator in (14) is negative. This confirms (9). The case of negative definite matrix $\mathbf{A}$ is treated similarly.

II. Distance evaluation. Using the Lagrange multipliers method we reduce the constrained optimization problem to the following system of algebraic equations

$$X - Y - \lambda \mathbf{A} X - \lambda B = \mathbf{O} \tag{15}$$

$$X - Y + \frac{1}{2} \mathbf{C} [\lambda_1, \ldots, \lambda_k]^T = \mathbf{O} \tag{16}$$

$$X^T \mathbf{A} X + 2 B^T X - 1 = 0 \tag{17}$$

$$\mathbf{C}^T Y = \mathbf{O}. \tag{18}$$

We introduce also a new variable responsible for the critical values of the distance function:

$$z - (X - Y)^T (X - Y) = 0. \tag{19}$$

Our aim is to eliminate all the variables from the system (15)–(19) except for z. We express first X and Y from (15) and (16) (hereinafter $\mathbf{I}$ stands for the identity matrix of an appropriate order):

$$X = -\mathbf{A}^{-1} B - \frac{1}{2\lambda} \mathbf{A}^{-1} \mathbf{C} [\lambda_1, \ldots, \lambda_k]^T \tag{20}$$

$$Y = -\mathbf{A}^{-1} B - \frac{1}{2\lambda} (\mathbf{A}^{-1} - \lambda \mathbf{I}) \mathbf{C} [\lambda_1, \ldots, \lambda_k]^T. \tag{21}$$

Then we substitute (21) into (18) with the aim to express $\lambda_1, \ldots, \lambda_k$ via λ. This can be performed with the aid of the following matrix

$$\mathbf{M} \stackrel{def}{=} \frac{1}{\lambda} \mathbf{C}^T \mathbf{A}^{-1} \mathbf{C} - \mathbf{C}^T \mathbf{C} = \mu \mathbf{C}^T \mathbf{A}^{-1} \mathbf{C} - \mathbf{G}, \tag{22}$$

where $\mathbf{G}$ is the Gram matrix of the columns $C_1, \ldots, C_k$ and $\mu \stackrel{def}{=} 1/\lambda$. Indeed, one has

$$\mathbf{M}[\lambda_1, \ldots, \lambda_k]^T = -2\mathbf{C}^T \mathbf{A}^{-1} B \tag{23}$$

and, provided that $\mathbf{M}$ is non-singular,

$$[\lambda_1, \ldots, \lambda_k]^T = -2\mathbf{M}^{-1}\mathbf{C}^T \mathbf{A}^{-1} B. \tag{24}$$

Now substitute (24) into (16) and then the obtained result into (19):

$$z - B^T \mathbf{A}^{-1} \mathbf{C} \mathbf{M}^{-1} \mathbf{G} \mathbf{M}^{-1} \mathbf{C}^T \mathbf{A}^{-1} B = 0. \tag{25}$$

Equation (25) is a rational one with respect to the variables μ and z.

To find an extra equation for these variables, let us transform (17) using (20) and (24)

$$0 = X^T \mathbf{A} X + 2B^T X - 1 = -B^T \mathbf{A}^{-1} B - 1 +$$
$$+ \mu B^T \mathbf{A}^{-1} \mathbf{C} \mathbf{M}^{-1}(\mu \mathbf{C}^T \mathbf{A}^{-1} \mathbf{C} - \mathbf{G} + \mathbf{G})\mathbf{M}^{-1}\mathbf{C}^T \mathbf{A}^{-1} B.$$

Using (22) and (25), the last equation takes the form

$$\Psi(\mu, z) \stackrel{def}{=} -1 + \mu z - B^T \mathbf{A}^{-1} B + \mu B^T \mathbf{A}^{-1} \mathbf{C} \mathbf{M}^{-1} \mathbf{C}^T \mathbf{A}^{-1} B = 0. \tag{26}$$

Therefore, the system (15)–(19) is reduced to (25)–(26). It can be verified that the left-hand side of (25) is just the derivative of that of (26) with respect to μ and, thus, it remains to eliminate μ from the system

$$\Psi(\mu, z) = 0, \ \Psi'_\mu(\mu, z) = 0.$$

This can be done with the help of discriminant – and that is the reason of its appearence in the statement of the theorem.

The Schur complement formula (13) helps once again in representing $\Psi(\mu, z)$ in the determinantal form:

$$\Psi(\mu, z) \equiv \frac{\begin{vmatrix} \mathbf{A} & B & \mathbf{C} \\ B^T & -1 + \mu z & \mathbf{O} \\ \mathbf{C}^T & \mathbf{O} & \frac{1}{\mu}\mathbf{G} \end{vmatrix}}{\begin{vmatrix} \mathbf{A} & \mathbf{C} \\ \mathbf{C}^T & \frac{1}{\mu}\mathbf{G} \end{vmatrix}} = \frac{\mu^k \begin{vmatrix} \mathbf{A} & B & \mathbf{C} \\ B^T & -1 + \mu z & \mathbf{O} \\ \mathbf{C}^T & \mathbf{O} & \frac{1}{\mu}\mathbf{G} \end{vmatrix}}{\det(\mathbf{A})\det(\mathbf{M})}. \tag{27}$$

III. Finding the nearest points on the surfaces. Once the real zero $z = z_\star$ of (10) is evaluated, one can reverse the elimination scheme from part II of the proof in order to find the corresponding points $X_\star$ and $Y_\star$ on the surfaces.

For $z = z_\star$, the polynomial in μ standing in the numerator of (27) has a multiple zero $\mu = \mu_\star$. Provided that the multiple zero is unique, it can be expressed rationally in terms of the coefficients of this polynomial (and in $z_\star$) with the aid of (7). We substitute this value into (22) then resolve the linear system (23) with respect to $\lambda_1, \ldots, \lambda_k$ and, finally, substitute the obtained numbers into (20) and (21).

However, this algorithm fails if for $\mu = \mu_\star$ the matrix $\mathbf{M}$ becomes singular. For explanation of the geometrical reason, one may recall that the distance between the surfaces may be attained not in a unique pair of points.

We avoid this case by imposing the simplicity restriction for the minimal zero of $\mathcal{F}(z)$ in the statement of the theorem. As a matter of fact, we are referring here to the empirical hypothesis that the conditions

$$\mathcal{D}_{x_1}(\mathcal{D}_{x_2}(g(x_1, x_2))) \neq 0 \text{ and } \mathcal{D}_{x_2}(\mathcal{D}_{x_1}(g(x_1, x_2))) \neq 0$$

are equivalent for the generic polynomial $g(x_1, x_2)$. For our particular case, the derivative of the determinant in the numerator of (27) with respect to z coincides with the denominator. □

Corollary 1. *If the system of columns $C_1, \ldots, C_k$ is an orthonormal one then, by transforming the determinant in (10), one can diminish its order: the expression under discriminant can be reduced into*

$$\begin{vmatrix} \mathbf{A}_1 - \mu \mathbf{C} \mathbf{C}^T & B_1 \\ B_1^T & -1 + \mu z \end{vmatrix}. \tag{28}$$

Example 1. Find the distance to the x_1-axis from the ellipsoid

$$7 x_1^2 + 6 x_2^2 + 5 x_3^2 - 4 x_1 x_2 - 4 x_2 x_3 - 37 x_1 - 12 x_2 + 3 x_3 + 54 = 0.$$

Solution. One can choose here $C_1 = [0, 1, 0]^T, C_2 = [0, 0, 1]^T$, then the determinant (28) takes the form

$$\begin{vmatrix} -7/54 & 1/27 & 0 & 37/108 \\ 1/27 & -1/9 - \mu & 1/27 & 1/9 \\ 0 & 1/27 & -5/54 - \mu & -1/36 \\ 37/108 & 1/9 & -1/36 & -1 + \mu z \end{vmatrix}.$$

Equation (10)

$$\mathcal{F}(z) = 516019098077413632 \, z^4 - 15034745857812486912 \, z^3 +$$
$$+ 95300876926947983328 \, z^2 - 421036780846089455856 \, z +$$
$$+ 237447832908365535785 = 0$$

has two real zeros: $z_1 = 0.05712805$ and $z_2 = 22.54560673$. Hence, the distance equals $\sqrt{z_1} \approx 0.23901475$.

Corollary 2. *The square of the distance from the origin* $X = \mathbf{O}$ *to the ellipsoid (1) coincides with the minimal positive zero of the equation*

$$\mathcal{F}(z) \overset{def}{=} \mathcal{D}_\mu \left(f(\mu)(\mu z - 1) - B_1^T q(\mathbf{A}_1, \mu) B_1 \right) = 0 \qquad (29)$$

provided that this zero is not a multiple one. Here $f(\mu) \overset{def}{=} \det(\mathbf{A}_1 - \mu\mathbf{I})$ *is the characteristic polynomial of the matrix* $\mathbf{A}_1$ *whereas* $q(\mathbf{A}_1, \mu)$ *stands for the adjoint matrix to the matrix* $\mathbf{A}_1 - \mu\mathbf{I}$.

Remark 1. For large n, one can compute $f(\mu)$ and $q(\mathbf{A}_1, \mu)$ simultaneously with the aid of the Leverrier–Faddeev method [7].

Remark 2. For the case $B_1 = \mathbf{O}$, one gets $\mathcal{F}(z) \equiv \mathcal{D}(f)\,[z^n f(1/z)]^2$. This corresponds to the well-known result that the distance to the ellipsoid $X^T \mathbf{A}_1 X = 1$ from its center coincides with the square root of the reciprocal of the largest eigenvalue of the matrix $\mathbf{A}_1$.

We exploit the result of the last corollary to elucidate the importance of the simplicity restriction imposed on the minimal positive zero for $\mathcal{F}(z)$; this assumption will also appear in the foregoing results.

Example 2. Find the distance from the origin to the ellipse

$$5/4\, x_1^2 + 5/4\, x_2^2 - 3/2\, x_1 x_2 - \alpha\, x_1 - \alpha\, x_2 + \alpha^2 - 1 = 0.$$

Here $\alpha > 0$ stands for parameter.

Solution. One can see that the given ellipse is obtained from the one centered at the origin by translation along its principal axis by the vector $[\alpha, \alpha]^T$. Let us investigate the dependence of the distance on α.
 Polynomial (10)

$$\mathcal{F}(z) = \frac{1}{16} \frac{(z - 2\alpha^2 + 4\alpha - 2)(z - 2\alpha^2 - 4\alpha - 2)(6z + 4\alpha^2 - 3)^2}{(\alpha - 1)^6 (\alpha + 1)^6}$$

possesses the zeros $z_1 = 2\alpha^2 + 4\alpha + 2$, $z_2 = 2\alpha^2 - 4\alpha + 2$, $z_3 = -2/3\,\alpha^2 + 1/2$, and, for any specialization of the parameter, the value d^2 will be among these values.
 Furthermore, $z_3 = \min\{z_1, z_2, z_3\}$ for $\alpha \in\,]0, \sqrt{3}/2]$. Nevertheless, for $\alpha \in\,]3/4, \sqrt{3}/2]$ the square of the distance is calculated by the formula $d^2 = z_2$.
 Explanation for this phenomenon is as follows: the multiple zero z_3 corresponds to the pair of points $[x_1, x_2]^T$ on ellipse. These points are real for $\alpha \leq 3/4$ and imaginary (complex-conjugate) for $\alpha > 3/4$.

4   Distance to a Quadric

Consider first the case of surfaces centered at the origin: $B_1 = \mathbf{O}$, $B_2 = \mathbf{O}$.

Theorem 2. *The surfaces $X^T \mathbf{A}_1 X = 1$ and $X^T \mathbf{A}_2 X = 1$ intersect iff the matrix $\mathbf{A}_1 - \mathbf{A}_2$ is not sign-definite. If this condition is not satisfied then the value d^2 coincides with the minimal positive zero of the equation*

$$\mathcal{F}(z) \stackrel{def}{=} \mathcal{D}_\lambda(\det(\lambda \mathbf{A}_1 + (z - \lambda)\mathbf{A}_2 - \lambda(z - \lambda)\mathbf{A}_1 \mathbf{A}_2)) = 0 \qquad (30)$$

provided that this zero is not a multiple one.

Remark 3. We failed to establish the authors of the intersection condition from the above theorem. However, this condition should be treated as "well-known" since it is contained as an exercise in the problem book [8]. The other assertion of Theorem 2 follows from

Theorem 3. *The surfaces (1) and (3) intersect iff among the real zeros of the equation*

$$\Phi(z) \stackrel{def}{=} \mathcal{D}_\lambda \left(\det \left(\begin{bmatrix} \mathbf{A}_2 & B_2 \\ B_2^T & -1 - Z \end{bmatrix} - \lambda \begin{bmatrix} \mathbf{A}_1 & B_1 \\ B_1^T & -1 \end{bmatrix} \right) \right) = 0$$

there are the values of different signs or 0. If this condition is not satisfied then the value d^2 coincides with the minimal positive zero of the equation

$$\mathcal{F}(z) \stackrel{def}{=} \qquad (31)$$

$$\stackrel{def}{=} \mathcal{D}_{\mu_1,\mu_2} \left(\det \left(\mu_1 \begin{bmatrix} \mathbf{A}_1 & B_1 \\ B_1^T & -1 \end{bmatrix} + \mu_2 \begin{bmatrix} \mathbf{A}_2 & B_2 \\ B_2^T & -1 \end{bmatrix} - \begin{bmatrix} \mathbf{A}_2 \mathbf{A}_1 & \mathbf{A}_2 B_1 \\ B_2^T \mathbf{A}_1 & B_2^T B_1 - \mu_1 \mu_2 z \end{bmatrix} \right) \right) = 0$$

provided that this zero is not a multiple one.

Proof. is sketched as it is similar to that of Theorem 1. Intersection condition is a result of the following considerations. Extrema of the function $X^T \mathbf{A}_2 X + 2B_2^T X - 1$ on the ellipsoid (1) are all of the similar sign iff the surfaces (1) and (3) do not intersect. We state the problem of finding the extremal **values** of $X^T \mathbf{A}_2 X + 2B_2^T X - 1$ subject to (1), then apply the Lagrange multipliers method and finally eliminate all the variables except for z from the obtained algebraic system coupled with the equation $X^T \mathbf{A}_2 X + 2B_2^T X - 1 - z = 0$.

To prove the second part of the theorem denote

$$\mathbf{M} \stackrel{def}{=} \mathbf{I} - \frac{1}{\lambda_1} \mathbf{A}_1^{-1} - \frac{1}{\lambda_2} \mathbf{A}_2^{-1}, \ Q \stackrel{def}{=} -\mathbf{A}_1^{-1} B_1 + \mathbf{A}_2^{-1} B_2$$

and transform the equations of the system (4) into

$$X = -\mathbf{A}_1^{-1} B_1 + \frac{1}{\lambda_1} \mathbf{A}_1^{-1} \mathbf{M}^{-1} Q, \ Y = -\mathbf{A}_2^{-1} B_2 - \frac{1}{\lambda_2} \mathbf{A}_2^{-1} \mathbf{M}^{-1} Q \qquad (32)$$

$$-B_j^T \mathbf{A}_j^{-1} B_j + \frac{1}{\lambda_j^2} Q^T \mathbf{M}^{-1} \mathbf{A}_j^{-1} \mathbf{M}^{-1} Q - 1 = 0 \text{ for } j \in \{1, 2\} \qquad (33)$$

$$z - Q^T \mathbf{M}^{-2} Q = 0. \qquad (34)$$

On multiplying equations (33) by λ_j and using (34), we get

$$- \lambda_1 B_1^T \mathbf{A}_1^{-1} B_1 - \lambda_2 B_2^T \mathbf{A}_2^{-1} B_2 - Q^T \mathbf{M}^{-1} Q - \lambda_1 - \lambda_2 + z = 0. \qquad (35)$$

It can be verified that the derivative of the left-hand side of (35) with respect to λ_j coincides with that one of (33). Substitution $\mu_1 = 1/\lambda_2$, $\mu_2 = 1/\lambda_1$ and the use of the Schur complement formula (13) enable one to reduce (35) to the determinantal representation from (31).   □

Example 3. Find the distance between the ellipsoids

$$7\,x_1^2 + 6\,x_2^2 + 5\,x_3^2 - 4\,x_1x_2 - 4\,x_2x_3 - 37\,x_1 - 12\,x_2 + 3\,x_3 + 54 = 0$$

$$\text{and } 189\,x_1^2 + x_2^2 + 189\,x_3^2 + 2\,x_1x_3 - x_2x_3 - 27 = 0$$

and establish the coordinates of their nearest points.

Solution. Intersection condition from Theorem 3 is not satisfied: the sixth-order polynomial $\Phi(z)$ has all its real zeros positive. To compute the discriminant (31) we represent it as the determinant (5) of the order $N = 16$. The twenty-fourth-order polynomial $\mathcal{F}(z)$, with integer coefficients of the orders up to 10^{188}, has eight positive zeros $z_1 \approx 1.35377, \ldots, z_8 \approx 111.74803$. Thus, the distance between the given ellipsoids equals $\sqrt{z_1} \approx 1.16351$.

For the obtained value of z_1, the polynomial in μ_1 and μ_2 from (31) possesses a multiple zero which can be expressed rationally in terms of z_1 with the aid of the minors of the determinant (5) by (8). Substitution of the obtained values $\lambda_1 \approx 5.75593$, $\lambda_2 \approx -0.45858$ into (32) yields the coordinates of the nearest points on the given ellipsoids:

$$X \approx [1.52039,\ 1.50986,\ 0.12623]^T,\ Y \approx [0.36100,\ 1.48490,\ 0.03152]^T.$$

Remark 4. It turns out that generically the degree of the polynomial $\mathcal{F}(z)$ is given by the following table

Formula	(10)	(29)	(30)	(31)
deg $\mathcal{F}(z)$	$2k$	$2n$	$n(n+1)$	$2n(n+1)$

Formulas from the third and the fourth column are valid on excluding the extraneous factor from $\mathcal{F}(z)$ (in the case of the fourth column the mentioned factor is responsible for the equivalence of the transfer from the representation (35) to that one of (31)).

5   Conclusions

We have treated the problem of distance evaluation between algebraic surfaces in $\mathbb{R}^n$ via inversion of the traditional approach:

$$\text{nearest points} \ \rightarrow \ \text{distance}.$$

This has been performed via introduction of an extra variable responsible for the critical values of distance function and application of the Elimination Theory methods. It happens that the discriminant is fully responsible for everything: with its help it is not only possible to deduce a univariate polynomial equation for the square of the distance but also to express (Theorem 3) the necessary and sufficient condition for the intersection of the surfaces.

The proposed approach might be especially useful for the optimization problems connected with the parameter dependent surfaces, for instance, for finding an ellipsoid approximating a set of points in $\mathbb{R}^n$.

References

1. Schneider, P.J., Eberly, D.H.: Geometric Tools for Computer Graphics. Elsevier, San Francisco (2003)
2. Lin, A., Han, S.-P.: On the distance between two ellipsoids. SIAM J. on Optimization 13, 298–308 (2002)
3. Uteshev, A.Y., Cherkasov, T.M.: The search for maximum of a polynomial. J. Symbolic Computation 25, 587–618 (1998)
4. Kalinina, E.A., Uteshev, A.Y.: Elimination Theory (in Russian). St. Petersburg State University (2002)
5. Bikker, P., Uteshev, A.Y.: On the Bézout construction of the resultant. J. Symbolic Computation 28, 45–88 (1999)
6. Horn, R.A., Johnson, C.R.: Matrix Analysis. Cambridge University Press, Cambridge (1986)
7. Faddeev, D.K., Faddeeva, V.N.: Computational Methods of Linear Algebra. Freeman, San Francisco (1963)
8. Proskuryakov, I.V.: Problems in Linear Algebra. Mir, Moscow (1978)

Robust Stability for Parametric Linear ODEs

Volker Weispfenning

University of Passau, D-94030 Passau, Germany
weispfen@uni-passau.de
http://www.fmi.uni-passau.de/algebra/staff/weispfen.php3

Abstract. The study of linear ordinary differential equations (ODEs) with parametric coefficients is an important topic in robust control theory. A central problem is to determine parameter ranges that guarantee certain stability properties of the solution functions. We present a logical framework for the formulation and solution of problems of this type in great generality. The function domain for both parametric functions and solutions is the differential ring D of complex exponential polynomials. The main result is a quantifier elimination algorithm for the first-order theory T of D in a language suitable for global and local stability questions, and a resulting decision procedure for T. For existential formulas the algorithm yields also parametric sample solution functions. Examples illustrate the expressive power and algorithmic strengh of this approach concerning parametric stability problems. A contrasting negative theorem on undecidability shows the boundaries of extensions of the method.

1 Introduction

The theory of systems of linear ordinary differential equations (ODEs) with constant coefficients is a long established classical field of mathematics. If the right hand sides of these equations are natural combinations of polynomials and exponential functions (**exponential polynomials**), then all solutions of such systems will also be functions of this type ([3]), and these solutions can be obtained by algebraic manipulations on the coefficients and the right hand sides of the equations. In this paper we investigate the question of **uniformity** in this theory: Suppose the coefficients of the system are given as abstract complex number parameters and the right hand sides as abstract function parameters ranging over exponential polynomials. Is it possible to describe the solution uniformly in these parameters?

Having in addition the aspects of robust control theory ([1]) in mind, we extend the question as follows: Is it in addition possible to describe the local and asymptotic behaviour of solution functions uniformly in these parameters. A very special case of this question is answered by the famous theorems of **Routh-Hurwitz** and of **Lienard-Chipart** [1] that give an explicit description of the set of number parameters that guarantee stability or asymptotic stability of solutions functions of **homogeneous** first-order linear systems of ODEs with constant coefficients.

V.G. Ganzha, E.W. Mayr, and E.V. Vorozhtsov (Eds.): CASC 2007, LNCS 4770, pp. 402–422, 2007.

In this paper we give a **positive answer to a much more general form of this question:** We admit arbitrary boolean combinations of homogeneous and inhomogeneous linear ODEs of arbitrary order and ask not only for solvability and stability of solutions - coded by existential quantification over function variables - , but admit both existential and universal quantification over function variables. Moreover we admit quantifications over complex number variables and side conditions on these number variables and their real and imaginary parts consisting of non-linear polynomial equations and inequalities. The present paper results from a question of **A. Weber** at CASC 2005, asking whether it is possible to include stability conditions in the framework presented there [16]. The main results of this paper were announced at the ACA-conference in Varna, 2006 (see http://www.math.bas.bg/artint/mspirid/ACA2006/). On this occasion, **Vladimir Gerdt** was highly interested in the results and provided valuable additional information on possible extensions of the method.

Of course this ambitious undertaking requires some additional information on the function parameters as the properties of solution function will obviously depend on them: Recall that all function parameters will range over the domain D **complex exponential polynomials,** i. e. complex polynomials in the independent real variable x and in $\exp(\lambda x)$ for arbitrary complex values of λ. They have a **unique representation** in the form

$$f := \sum_{\alpha \in S} p_\alpha(x) \exp(\alpha x)$$

with non-zero complex polynomials $p_\alpha(x)$ and S a finite - possibly empty - set of complex numbers. We call S the **spectrum** $\mathrm{spec}(f)$ of f, $|S|$ the **specsize** $specsize(f)$ and the maximal degree of all $p_\alpha(x)$ the **degree** $deg(f)$ of f. In view of the uniqueness of this representation every $f \in D$ is uniquely determined by a finite set of complex numerical data, namely the elements of $\mathrm{spec}(f)$ and the coefficients of all polynomials $p_\alpha(x)$ for $\alpha \in \mathrm{spec}(f)$.

We enlarge our formal language for the problem description by predicates bounding the specsize of f, i. e. the size of the spectrum of f i.e. the number of summands in this sum, and the degree of f, i. e. the maximal degree of all polynomials $p_\alpha(x)$. In addition we introduce function-symbols extracting from a function variable f formal expressions for the elements $\alpha \in S$ and the coefficients of all polynomials $p_\alpha(x)$. These informations are then sufficient in order to complete our task: On the one hand the language allows to formulate the required stability conditions. On the other hand we obtain an **algorithmic quantifier elimination procedure** in this framework. It provides to every parametric question expressible in our framework quantifier-free necessary and sufficient conditions on the number and function parameters in the question for a positive answer.

The only **restriction in this procedure** is that all quantified function variables and all function parameters have bounded specsize and bounded degree. We express this restriction by saying that we have a **constrained quantifier elimination procedure for** D.

For purely existential formulas our quantifier elimination procedure works alternatively also for unconstrained quantified function variables and constrained function parameters. In fact this is also true for the wider class of F-existential formulas, where additional arbitrary number quantifiers may occur. In this case the classical theory of linear ODEs provides corresponding resulting bounds on the specsize and degree of quantified function variables. This result can be construed as a strong generalization of the classical criteria by **Routh-Hurwitz** and **Lienard-Chipart** in terms of the positivity of certain coefficients and determinants concerning the stability of solutions of parametric homogeneous first-order systems of ODEs [1].

The **proof idea** of this constrained quantifier elimination procedure is a coding a quantifier referring to a constrained function variable by a block of quantifiers referring to complex numbers, i. e. the corresponding numerical data of the function. Due to constraints on the function variable and the function parameters the number of this numerical data, and hence the size of this quantifier block is explicitly bounded. At this point a suitable extension of an arbitrary quantifier elimination for the ordered fields of reals to the field of complex numbers with real and imaginary parts completes the quantifier elimination.

A reduction of certain stability problems for systems of linear constant coefficient ODEs and PDEs to problems of real quantifier elimination has already been performed successfully in [10,11]. Other application of real quantifier elimination methods to control theory are treated in [9,2]. While the present paper is concerned only with ODEs, the framework considered is much larger by allowing arbitrary boolean combinations of parametric ODEs, both existential and universal quantification and additonal predicates and functions on the structure of exponential polynomials.

An immediate consequence of the quantifier elimination procedure is a decision procedure for the domain D and constrained closed formulas in this language. A detailed analysis of the quantifier elimination procedure for F-existential formulas yields in addition **explicit parametric solutions** to existential questions. In other words we have a **constrained quantifier elimination procedure with answers** in the sense of [15,8].

The exact formulation of these results appears in Theorems 2,4,6 and their corollaries. Applications to various types of stability problems are described in section 3. In section 4 we illustrate the results by two classical examples, the harmonic oscillator with external force and a parametric linear first-order system. Section 5 is concerned with the proofs of Theorems 2 and 4. Section 6 is devoted to the computation of solutions to parametric existential problems uniformly in the parameters. Finally we show in section 7 that even seemingly simply problems on ODEs with parametric non-constant coefficients turn out to be algorithmically undecidable in the domain of holomorphic functions. The last section draws some conclusions and specifies some open problems.

2   The Formal Framework

We begin by reviewing some properties of our function domain D of exponential polynomials. Every $f \in D$ is a complex polynomial in the independent real variable x and in $\exp(\lambda x)$ for arbitrary complex values of λ. Thus D forms ring of complex-valued functions under pointwise operations. Since the derivative f' of an $f \in D$ is again in D, D forms in fact a differential ring. Clearly each such f has a representation in the form

$$f := \sum_{\alpha \in S} p_\alpha(x) \exp(\alpha x)$$

with non-zero complex polynomials $p_\alpha(x)$ and S a finite (possibly empty) set of complex numbers. The empty sum represents the zero exponential polynomial. This representation is **unique** if the summands are ordered by increasing $\alpha \in \mathbb{R}^2$ wrt. the lexicographical order.

To **prove uniqueness** it suffices to show that for a non-empty sum as above

$$f := \sum_{\alpha \in S} p_\alpha(x) \exp(\alpha x) \neq 0.$$

In fact the zero-set of f as holomorphic function is countable and has no finite accumulation point in $\mathbb{C}$.

Let $\emptyset \neq S = \{\alpha_1, \ldots, \alpha_n\}$ with $\alpha_1 <_{lex} \ldots <_{lex} \alpha_n$, and let $p_{\alpha_i}(x) = \sum_{k=0}^{d} a_{i,j} x^j$ with $a_{i,j} \in \mathbb{C}$ not all zero. Recall that the power-series expansion of $\exp(\alpha_i x)$ is $\sum_{k=0}^{\infty} (\alpha_i^k / k!) x^k$. Setting all coefficients of the resulting power-series expansion of f to zero leads to an infinite system of homogeneous linear equations for the coefficients $a_{i,j}$ regarded as a column vector

$$a_{1,0}, \ldots, a_{n,0}, a_{1,1}, \ldots, a_{n,1}, \ldots, a_{1,d}, \ldots, a_{n,d}$$

of length $n(d+1)$. The $\infty \times n(d+1)$- matrix of this system has the rows

$$(1, \ldots, 1, 0, \ldots, 0)$$

$$(\alpha_1, \ldots, \alpha_n, 1, \ldots, 1, 0, \ldots, 0)$$

$$(\alpha_1^2/2!, \ldots, \alpha_n^2/2!, \alpha_1, \ldots, \alpha_n, 1, \ldots, 1, 0, \ldots, 0)$$

$$(\alpha_1^3/3!, \ldots, \alpha_n^3/3!, \alpha_1^2/2!, \ldots, \alpha_n^2/2!, \alpha_1, \ldots, \alpha_n, 1, \ldots, 1, 0, \ldots, 0)$$

$$\cdots$$

$$\cdots$$

Here each group of similar symbols except the group of zeros has length n.

After multiplying the i-th row of this matrix ($1 \leq i \leq n$) by $i!$, we see that the first n columns - restricted to their first n rows - form a Vandermonde-matrix; hence the first n columns are linearly independent. Similarly we may multiply the i-th row of this matrix ($2 \leq i \leq n+1$) by $(i-1)!$; then the second n columns - restricted to row 2 to $n+1$ - again form a Vandermonde-matrix; so also the second n columns are linearly independent. Continuing in this way in groups of n columns, until the d-th group, we see that each group of columns is linearly independent. By the upper triangular block structure of the zeros in this matrix, this implies that the set of all columns is in fact linearly independent. Thus the linear system (even restricted to the first $n+d$ rows) has only the trivial solution, contradicting our hypothesis that not all $a_{i,j}$ are zero. This proves the uniqueness.

It will be convenient for us to slightly modify this representation of functions $f \in D$: We call the representation

$$f := \sum_{\alpha \in S'} p_\alpha(x) \exp(\alpha x)$$

with complex polynomials $p_\alpha(x)$ and S' a finite set of complex numbers the **normal form** of f, if $0 \in S'$ and all polynomials except possibly $p_0(x)$ are non-zero. Notice that the uniqueness result above carries over to the normal form representation: Indeed if 0 is in the spectrum of f, then both representations are identical, and otherwise $p_0(x) = 0$. We call S' the **extended spectrum** of f. This extended spectrum is then stable under the formation of derivatives.

The operations $-, +, \cdot$ on normal forms work similar as for polynomials by collecting polynomial coefficients of identical expressions $\exp(\alpha x)$. The derivative of f has normal form

$$f' := \sum_{\alpha \in S'} (\alpha p_\alpha(x) + p'_\alpha(x))(\exp(\alpha x).$$

Similarly higher derivatives $f^{(j)}$ of f have normal form

$$f^{(j)} := \sum_{\alpha \in S'} p_{\alpha,j}(x) \exp((\alpha x),$$

where

$$p_{\alpha,j}(x) := (\sum_{i=0}^{j} \binom{j}{i} \alpha^{j-i} p_\alpha^{(i)})(x)$$

So the kth-coefficient of $p_{\alpha,j}(x)$ is $\sum_{i=0}^{j} \binom{j}{i} \alpha^{j-i} k(k-1) \cdots (k-i+1) a_{\alpha,k+i}$, where $a_{\alpha,k}$ is the k-th coefficient of p_α.

We call $S := \{\alpha \in S' \mid p_\alpha \neq 0\}$ the **spectrum** $\mathrm{spec}(f)$ of f, and $|S|$ the **specsize** $specsize(f)$ and the maximal degree of all $p_\alpha(x)$ the **degree** $deg(f)$ of f.

The **logical framework** for $(\mathbb{C}, D)$ as a two-sorted structure for a first-order language L is as follows:

We have **two sorts** of variables, the F-variables ranging over D and the N-variables ranging over $\mathbb{C}$. In the N-sort we have constants for all rational numbers and for $I := \sqrt{-1}$, the ring operations $+, -, \cdot$, the operations $\Re$ and $\Im$ (for real and imaginary parts), and the order relation (restricted to $\mathbb{R}$). In the F-sort we have the constant 1 and the operations $+, -, '$ and for every natural number b the unary predicates $Specsize_b(y)$ and $Deg_b(y)$. In addition we have mixed function symbols for scalar multiplication of an N-term with an F-term, and for all natural numbers i, j unary function-symbols $spec_i(y)$ and $spec_{i,j}(y)$ mapping F-terms into N-terms.

Semantics: $Specsize_b(y)$ holds iff the number of elements in the spectrum of y is at most b; $Deg_b(y)$ holds iff the maximal degree of all polynomials p_α with α in the spectrum of y is at most b. $spec_i(y)$ denotes the i-th element of the spectrum of y in the lexicographical order. $spec_{i,j}(y)$ denotes the j-th coefficient of the polynomial $p_\alpha(x)$ belonging to $\alpha = spec_i(y)$ in the unique representation of y. In the exceptional case $i > specsize(y)$ we put both values to zero. The order relation applied to complex numbers holds if and only if both arguments are real and satisfy the natural order relation as reals.

Pure N-Terms are expressions obtained from N-variables and N-constants by superposition of the operations $+, -, \cdot, \Re, \Im$. Since they represent complex numbers, they can be rewritten modulo semantic equality as multivariate polynomials with Gaussian rational coefficients in N-variables and their real parts and imaginary parts.

F-**Terms** are expressions of the form

$$\sum_{(i,j)\in M} t_{i,j} \cdot y_i^{(j)} + t_0 \cdot 1$$

where $t_{i,j}$ and t_0 are pure N-terms and $y_i^{(j)}$ are F-variables and their iterated derivatives. We refer to the number $|M|$ as the **additive length** of this term.

"Linear combinations" of F-terms and their iterated derivatives of the form $t = \sum_{m=1}^n s_m \cdot t_m^{j_m}$ with pure N-terms s_m and F-terms t_m are written in the obvious way as an F-term by collecting coefficients of identical variable-derivatives $y_i^{(j)}$. In particular $spec_h(t)$ and $spec_{h,k}(t)$ can be expressed as terms in $spec_h(y_i^{(j)})$ and $spec_{h,k}(y_i^{(j)})$. A similar remark applies to the predicates $Specsize_q(t)$ and $Deg_q(t)$: Under the hypothesis that the formula $\bigwedge_{m=1}^n Specsize_b(t_m) \wedge Deg_b(t_m)$ holds for a fixed natural number b, one can express for every natural number q $Specsize_q(t)$ and $Deg_q(t)$ by an equivalent quantifier-free formula. In particular for $q > b$, $Deg_q(t)$ is equivalent to "false" and for $q > bn$, $Specsize_q(t)$ is equivalent to "false".

Arbitrary N-Terms are obtained from pure N-Terms and expressions of the form $spec_i(s)$ and $spec_{i,j}(s)$ with F-terms s by superposition of the N-operations $+, -, \cdot, \Re, \Im$.

Atomic formulas are equations $s = t$ between two F-terms s, t, (F-**equations**) or equations $s = t$ or inequalities $s < t$ between two N-terms s, t, (N-**equations** and N-**inequalities**) or finally predicates $Specsize_s(t)$ and $Deg_d(t)$ for an F-term t.

The first type represents **parametric inhomogeneous implicit linear ODEs**, the second type **complex polynomial equations of inequalities** and the third type **restraints on function parameters**. We refer to F-equations and predicates of the form $Specsize_s(t)$ and $Deg_d(t)$ as **atomic F-formulas** and to N-equations and N-inequalities as **atomic N-formulas**. We call an atomic N-formula **pure** if both terms occurring in it are pure N-terms.

Quantifier-free formulas are arbitrary combinations of atomic formulas by $\wedge$ (and), $\vee$ (or), $\neg$ (not). In the formation of **arbitrary formulas** we allow in addition quantification $\exists y$, $\forall y$ over F-variables and over N-variables $\exists \xi$,

$\forall\xi$ - **provided** ξ does not occur in any atomic formula of the F-sort. **Constrained formulas** are formulas in which each F-quantifier is constrained by two explicit natural numbers s and d in the following way:

An existential F-quantifier occurs only in the form

$$\exists y(Specsize_s(y) \ \wedge \ Deg_d(y) \ \wedge \ \varphi)$$

and a universal F-quantifier occurs only in the form

$$\forall y((Specsize_s(y) \ \wedge \ Deg_d(y)) \ \Longrightarrow \ \varphi),$$

where φ is a formula. We will also use the short-hand notation

$$\exists_{s,d}y(\varphi)$$

and

$$\forall_{s,d}y(\varphi),$$

respectively, for these formulas.

Pure N-formulas are formulas containing as atomic subformulas only equations and inequalities between pure N-terms.

We will need the following fact that is a straightforward extension of real quantifier elimination (QE):

Theorem 1. *There is an algorithm assigning to every pure N-formula an equivalent quantifier-free pure N-formula.*

Proof. Let φ be a pure N-formula. We replace every occurence of a N-variable ξ in φ by $\xi_1 + I \cdot \xi_2$, with two new variables ξ_1 and ξ_2. If a corresponding quantifier $Q\xi$ occurs in φ we replace it by two new corresponding quantifiers $Q\xi_1 Q\xi_2$. Here we consider ξ_1 and ξ_2 as variables ranging over the real numbers. Next we split every equation and inequality occuring as subformula of the resulting formula φ_1 into its real and imaginary part; thus we obtain a new formula φ_2 in the language of the ordered field of real numbers. Applying any real quantifier elimination algorithm (see e.g. [4,7,6]) to φ_2 we obtain an equivalent quantifier-free formula φ_3 in the language of the ordered field of real numbers. Replacing in φ_3 every occurence of ξ_1 by $\Re(\xi)$ and every occurence of ξ_2 by $\Im(\xi)$, we obtain a quantifier-free pure N-formula φ_4 equivalent to φ in the complex numbers. $\square$

Using this algorithm as a tool we will obtain our first **main theorem**:

Theorem 2. *There is an algorithmic QE-procedure for $(\mathbb{C}, D)$ in this language for constrained input formulas, provided all function-parameters in the input formula are constrained. So for a constrained formula $\varphi(\eta_1, \ldots, \eta_m, u_1, \ldots u_n)$ and natural numbers s, d one can compute a quantifier-free formula $\varphi'_{s,d}(\eta_1, \ldots, \eta_m, u_1, \ldots u_n)$ such that the following holds:*

$$\bigwedge_{i=1}^{n} (Specsize_s(u_i) \ \wedge \ Deg_d(u_i)) \ \longrightarrow$$

$$(\varphi(\eta_1, \ldots, \eta_m, u_1, \ldots u_n) \ \longleftrightarrow \ \varphi'_{s,d}(\eta_1, \ldots, \eta_m, u_1, \ldots u_n))$$

The **constrained elementary theory** of $(\mathbb{C}, D)$ consists of all constrained formulas without parameters that are true in $(\mathbb{C}, D)$.

Corollary 1. *The constrained elementary theory of $(\mathbb{C}, D)$ in this language is decidable.*

Proof. In order to decide the validity of a constrained sentence (i. e. a constrained formula without parameters) in $(\mathbb{C}, D)$ one first applies the algorithmic QE-procedure for $(\mathbb{C}, D)$. This reduces the decision problem to quantifier-free sentences. Since the only function constants we have are constant functions with integer values, the decision of atomic sentences in $(\mathbb{C}, D)$ is simple arithmetic. $\square$

Notice that the corollary remains valid, if we include an F-constant X for the independent variable in our language.

An F-existential formula is a formula of the form

$$\exists y_1 \ldots \exists y_k (\varphi(\eta_1, \ldots, \eta_m, y_1, \ldots, y_k, u_1, \ldots u_n),$$

where φ contains no F-quantifiers, but may contain arbitrary N-quantifiers. Notice that here the F-quantifiers are unconstrained.

Then the classical theory of ODEs with constant coefficients provides as a consequence of our first main theorem the following second main theorem:

Theorem 3. *There is an algorithmic QE-procedure for $(\mathbb{C}, D)$ and F-existential input formulas in this language, provided all function-parameters in the input formula are constrained. So for a given F-existential formula $\varphi(\eta_1, \ldots, \eta_m, u_1, \ldots u_n)$ and natural numbers s, d one can compute a quantifier-free formula $\varphi'_{s,d}(\eta_1, \ldots, \eta_m, u_1, \ldots u_n)$ such that the following holds:*

$$\bigwedge_{i=1}^{n} (Specsize_s(u_i) \wedge Deg_d(u_i)) \longrightarrow$$

$$(\varphi(\eta_1, \ldots, \eta_m, u_1, \ldots u_n) \longleftrightarrow \varphi'_{s,d}(\eta_1, \ldots, \eta_m, u_1, \ldots u_n))$$

The F-**existential theory** of $(\mathbb{C}, D)$ consists of all F-existential formulas without parameters that are true in $(\mathbb{C}, D)$. Then as above we have a corresponding corollary:

Corollary 2. *The F-existential theory theory of $(\mathbb{C}, D)$ in this language is decidable.*

3   Expressive Power of the Framework

What kind of properties of exponential polynomials can be expressed in the given language?

We begin by considering the **expressive power of pure N-formulas:** Since the **absolute value** satisfies $|x|^2 = \Re(x)^2 + \Im(x)^2$ and **conjugation** is given by $\bar{x} = \Re(x) - \Im(x)$, both operations are definable by quantifier-free pure N-formulas. Hence their use can be coded by pure N-formulas.

Next, we illustrate the **expressive power of formulas in general:**

A straightforward instance are **initial value conditions on F-variables** y : and their higher derivatives $y^{(j)}$:

Under the hypothesis $specsize_s(y)$ the term $\sum_{i=1}^s spec_{i,0}(y)$ describes the initial value $y(0)$. The terms $\sum_{i=1}^b spec_{i,1}(y)$ and $\sum_{i=1}^b 2spec_{i,2}(y)$ describe the initial values $y'(0)$ and $y''(0)$, respectively. Similarly for higher derivatives.

The most important properties expressible in our framework are **global and local stability properties:**

The (global) **asymptotic stability** of an F-variable y constrained by numbers s, d (with repect to the zero function as equilibrium) can be expressed by the following quantifier-free formula:

$$Asympstab_{s,d}(y): \ Specsize_s(y) \ \wedge \ Deg_d(y) \ \wedge$$

$$\bigwedge_{i=1}^{s} \Re(spec_i(y)) < 0$$

Similarly the (global) **stability** of an F-variable y constrained by numbers s, d (with repect to the zero function as equilibrium) can be expressed by the quantifier-free formula:

$$Stab_{s,d}(y): \ Specsize_s(y) \ \wedge \ Deg_s(y) \ \wedge$$

$$\bigwedge_{i=1}^{s} (\Re(spec_i(y)) < 0 \ \vee \ (\Re(spec_i(y)) = 0 \ \wedge \ \bigwedge_{j=1}^{d} spec_{i,j}(y) = 0))$$

When one considers global stability of solutions of inhomogeneous linear ODEs one wants to express stability of a solution y w.r.t. the given right-hand side function z by a quantifier-free formula. This can be done as follows:

$$Asympstab_{s,d}(y,z): \ Specsize_s(y) \ \wedge \ Deg_d(y) \ \wedge \ Specsize_s(z) \ \wedge \ Deg_d(z) \ \wedge$$

$$\bigwedge_{i=1}^{s} (\Re(spec_i(y)) \geq 0 \ \longrightarrow$$

$$\bigvee_{j=1}^{s} spec_i(y) = spec_j(z) \ \wedge \ \bigwedge_{k=1}^{d} (spec_{i,k}(z) = 0 \implies spec_{i,k}(y) = 0)$$

$$Stab_{s,d}(y,z): \ Specsize_s(y) \ \wedge \ Deg_d(y) \ \wedge \ Specsize_s(z) \ \wedge \ Deg_d(z) \ \wedge$$

$$\bigwedge_{i=1}^{s} (\Re(spec_i(y)) > 0 \ \longrightarrow \ \bigvee_{j=1}^{s} spec_i(y) = spec_j(z) \ \wedge$$

$$\bigwedge_{i=1}^{s} (\Re(spec_i(y)) = 0 \longrightarrow (\bigwedge_{k=1}^{d} spec_{i,k}(y) = 0 \vee$$

$$\bigvee_{j=1}^{s} (spec_i(y) = spec_j(z) \wedge \bigwedge_{k=1}^{d} (spec_{i,k}(z) = 0 \implies spec_{i,k}(y) = 0)$$

Global stability and asymptotic stability - coded by the formulas above - refer to the long time behaviour of a function. In practice - even if a function is asymptotically stable - it may still have very large values (e. g. amplitudes in case of a damped oscillator) for small positive argument values. This can be technically undesirable or even dangerous. So it is desirable to have in addition formulas that express **local stability** on a compact or right semiinfinite interval: Next we show how parametric bounds on function values within a parametric interval can be expressed in the given framework:

Let F be a constrained set of functions in D and assume we have fixed $s, d \in \mathbb{N}$ with $specsize(f) \le s$ and $deg(f) \le d$ for all $f \in F$.

Notice that for an interval $[a, b]$ in $\mathbb{R}$ and for $x \in [a, b], \alpha \in \mathbb{C}$ the function values $|exp(\alpha x)|$ are bounded by $\max(\exp(\pm|\Re(\alpha)a|), \exp(\pm|\Re(\alpha)b|))$.

Hence for an exponential polynomial $f \in F$ in normal form

$$f := \sum_{\alpha \in S} p_\alpha(x) \exp(\alpha x)$$

with $|S| \le s$, and $deg(p_\alpha) \le s$ for all $\alpha \in S$, we obtain for $[a, b]$ in $\mathbb{R}$ and for $x \in [a, b]$, the bound

$$|f(x)| \le \sum_{\alpha \in S} \max(|p_\alpha(x)| \ |x \in [a, b]) \cdot \max(\exp(\pm|\Re(\alpha)a|), \exp(\pm|\Re(\alpha)b|)).$$

For real variables a, b, c and a function variable f constrained by s, d we can now indeed find for arbitrary fixed $n \in \mathbb{N}$ a formula $\beta_{s,d,n}(a, b, c, f)$ of our framework that implies the fact that $|f(x)|$ is bounded by the real parameter c on the parametric interval $[a, b]$.

For this purpose we use the following well-known upper and lower bounds for $\exp(r)$ for arbitrary fixed $n \in \mathbb{N}$ and arbitrary real value of r:

$$(1 + \frac{r}{n})^n < \exp(r) < (1 + \frac{-r}{n})^{-n}.$$

Using these bounds instead of the value of $\exp(r)$ above and by expressing $\max(|p_\alpha(x)| \ |x \in [a, b])$ by universal quantification over a real variable ranging in $[a, b]$, it is now easy to find such a formula $\beta_{s,d,n}(a, b, c, f)$ explicitly. In a similar manner one can find a corresponding formula $\beta_{s,d,n}(a, \infty, c, f)$ for the case of an upper unbounded parametric interval $[a, \infty)$.

Notice that for fixed s, d and increasing values of $n \in \mathbb{N}$ the formulas $\beta_{s,d,n}(a, b, c, f)$ and $\beta_{s,d,n}(a, \infty, c, f)$ get weaker and weaker, since the approximation of the exponential function used therein gets better and better. For all values of s, d and n these formulas underestimate the true range of the parameter c that bound the function values of y on the interval $[a, b]$ or $[a, \infty)$. In other words, whenever these formulas hold for given parameter values and a given function y, then y has the respective local stability behaviour.

In **summary** this section shows that all the properties of exponential polynomials mentioned above can be expressed by quantifier-free formulas in our language. Hence these properties - in arbitrary boolean combinations - can be required of some or all constrained quantified function variables in an input formula φ of the QE procedure. Then for every specialization of the number parameters $\eta_1, \ldots, \eta_m$ and the function parameters $u_1, \ldots, u_n$ of φ to concrete values, the corresponding quantifier-free formula φ' can be easily evaluated to 'true' or 'false'. Thus we have an **answer** to problems expressed by φ that is **uniform in the parameters**.

4 Examples

As a **first example** we consider the differential equation of an harmonic oscillator with external force $f(x)$ [3], section 2.6. It is modelled by the equation

$$my'' + cy' + ky = f(x),$$

where m denotes the mass of the oscillating body, c is the damping constant, and k is the spring constant.

We assume that the external force is of the form $\gamma \exp(\omega x)$. So we can model $f(x)$ by a function variable u with the constraint $Specsize_1(u) \wedge Deg_0(u)$. Moreover for solutions y we fix constraining number $s = 2$, $d = 1$. Then we can model the question of the existence of stable or asymptotically stable solutions, respectively, by the following two formulas:

$$\exists y(my'' + cy' + ky = u \ \wedge \ y \neq 0 \wedge \ Stab_{2,1}(y))$$

$$\exists y(my'' + cy' + ky = u \ \wedge \ y \neq 0 \wedge \ Asympstab_{2,1}(y))$$

Following the proof of Theorem 1, we replace the function quantifier $\exists y$ by a block of number quantifiers, taking into account the fact that one add conjunctively the constraints $Specsize_2(y)$ and $Deg_1(y)$. This is achieved by the "Ansatz"

$$u = \gamma \exp(\omega x)$$

with $\omega = spec_1(u)$, $\gamma = spec_{1,0}(u)$,

$$y = (\xi_0 + \xi_1 x) \exp(\alpha x) + (\eta_0 + \eta_1 x) \exp(\beta x).$$

By computing y' and y'' in this "Ansatz" and by comparing "coefficients" in the resulting equation between exponential polynomials, we obtain the equivalent quantifier-free formula φ' in the unknown N-variables $\xi_0, \xi_1, \eta_0 . \eta_1, \alpha, \beta$ and the N-parameters m, c, k, γ, ω. The elimination of the N-quantifiers in the formula

$$\exists(\xi_0, \xi_1, \eta_0 . \eta_1, \alpha, \beta)(\varphi')$$

yields then the desired quantifier-free formula $\varphi''(m, c, k, \gamma, \omega)$.

Our **second example** concerns a homogeneous linear 2×2 system of ODEs with one number parameter d:

$$y_1' = -3y_1 - 4y_2, \ y_2' = 2y_1 + dy_2$$

For simplicity we delete the case of identical eigenvalues. Accordingly we put

$$\varphi: \ y_1' = -3y_1 - 4y_2 \ \wedge \ y_2' = 2y_1 + dy_2 \ \wedge \ d^2 + 6d - 23 \neq 0 \ \wedge \ Asympstab_{2,0}(y_2)$$

and we let

$$\exists_{2,0}y_2 \exists_{2,0}y_1(\varphi)$$

be our input formula.

Then the innermost quantifier is eliminated by transforming the formula $\exists_{2,0}y_1(\varphi)$ into

$$\exists(\xi_1, \xi_2, \eta_{1,0}\eta_{2,0}((\xi_1 \neq spec_1(y_2), spec_2(y_2) \ \wedge \ \xi_2 \neq spec_1(y_2), spec_2(y_2) \ \wedge$$

$$\eta_{1,0}\xi_1 = -3\eta_{1,0} \ \wedge \ \eta_{2,0}\xi_2 = -3\eta_{2,0} \ \wedge \ spec_{1,0}(y_2) = 0 \ \wedge \ spec_{2,0}(y_2) = 0) \ \vee$$

$$(\xi_2 = spec_1(y_2) \ \wedge \ \eta_{1,0}\xi_1 = -3\eta_{1,0} \ \wedge \ \eta_{2,0}\xi_2 = -3\eta_{2,0}(y_2) - 4spec_{1,0}(y_2) \ \wedge$$

$$spec_2(y_2) = 0) \ \vee \ (\xi_1 = spec_2(y_2) \ \wedge$$

$$\eta_{2,0}\xi_2 = -3\eta_{2,0} \ \wedge \ \eta_{1,0}\xi_1 = -3\eta_{1,0} - 4spec_{2,0}(y_2) \ \wedge \ spec_1(y_2) = 0) \ \vee$$

$$(\xi_1 = spec_1(y_2) \ \wedge \ \xi_2 \neq spec_2(y_2) \ \wedge \ \eta_{1,0}\xi_1 = -3\eta_{1,0} - 4spec_{1,0}(y_2) \ \wedge$$

$$\eta_{2,0}\xi_2 = -3\eta_{2,0} \ \wedge \ spec_2(y_2) = 0) \ \vee \ (\xi_2 = spec_2(y_2) \ \wedge \ \xi_1 \neq spec_1(y_2) \ \wedge$$

$$\eta_{2,0}\xi_2 = -3\eta_{2,0}(y_2) - 4spec_{2,0}(y_2) \ \wedge \ \eta_{1,0}\xi_1 = -3\eta_{1,0} \ \wedge \ spec_1(y_2) = 0) \ \vee$$

$$(\xi_1 = spec_1(y_2) \ \wedge \ \xi_2 = spec_2(y_2) \ \wedge \ \eta_{1,0}\xi_1 = -3\eta_{1,0} - 4spec_{1,0}(y_2) \ \wedge$$

$$\eta_{2,0}\xi_2 = -3\eta_{2,0} - 4spec_{2,0}(y_2)).$$

At this point the N-quantifier block $\exists(\xi_1, \xi_2, \eta_{1,0}\eta_{2,0})$ can be eliminated by Theorem 1 yielding a quantifier-free formula $\varphi'(y_2, d, spec_1(y_2), spec_2(y_2), spec_{1,0}(y_2), spec_{2,0}(y_2))$

The remaining F-quantifier elimination in the formula $\exists_{2,0}y_2(\varphi')$ is now performed in a similar way. The output is then a quantifier-free formula $\varphi''(d)$. By the classical theory this formula will be equivalent to the Routh-Hurwitz condition on the polynomial $X^2 + (3 - d)X + (8 - 3d)$.

5   Proof of the Two Main Theorems

For the proof of Theorem 2 we have to exhibit a quantifier elimination algorithm for D in the given language. For this purpose it suffices to show how to eliminate a single existential quantifier $\exists$ in front of a N-variable ξ (a number-quantifier) and a single constrained existential quantifier $\exists_{s,d}$ in front of an F-variable y (a constrained function-quantifier).

For the **case of a number-quantifier** consider a formula $\exists\xi(\varphi)$, where φ is quantifier-free. By our definition of atomic formulas, the bound variable ξ does not occur in φ in an F-equation. We replace temporarily each subterm of the form $spec_i(t)$ or $spec_{i,j}(t)$ with an F-term t in each non-pure atomic N-subformula of φ by a new N-variable $\eta_{t,i}$ and $\eta_{t,i,j}$, respectively. Then the resulting formula $\exists\xi(\varphi_1)$ is a pure N-formula. By theorem 1 there is a quantifier elimination algorithm assigning to this formula a quantifier-free pure N-formula φ_2 that is equivalent to the given formula in D. Back-substituting the original terms $spec_i(t)$ for $\eta_{t,i}$ and $spec_{i,j}(t)$ for $\eta_{t,i,j}$, respectively, we obtain a quantifier-free formula φ_3 equivalent to $\exists\xi(\varphi)$ in d.

The **case of a constrained function-quantifier** is considerably more subtle: Consider a formula $\exists_{s,d}y(\varphi)$, where φ is a quantifier-free formula. containing at most the free F-variables (function-parameters) $z_1, \ldots, z_m$.

We are going to exhibit an equivalent F-quantifier-free formula where the constrained function quantifier $\exists_{s,d}y$ has been replaced by a block of existential number quantifiers preceeding a quantifier-free formula φ'. At this point we use the elimination of number quantifers described in Theorem 1 in order to obtain an equivalent quantifier-free formula. The number of new number quantifiers introduced in this way depends on s, d and on the maximal order of the derivatives of y occuring in φ. The new number-variables will semantically represent elements of the spectrum of a "solution function" y and coefficients of the polynomials occuring in the normal form of y, i. e, the numerical data determining the function y.

We may assume that each F-equation occuring in φ has been normalized wrt. the quantified F-variable y to the form $t = u$, where t is an F-term containing no F-variable except y and u is an F-term not containing the F-variable y (but possibly other variables). Let n be an upper bound on the additive length of the F-terms u. In view of our hypothesis on the function parameters and the remarks on F-terms we can assume that for all these F-terms u holds $Deg_d(u)$ and $Specsize_{ns}(u)$. Let m be an upper bound on the order of y in all terms t, i.e. assume $j \leq m$ for all derivatives $y^{(j)}$ occuring in some t.

By our remarks on F-terms we may assume that an occurrence of the F-variable y in a subterm of the form $spec_i(t)$ or $spec_{i,j}(t)$ happens only if $t = y$. Similarly we may assume that an occurrence of the F-variable y in a subformula of the form $Specsize_q(t)$ or $Deg_q(t)$ for a natural number q and an F-term t happens only if $t = y$. Moreover in view of the constraint $Specsize_s(y) \wedge Deg_d(y)$ on y, we may in addition assume that all indices i, j, q occurring in $spec_i(y)$, $spec_{i,j}(y)$, $Specsize_q(y)$, $Deg_q(y)$ are natural numbers smaller or equal to s, and that all indices occurring in $spec_{i,j}(y)$, $Deg_q(y)$ are natural numbers smaller or equal to d.

Next we introduce $s + s(d+1)$ many new N-variables ξ_i, $\eta_{i,j}$, where $1 \leq i \leq c$, $0 \leq j \leq d$. Then we write the given formula equivalently in the form

$$\exists \xi_1 \ldots \exists \xi_c \exists \eta_{1,0} \ldots \exists \eta_{c,d} \exists y (Specsize_s(y) \wedge Deg_d(y) \wedge$$

$$\bigwedge_{i=1}^{s} (\text{spec}_i(y) = \xi_i \wedge \bigwedge_{j=0}^{d} spec_{i,j}(y) = \eta_{i,j} \wedge \varphi)$$

Next notice that by our hypothesis and the remarks on F-terms we have $Specsize_s(t)$ and $Deg_d(t)$ for every term t on the left hand side of an F-equation $t = u$ in φ. Hence we can rewrite each F-equation $t = u$ with $t = \sum_{h=0}^{m} t_h \cdot y^{(h)} + t_0 \cdot 1$, where all t_h all pure N-terms, in the formula above equivalently by "matching the numerical data of both sides" in the following formula:

$$\bigvee_{1 \leq s_1 < \ldots < s_{ns}} (\bigwedge_{r=1}^{ns} (\xi_{s_r} = spec_r(u) \wedge$$

$$\bigwedge_{k=0}^{d} \left(\sum_{h=0}^{m} \sum_{i=0}^{h} t_h \xi_{s_r}^{h-i} \binom{h}{i} k(k-1)\cdots(k-i+1) \cdot \eta_{s_r,k+i} = spec_{r,k}(u) \right) \wedge$$

$$(\bigwedge_{1 \leq q \leq c, q \neq s_1, \ldots, s_{ns}} \bigwedge_{k=0}^{d} \eta_{q,k} = 0)$$

Notice that the variables ξ_i can occur non-linearly in this formula.

Next we consider occurences of the variable y in predicates of the form $Specsize_q(t + u)$ and $Deg_q(t + u)$, where the F-terms t and u are as above. Then for $q \geq (n+1)s$ the first formula can clearly be replaced by 'true', and for $q \geq d$ the second formula can be replaced by 'true'. For smaller values of s the remarks in section 2 show that these formulas can be equivalently replaced by quantifier-free formulas involving in addition the variables $\xi_i, \eta_{i,j}$, but not the variable y.

Finally we consider occurences of the variable y in an N-equation or an N-inequality ψ. Here such occurences can only be inside subterms of the form $spec_i(t)$ and $spec_{i,j}(t)$ for f-terms t. By the remarks in section 2 ψ can then be rewritten equivalently as a quantifier-free formula not containing y but instead the n-variables ξ_i and $\eta_{i,j}$.

We are now in a position to rewrite the given formula equivalently in the form

$$\exists \xi_1 \ldots \exists \xi_c \exists \eta_{1,0} \ldots \exists \eta_{c,d} \exists_{s,d} y (Specsize_c(y) \wedge Deg_d(y) \wedge$$

$$\bigwedge_{i=1}^{s} (\text{spec}_i(y) = \xi_i \wedge \bigwedge_{j=0}^{d} spec_{i,j}(y) = \eta_{i,j} \wedge \varphi'),$$

where φ' is a quantifier-free formula containing no occurence of the variable y. The range of the quantifier $\exists_{s,d} y$ can now be restricted to that part of the formula that may contain occurences of the variable y, namely the formula

$$\exists y(Specsize_s(y) \,\wedge\, Deg_d(y) \,\wedge\, \bigwedge_{i=1}^{s}(\mathrm{spec}_i(y) = \xi_i \,\wedge\, \bigwedge_{j=0}^{d} spec_{i,j}(y) = \eta_{i,j})$$

But this formula is obviously equivalent to 'true.' In this way we have now completely eliminated the quantifier $\exists_{s,d}y$ in favour of new number quantifiers

$$\exists \xi_1 \ldots \exists \xi_s \exists \eta_{1,0} \ldots \exists \eta_{s,d}$$

By the elimination method for number quantifiers in Theorem 1, these can now be eliminated, yielding a final quantifier-free formula equivalent to the given one.

Proof of Theorem 4

Let an F-existential formula of the form

$$\exists y_1 \ldots \exists y_k (\varphi(\eta_1, \ldots, \eta_m, y_1, \ldots, y_k, u_1, \ldots u_n),$$

be given, where φ contains no F-quantifiers, but may contain arbitrary N-quantifiers. Then by applying Theorem 1 to the formula φ we may assume without restriction that φ is already quantifier-free.

We **claim** that under the given constraints s, d on the F-parameters u_i the F-quantifiers $\exists y_i$ can be equivalently replaced by constrained F-quantifiers $\exists_{s',d'} y_i$ for suitably computed bounds s', d'. In other words: Any k-tuple of solution functions f_i of $\varphi(y)$ can be replaced by a tuple of solution function g_i with $Specsize(g_i) \leq s'$ and $Deg(g_i) \leq d'$. Let p be a bound on the order j of derivatives $y_h^{[j]}$ occuring in φ. Let q be the highest index of a predicate $Specsize_i(y_h)$ or $Deg_i(y_h)$ or a function expression of type $spec_i(y_h)$ or $spec_{i,j}(y_h)$ occuring in φ and let r be the maximum of q, s, d.

Then we **claim specifically** that we can take

$$s' := pk + nr, \quad d' := pk + r$$

If the claim holds, then we are back in the situation of Theorem 2 and can apply the quantifer elimination algorithm described there.

It remains to prove the **claim:** For this purpose we may assume - by passing to a disjunctive normal form - that φ is actually a conjunction of atomic formulas and negated atomic formulas involving the variables y_h. By the remarks in section 2 we may assume that the atomic subformulas of φ are of the form $t = u$ with F-terms t, u as above or $Specsize_q(y_h)$ or $Deg_r(y_h)$ or N-equations or N-inequalities in which the variables y_h occur in the forms $spec_i(y_h)$ and/or $spec_{i,k}(y_h)$. Fix values in D for all F-parameters occuring in φ and fix values in $\mathbb{C}$ for all N-parameters occuring in φ. Assume moreover that for some fixed values of y_h in D φ holds in D. Then if $Specsize_{s'}(y_h) \,\wedge\, Deg_{d'}(y_h)$ holds in D there is nothing to prove.

Notice that every atomic subformulas of φ of the form $t = u$ with F-terms t, u such that u contains no $y_h^{(j)}$ now represents a linear ordinary differential equation with constant complex coefficients in the unknown functions y_h. Moreover if this equation does actually contain some higher derivative of the variable y_h with non-vanishing coefficient then the order of y_h in this equation is bounded by $ord(y_h, t)$ and hence by p.

The totality of all such equations that occur unnegated in the conjunction φ forms a system of linear ODES with constant coefficients for the unknown functions $y_1, \ldots, y_k$. This system may be underdetermined or determined or overdetermined. In the latter two cases the classical theory of linear ODEs [3] entails that $specsize(y_h) \leq pk + ns$ and $deg(y_h) \leq pk + d$. So we have actually satisfied all the constraint quantifiers $\exists_{s',d'} y_h$. If the system is underdetermined we may have to replace the solution functions y_h by others of smaller specsize and/or degree. For this purpose we replace the undetermined functions y_h by functions of smallest possible specsize and degree in view of the side conditions expressed by the negated predicates occuring in the conjunction and possible conditions on $spec_i(y_h)$ and $spec_{i,j}(y_h)$. The remaining functions y_i can now be determined by the classical theory and the resulting solution functions will be constrained by s' and d' as required. So again we have satisfied the constrained quantifiers $\exists_{s',d'} y_h$ □

6 Computing Parametric Solutions

The quantifier elimination for F-existential formulas described in Theorem 4 provides in particular for purely existential formulas of the form

$$\exists \xi_1 \ldots \exists \xi_r \exists y_1 \ldots \exists y_r (\varphi)$$

with a quantifier-free formula φ a necessary and sufficient test in the (constrained) parameters in φ for the existence of solutions $\xi_1, \ldots, \xi_r, y_1, \ldots, y_s$ for φ. In practice one is, however, also interested in computing explicit sample solutions. In view of the presence of parameters such solution can only be described by formal expressions in the parameters. The problem of finding a finite quantifier-free cases distinction on the parameters and corresponding formal expressions in the parameters that describe a solution in each case is known as **extended quantifier elimination** or **quantifier elimination with answers** [15,8].

Based on the structure of the proofs of Theorem 1 and Theorem 4 we are going to describe such an extended quantifier elimination procedure in our setting. The result is as follows:

Theorem 4. *For every existential input formula*

$$\exists \xi_1 \ldots \exists \xi_r \exists y_1 \ldots \exists y_s (\varphi)$$

with number parameters $\eta_1, \ldots, \eta_m$ and function parameters $u_1, \ldots, u_n$ and all natural numbers c, d, one can compute a finite system of tuples consisting of a *quantifier-free formulas ψ_i in the parameters, formal number expressions $\gamma_{,j}i$ in these parameters and formal function expressions $t_{i,j}$ in these parameters ($i \in I$) such that in $(\mathbb{C}, D)$ the following holds:*

$$\bigwedge_{i=1}^{n} (Specsize_c(u_i) \wedge Deg_d(u_i)) \longrightarrow$$

$$(\bigvee_{i \in I} \psi_i \wedge \bigwedge_{i \in I} (\exists \xi_1 \dots \exists \xi_r \exists y_1 \dots \exists y_s(\varphi) \wedge \psi_i) \longrightarrow$$

$$\varphi[\gamma_{i,1}/\xi_1, \dots, \gamma_{i,r}/\xi_r, t_{i,1}/y_1, \dots, t_{i,s}/y_s])$$

In short, we have an algorithm for constrained extended quantifier elimination for existential formulas in D.

As a consequence we get the equivalence

$$\exists \xi_1 \dots \exists \xi_r \exists y_1 \dots \exists y_s(\varphi) \longleftrightarrow \bigvee_{i \in I} (\psi_i \wedge \varphi[\gamma_{i,1}/\xi_1, \dots, \gamma_{i,r}/\xi_r, t_{i,1}/y_1, \dots, t_{i,s}/y_s]).$$

It remains to describe the formal number-expressions $\gamma_{i,j}$ and the formal-function expressions $t_{i,j}$ above in greater detail.

By splitting complex variables in their real and imaginary parts, we may assume that the number quantifiers refer actually to real variables. These can be eliminated by an extended real quantifier elimination based on cylindrical algebraic decomposition as described in [12,13]. This conforms with the description above, where the N-expressions $\gamma_{i,j}$ are nested occurences of expressions describing some real zero in increasing order of a univariate real polynomial with parametric coefficients. These expressions can also occur in arithmetic means and combined with addition or subtraction of 1.

The F-expressions are then of a form corresponding to normal forms of elements of D, i.e. of the form

$$\sum_{\alpha \in S} p_\alpha(x) \exp(\alpha x),$$

where S is a finite set of N-expressions as above and for each $\alpha \in S$ $p_\alpha(x)$ is a univariate polynomial expression in x with coefficients that are N-expressions as above.

Notice that for both the N-expressions $\gamma_{i,j}$ and the F-expressions $t_{i,j}$ a natural substitution of these expressions for N-variables and F-variables, respectively, in a quantifier-free formula will again yield in a natural way a quantifier-free formula.

The **proof** of the theorem proceeds by induction on the number s of function-quantifiers in the given existential formula.

If $s = 0$ then all quantified N-variables occur only in N-equations and N-inequalities. By "splitting complex number variables in real and imaginary parts" (as in the proof of Theorem 1) we may assume that these N-variables are already real variables. At this point the extended real quantifier elimination based on cylindrical algebraic decomposition as described in [12] provides the required N-expressions for the newly introduced real variables. Combining these as real and imaginary parts then yields the required N-expressions for the original N-variables satisfying the given formula φ. Notice that these expressions may involve the F-parameters occuring in φ in the form of N-terms $spec_i(t)$ and $spec_{i,j}(t)$.

In the case $s = 1$ we have a single F-quantifier $\exists y$. Removing this quantifier, we may assume by the induction assumption that extended quantifier elimination has already been performed for the remaining existential number quantifiers. Next we consider each quantifier-free formula ψ_i and $\varphi[\gamma_1/\xi_1, \ldots, \gamma_r/\xi_r]$ occuring in the output separately. Then the given input formula

$$\exists \xi_1 \ldots \exists \xi_r \exists y(\varphi)$$

is by induction assumption under the constraint hypothesis equivalent to

$$\bigvee_{i \in I} \exists y(\psi_i \wedge \varphi[\gamma_{i,1}/\xi_1, \ldots, \gamma_{i,r}/\xi_r]).$$

By the proof of the main theorem, the F-quantifier $\exists y$ can be equivalently replaced by a block of new existential quantifiers wrt. new N-variables representing the objects $spec_i(y)$ and $spec_{i,j}(y)$. By the previous case extended quantifier elimination can be performed eliminating these quantifiers by a finite system of case distinctions and corresponding N-expressions. By combining the N-expressions for all these variables into an F-expression of the form

$$t := \sum_{\alpha \in S} p_\alpha(x) \exp(\alpha x)$$

with the specified values α corresponding to the N-expressions for $spec_i(t)$ and the coefficients of the polynomials $p_\alpha(x)$ corresponding to the N-expressions for $spec_{i,j}(t)$, we obtain the desired extended quantifier elimination of the F-quantifier $\exists y$.

Finally the case $s > 1$ of several F-quantifiers is handled by a simple induction on s, using the case $s = 1$ in the induction step.

This completes the proof.

7   Warning About Variants

It is tempting to conjecture that the present method can be extended to a logical framework inculding linear differential equations with **parametric coefficients that are not necessarily differential constants**. This is indeed possible in domains of germs of meromorphic functions as shown in [16], **when no stability conditions are present**. In fact the present study results from a question of **A. Weber** at CASC 2005, asking whether it is possible to include stability conditions in that framework.

If, however, the present framework for the domain $(\mathbb{C}, D)$ even without stability conditions is extended to include a function constant for the independent variable X then we arrive at an **algorithmically undecidable situation**. In particular no algorithmic quantifier elimination is possible in that extended framework.

This is not only the case for the present semantic domain $(\mathbb{C}, D)$ of exponential polynomials, but in fact for any reasonable differential subring of the ring of complex formal power series:

Theorem 5. *Let K be a subfield of $\mathbb{C}$ and let R be a differential subring of $K[[X]]$ in the language $L := \{0, 1, X, +, -, \cdot,'\}$ of linear homogeneous ODEs and of polynomial equations over K. Here X is a language constant for the independent variable. Then the existential theory of R is undecidable. In particular there cannot be an algorithmic quantifier elimination for existential L-formulas in R.*

Proof. The linear differential equation $X \cdot y' = \xi \cdot y$ with number parameter ξ has a non-trivial solution y in R iff $\xi \in \mathbb{N}_0$: Indeed if such a solution y is given by a power series $\sum_{n=0}^{\infty} a_n X^n$, with derivative $\sum_{n=0}^{\infty} na_{n+1}X^n$, then the given differential equation implies by coefficient comparison

$$\xi a_0 = 0, \xi a_1 = a_1, \xi a_2 = 2a_2, \xi a_3 = 3a_3, \ldots$$

Under the hypothesis that for some $i \in \mathbb{N}_0$ $a_i \neq 0$ this implies that $\xi = i \in \mathbb{N}_0$ and that the solution $y = a_i X^i$.

Hence the non-trivial solvability of the given parametric linear differential equation given by the existential formula

$$\varphi(\xi) := \exists y(y \neq 0 \ \wedge \ X \cdot y' = \xi \cdot y \ \wedge \xi' = 0)$$

codes the semiring $\mathbb{N}_0$ of natural numbers inside the ring R. As a consequence every existential sentence about $\mathbb{N}_0$ can be coded into an existential formula about R of the form

$$\exists \xi_1 \ldots \exists \xi_k (\bigwedge_{i=1}^{k} \varphi(\xi_i) \ \wedge \ \psi),$$

where ψ is a quantifier-free formula about R. By the negative solution of Hilbert's tenth problem the existential theory of $\mathbb{N}_0$ is undecidable; by the coding above, this proves the theorem. $\square$

It appears that weaker versions of this theorem have been known to some algebraic model theorists (A. Macintyre, personal communication). Whether the theorem holds also in a language without the constant X is an open problem. The use of this constant can at any rate be avoided, when R is regarded as a differential valuation ring with the natural valuation v, since X is the only solution of the condition $x' = 1 \ \wedge v(x) > 0$. Alternatively, X can be defined in our given language of the domain $(\mathbb{C}, D)$ by the formula

$$Specsize_1(x) \ \wedge \ Deg_1(x) \wedge \ spec_1(x) = 0 \ \wedge \ spec_{1,0}(x) = 0 \ \wedge \ spec_{1,1}(x) = 1.$$

So if we allow in the language of $(\mathbb{C}, D)$ also function terms as coefficients of F-variables and their derivatives, then the existential theory of $(\mathbb{C}, D)$ becomes undecidable and algorithmic quantifier elimination becomes impossible.

8 Conclusions and Open Problems

We have presented an algorithmic constrained quantifier elimination algorithm and a decision algorithm for the domain $(\mathbb{C}, D)$ of complex exponential polynomials in a language including arbitrary polynomial relations for complex number variables, linear ODEs with complex number parameters for function variables and additional predicates specifying the properties of a function variable as exponential polynomial. Moreover the quantifier elimination algorithm can also be extended to produce solutions for existential parametric problems uniformly in the number and function parameters. The chosen language allows the formulation of global and local stability properties of function terms, as well as evaluation at point zero. Hence it appears to be well suited to the solution of parametric stability problems in robust control theory [1]. Examples demonstrate the scope of the method. A negative theorem on undecidability shows a limitation for extensions of this approach.

An implementation of the methods is in principle possible e. g. in REDLOG [7], where a differential algebra context is already present. In order to be of practical relevance the basic algorithms presented here have to be optimized in several more or less obvious ways. The question of the asymptotic complexity of the QE algorithm also remains open. Since it is based on real QE, it must be at least doubly exponential in the worst case [14,5]. It is also rather obvious that an upper complexity bound will depend on the constraint bound b on the function parameters. For fixed value of b, I conjecture that the algorithms are elementary recursive, i. e. bounded in complexity by a fixed number of iterated exponential functions. The number of these interations may well depend on b.

References

1. Ackermann, J.: Robust Control. Communication and Control Engineering (1993)
2. Anai, H., Hara, S.: Fixed-structure robust controller synthesis based on sign definite condition by a special quantifier elimination. In: Proceedings of ACC2000 (to appear)
3. Braun, M.: Differential equations and their applications. In: Applied Mathematical Sciences, 3rd edn., Springer, Heidelberg (1983)
4. Brown, C.W.: Improved projection for cylindrical algebraic decomposition. J. Symb. Computation 32(5), 447–465 (2001)
5. Davenport, J.H., Heintz, J.: Real quantifier elimination is doubly exponential. Journal of Symbolic Computation 5(1–2), 29–35 (1988)
6. Dolzmann, A., Gilch, L.A.: Generic hermitian quantifier elimination. In: Buchberger, B., Campbell, J.A. (eds.) AISC 2004. LNCS (LNAI), vol. 3249, pp. 80–92. Springer, Heidelberg (2004)

7. Dolzmann, A., Seidl, A.: Redlog – first-order logic for the masses. Journal of Japan Society for Symbolic and Algebraic Computation 10(1), 23–33 (2003)
8. Dolzmann, A., Sturm, T., Weispfenning, V.: Real quantifier elimination in practice. In: Matzat, B.H., Greuel, G.-M., Hiss, G. (eds.) Algorithmic Algebra and Number Theory, pp. 221–247. Springer, Berlin (1998)
9. Dorato, P., Yang, W., Abdallah, C.: Robust multi-object feedback design by quantifier elimination. J. Symb. Computation 24(2), 153–159 (1997)
10. Hong, R., Liska, H., Steinberg, S.: Testing stability by quantifier elimination. J. Symb. Comp. 24(2), 161–187 (1997)
11. Liska, R., Steinberg, S.: Applying quantifier elimination to stability analysis of difference schemes. The Computer Journal 36, 497–509 (1993)
12. Seidl, A.: Cylindrical Decomposition under Application-Oriented Paradigms. PhD thesis, FMI, Univ. Passau (2006)
13. Seidl, A., Sturm, T.: A generic projection operator for partial cylindrical algebraic decomposition. In: Sendra, R. (ed.) Proceedings of the 2003 International Symposium on Symbolic and Algebraic Computation (ISSAC 03), Philadelphia, Pennsylvania, pp. 240–247. ACM Press, New York (2003)
14. Weispfenning, V.: The complexity of linear problems in fields. Journal of Symbolic Computation 5(1–2), 3–27 (1988)
15. Weispfenning, V.: Parametric linear and quadratic optimization by elimination. Technical Report MIP-9404, FMI, Universität Passau, D-94030 Passau, Germany (April 1994)
16. Weispfenning, V.: Solving linear differential problems with parameters. In: Ganzha, V.G., Mayr, E.W., Vorozhtsov, E.V. (eds.) CASC 2005. LNCS, vol. 3718, pp. 469–488. Springer, Heidelberg (2005)

Symbolic and Algebraic Methods for Linear Partial Differential Operators

Franz Winkler and Ekaterina Shemyakova

RISC, Linz

(Plenary Talk)

Abstract. The solution of Partial Differential Equations (PDEs) is one of the most important problems of mathematics, and has an enormous area of applications. One of the methods for extending the range of analytically solvable PDEs consists in transformations of PDEs and the corresponding transformations of their solutions. Thus, based on the fact that a second-order equation can be solved if one of its factorizations is known, the famous method of Laplace Transformations suggests a certain sequence of transformations of a given equation. Then, if at a certain step in this transformation process an equation becomes factorizable, an analytical solution of this transformed equation — and then of the initial one — can be found.

The aim of this talk is a description of some old and new developments and generalizations of analytical approaches to the solution of PDEs and the corresponding algebraic theory of differential operators. Recently we have introduced the notion of *obstacle* for the factorization of a differential operator, i.e. conditions preventing a given operator from being factorizable. These obstacles give rise to a *ring of obstacles* and furthermore to a classification of operators w.r.t. to their factorization properties. From obstacles we can also get (Laplace) invariants of operators w.r.t. to certain (gauge) transformations. We have shown how such systems of invariants can be extended to full systems of invariants for certain low order operators. Another related problem is the description of the structure of families of factorizations. For operators of order 3 it has been shown that a family of factorizations depends on at most 3 or 2 parameters, each of these parameters being a function on one variable.

V.G. Ganzha, E.W. Mayr, and E.V. Vorozhtsov (Eds.): CASC 2007, LNCS 4770, p. 423, 2007.
© Springer-Verlag Berlin Heidelberg 2007

A New Scheme for Deniable/Repudiable Authentication

Song Y. Yan[1,2], Carsten Maple[1], and Glyn James[3]

[1] Institute for Research in Applicable Computing, University of Bedfordshire, Park Square Luton, LU1 3JU, UK
song.yan@beds.ac.uk
[2] School of Computer Science and Engineering, South China University of Technology, Guangzhou 510006, China
[3] Department of Mathematical Sciences, Coventry University, CV1 5FB, UK

Abstract. One of the most important features of authentication is the non-repudiation property, implemented by digital signatures. This useful feature of authentication may, however, not be necessary in some cases, such as e-voting, and should indeed be avoided. In this paper, by a combined use of public-key encryption, digital signatures, coding, quadratic residues, and randomness, a new scheme for deniable/repudiable authentication is proposed and analyzed, and a complete example of the scheme is given. The security of the scheme is based on the intractability of the quadratic residuosity problem.

Keywords: Deniable/repudiable authentication, intractability, quadratic residuosity problem.

1 Introduction

One of the most important features of authentication is the non-repudiation property, implemented by digital signatures. According to Hellman [4], a true digital signature must be a number (so it can be sent in electronic form) that is easily recognized by the receiver as validating the particular message received, and which could only have been generated by the sender. Remarkably enough, this important feature of digital signatures is in fact the main motivation for Hellman and his colleagues to develop the public-key cryptography (see [2], [3] and [4]) since the main application of public-key cryptography is for digital signatures and authentication rather than for encryption. Notice however that the requirement for digital signature to be *easily recognized by the receiver* may not be necessary in privacy-preserved authentication systems, such as e-voting systems. Traditionally, once the signer, say e.g. Alice, signs a document, she cannot deny her signature later, since the signature was generated only by using her private key. This marvellous feature of authentication is, however, not necessary and should be avoided whenever possible in some e-systems. For example, in an e-voting system, suppose Alice votes Bob but not John. As the voting is confidential, Alice does not want to disclose her voting to anyone regardless of whether or

V.G. Ganzha, E.W. Mayr, and E.V. Vorozhtsov (Eds.): CASC 2007, LNCS 4770, pp. 424–432, 2007.

not she actually votes Bob or not. By a combined use of public-key encryption, digital signatures, coding and randomness, this paper proposes a new scheme for deniable/repudiable authentication suitable for e-voting, in which the author of a vote can deny the authorship of the vote even if she actually did the vote. Of course, if needed and if she wishes, Alice can at any time show a restricted number of individuals, whilst still keeping the information from others, that she is the author of the vote. The security of the scheme is based on the quadratic residuosity problem, which is believed to be intractable. In the next sections, the intractability of the quadratic residuosity problem will be introduced, then the deniable/repudiable authentication scheme will be proposed and discussed, followed by an illustrative example. The security analysis and the related work will also be addressed.

2   The Deniable Scheme

A positive integer $a > 1$ is a quadratic residue modulo N, denoted by $a \in Q_N$, if $\gcd(a, N) = 1$ and if there exists a solution x to the congruence $x^2 \equiv a \pmod{N}$; otherwise, it is a quadratic non-residue modulo N, denoted by $a \in \overline{Q}_N$. The quadratic residuosity problem (QRP) is to decide whether a is a quadratic residue or quadratic non-residue modulo N. That is, to decide whether a is a square or pseudo-square modulo N. If $N = p$ is an odd prime, by Euler's criterion [9] a is a quadratic residue modulo p if and only if $a^{(p-1)/2} \equiv 1 \pmod{p}$, which is in fact easy to decide. However, if N is an odd composite, then one needs to know the prime factorization of N which, as it is well known, is intractable. Consequently it is hard to decide whether or not $a \in Q_N$ when N is a large odd composite, since obviously, a is a quadratic residue modulo N if and only if it is quadratic residue modulo *every prime* dividing N [7]. Define the Legendre symbol, $(\frac{a}{p})$ with p odd prime, as follows:

$$\left(\frac{a}{p}\right) = \begin{cases} 1, & a \in Q_N \\ -1, & a \in \overline{Q}_N \end{cases}$$

Also define the Jacobi symbol, $(\frac{a}{N})$ with N odd composite, as follows:

$$\left(\frac{a}{N}\right) = \left(\frac{a}{p_1}\right)^{\alpha_1} \left(\frac{a}{p_2}\right)^{\alpha_2} \cdots \left(\frac{a}{p_k}\right)^{\alpha_k}$$

when $N = p_1^{\alpha_1} p_2^{\alpha_2} \cdots p_k^{\alpha_k}$. Obviously, if $(\frac{a}{N}) = -1$, then $(\frac{a}{p_i}) = -1$ for some i, and a is a quadratic non-residue modulo N. However, it may well be possible that a is a quadratic non-residue modulo N even if $(\frac{a}{N}) = 1$, since

$$a \in Q_N \implies \left(\frac{a}{N}\right) = 1,$$

$$a \in Q_N \overset{?}{\Longleftarrow} \left(\frac{a}{N}\right) = 1,$$

$$a \in \overline{Q}_N \Longleftarrow \left(\frac{a}{N}\right) = -1.$$

Let $J_N = \{a \in \mathbb{Z}_N^* : (\frac{a}{N}) = 1\}$. Then $\tilde{Q}_N = J_N - Q_N$. Thus $\tilde{Q}_N$ is the set of all pseudosquares modulo N; it contains those elements of J_N that do not belong to Q_N. Thus QRP may be stated as follows. Given an odd composite N and an integer $a \in J_N$, decide whether or not $a \in Q_N$.

Example 1. Let $N = 21$, then

$$\mathbb{Z}_{21}^* = \{1, 2, 4, 5, 8, 10, 11, 13, 16, 17, 19, 20\},$$

$$J_{21} = \{1, 4, 5, 16, 17, 20\} = Q_{21} \cup \tilde{Q}_{21},$$

$$\overline{Q}_{21} = \mathbb{Z}_{21}^* - J_{21} = \{2, 8, 10, 11, 13, 19\}.$$

Now, both Q_{21} and $\tilde{Q}_{21}$ are mixed in J_{21}. To distinguish $\tilde{Q}_{21}$ from Q_{21}, the only method is to factor $21 = 3 \times 7$ and find Q_3 and Q_7, and compute $Q_{21} = Q_3 \cap Q_7$. From Page 150 of [9], we know that $Q_3 = \{1, 4, 10, 13, 16, 19\}$ and $Q_7 = \{1, 2, 4, 8, 11, 16\}$. Thus $Q_{21} = Q_3 \cap Q_7 = \{1, 4, 16\}$. Hence, $\tilde{Q}_{21} = \{5, 17, 20\}$. That is, $\tilde{Q}_{21}$ has been successfully distinguished from Q_{21}. This process will be, however, very difficult to perform if N is large since N needs to be factored first.

As indicated in Example 1, to decide the quadratic residuosity, the only method we know is to factor N, which is intractable for large N; the fastest factoring method, the Number Field Sieve (NFS), runs in time

$$\mathcal{O}(\exp(c(\log N)^{1/3}(\log \log N)^{2/3})),$$

where $c = (64/9)^{1/3} \approx 1.92$ if a general version of NFS is used or $c = (32/9)^{1/3} \approx 1.53$ if a special version of NFS is used.

In our deniable/repudiable authentication scheme, we first generate a conventional non-deniable/non-repudiable digital signature S using a standard digital signature system. Then we convert S into a deniable/repudiable signature S':

$$S \qquad \Longrightarrow \qquad S'$$
$$\Uparrow \qquad\qquad\qquad \Uparrow$$
$$\text{Non-deniable signature} \qquad \text{Deniable signature}$$

On receiving the deniable signature, the receiver obviously knows there is a signature attached to the document, say a vote, but he cannot verify who is the author of the vote, as the vote is privacy-preserved by the deniable signature, which cannot be verified directly by the sender's public-key. There are several nontrivial ways to implement the deniable/repudiable authentication scheme; one of the ways was developed by Aumann and Rabin [1] and based on the intractability of the Modular Square Root Problem. Our method, however, is based on the intractability of the Quadratic Residuosity Problem, which may be described as follows:

Algorithm 1. *A new deniable/repudiable authentication scheme based on a combined use of public-key encryption, digital signatures, coding, quadratic residues, and randomness, suitable for e-voting environment.*

[1] *Generate the non-deniable signature $S = \{s_1, s_2, \cdots\}$.* The voter, say, e.g., Bob, uses his private key d to generate his digital signature S on his vote M: $S = M^d \pmod{N}$, where N is a product of two large prime numbers. This signature can be verified by using Bob's public-key e: $M = S^e \pmod{N}$, where $ed \equiv 1 \pmod{\phi(N)}$.

[2] $S \implies (S', B)$: *Obtain the deniable signature S' by adding randomness to S.*

    [2-1] $S \implies S'$: *Randomize the signature.* Randomly add some extra digits (noise) to the digital signature S to get a corresponding randomized digital signature $S' = \{s'_1, s'_2, \cdots\}$.

    [2-2] $S' \implies B$: *Generate a bit string $B = \{b_1, b_2, \cdots\}$ for the randomized signature S'.* Generate a binary string B, with each bit in B corresponding to a digit in S', assign 1 to the bit if the corresponding digit appears in both S and S', otherwise, assign 0 to the bit.

    [2-3] $B \implies X$: *Generate a random mixed string X of squares and pseudo-squares.* Generate a string of integers $X = \{x_1, x_2, \cdots\}$ which are the mixed squares and pseudo-squares, based on the quadratic residuosity problem [9] (it can also even be based on the kth power residuosity problem (kPRP) [8]).

    To get the X string, we choose $N = pq$ with p, q prime. Find a pseudo random square $y \in \mathbb{Z}_N^*$ such that $y \in \overline{Q}_N$ and $(\frac{y}{N}) = 1$. That is, $y \in \tilde{Q}_N$. (N, y) can be made public, but p, q must be kept as a secret. Choose at random the number r_i and compute

$$x_i \equiv \begin{cases} r_i^2 \bmod N & \text{if } b_i = 0 \text{ (random square)} \\ yr_i^2 \bmod N & \text{if } b_i = 1 \text{ (pseudo random square)} \end{cases}$$

    [2-4] Send $\{(S', X), E\}$ to the Election Centre, where (S', X) is the repudiable digital signature and E the e-vote. The author of the e-vote can deny his authorship of the vote, since S' is a random string of digits which is different from S and from which it is not easy to get back to S unless the quadratic residuosity problem can be solved in polynomial-time.

    [2-5] $S' \implies B$: To verify the signature if needed only the author of the vote, who knows the trap-door information (i.e., the prime factors p and q of the composite modulus N), can recover B from X.

    To get the string $B = b_1, b_2, \cdots$ from the string $X = x_1, x_2, \cdots$, one may perform the following operations:

$$b_i \equiv \begin{cases} 0, \text{ if } e_i^p = e_i^q = 1 \\ 1, \text{ otherwise} \end{cases}$$

where

$$e_i^p = \left(\frac{x_i}{p}\right), \quad e_i^q = \left(\frac{x_i}{q}\right).$$

    [2-6] $\{B, S'\} \implies S$: Remove the noise from S' to get S according to B.

Example 2. We demonstrate the above idea in Algorithm 1 in this example for an e-voting system.

[1] *Generate the non-repudiable signature.* The voter, say, e.g., Bob, uses his private key to generate his digital signature S for his vote:

Bob Newman $\implies$ 0215020014052313013 $\implies$ 42527067843532368

$\Uparrow$ $\Uparrow$ $\Uparrow$

Name for Signature   Numerical Form (M)   Digital Signature (S)

This digital signature was generated and verified by

$$S \equiv M^e \equiv 215020014052313013^7$$
$$\equiv 42527067843532368 \ (\mathrm{mod}\ 1832970702926065247)$$

$$M \equiv S^d \equiv 42527067843532368^{785558871522137263}$$
$$\equiv 215020014052313013 \ (\mathrm{mod}\ 1832970702926065247)$$

with

$$ed \equiv 1 \ (\mathrm{mod}\ \phi(1832970702926065247)).$$

[2] $S \implies (S', B)$: *Add repudiable feature to the non-repudiable signature.*

[2-1] $S \implies S'$: *Randomize the signature.* Randomly add some extra digits into the digital signature S to get a corresponding pseudo digital signature S'.

S   4   2   5   2   7 0 6   7 8   4 3 5   3 2   3 6   8

$\Updownarrow \Uparrow$

S' 7 9 1 4 8 5 2 1 4 5 3 2 2 8 9 1 7 0 6 3 7 8 9 4 3 5 9 1 3 2 3 2 1 7 3 6 6 8

[2-2] $S' \implies B$: *Generate a bit string $B = \{b_1, b_2, \cdots\}$ for the randomized signature S'.* Generate a binary string B with each bit in B corresponding to a digit in S', assign 1 to the bit if the corresponding digit appears in both S and S', otherwise, assign 0 to the bit.

S' 7 9 1 4 8 5 2 1 4 5 3 2 2 8 9 1 7 0 6 3 7 8 9 4 3 5 9 1 3 2 3 2 1 7 3 6 6 8

$\Updownarrow \Uparrow$

B 0 0 0 1 0 0 1 0 0 1 0 0 1 0 0 0 1 1 1 0 1 1 0 1 1 1 0 0 0 0 1 1 0 0 1 1 0 1

[2-3] $B \implies X$: Generate a random mixed string X of squares and pseudo-squares: Generate a string of integers $X = \{x_1, x_2, \cdots\}$ which are the mixed squares and pseudo squares. based on the quadratic residuosity problem.

Choose a random pseudo square $y = 1234567 \in \mathbb{Z}_N^*$ such that $y \in \overline{Q}_N$ and $(\frac{y}{N}) = 1$. That is, $y \in \tilde{Q}_N$. (N, y) can be made public, but p, q must be kept as a secret.

We just calculate the values for x_3 and x_4; all the rest can be calculated in exactly the same way.

Choose at random the number $r_3 = 8194920765$ and $r_4 = 1740298374$, and compute (note that $b_3 = 0, b_4 = 1$):

$$x_3 \equiv r_3^2 \equiv 8194920765^2$$
$$\equiv 1169781039289836333 \pmod{1832970702926065247}$$

$$x_4 \equiv yr_4^2 \equiv 1234567 \cdot 1740298374^2$$
$$\equiv 287064024006224109 \pmod{1832970702926065247}$$

[2-4] Send $\{(S', X), E\}$ to the Election Centre, where (S', X) is the deniable/repudiable digital signature and E the e-vote. The author of the e-vote can deny his authorship of the vote, since S' is a random string of digits which is different from S.

[2-5] $S' \implies B$: To verify the signature if needed, only the author of the vote who knows the trap-door information (the composite modulus N) can show the Election Centre that he is the author of the vote, anyone else should not be able to verify the authorship of the vote.

We just show how to recover S_3' and S_4' by computing b_3 and b_4 from x_3 and x_4 as follows (all the rest are performed in exactly the same way): Since

$$e_3^p = \left(\frac{x_3}{p}\right) = \left(\frac{1169781039289836333}{1353874987}\right) = 1,$$

$$e_3^q = \left(\frac{x_3}{q}\right) = \left(\frac{1169781039289836333}{1353869981}\right) = 1,$$

$$e_4^p = \left(\frac{x_4}{p}\right) = \left(\frac{287064024006224109}{1353874987}\right) = -1,$$

$$e_4^q = \left(\frac{x_4}{q}\right) = \left(\frac{287064024006224109}{1353869981}\right) = -1,$$

then

$$b_3 = 0, \text{ as } e_3^p = e_3^q = 1,$$

$$b_4 = 1, \text{ as } e_4^p = e_4^q = -1.$$

On the other hand, for the cryptanalyst who does not know the factorization of $N = pq$, then the following two pieces of computation must be performed

$$e_3^N = \left(\frac{x_3}{N}\right) = \left(\frac{1169781039289836333}{1832970702926065247}\right) = 1,$$

$$e_4^N = \left(\frac{x_4}{N}\right) = \left(\frac{287064024006224109}{1832970702926065247}\right) = 1,$$

So, he cannot distinguish squares from pseudo squares. Thus, he cannot remove the noise correctly, as indicated in the next step.

Table 1. The Complete List of Computation Results in Example 2

i	S_i	S'_i	B_i	r_i	x_i	e_i^p	e_i^q	e_i^N	B_i	S'_i	S_i
1		7	0	5984746251	990744333263315308	1	1	1	0	7	
2		9	0	8321764556	1431849317233462997	1	1	1	0	9	
3		1	0	8194920765	1169781039289836333	1	1	1	0	1	
4	4	4	1	1740298374	287064024006224109	-1	-1	1	1	4	4
5		8	0	9827365106	1262628374470998392	1	1	1	0	8	
6		5	0	1029837512	1060565301122350144	1	1	1	0	5	
7	2	2	1	6473682901	150184231223858133	-1	-1	1	1	2	2
8		1	0	8376453217	512081785399169703	1	1	1	0	1	
9		4	0	2274645280	1508069743974147906	1	1	1	0	4	
10	5	5	1	427419669081	370940926090515520	-1	-1	1	1	5	5
11		3	0	321110693270	143409937148888162	1	1	1	0	3	
12		2	0	746753830538	272413385043405128	1	1	1	0	2	
13	2	2	1	474256143563	1334682782008751010	-1	-1	1	1	2	2
14		8	0	32062222085	1522491389265208905	1	1	1	0	8	
15		9	0	604305613921	688901773803769184	1	1	1	0	9	
16		1	0	558458718976	660632574793168744	1	1	1	0	1	
17	7	7	1	722974121768	26578238445673413	-1	-1	1	1	7	7
18	0	0	1	845473509473	1162153979472023243	-1	-1	1	1	0	0
19	6	6	1	343633073697	1723201744301369246	-1	-1	1	1	6	6
20		3	0	676470788342	593670337217803932	1	1	1	0	3	
21	7	7	1	155590763466	1760530014633994358	-1	-1	1	1	7	7
22	8	8	1	429392673709	1131678843033095552	-1	-1	1	1	8	8
23		9	0	525428510973	404751391738164577	1	1	1	0	9	
24	4	4	1	272600608981	1312583620580057586	-1	-1	1	1	4	4
25	3	3	1	219760099374	401697423373942699	-1	-1	1	1	3	3
26	5	5	1	675982933766	466825511754271815	-1	-1	1	1	5	5
27		9	0	146486307198	1483148054306829822	1	1	1	0	9	
28		1	0	920624947349	1137384662786502224	1	1	1	0	1	
29		3	0	644031395307	829658753701141607	1	1	1	0	3	
30		2	0	453747019461	1589404976685991740	1	1	1	0	2	
31	3	3	1	812920457916	640910416677721889	-1	-1	1	1	3	3
32	2	2	1	960498834085	7965085299907034044	-1	-1	1	1	2	2
33		1	0	88430571674	512987907856818574	1	1	1	0	1	
34		7	0	39169594160	60628309989493861	1	1	1	0	7	
35	3	3	1	797179490457	573155683439910649	-1	-1	1	1	3	3
36	6	6	1	310075487163	284578868772975450	-1	-1	1	1	6	6
37		6	0	259811952655	1271636448682763003	1	1	1	0	6	
38	8	8	1	745580037409	1355861018555287891	-1	-1	1	1	8	8

[2-6] $\{B, S'\} \implies S$: Remove the noise from S' to get S. Since $b_3 = 0$, its corresponding digit 1 in S'_3 is noise and should be removed from S'. However, as $b_4 = 1$, its corresponding digit 4 should remain in S'. Clearly, after removing all the noise from S', S' will eventually become S, the true digital signature. A complete list of computation results in this example is given in table 1.

Clearly, anyone who can solve the quadratic residuosity problem (or the kth power residuosity problem) can distinguish the pseudo squares from square (or the pseudo kth powers from the kth powers), and hence can verify the digital signature and the authorship of the e-vote. But as everybody knows, solving QRP/kPRP is intractable, thus the author can deny his authorship of an e-vote, regardless of whether or not he actually did vote.

3 Conclusions

In this paper, a new scheme, based on a combined use of public-key encryption, digital signatures, coding, quadratic residues and randomness, of deniable/repudiable authentication/authorization is proposed and discussed. The scheme is well suited for such e-systems as e-voting where the privacy of the signature needs to be preserved. The security of the scheme is based on the intractability of the quadratic residuosity problem (QSP). With some extension, it can also be based on the kth power residuosity problem (kPSP) [8]. It is interesting to note that the quadratic residuosity problem has been used for probabilistic encryption (see [5] and [6]) since 1984, but to best of our knowledge, it is the first time to be used for authentication/authorization, particularly for deniable/repudiable authentication/authorization. Similar scheme was developed by Aumann and Rabin [1], but based on a different intractable number-theoretic problem, the modular square root problem.

Acknowledgements. The authors would like to thank the three anonymous referees for their helpful comments, suggestions and corrections over the earlier draft paper.

References

1. Aumann, Y., Rabin, M.O.: Authentication, Enhanced Security and Error Correcting Codes. In: Krawczyk, H. (ed.) CRYPTO 1998. LNCS, vol. 1462, pp. 299–303. Springer, Heidelberg (1998)
2. Diffie, W., Hellman, E.: New Directions in Cryptography. IEEE Transactions on Information Theory 22(5), 644–654 (1976)
3. Diffie, W., Hellman, M.E.: Privacy and Authentication: An Introductionp to Cryptography. Proceedings of the IEEE 67(3), 393–427 (1979)
4. Hellman, M.E.: An Overview of Public Key Cryptography. IEEE Communications Magazine, Landmark 10 Articules, 50th Anniversary Commemorative Issue, 42–49 (2002)

5. Goldwasser, S.: The Search for Provably Secure Cryptosystems. In: Pomerance, C. (ed.) Cryptology and Computational Number Theory. Proceedings of Symposia in Applied Mathematics, vol. 42, American Mathematical Society (1990)
6. Goldwasser, S., Micali, S.: Probabilistic Encryption. Journal of Computer and System Sciences 28, 270–299 (1984)
7. McCurley, K.S.: Odds and Ends from Cryptology and Computational Number Theory. In: Pomerance, C. (ed.) Cryptology and Computational Number Theory. Proceedings of Symposia in Applied Mathematics, vol. 42, pp. 49–74. American Mathematics Society (1990)
8. Yan, S.Y.: A New Cryptographic Scheme based on the kth Power Residuosity Problem. In: 15th British Colloquium for Theoretical Computer Science (BCTCS15), 14-16 April 1999, Keele University (1999)
9. Yan, S.Y.: Number Theory for Computing, 2nd edn. Springer, Heidelberg (2002)

An Algebraic-Numeric Algorithm for the Model Selection in Kinetic Networks

Hiroshi Yoshida[1,*], Koji Nakagawa[2], Hirokazu Anai[3], and Katsuhisa Horimoto[2,*]

[1] Faculty of Mathematics, Organization for the Promotion of Advanced Research, Kyushu
University, Hakozaki 6-10-1, Higashi-ku, Fukuoka 812-8581 Japan
Tel/Fax.: +81-92-642-7396
phiroshi@math.kyushu-u.ac.jp

[2] Computational Biology Research Centre (CBRC), National Institute of Advanced Industrial
Science and Technology (AIST), Aomi 2-42, Koto-ku, Tokyo 135-0064, Japan
* Tel.: +81-3-3599-8711; Fax.: +81-3-3599-8081
{nakagawa-koji,k.horimoto}@aist.go.jp

[3] IT Core Laboratories, FUJITSU LABORATORIES LTD./CREST, JST., Kamikodanaka 4-1-1,
Nakahara-ku, Kawasaki 211-8588, Japan
anai@jp.fujitsu.com

Abstract. We propose a novel algorithm to select a model that is consistent with
the time series of observed data. In the first step, the kinetics for describing a biological phenomenon is expressed by a system of differential equations, assuming
that the relationships between the variables are linear. Simultaneously, the time
series of the data are numerically fitted as a series of exponentials. In the next
step, both the system of differential equations with the kinetic parameters and the
series of exponentials fitted to the observed data are transformed into the corresponding system of algebraic equations, by the Laplace transformation. Finally,
the two systems of algebraic equations are compared by an algebraic approach.
The present method estimates the model's consistency with the observed data and
the determined kinetic parameters. One of the merits of the present method is that
it allows a kinetic model with cyclic relationships between variables that cannot
be handled by the usual approaches. The plausibility of the present method is illustrated by the actual relationships between specific leaf area, leaf nitrogen and
leaf gas exchange with the corresponding simulated data.

1 Introduction

The knowledge-based approach to constructing a biological network model is recognized as one of the most promising approaches [4]. In this approach, the causal relations between biological molecules are described as a directed graph, based on the gene
interaction information collected from a large number of previous reports. Since each
relation identified by experimental studies is regarded as strong evidence for the existence of edges in the network model, biological network models have been constructed
for various biological phenomena by a knowledge-based approach. On the other hand,
it is well-known that the relationships between the molecules in a living cell change

---

* Corresponding authors.

V.G. Ganzha, E.W. Mayr, and E.V. Vorozhtsov (Eds.): CASC 2007, LNCS 4770, pp. 433–447, 2007.

dynamically, depending on the cellular environment. Thus, the molecular relationships in the literature represent the responses to the different conditions in the experimental studies, and in the network model generated from the biological knowledge, the consistency of the model with the data observed by experimental studies must be considered carefully. Actually, several distinctive models of the relationship between molecules for a biological phenomenon can be obtained from the large amount of information in the literature [2, 5]. In these cases, a model that is consistent with the data observed under particular conditions should be selected from the candidate models.

The consistency of a model with the observed data first reminds us of the identifiability problem in the compartmental models for tracer kinetics [1, 5, 6]. In the compartmental models, the unknown parameters are estimated from tracer data in the accessible pools. The identifiability problem addresses the issue of whether the unknown parameters can be determined uniquely or non-uniquely from the tracer data. This issue has usually been solved through the transformation of differential equations into algebraic equations, by the Laplace transformation. Although a systematic algorithm for the identifiability problem was proposed [3], its application is limited to the unrealistic context of an error–free model structure and noise–free tracer data. Thus, it still seems to be difficult to solve the identifiability problem for actually observed data, in spite of the mathematical studies.

The issue of the consistency of a model with the observed data is also well known in statistics, as the test for causal hypotheses by using the observed data. The origin of the test for causal hypotheses is attributed to path analysis [12]. Unfortunately, the importance of this cornerstone research has been ignored for a long time, but the natural extension of the path analysis has been established as the well-known structural equation model (SEM) [8]. Indeed, the SEM has been utilized recently in various fields, in accordance with increased computer performance. However, the SEM without any latent variables, which is the natural form for applying the SEM to the biological networks, frequently faces difficulty in the numerical calculation of the maximum likelihood for the observed data. To overcome the difficulty of this calculation, the d-sep test [11] has been developed, based on the concept of d-separation in a directed acyclic graph [10]. Notice that the graph consistency with the data in the d-sep test can consider only the directed acyclic graph (DAG), without any cyclic relationships.

In this study, we propose a new method for selecting models, by estimating the consistency of a kinetic model with the time series of observed data. Our method is described in the following section. First, the kinetics for describing a biological phenomenon is expressed by a system of differential equations, assumed that the relationships between the variables are linear. Simultaneously, the time series of the data are numerically fitted as a series of exponentials. Next, the differential equations with the kinetic parameters and the series of exponentials fitted to the observed data are both transformed into the corresponding system of algebraic equations, by the Laplace transformation. Finally, the two systems of algebraic equations are compared by an algebraic approach. Thus, the present method estimates the model's consistency with the observed data and the determined kinetic parameters. In §3, the plausibility of the present method is illustrated by the actual relationships between specific leaf area, leaf

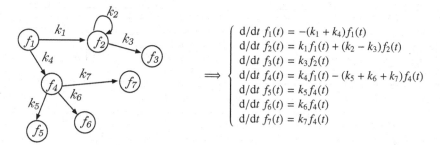

Fig. 1. Correspondence between a network and a system of differential equations. By assuming a linear relation between the variables, the kinetics of chemicals $f_1, f_2, \ldots$ in the left graph can be described by the system of differential equations on the right side.

nitrogen and leaf gas exchange [9], with the corresponding data generated by the differential equations for the relationships. Furthermore, the merits and pitfalls of the present method are discussed. In particular, one of the merits of the present method is that it allows a kinetic model with cyclic relationships between variables that cannot be handled by the usual approaches.

2 Methods

The aim of this paper is to select the model most consistent with the given sampling data. In this section, we propose a method to perform this selection, where the model is described as a network. The network addressed in this paper designates the kinetics of chemicals, which can be described by a system of differential equations, as seen in Fig. 1.

First, we will show the overview of our method by a schematic illustration. We will then provide an explanation for the Laplace transformations of model formulae and sampling data over the time domain, as preparation for the model selection over the Laplace domain. Lastly, we describe a procedure to estimate the model consistency with the definition of *consistency measure*.

2.1 Overview

The overview of our method is schematically illustrated in Fig. 2. The point is that we perform the model selection over the Laplace domain. Therefore, both the model formulae and sampling data must be transformed into functions over the Laplace domain. Suppose that the model formulae are $\{d/dt\, h(t) = -k_1 h(t),\ d/dt\, f(t) = k_1 h(t) - k_2 f(t)\}$ and the sampling data are fitted to $h_o(t) = \beta_0 \exp(-\alpha_0 t)$, $f_o(t) = \beta_1 \exp(-\alpha_1 t) + \beta_2 \exp(-\alpha_2 t)$. The Laplace-transformed formulae of the model formula: $L[f(t)](s)$ and the fitted function: $L[f_o(t)](s)$ are rational functions in s, as seen in the middle row of Fig. 2. Let *comp* denote the set of polynomials obtained by matching the coefficients in s of $L[f(t)](s)$ and $L[f_o(t)](s)$ over the Laplace domain, in which every element is equal to zero when $L[f(t)](s)$ is exactly identical to $L[f_o(t)](s)$ in s. Then we have adopted the smallest sum-square value of the elements in *comp* as a *consistency measure* between the model

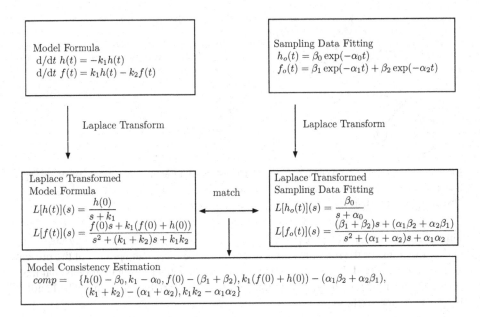

Fig. 2. Overview of our method. The top row designates the model formulae and the sampling data over the time domain, and the middle row designates their Laplace transformations. *comp* denotes the set of polynomials derived by matching the coefficients in s of $L[f(t)](s)$ and $L[fo(t)](s)$ over the Laplace domain, which is zero when the model and sampling data are completely consistent with each other.

and the sampling data, because this value is zero in the case of $L[f(t)](s) = L[fo(t)](s)$. We shall mention the formal procedure and definitions concretely in the following subsections.

2.2   Preparations: Transformation into Laplace Domain

Model Formula. Suppose that the model formulae are described over the time domain as the following system of differential equations:

$$\frac{\mathrm{d}f_i(t)}{\mathrm{d}t} = F_i(\vec{f}, \vec{k}), \tag{2.1}$$

where $\vec{f} = \{f_1, f_2, \ldots, f_n\}$ and $\vec{k} = \{k_1, k_2, \ldots, k_m\}$. $F_i(\vec{f}, \vec{k})$ can be determined in accordance with the network representing the model, and $\vec{k}$ denotes the kinetic constants between the chemicals. We transform this system of differential equations into the system of algebraic equations over the Laplace domain, and solve the equations in $L[f_i(t)](s)$ ($i = 1, 2, \ldots, n$). Notice that in this paper, we deal only with an autonomous system of differential equations, but in the framework of the Laplace transformation, we can deal with differential equations containing external forces or 'convolutions' of complex functions, as long as the Laplace-transformed algebraic equations can explicitly be solved in $L[f_i(t)](s)$ ($i = 1, 2, \ldots, n$).

Sampling Data Fitting. In this paper, we need the Laplace transformation of the sampling data, because we perform the model selection over the Laplace domain. Let $fo_i(t)$ denote the sampling data corresponding to $f_i(t)$ derived theoretically. By using non-linear regression (via Maple 10 Global Optimization toolbox, ©MapleSoft), $fo_i(t)$ is expressed in terms of a series of exponentials, according to [6], as follows:

$$fo_i(t) = \beta_0 + \sum_{i=1}^{k} \beta_i \exp(-\alpha_i t), \qquad (2.2)$$

where k is the number of distinct exponentials determined by $f_i(t)$, and β_0 is zero in the case of the non-existence of a constant term within $f_i(t)$. $fo_i(t)$ thus fitted is changed into the Laplace-transformed data as follows:

$$L[fo_i(t)](s) = \frac{\beta_0}{s} + \sum_{i=1}^{k} \frac{\beta_i}{s + \alpha_i}, \qquad (2.3)$$

where L denotes the Laplace transformation.

2.3 Estimation of Model Consistency

Consistency Measure. To evaluate the consistency of the model with the sampling data, here we define two *consistency measures*. If the model is completely consistent with the sampling data and the data lack noise and inaccuracies, then $L[f_i(t)](s) = L[fo_i(t)](s)$ ($i = 1, 2, \ldots, n$) holds. This fact has led us to the following definitions of consistency measure:

Let *comp* denote the set of polynomials obtained by matching the coefficients of $L[f(t)](s)$ and $L[fo(t)](s)$ over the Laplace domain, in which every element is zero in the case of $L[f_i(t)](s) = L[fo_i(t)](s)$ ($i = 1, 2, \ldots, n$); that is, when Formula $L[f_i(t)](s) = L[fo_i(t)](s)$ is an identity in s.

The first consistency measure (in short, $CM1$) of the model is defined as the smallest sum-square value of the elements in *comp* under the following constraint:

$$k_1 > 0, k_2 > 0, \ldots, k_m > 0. \qquad (2.4)$$

In order to obtain the smallest value, we have utilized the least squares method using the following equations:

$$\frac{\partial}{\partial k_1} g(\vec{k}) = 0, \frac{\partial}{\partial k_2} g(\vec{k}) = 0, \cdots, \frac{\partial}{\partial k_m} g(\vec{k}) = 0, \qquad (2.5)$$

where $g(\vec{k})$ is the sum-square value of the elements in *comp*. It should be noted that in this paper we deal only with the case that the ideal associated with the set of polynomials in (2.5) is zero-dimensional. Then, we survey all of the possible candidates of the minimum by calculating *all* of the real positive roots of the system of algebraic equations (2.5). Several methods and tools exist to calculate all real roots of algebraic equations adjoined by a zero-dimensional ideal. Here we employed 'NSolve' in Mathematica 5.2 (Wolfram Research Inc.), which computes the desired roots efficiently.

Remark 1. If *comp* is a positive dimensional, then we can always perturb the set of polynomials in *comp* in order to obtain a zero-dimensional variety. Although here we cannot discuss the stability and convergency issues related to such perturbations, it is an important research issue on its own light (see [7] for an example).

In this paper, we have calculated the other consistency measure (in short, *CM2*) as the smallest $g(\vec{k})$ under the following constraint:

$$k_1 \geq 0, k_2 \geq 0, \ldots, k_m \geq 0. \tag{2.6}$$

The difference between Constraints: (2.4) and (2.6) is that one takes account of the zero value of the kinetic constants $\vec{k}$, corresponding to the non-existence of edges in the network. This account yields a finer model selection where all of the subnetworks of the presupposed network are also considered. We can calculate the smallest value of $g(\vec{k})$ under Constraint (2.6), using the following recursive definition:

Let $MinimumValue(q(\vec{l}))$ denote the *minimum value* of function q with variables: $\vec{l} = \{l_1, l_2, \ldots, l_m\}$ by the following procedure:

1. If the cardinality of $\vec{l}$, namely m, is zero, then the *minimum value* is infinity.
2. Otherwise, let v_0 denote the minimum value of q under Constraint (2.4) via 'NSolve.' Furthermore, let v_i ($i = 1, 2, \ldots, m$) denote the value calculated by $MinimumValue(q(\vec{l_i}))$, where $\vec{l_i}$ is the vector: $\{l_1, l_2, \ldots, l_{i-1}, 0, l_{i+1}, \ldots, l_m\}$.
3. The *minimum value* is the smallest value among $v_0, v_1, \ldots, v_m$.

Model Selection. Using the consistency measure defined in §2.3, we performed a model selection. We, first, calculated the consistency measures among all of the combinations of the presupposed models with the sampling data. Next, we arranged the combinations of the models with the data in ascending order by the consistency measure. Last, we estimated the most consistent model having the first element (the smallest value).

3 Results and Discussion

3.1 Preparations: Transformation into Laplace Domain

Model Formula. We analyzed the models for a relationship between specific leaf area, leaf nitrogen, and leaf gas exchange in botany [9]. In the original paper, six models for the kinetics of four biomolecules are listed, and the consistency of the models with the observed data, which are composed of various properties of the molecules, rather than time series data, are tested by the d-sep test. In this paper, four of the six original models (models A, B, C, and D) and one model (model E) modified from the original one are considered, to show how cyclic relationships can be handled. The models considered in this paper are shown in Fig. 3. Each model expressed the relationship between four biomolecules, SLA, N, A, and G. According to the definition in §2.2, each relationship

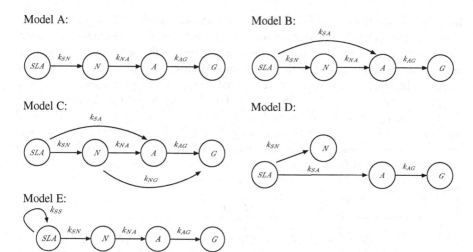

Model A:

Model B:

Model C:

Model D:

Model E:

Fig. 3. Models analyzed in the present study. In the above models, the causal relationships between molecules are denoted by arrows. The molecules corresponding to the variables, denoted within the circles, are SLA, N, A, and G, and the kinetic parameters, denoted over the arrows, are k_{SN}, k_{NA}, k_{AG}, k_{SA}, k_{NG}, and k_{SS}.

between the variables is assumed to be linear, and then the differential equations for the five models can be formulated as follows:

Model A:

$$\begin{cases} \mathrm{d}/\mathrm{d}t\, SLA(t) = -k_{SN}\, SLA(t), \\ \mathrm{d}/\mathrm{d}t\, N(t) = k_{SN}\, SLA(t) - k_{NA}\, N(t), \\ \mathrm{d}/\mathrm{d}t\, A(t) = k_{NA}\, N(t) - k_{AG}\, A(t), \\ \mathrm{d}/\mathrm{d}t\, G(t) = k_{AG}\, A(t). \end{cases} \qquad (3.1)$$

Model B:

$$\begin{cases} \mathrm{d}/\mathrm{d}t\, SLA(t) = -(k_{SN} + k_{SA})\, SLA(t), \\ \mathrm{d}/\mathrm{d}t\, N(t) = k_{SN}\, SLA(t) - k_{NA}\, N(t), \\ \mathrm{d}/\mathrm{d}t\, A(t) = k_{SA}\, SLA(t) + k_{NA}\, N(t) - k_{AG}\, A(t), \\ \mathrm{d}/\mathrm{d}t\, G(t) = k_{AG}\, A(t). \end{cases} \qquad (3.2)$$

Model C:

$$\begin{cases} \mathrm{d}/\mathrm{d}t\, SLA(t) = -(k_{SN} + k_{SA})\, SLA(t), \\ \mathrm{d}/\mathrm{d}t\, N(t) = k_{SN}\, SLA(t) - (k_{NA} + k_{NG})\, N(t), \\ \mathrm{d}/\mathrm{d}t\, A(t) = k_{SA}\, SLA(t) + k_{NA}\, N(t) - k_{AG}\, A(t), \\ \mathrm{d}/\mathrm{d}t\, G(t) = k_{AG}\, A(t) + k_{NG}\, N(t). \end{cases} \qquad (3.3)$$

Model D:

$$\begin{cases} \mathrm{d}/\mathrm{d}t\, SLA(t) = -(k_{SN} + k_{SA})\, SLA(t), \\ \mathrm{d}/\mathrm{d}t\, N(t) = k_{SN}\, SLA(t), \\ \mathrm{d}/\mathrm{d}t\, A(t) = k_{SA}\, SLA(t) - k_{AG}\, A(t), \\ \mathrm{d}/\mathrm{d}t\, G(t) = k_{AG}\, A(t). \end{cases} \qquad (3.4)$$

Model E:

$$\begin{cases} \mathrm{d}/\mathrm{d}t\, SLA(t) = (k_{SS} - k_{SN})\, SLA(t), \\ \mathrm{d}/\mathrm{d}t\, N(t) = k_{SN}\, SLA(t) - k_{NA}\, N(t), \\ \mathrm{d}/\mathrm{d}t\, A(t) = k_{NA}\, N(t) - k_{AG}\, A(t), \\ \mathrm{d}/\mathrm{d}t\, G(t) = k_{AG}\, A(t). \end{cases} \qquad (3.5)$$

In the above equations, k_{SN}, k_{NA}, k_{AG}, k_{SA}, k_{NG}, and k_{SS} are the kinetic parameters between the molecules. Notice that the relationships between the molecules in the actual kinetics cannot be expressed by the above equations. In the actual case, some relationships are non-linear, such as the well-known Michaelis–Menten kinetics in enzyme reactions. In the present study, we have adopted the relationships between molecules as typical ones, but do not consider the details of the kinetics between molecules.

According to the definitions in §2.2, we transform the above systems of differential equations of (3.1)–(3.5) into the system of algebraic equations over the Laplace domain, and solve the equations for the five models. For instance, the solution to the system of differential equations for Model A is expressed over the Laplace domain, as follows:

$$
\begin{cases}
L[SLA(t)](s) = \dfrac{SLA(0)}{s + k_{SN}}, \\[2mm]
L[N(t)](s) = \dfrac{N(0)\,s + N(0)\,k_{SN} + k_{SN}\,SLA(0)}{s^2 + (k_{SN} + k_{NA})\,s + k_{NA}\,k_{SN}}, \\[2mm]
L[A(t)](s) = (A(0)\,s^2 + (k_{NA}\,N(0) + A(0)\,k_{SN} + A(0)\,k_{NA})\,s + k_{NA}\,N(0)\,k_{SN} \\
\qquad + k_{NA}\,k_{SN}\,SLA(0) + A(0)\,k_{NA}\,k_{SN})/(s^3 + (k_{AG} + k_{SN} + k_{NA})\,s^2 + (k_{AG}\,k_{SN} \\
\qquad + k_{AG}\,k_{NA} + k_{NA}\,k_{SN})\,s + k_{AG}\,k_{NA}\,k_{SN}), \\[2mm]
L[G(t)](s) = (G(0)\,s^3 + (G(0)\,k_{AG} + G(0)\,k_{NA} + k_{AG}\,A(0) + G(0)\,k_{SN})\,s^2 \\
\qquad + (G(0)\,k_{AG}\,k_{NA} + k_{AG}\,k_{NA}\,N(0) + G(0)\,k_{AG}\,k_{SN} + k_{AG}\,A(0)\,k_{NA} \\
\qquad + k_{AG}\,A(0)\,k_{SN} + G(0)\,k_{NA}\,k_{SN})s + k_{AG}\,k_{NA}\,N(0)\,k_{SN} + k_{AG}\,k_{NA}\,k_{SN}\,SLA(0) \\
\qquad + G(0)\,k_{AG}\,k_{NA}\,k_{SN} + k_{AG}\,A(0)\,k_{NA}\,k_{SN})/(s^4 + (k_{AG} + k_{SN} + k_{NA})\,s^3 \\
\qquad + (k_{AG}\,k_{SN} + k_{AG}\,k_{NA} + k_{NA}\,k_{SN})\,s^2 + s\,k_{AG}\,k_{NA}\,k_{SN}).
\end{cases}
$$

$$(3.6)$$

In the above equations, the initial values for each molecule are denoted by $SLA(0)$, $N(0)$, $A(0)$, and $G(0)$.

Sampling Data Fitting. To estimate the consistency of the above equations derived from the models with the data, we should presuppose the equations for the sampling data. For this purpose, first, a series of exponentials with parameters are set. For instance, the equations for fitting to the data in Model A are expressed as follows:

$$
\begin{cases}
SLA_O(t) = \beta_{SLA,1}\exp(-\alpha_{SLA,1}t), \\
N_O(t) = \beta_{N,1}\exp(-\alpha_{N,1}t) + \beta_{N,2}\exp(-\alpha_{N,2}t), \\
A_O(t) = \beta_{A,1}\exp(-\alpha_{A,1}t) + \beta_{A,2}\exp(-\alpha_{A,2}t) + \beta_{A,3}\exp(-\alpha_{A,3}t), \\
G_O(t) = \beta_{G,1}\exp(-\alpha_{G,1}t) + \beta_{G,2}\exp(-\alpha_{G,2}t) + \beta_{G,3}\exp(-\alpha_{G,3}t) + \beta_{G,4}.
\end{cases}
$$

$$(3.7)$$

Then, the corresponding algebraic equations are obtained by the Laplace transformation. The corresponding algebraic equations in Model A are as follows:

$$
\begin{cases}
L[SLA_O(t)](s) = \dfrac{\beta_{SLA,1}}{s + \alpha_{SLA,1}}, \\[2mm]
L[N_O(t)](s) = \dfrac{\beta_{N,1}}{s + \alpha_{N,1}} + \dfrac{\beta_{N,2}}{s + \alpha_{N,2}}, \\[2mm]
L[A_O(t)](s) = \dfrac{\beta_{A,1}}{s + \alpha_{A,1}} + \dfrac{\beta_{A,2}}{s + \alpha_{A,2}} + \dfrac{\beta_{A,3}}{s + \alpha_{A,3}}, \\[2mm]
L[G_O(t)](s) = \dfrac{\beta_{G,1}}{s + \alpha_{G,1}} + \dfrac{\beta_{G,2}}{s + \alpha_{G,2}} + \dfrac{\beta_{G,3}}{s + \alpha_{G,3}} + \dfrac{\beta_{G,4}}{s}.
\end{cases}
$$

$$(3.8)$$

Notice that the parameters in the above equations are estimated by numerically fitting them to the data.

3.2   Estimation of Model Consistency

Data Generation for Simulation. In the present study, we have no actual data for the molecules in the models, and thus we need to generate the time series of data for the constituent molecules for the simulation study, before the model consistency estimation. Notice that, if the data for the constituent molecules in the models are actually observed, then this process is not necessary. First, the system of differential equations of (3.1)–(3.5) is solved over the time domain. For instance, the solution of the Model A is expressed as follows:

$$
\begin{cases}
SLA(t) = SLA(0)\ \exp(-k_{SN}\,t), \\[4pt]
N(t) = \left(N(0) - \dfrac{SLA(0)\,k_{SN}}{k_{NA} - k_{SN}}\right)\exp(-k_{NA}\,t) + \dfrac{SLA(0)\,k_{SN}}{k_{NA} - k_{SN}}\exp(-k_{SN}\,t), \\[10pt]
A(t) = \dfrac{k_{NA}\,(k_{SN}\,SLA(0) - k_{NA}\,N(0) + N(0)\,k_{SN})}{(k_{NA} - k_{SN})(k_{NA} - k_{AG})}\exp(-k_{NA}\,t) \\[10pt]
\qquad + \dfrac{k_{NA}\,k_{SN}\,SLA(0)}{(k_{AG} - k_{SN})(k_{NA} - k_{SN})}\exp(-k_{SN}\,t) \\[10pt]
\qquad + \left(A(0) + \dfrac{k_{NA}\,(-k_{SN}\,SLA(0) + k_{AG}\,N(0) - N(0)\,k_{SN})}{(k_{AG} - k_{SN})(k_{NA} - k_{AG})}\right)\exp(-k_{AG}\,t), \\[10pt]
G(t) = \dfrac{k_{AG}\,(-k_{SN}\,SLA(0) + k_{NA}\,N(0) - N(0)\,k_{SN})}{(k_{NA} - k_{AG})(k_{NA} - k_{SN})}\exp(-k_{NA}\,t) \\[10pt]
\qquad + \dfrac{k_{AG}\,(-k_{NA} + k_{AG})\,SLA(0)\,k_{NA}}{(k_{NA} - k_{AG})(k_{AG} - k_{SN})(k_{NA} - k_{SN})}\exp(-k_{SN}\,t) \\[10pt]
\qquad + \left(-A(0) + \dfrac{(k_{SN}\,SLA(0) - k_{AG}\,N(0) + N(0)\,k_{SN})\,k_{NA}}{(k_{NA} - k_{AG})(k_{AG} - k_{SN})}\right)\exp(-k_{AG}\,t) \\[10pt]
\qquad + SLA(0) + N(0) + A(0) + G(0).
\end{cases}
$$
$$(3.9)$$

In the above equations, we have no information about the actual values of the kinetic parameters and their initial values. Thus, we set them as follows: $k_{SN} = 1$, $k_{NA} = 0.1$, $k_{AG} = 0.5$, $k_{NG} = 0.2$, $k_{SA} = 0.4$, and $k_{SS} = 0.7$ for the kinetic parameters, and $SLA(0) = 10$, $N(0) = 7$, $A(0) = 3$, and $G(0) = 1$ for the initial values. By using the above values, the differential equations of (3.9) are simulated from $t = 0$ to 100 with intervals of 1. Then, we obtain the time series of data for each molecule at 101 sample points. We then numerically estimate the parameters by fitting the equations of (3.7) over the time domain to the above-generated data by the Maple 10 Global Optimization tool (©MapleSoft). In Fig. 4, the sampling data at 101 points and the corresponding equations (fitted curve) are plotted in Model A, together with the given and estimated parameters. Notice that, besides the estimation, all of the parameters in (3.7) can be exactly obtained from the given values for the kinetic parameters and the initial values in (3.9). In the present case, it is natural that the estimated values of the parameters are quite consistent with the given values of the parameters for generating the data.

Consistency Measure. As the first step for the model consistency estimation, we construct a set of polynomials, *comp*, from the algebraic equations of (3.6) for the models and those of (3.8) for the sampling data. The following equations are *comp* for Model A:

$$comp = \{ k_{SN} - \alpha_{SLA,1},$$
$$k_{NA} k_{SN} - \alpha_{N,1} \alpha_{N,2},$$
$$k_{SN} + k_{NA} - \alpha_{N,2} - \alpha_{N,1},$$
$$N(0) k_{SN} + k_{SN} S LA(0) - \beta_{N,1} \alpha_{N,2} - \beta_{N,2} \alpha_{N,1},$$
$$k_{AG} + k_{SN} + k_{NA} - \alpha_{A,1} - \alpha_{A,3} - \alpha_{A,2},$$
$$k_{AG} k_{SN} + k_{AG} k_{NA} + k_{NA} k_{SN} - \alpha_{A,1} \alpha_{A,3} - \alpha_{A,1} \alpha_{A,2} - \alpha_{A,2} \alpha_{A,3},$$
$$k_{NA} N(0) + A(0) k_{SN} + A(0) k_{NA} - \beta_{A,1} \alpha_{A,3} - \beta_{A,1} \alpha_{A,2} - \beta_{A,2} \alpha_{A,3} - \beta_{A,2} \alpha_{A,1}$$
$$-\beta_{A,3} \alpha_{A,2} - \beta_{A,3} \alpha_{A,1},$$
$$k_{NA} N(0) k_{SN} + k_{NA} k_{SN} S LA(0) + A(0) k_{NA} k_{SN} - \beta_{A,1} \alpha_{A,2} \alpha_{A,3} - \beta_{A,2} \alpha_{A,1} \alpha_{A,3}$$
$$-\beta_{A,3} \alpha_{A,1} \alpha_{A,2},$$
$$k_{AG} + k_{SN} + k_{NA} - \alpha_{G,2} - \alpha_{G,1} - \alpha_{G,3},$$
$$k_{AG} k_{NA} k_{SN} - \alpha_{A,1} \alpha_{A,2} \alpha_{A,3}, k_{AG} k_{NA} k_{SN} - \alpha_{G,1} \alpha_{G,2} \alpha_{G,3},$$
$$k_{AG} k_{SN} + k_{AG} k_{NA} + k_{NA} k_{SN} - \alpha_{G,1} \alpha_{G,2} - \alpha_{G,2} \alpha_{G,3} - \alpha_{G,1} \alpha_{G,3},$$
$$k_{AG} k_{NA} N(0) k_{SN} + k_{AG} k_{NA} k_{SN} S LA(0) + G(0) k_{AG} k_{NA} k_{SN} + k_{AG} A(0) k_{NA} k_{SN}$$
$$-\beta_{G,4} \alpha_{G,1} \alpha_{G,2} \alpha_{G,3},$$
$$G(0) k_{AG} k_{NA} + k_{AG} k_{NA} N(0) + G(0) k_{AG} k_{SN} + k_{AG} A(0) k_{NA} + k_{AG} A(0) k_{SN}$$
$$+G(0) k_{NA} k_{SN} - \beta_{G,3} \alpha_{G,1} \alpha_{G,2} - \beta_{G,1} \alpha_{G,2} \alpha_{G,3} - \beta_{G,2} \alpha_{G,1} \alpha_{G,3}$$
$$-\beta_{G,4} \alpha_{G,1} \alpha_{G,3} - \beta_{G,4} \alpha_{G,2} \alpha_{G,3} - \beta_{G,4} \alpha_{G,1} \alpha_{G,2},$$
$$G(0) k_{AG} + G(0) k_{NA} + k_{AG} A(0) + G(0) k_{SN} - \beta_{G,2} \alpha_{G,1} - \beta_{G,2} \alpha_{G,3} - \beta_{G,3} \alpha_{G,1}$$
$$-\beta_{G,3} \alpha_{G,2} - \beta_{G,1} \alpha_{G,3} - \beta_{G,4} \alpha_{G,1} - \beta_{G,4} \alpha_{G,2} - \beta_{G,4} \alpha_{G,3} - \beta_{G,1} \alpha_{G,2}\}.$$

In the *comp*, the parameters and the initial values can be expressed as numerical values by the sample data fitting. Thus, only the set of kinetic parameters in the model remains as the unknown parameters in the *comp*. In the following section, we will estimate the kinetic parameters under the constraints in equations (2.4) and (2.6), and will select the model by considering the smallest value of $g(\vec{k})$, the sum-square value of the elements in *comp*.

Model Selection. The model selections by estimating the consistency of the models with the simulated data under the two constraints of equations (2.4) and (2.6) are shown in Table 1. In the first column, the query models, from which the simulated data are generated, are listed, and the models with consistencies that are estimated for the query model are listed in the second column. In the following column, the smallest values of the consistency measure are sorted in ascending order, and the corresponding kinetic measures are listed. As easily seen in this table, the present method has successfully identified the query models. Indeed, all of the models and four of the five models under the two constraints of (2.4) and (2.6) are correctly selected in Table 1, respectively. In addition to the successful selection, the characteristic features for the model selection are observed in the selections by the two constraints. The details of the features are as follows.

As for the selection under the constraint of (2.4), all of the models are clearly selected. By each query model, the corresponding models show the smallest consistency measure ($CM1$) in the constraint of (2.4). For example, when the query model is Model A, the corresponding value for the model consistency for Model A is 1.34×10^{-11}, which

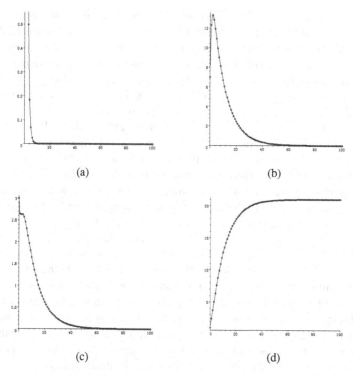

Fig. 4. Sample data for numerical fitting (circles), together with fitted curves (solid lines). The data were generated by numerical calculation from the differential equations (3.9), and the curves were fitted by commercial software (see details in the text). The given and estimated parameters are as follows: $\alpha_{SLA,1}$, 1 (given) and 1.00 (estimated); $\beta_{SLA,1}$, 10 and 10.0; $\alpha_{N,1}$, 1/10 and 0.100; $\alpha_{N,2}$, 1 and 1.00; $\beta_{N,1}$, 163/9 and 18.1; $\beta_{N,2}$, 100/9 and 11.1; $\alpha_{A,1}$, 1/10 and 0.100; $\alpha_{A,2}$, 1/2 and 0.500; $\alpha_{A,3}$, 1 and 1.00; $\beta_{A,1}$, 163/36 and 4.53; $\beta_{A,2}$, $-15/4$ and -3.75; $\beta_{A,3}$, 20/9 and 2.22; $\alpha_{G,1}$, 1/10 and 0.100; $\alpha_{G,2}$, 1/2 and 0.500; $\alpha_{G,3}$, 1 and 1.00; $\beta_{G,1}$, $-815/36$ and -22.6; $\beta_{G,2}$, 15/4 and 3.75; $\beta_{G,3}$, $-10/9$ and -1.11; $\beta_{G,4}$, 21 and 21.0. Each figure corresponds to the four variables (molecules) in the model: (a) SLA, (b) N, (c) A, (d) G.

is the smallest among the values of the five models. The magnitude is slightly smaller than 1.36×10^{-11} for Model E. Interestingly, the parameter value for k_{SS} in Model E is estimated to be nearly zero, 1.40×10^{-6}, and when k_{SS} is zero, Model E is identical to Model A. In the remaining models, the parameters cannot be estimated under the constraint of (2.4). In the other query models, the model corresponding to the query model shows the smallest values for the model consistency, and the remaining models show relatively large values or no values, due to the constraint of (2.4). In particular, Model E, in which a cyclic relationship is included, is successfully selected from the other models, especially Model A, which differs from Model E, only in the cyclic part. Furthermore, in all cases, the values of the kinetic parameters are estimated to be equal to the values that are set for the data generation. Thus, the model selection by using the constraint of (2.4) has completely succeeded in all of the models.

Four of the five models are successfully selected under the constraint of (2.6). In the model selection for Model A, Model C is selected. However, Models C, A, E, and B show small values for the consistency measure ($CM2$). Furthermore, three models, Models C, D, and B, become the same form as Model A, by considering the values of the kinetic parameters. Indeed, k_{NG} and k_{SA} in Model C and k_{SA} in Model B are estimated to be exactly zero values, and k_{SS} in Model E is estimated to be a very small value, 1.40×10^{-6}. A similar situation is also found when the query model is Model B. In this case, while the model showing the smallest value is Model B, a similar value is also found in Model C. However, the value of k_{NG} is estimated to be exactly zero, and this indicates that Model C, with the estimated values for kinetic parameters, is the same form as Model B. Thus, the constraint of (2.6) effectively excludes the false relationship between the molecules by estimating the values of the kinetic parameters. As for the model selection for Model E, the small value appears only in the query model, and the relatively large values appear in the other models. In the models with the large values, the $CM2$ values in Models A, B, and C are relatively smaller than the $CM2$ value in Model D. Interestingly, the former models share common chain relationships between SLA, N, A, and G with Model E, as seen in Fig 3, while the latter model is a distinctive form from Model E. Even in the inconsistent models, $CM2$ may reflect the similarity of the model form between the query and the estimated models. At any rate, the model selection under the constraint of (2.6) also has succeeded in all of the models.

In summary, the present model selection algorithm shows high performance under the constraints of both (2.4) and (2.6). The constraint of (2.4) focuses on only the selection of a model consistent with the data by a simple algorithm, and the constraint of (2.6) focuses on finer model selection, with the exclusion of false relationships, by a slightly and complicated algorithm. Thus, the algorithm with the constraint of (2.4) is useful to select a model consistent with the data among many candidate models, and that with the constraint of (2.6) is effective to select a model among the candidate models including similar forms.

3.3 Discussion

We have proposed a method for selecting a model that is the most consistent with the data in the present study. In small but distinctive networks, our algorithm has successfully selected the query model, from which the sampling data are generated. The present study partly exploits the previous studies of Cobelli et al. [5, 6] about the relationship between observational parameters and model parameters over the Laplace domain. In these studies, they dealt with the case of differential equations adjoined by a higher dimensional ideal to survey whether the model parameters themselves can be determined uniquely or non-uniquely. In our work, the combination of the transformation of equations over the Laplace domain with the numerical fitting to the observed data enables us to estimate the model's consistency with the data as well as with the values of the kinetic parameters. Although the robustness for data including noise should be further tested, our algorithm is expected to be feasible for actual biological issues regarding the selection of a kinetics model.

The scalability of the present algorithm also remains to be tested. Actually, the present model selection algorithm required several hours for one model. In addition,

Table 1. Model selections. The five models in Fig. 3 were examined for the model selection and the determination of kinetic parameters with the simulated data by the two constraints (see the details in the text). The 'query' and 'estimated' indicate the model from which the simulated data are generated, and the model the consistency of which is estimated by the corresponding query model, respectively.

$CM1$, under the constraint (2.4)

Query	estimated	smallest	k_{SN}	k_{NA}	k_{AG}	k_{NG}	k_{SA}	k_{SS}
A	A	1.34×10^{-11}	1.00	0.100	0.500	-	-	-
	E	1.36×10^{-11}	1.00	0.100	0.500	-	-	1.40×10^{-6}
	D	×	×	-	×	-	×	-
	B	×	×	×	×	-	×	-
	C	×	×	×	×	×	×	-
B	B	4.20×10^{-11}	1.00	0.100	0.500	-	0.400	-
	A	20.1	1.19	0.167	0.637	-	-	-
	D	×	×	-	×	-	×	-
	E	×	×	×	×	-	-	×
	C	×	×	×	×	×	-	-
C	C	2.78×10^{-9}	1.00	0.100	0.500	0.200	0.400	-
	B	0.558	0.994	0.160	0.913	-	0.408	-
	A	28.0	1.19	0.213	1.17	-	-	-
	D	×	×	-	×	-	×	-
	E	×	×	×	×	-	×	×
D	D	1.83×10^{-14}	1.00	-	0.500	-	0.400	-
	A	576	1.02	3.98	0.623	-	-	-
	E	×	×	×	×	-	-	×
	B	×	×	×	×	-	×	-
	C	×	×	×	×	×	×	-
E	E	9.26×10^{-11}	1.00	0.100	0.500	-	-	0.700
	A	1.46	0.702	0.0564	0.367	-	-	-
	D	×	×	-	×	-	×	-
	B	×	×	×	×	-	×	-
	C	×	×	×	×	×	×	-

$CM2$, under the constraint (2.6)

Query	estimated	smallest	k_{SN}	k_{NA}	k_{AG}	k_{NG}	k_{SA}	k_{SS}
A	C	7.66×10^{-12}	1.00	0.100	0.500	0*	0*	-
	A	1.34×10^{-11}	1.00	0.100	0.500	-	-	-
	E	1.36×10^{-11}	1.00	0.100	0.500	-	-	1.40×10^{-6}
	B	1.35×10^{-10}	1.00	0.100	0.500	-	0*	-
	D	1.68	0*	-	0.435	-	0.0439	-
B	B	4.20×10^{-11}	1.00	0.100	0.500	-	0.400	-
	C	6.44×10^{-11}	1.00	0.100	0.500	0*	0.400	0*
	E	20.1	1.19	0.167	0.637	-	-	-
	A	20.1	1.19	0.167	0.637	-	-	-
	D	1050	0*	-	0.0351	-	1.97	-
C	C	2.78×10^{-9}	1.00	0.100	0.500	0.200	0.400	-
	B	0.558	0.994	0.160	0.913	-	0.408	-
	D	23.9	0*	-	1.22	-	0.418	-
	E	28.0	1.19	0.213	1.17	-	-	0*
	A	28.0	1.19	0.213	1.17	-	-	-
D	D	1.83×10^{-14}	1.00	-	0.500	-	0.400	-
	E	358.	1.13	3.63	0.285	-	-	-
	B	399.	1.10	3.74	0.395	-	0*	-
	C	434.	1.18	3.43	0.454	0.528	0*	-
	A	576.	1.02	3.98	0.623	-	-	-
E	E	9.26×10^{-11}	1.00	0.100	0.500	-	-	0.700
	C	1.46	0.702	0.0564	0.367	0*	-	-
	A	1.46	0.702	0.0564	0.367	-	0*	-
	B	1.46	0.702	0.0564	0.367	-	0*	-
	D	2.57	0*	-	0.258	-	0.0284	-

0*: *exact* zero value.

-: no corresponding parameters.

×: no real positive solutions.

the limit of the nodes and edges in the tested network approximately ranged within 10 edges between 10 nodes. However, the present algorithm over the Laplace domain may overcome the issue of scalability. In a local network within a large-scale network, the relationships of the molecules in the local network with those outside of it are regarded as inputs from the outside, and the variables corresponding to the inputs may easily be eliminated, if the relationships are treated over the Laplace domain. Indeed, we have successfully eliminated the *unnecessary* variables to estimate the parameter values in complex compartmental models for Parkinson's disease by PET measurements [13]. If the *unnecessary* variables in the local network can be eliminated, then the present algorithm can be applied to estimate the model's consistency. Thus, the iteration of the elimination and the consistency estimation may be applicable for the consistency estimation, even in a large-scale network model. Further examinations of the present algorithm for a large-scale network and for noisy data will appear in the near future.

4   Conclusion

In the present model selection, an algebraic manipulation of the differential equations over the Laplace domain, formulated based on the assumption of linear relationships between the variables, is combined with the numerical fitting of the sampling data. The performance of our approach is illustrated with simulated data, in the distinctive forms of models, one of which includes a cyclic relationship hitherto unavailable in previous methods. Although some further examinations of the present method are necessary, especially of the analyzed data and its robustness with noise, the extension of our approach to a large-scale network is promising.

Acknowledgments

H. Y. and K. H. were partly supported by a Grant-in-Aid for Scientific Research on Priority Areas "Systems Genomics" (grant 18016008), by a Grant-in-Aid for Scientific Research (grant 19201039) and by a Grant-in-Aid for Young Scientists (B) (grant 19790881) from the Ministry of Education, Culture, Sports, Science and Technology of Japan (MEXT). This study was supported in part by the New Energy and Industrial Technology Development Organization (NEDO) of Japan, by Core Research for Evolutional Science and Technology (CREST), by Program for Improvement of Research Environment for Young Researchers from Special Coordination Funds for Promoting Science and Technology (SCF) commissioned by Japan Science and Technology Agency (JST,MEXT).

References

[1] Audoly, S., D'Angiò, L., Saccomani, M.P., Cobelli, C.: Global identifiability of linear compartmental models — A computer algebra algorithm. IEEE Trans. Biomed. Eng. 45, 36–47 (1998)
[2] Bisits, A.M., Smith, R., Mesiano, S., Yeo, G., Kwek, K., MacIntyre, D., Chan, E.C.: Inflammatory aetiology of human myometrial activation tested using directed graphs. PLoS Comput. Biol. 1, 132–136 (2005)

[3] Buchberger, B.: An Algorithmic Criterion for the Solvability of a System of Algebraic Equations. In: Buchberger, B., Winkler, F. (eds.) Gröbner Bases and Applications. London Mathematical Society Lecture Notes Series, vol. 251, pp. 535–545. Cambridge University Press, Cambridge (1998)

[4] Calvano, S.E., Xiao, W., Richards, D.R., Felciano, R.M., Baker, H.V., Cho, R.J., Chen, R.O., Brownstein, B.H., Cobb, J.P., Tschoeke, S.K., Miller-Graziano, C., Moldawer, L.L., Mindrinos, M.N., Davis, R.W., Tompkins, R.G., Lowry, S.F.: Inflammation and Host Response to Injury Large Scale Collab. Res. Program: A network-based analysis of systemic inflammation in humans. Nature 437, 1032–1037 (2005)

[5] Cobelli, C., Foster, D., Toffolo, G.: Tracer Kinetics in Biomedical Research: From Data to Model. Kluwer Academic/Plenum Publishers (2000)

[6] Cobelli, C., Toffolo, G.: Theoretical aspects and practical strategies for the identification of unidentifiable compartmental systems. ch. 8, pp. 85–91. Pergamon Press, Oxford (1987)

[7] Hanzon, B., Jibetean, D.: Global minimization of a multivariate polynomial using matrix methods. Journal of Global Optimization 27, 1–23 (2003)

[8] Joreskog, K.G.: A general method for analysis of covariance structures. Biometrika 57, 239–251 (1970)

[9] Meziane, D., Shipley, B.: Direct and Indirect Relationships Between Specific Leaf Area, Leaf Nitrogen and Leaf Gas Exchange. Effects of Irradiance and Nutrient Supply. Annals of Botany 88, 915–927 (2001)

[10] Pearl, J.: Probabilistic Reasoning in Intelligent Systems: Networks of Plausible Inference. Morgan Kaufmann Publishers, San Francisco (1988)

[11] Shipley, B.: A new inferential test for path models based on directed acyclic graphs. Structural Equation Modeling 7, 206–218 (2000)

[12] Wright, S.: The method of path coefficients. Ann. Math. Statist. 5, 161–215 (1934)

[13] Yoshida, H., Nakagawa, K., Anai, H., Horimoto, K.: Exact parameter determination for Parkinson's disease diagnosis with PET using an algebraic approach. In: Anai, H., Horimoto, K., Kutsia, T. (eds.) Algebraic Biology 2007. LNCS, vol. 4545, pp. 110–124. Springer, Heidelberg (2007)

On the Representation of the Differential Operator in Bases of Periodic Coiflets and It's Application

Anna Deytseva

Grodno State University, Grodno, Belarus

Abstract. In the present paper multiresolutional representation of differential operator in basis of periodized coiflets was constructed. The properties of differential operator coefficients have been investigated. Taking into consideration the behavior of Coifman scaling function, in considered multiresolutional representation of differential operator, coefficients of wavelet-approximation are substituted by evaluations of the function at dyadic points. For sufficiently smooth function the convergence rate of the considered approximations to derivative is stated. For 1-periodic function the formula of numeric differentiation is obtained, and also the error estimate is stated. The application of multiresolution representation of differential operator for numerical solution of ordinary differential equation with periodic boundary conditions is considered.

1   Introduction

The theory of wavelet analysis has grown explosively in the last two decades. The terminology "wavelet" was first introduced in 1984 by A. Grossmann and J. Morlet. Families of functions

$$\psi_{j,k}(x) = 2^{j/2} \psi \left(2^j x - k \right),$$

$j, k \in \mathbb{Z}$, derived from a single function $\psi(x)$ by dilation and translation, which form a basis for $L_2(\mathbb{R})$, are called wavelets.

In 1988, I. Daubechies made an important contribution in wavelet theory. She introduced a class of compactly supported orthonormal wavelet systems with vanishing moments for the wavelet function. Her work has stimulated further study of wavelets and it's applications, and, in particular, application of wavelets in calculus of approximations theory. Thus, in work of G. Beylkin, R. Coifman and V. Rokhlin the fast algorithms of estimating some operators (including differentiatial operator) application on arbitrary function were obtained [1]. At that, matrix representation of the operator and vector representation of function in wavelet basis were used. In the consequent work G. Beylkin has considered in detail the task of building matrix representation of operators in Daubechies wavelet basis [2]. Unlike Daubechies wavelets, Coifman wavelet system have vanishing moments not only for the wavelet functions, but also for

V.G. Ganzha, E.W. Mayr, and E.V. Vorozhtsov (Eds.): CASC 2007, LNCS 4770, pp. 448–457, 2007.

the scaling functions [3,4]. The investigation of the approximation of differential operator in Coifman wavelet basis was conducted in [5]. But for the study of periodic functions use of periodic wavelets is quite natural. The goal of this article is to research the approximation of differential operator in basis of periodized coiflets, and it's application for numerical solution of differantial equation.

2   Periodized Wavelets

In [4] it was shown that on the basis of scaling function φ and corresponding wavelet ψ, generating multiresolution analysis on space $L_2(\mathbb{R})$, multiresolutional representation for the space $L_2([0,1])$ can be obtained.

Multiresolution analysis on the space $L_2([0,1])$ is a chain of closed subspaces

$$V_0^{per} \subset V_1^{per} \subset \ldots \subset L_2([0,1]),$$

such that

$$L_2([0,1]) = \overline{\bigcup_{j \in \mathbb{N}_0} V_j^{per}},$$

$\mathbb{N}_0 = \{0,1,2,\ldots\}$.

By defining W_j^{per} as an orthogonal complement of V_j^{per} in V_{j+1}^{per}:

$$V_{j+1}^{per} = V_j^{per} \oplus W_j^{per},$$

the space $L_2([0,1])$ is represented as a direct sum

$$L_2([0,1]) = V_J^{per} \overset{+\infty}{\underset{j=J}{\oplus}} W_j^{per},$$

$J \in \mathbb{N}_0$.

At this orthonormal basis of the scaling space V_j^{per} is formed by the system of functions $\left\{\varphi_{j,k}^{per}\right\}_{k=0}^{2^j-1}$

$$\varphi_{j,k}^{per}(x) = \sum_{l \in \mathbb{Z}} \varphi_{j,k}(x+l), \tag{1}$$

$$\varphi_{j,k} = 2^{j/2}\varphi\left(2^j x - k\right). \tag{2}$$

Orthonormal basis of the detailing space W_j^{per} is formed by the system of functions $\left\{\psi_{j,k}^{per}\right\}_{k=0}^{2^j-1}$

$$\psi_{j,k}^{per}(x) = \sum_{l \in \mathbb{Z}} \psi_{j,k}(x+l),$$

$$\psi_{j,k} = 2^{j/2}\psi\left(2^j x - k\right).$$

3   Operator of Differentiation for Periodic Functions

Let $\left\{V_j^{per}\right\}_{j\in\mathbb{N}_0}$ – multiresolution analysis on $L_2\left([0,1]\right)$, generated by Coifman scaling function $\varphi \in L_2(\mathbb{R})$ of the order $L = 2K$, $K \in \mathbb{N}$, $\psi \in L_2(\mathbb{R})$ – corresponding coiflet. Functions φ and ψ – are of real value, have compact supports and following features [3,4]:

$$\varphi(x) = \sqrt{2} \sum_{k=-L}^{2L-1} h_k \varphi(2x - k), \tag{3}$$

$$\int_{\mathbb{R}} \varphi(x)\, dx = 1,$$

$$\int_{\mathbb{R}} x^l \varphi(x)\, dx = 0, \tag{4}$$

$l = \overline{1, L-1}$;

$$\int_{\mathbb{R}} x^l \psi(x)\, dx = 0,$$

$l = \overline{0, L-1}$.

Let us make a general observation about the representation of differential operator $T = \frac{d^n}{dx^n}$ in wavelet bases. In accordance with conception of multiresolution analysis the arbitrary function $f \in L_2\left([0,1]\right)$ can be represented as limit of successive approximations $P_j^{per} f \in V_j^{per}$, when $j \to +\infty$, defined as following:

$$P_j^{per} f(x) = \sum_{k=0}^{2^j-1} \alpha_{j,k} \varphi_{j,k}^{per}(x), \tag{5}$$

where

$$\alpha_{j,k} = \left\langle f, \varphi_{j,k}^{per} \right\rangle = \int_0^1 f(x)\varphi_{j,k}^{per}(x)\, dx = \int_{\mathbb{R}} f(x)\varphi_{j,k}(x)\, dx, \tag{6}$$

$\varphi_{j,k}^{per}$ and $\varphi_{j,k}$ are given by (1) and (2) correspondingly. Then

$$T P_j^{per} f(x) = \sum_{k=0}^{2^j-1} \alpha_{j,k} T \varphi_{j,k}^{per}(x)$$

and finding projection on the scaling space V_j^{per} we'll obtain

$$T_j^{per} f(x) = P_j^{per} T P_j^{per} f(x) = \sum_{k=0}^{2^j-1} \alpha_{j,k} \sum_{k'=0}^{2^j-1} \left\langle T \varphi_{j,k}^{per}(x), \varphi_{j,k'}^{per}(x) \right\rangle \varphi_{j,k'}^{per}(x) =$$

$$= \sum_{k'=0}^{2^j-1} \left(\sum_{k=0}^{2^j-1} \left\langle T\varphi_{j,k}^{per}(x), \varphi_{j,k'}^{per}(x) \right\rangle \alpha_{j,k} \right) \varphi_{j,k'}^{per}(x),$$

$x \in [0,1]$.

Further on it will be shown that for sufficiently smooth 1-periodic function f the following convergence takes place

$$T_j^{per} f \to Tf, \text{ when } j \to +\infty.$$

In other word, for sufficiently great $j \in \mathbb{N}$ the action of the differential operator T on the function f is translated into the action of the matrix $A_T = \left\{ \left\langle T\varphi_{j,k}^{per}(x), \varphi_{j,k'}^{per}(x) \right\rangle \right\}_{k,k'=0}^{2^j-1}$ on the sequence $\{\alpha_{j,k}\}_{k=0}^{2^j-1}$.

Let consider the coefficients of the matrix A_T

$$\left\langle T\varphi_{j,k}^{per}(x), \varphi_{j,k'}^{per}(x) \right\rangle = \int_0^1 T\varphi_{j,k}^{per}(x)\varphi_{j,k'}^{per}(x)\, dx = \int_0^1 \frac{d^n}{dx^n} \left(\varphi_{j,k}^{per}(x) \right) \varphi_{j,k'}^{per}(x)\, dx =$$

$$= \int_0^1 \frac{d^n}{dx^n} \left(\sum_{l \in \mathbb{Z}} \varphi_{j,k}(x+l) \right) \sum_{l' \in \mathbb{Z}} \varphi_{j,k'}(x+l')\, dx =$$

$$= \int_0^1 \frac{d^n}{dx^n} \left(2^{j/2} \sum_{l \in \mathbb{Z}} \varphi\left(2^j x + 2^j l - k\right) \right) 2^{j/2} \sum_{l' \in \mathbb{Z}} \varphi\left(2^j x + 2^j l' - k'\right)\, dx =$$

$$= \int_0^1 2^{jn} 2^j \sum_{l \in \mathbb{Z}} \varphi^{(n)}\left(2^j x + 2^j l - k\right) \sum_{l' \in \mathbb{Z}} \varphi\left(2^j x + 2^j l' - k'\right)\, dx.$$

Further on we will consequently use argument substitution $y = x + l$, suppose $l' - l = m$ and make substitution $2^j y - k = z$

$$\sum_{l \in \mathbb{Z}} \sum_{l' \in \mathbb{Z}} 2^{jn} 2^j \int_l^{l+1} \varphi^{(n)}\left(2^j y - k\right) \varphi\left(2^j y + 2^j (l' - l) - k'\right)\, dy =$$

$$= 2^{jn} \sum_{m \in \mathbb{Z}} 2^j \sum_{l \in \mathbb{Z}} \int_l^{l+1} \varphi^{(n)}\left(2^j y - k\right) \varphi\left(2^j y + 2^j m - k'\right)\, dy =$$

$$= 2^{jn} \sum_{m \in \mathbb{Z}} 2^j \int_{\mathbb{R}} \varphi^{(n)}\left(2^j y - k\right) \varphi\left(2^j y + 2^j m - k'\right)\, dy =$$

$$= 2^{jn} \sum_{m \in \mathbb{Z}} \int_{\mathbb{R}} \varphi^{(n)}(z) \varphi \left(z + 2^j m - k' + k \right) dz = 2^{jn} \sum_{m \in \mathbb{Z}} r^{(n)}_{k'-k-2^j m},$$

where

$$r^{(n)}_m = \left\langle \varphi(x - m), \varphi^{(n)}(x) \right\rangle = \int_{\mathbb{R}} \varphi(x - m) \varphi^{(n)}(x) \, dx, \tag{7}$$

$k, k' = \overline{0, 2^j - 1}$, $m \in \mathbb{Z}$, $j \in \mathbb{N}_0$.

Thus, for arbitrary function $f \in L_2([0,1])$ operator $T_j^{per} f$ is defined by relation

$$T_j^{per} f(x) = 2^{jn} \sum_{k'=0}^{2^j-1} \sum_{k=0}^{2^j-1} \alpha_{j,k} r^{(n),per}_{k'-k} \varphi^{per}_{j,k'}(x), \tag{8}$$

$$r^{(n),per}_l = \sum_{m \in Z} r^{(n)}_{l-2^j m}, \tag{9}$$

$l = \overline{-2^j + 1, 2^j - 1}$, $j \in \mathbb{N}_0$, $x \in [0,1]$, coefficients of wavelet-approximation $\alpha_{j,k}$ calculated via formulas (6).

As Coifman scaling function has no analytical definition, direct calculation of the differential operator coefficients via formulas (7) is not possible. In [5] it was proved that coefficients $r^{(n)}_m$ can be obtained via solving system of linear algebraic equations.

Theorem 1. *[5] Let φ – scaling Coifman function of the order $L = 2K$, $K \in \mathbb{N}$, $L \geq n$ and the right hand member integral (7) exists, then coefficients $r^{(n)}_m$, defined by expression (7), satisfy the following set of linear algebraic equations*

$$r^{(n)}_m = 2^n \left(r^{(n)}_{2m} + \frac{1}{2} \sum_{l=1}^{3K} a_{2l-1} \left(r^{(n)}_{2m+2l-1} + r^{(n)}_{2m-2l+1} \right) \right),$$

$m \in \mathbb{Z}$, *where coefficients a_l are calculated via formulas*

$$a_l = \sum_{k=-L}^{2L-1-l} h_k h_{k+l},$$

$l = \overline{1, 3L - 1}$, *coefficients h_k, $k = \overline{-L, 2L - 1}$ satisfy the refinement equation (3).*

Coefficients $r^{(n)}_m$ satisfy the following conditions

$$r^{(n)}_m = 0, \quad |m| > 3L - 2;$$

$$\sum_{m \in \mathbb{Z}} m^l r^{(n)}_m = (-1)^n n! \delta_{l,n},$$

$$\delta_{l,n} = \begin{cases} 0, l \neq n; \\ 1, l = n, \end{cases} \quad l = \overline{0, L-1}.$$

If $\varphi', \ldots, \varphi^{(n)} \in L_1(\mathbb{R})$, then

$$r_m^{(n)} = (-1)^n r_{-m}^{(n)},$$

$m = \overline{1, 3L - 2}$, $n \in \mathbb{N}$.

Coefficients $r^{(n)}$ for Coiflets with 6 vanishing moments for the first and second derivate are given in Table 1. They were calculated using standard functions of MATHEMATICA system, along with functions of package Wavelet Explorer. Table 1 displays only the coefficients $\{r_m^n\}_{m=0}^{16}$ since $r_m^n = (-1)^n r_{-m}^n$.

Table 1. Coefficients $r^{(1)}$ and $r^{(2)}$ for Coiflets with 6 vanishing moments

m	Coefficients $r_m^{(1)}$	Coefficients $r_m^{(2)}$
0	0	-3.6498559820
1	-0.85870133045	2.2864769185
2	0.27014190725	-0.6265697317
3	-0.08137436347	0.2138072082
4	0.01908146761	-0.0585392225
5	-0.00312072218	0.0109121919
6	0.00033051564	-0.1256216893E-2
7	-0.00002564437	0.1075566274E-3
8	1.76453581106E-6	-1.0658859563E-5
9	7.68664190049E-8	-5.0136014077E-8
10	-3.72285469962E-9	-8.7968664067E-9
11	-6.52539388499E-10	4.3680701748E-9
12	-7.26069305181E-12	5.7729729291E-11
13	2.42680606269E-14	-8.1065865439E-14
14	-1.71449741989E-16	2.2943344187E-15
15	6.37055338073E-21	-4.2560564669E-20
16	1.67232995005E-27	-8.8992880760E-25

4   Formula of Numerical Differentiation for Periodic Function

In this section, using multiresolution representation of the differential operator, we obtain the formula of numeric differentiation for 1-periodic function. For the first we will state the convergence rate of the approximations (8) to the n-th derivative of the function f.

Theorem 2. Let φ, ψ – Coifman scaling function and coiflet of the order $L = 2K, K \in \mathbb{N}$ correspondingly, and right hand member integral (7) exists. Let $f \in C^{n+L}(\mathbb{R})$ – 1-periodic function. Then sequence $\{T_j^{per}f\}_{j \in \mathbb{N}_0}$, defined by (8), convergence to $f^{(n)}$ and for every $x \in [0,1]$ the following inequality is valid

$$\left| f^{(n)}(x) - T_j^{per} f(x) \right| \leq 2^{-j(L-n)} \left(C_1 + C_2 \cdot 2^{-jn} \right),$$

where constants C_1 and C_2 do not depend on j; $n < L$, $n \in \mathbb{N}$.

As Coifman scaling function has vanishing moments, coefficients $\alpha_{j,k}$ may be approximated by evaluations of the function $f(x)$ at dyadic points:

$$\alpha_{j,k} \approx 2^{-j/2} f \left(\frac{k}{2^j} \right), \tag{10}$$

$k = \overline{0, 2^j - 1}$, $j \in \mathbb{N}_0$.

Taking this fact into consideration, we'll substitute the approximating coefficients $\alpha_{j,k}$ with the corresponding approximations (10) in defining operator $T_j^{per} f$ and shall define operator $T_j^{per,s} f : L_2([0,1]) \to V_j^{per}$ as following

$$T_j^{per,s} f(x) = 2^{jn} \sum_{k'=0}^{2^j-1} \sum_{k=0}^{2^j-1} s_{j,k} r_{k'-k}^{(n),per} \varphi_{j,k'}^{per}(x), \tag{11}$$

where

$$s_{j,k} = 2^{-j/2} f \left(\frac{k}{2^j} \right),$$

coefficients $r_{k-k'}^{(n),per}$ are given by (9), $j \in \mathbb{N}_0$, $x \in [0,1]$.

Corollary 1. *Let the conditions of the theorem 2 to be satisfied. Then for every $x \in [0,1]$ the following inequality is valid*

$$\left| f^{(n)}(x) - T_j^{per,s} f(x) \right| \leq 2^{-j(L-n)} \left(C_3 + C_4 \cdot 2^{-jn} \right),$$

where operator $T_j^{per,s} f$ is defined by relation (11), $j \in \mathbb{N}_0$; constants C_3 and C_4 do not depend on j; $n < L, n \in \mathbb{N}$.

We have to mention, that n-th derivative of the function f can also be represented as limit of successive approximations $P_j^{per} f$ when $j \to +\infty$, defined by

$$P_j^{per} f^{(n)}(x) = \sum_{k'=0}^{2^j-1} \alpha_{j,k'}^{(n)} \varphi_{j,k}^{per}(x), \tag{12}$$

$$\alpha_{j,k'}^{(n)} = \int_0^1 f^{(n)}(x) \varphi_{j,k'}^{per}(x) \, dx$$

at this for sufficiently smooth function f the coefficients $\alpha_{j,k'}^{(n)}$ may be approximated by evaluations of the function $f^{(n)}(x)$ at dyadic points:

$$\alpha_{j,k}^{(n)} \approx 2^{-j/2} f^{(n)} \left(\frac{k}{2^j} \right).$$

Thus, comparing the approximations (11) and (12) we obtain the following *formula of numerical differentiation for 1-periodic function f*:

$$f^{(n)}\left(\frac{k'}{2^j}\right) \approx 2^{jn} \sum_{k=0}^{2^j-1} f\left(\frac{k}{2^j}\right) r_{k'-k}^{(n),per}, \tag{13}$$

coefficients $r_{k-k'}^{(n),per}$ are given by (9), $k, k' = \overline{0, 2^j - 1}$, $j \in \mathbb{N}_0$.

The error estimate of this approximation is defined with the following corollary.

Corollary 2. *Let the conditions of the theorem 2 to be satisfied, then the following inequality is valid*

$$\left| f^{(n)}\left(\frac{k'}{2^j}\right) - 2^{jn} \sum_{k=0}^{2^j-1} f\left(\frac{k}{2^j}\right) r_{k'-k}^{(n),per} \right| \leq 2^{-j(L-n-1)}\left(C_5 + 2^{-j}C_6 + C_7 2^{-j(n+1)}\right),$$

$k' = \overline{0, 2^j - 1}$, $j \in \mathbb{N}_0$, *coefficients* $r_{k-k'}^{(n),per}$ *are given by (9) constants* C_5, C_6 *and* C_7 *do not depend on* j; $n + 1 < L$, $n \in \mathbb{N}$.

Remark 1. Formula of the numerical differentiation (13) one can rewrite in matrix form

$$F^{(n)} = D^{(n)}F, \tag{14}$$

where

$$F = col\left[f(x_0), f(x_1), \ldots, f(x_{2^j-1})\right],$$

$$F^{(n)} = col\left[f^{(n)}(x_0), f^{(n)}(x_1), \ldots, f^{(n)}(x_{2^j-1})\right],$$

$$x_k = \frac{k}{2^j}, \quad k = \overline{0, 2^j - 1}, \quad j \in \mathbb{N}_0,$$

$$D^{(n)} = \left[d_{k,l}^{(n)}\right]_{k,l=0}^{2^j-1}, \quad d_{k,l}^{(n)} = r_{k-l}^{(n),per}. \tag{15}$$

5   Example

In this section we present the results of numerical experiments in which we compute, using periodic coifelts, solution of the equation

$$y''(x) + p(x)y'(x) + q(x)y(x) = r(x),$$

$x \in [0, 1]$, where $y(x)$ - unknown 1-periodic function.

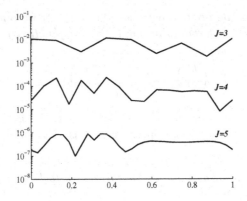

Fig. 1. Relative error of the multiscale solution for the example problem with $J = 3, 4, 5$

Discretizing this problem on a staggered grid using the formula of the numerical differentiation, we obtain the following system of linear algebraic equations

$$BY = R,$$

where

$$B = D^{(2)} + PD^{(1)} + QE,$$

$$Y = col\,[y(x_0), y(x_1), \ldots, y(x_{2^J-1})],$$

$$R = col\,[r(x_0), r(x_1), \ldots, r(x_{2^J-1})],$$

$$PD^{(1)} = \left[p(x_k)\,d^{(1)}_{k,l}\right]_{k,l=0}^{2^J-1}, \quad QE = [q(x_k)\,\delta_{k,l}]_{k,l=0}^{2^J-1},$$

$$\delta_{k,l} = \begin{cases} 0, & k \neq l; \\ 1, & k = l, \end{cases} \quad x_k = \frac{k}{2^J},$$

matrix $D^{(2)}$ and coefficients $d^{(1)}_{k,l}$ are given by the formulas (15). Here $J \in \mathbb{N}_0$ is a fixed scale of the resolution.

We illustrate the accuracy of such a computation by the following example. Let

$$p(x) = x + 1, \qquad q(x) = e^x,$$

$$r(x) = \left(-4\pi^2 \sin 2\pi x + 4\pi^2 \cos^2 2\pi x + 2\pi(x+1) \cos 2\pi x + e^x\right) e^{\sin 2\pi x}.$$

The periodic solution of such equation is

$$y(x) = e^{\sin 2\pi x}.$$

The relative error of the approximate solutions, obtained using Coiflets with 6 vanishing moments, for various scales of the resolution J is shown in Figure 1.

Thus, by numerical experiments we provide results which corroborate the theory.

References

1. Beylkin, G., Coifman, R., Rokhlin, V.: Fast Wavelet Transforms and Numerical Algorithm I. Commun. Pure and Appl. Math. 44, 141–183 (1991)
2. Beylkin, G.: On the representation of operators in bases of compactly supported wavelets. SIAM J. Numer. Anal. 6(6), 1716–1740 (1992)
3. Daubechies, I.: Orthonormal bases of compactly supported wavelets II. Variation on a theme. SIAM J. Math. Anal. 24(2), 499–519 (1993)
4. Daubechies, I.: Ten lectures on wavelets. SIAM, Philadelphia (1992)
5. Deytseva, A.: On the representation of the differential operators in bases of coiflets. In: Proceedings of CASTR 2007, pp. 52–58 (2007)

Author Index

Lecture Notes in Computer Science

Sublibrary 1: Theoretical Computer Science and General Issues

For information about Vols. 1– 4431
please contact your bookseller or Springer

Vol. 4613: F.P. Preparata, Q. Fang (Eds.), Frontiers in Algorithmics. XI, 348 pages. 2007.

Vol. 4600: H. Comon-Lundh, C. Kirchner, H. Kirchner (Eds.), Rewriting, Computation and Proof. XVI, 273 pages. 2007.

Vol. 4599: S. Vassiliadis, M. Berekovic, T.D. Hämäläinen (Eds.), Embedded Computer Systems: Architectures, Modeling, and Simulation. XVIII, 466 pages. 2007.

Vol. 4598: G. Lin (Ed.), Computing and Combinatorics. XII, 570 pages. 2007.

Vol. 4596: L. Arge, C. Cachin, T. Jurdziński, A. Tarlecki (Eds.), Automata, Languages and Programming. XVII, 953 pages. 2007.

Vol. 4595: D. Bošnački, S. Edelkamp (Eds.), Model Checking Software. X, 285 pages. 2007.

Vol. 4590: W. Damm, H. Hermanns (Eds.), Computer Aided Verification. XV, 562 pages. 2007.

Vol. 4588: T. Harju, J. Karhumäki, A. Lepistö (Eds.), Developments in Language Theory. XI, 423 pages. 2007.

Vol. 4583: S.R. Della Rocca (Ed.), Typed Lambda Calculi and Applications. X, 397 pages. 2007.

Vol. 4580: B. Ma, K. Zhang (Eds.), Combinatorial Pattern Matching. XII, 366 pages. 2007.

Vol. 4576: D. Leivant, R. de Queiroz (Eds.), Logic, Language, Information and Computation. X, 363 pages. 2007.

Vol. 4547: C. Carlet, B. Sunar (Eds.), Arithmetic of Finite Fields. XI, 355 pages. 2007.

Vol. 4546: J. Kleijn, A. Yakovlev (Eds.), Petri Nets and Other Models of Concurrency – ICATPN 2007. XI, 515 pages. 2007.

Vol. 4545: H. Anai, K. Horimoto, T. Kutsia (Eds.), Algebraic Biology. XIII, 379 pages. 2007.

Vol. 4533: F. Baader (Ed.), Term Rewriting and Applications. XII, 419 pages. 2007.

Vol. 4528: J. Mira, J.R. Álvarez (Eds.), Nature Inspired Problem-Solving Methods in Knowledge Engineering, Part II. XXII, 650 pages. 2007.

Vol. 4527: J. Mira, J.R. Álvarez (Eds.), Bio-inspired Modeling of Cognitive Tasks, Part I. XXII, 630 pages. 2007.

Vol. 4525: C. Demetrescu (Ed.), Experimental Algorithms. XIII, 448 pages. 2007.

Vol. 4514: S.N. Artemov, A. Nerode (Eds.), Logical Foundations of Computer Science. XI, 513 pages. 2007.

Vol. 4513: M. Fischetti, D.P. Williamson (Eds.), Integer Programming and Combinatorial Optimization. IX, 500 pages. 2007.

Vol. 4510: P. Van Hentenryck, L.A. Wolsey (Eds.), Integration of AI and OR Techniques in Constraint Programming for Combinatorial Optimization Problems. X, 391 pages. 2007.

Vol. 4507: F. Sandoval, A.G. Prieto, J. Cabestany, M. Graña (Eds.), Computational and Ambient Intelligence. XXVI, 1167 pages. 2007.

Vol. 4501: J. Marques-Silva, K.A. Sakallah (Eds.), Theory and Applications of Satisfiability Testing – SAT 2007. XI, 384 pages. 2007.

Vol. 4497: S.B. Cooper, B. Löwe, A. Sorbi (Eds.), Computation and Logic in the Real World. XVIII, 826 pages. 2007.

Vol. 4494: H. Jin, O.F. Rana, Y. Pan, V.K. Prasanna (Eds.), Algorithms and Architectures for Parallel Processing. XIV, 508 pages. 2007.

Vol. 4493: D. Liu, S. Fei, Z. Hou, H. Zhang, C. Sun (Eds.), Advances in Neural Networks – ISNN 2007, Part III. XXVI, 1215 pages. 2007.

Vol. 4492: D. Liu, S. Fei, Z. Hou, H. Zhang, C. Sun (Eds.), Advances in Neural Networks – ISNN 2007, Part II. XXVII, 1321 pages. 2007.

Vol. 4491: D. Liu, S. Fei, Z.-G. Hou, H. Zhang, C. Sun (Eds.), Advances in Neural Networks – ISNN 2007, Part I. LIV, 1365 pages. 2007.

Vol. 4490: Y. Shi, G.D. van Albada, J.J. Dongarra, P.M.A. Sloot (Eds.), Computational Science – ICCS 2007, Part IV. XXXVII, 1211 pages. 2007.

Vol. 4489: Y. Shi, G.D. van Albada, J.J. Dongarra, P.M.A. Sloot (Eds.), Computational Science – ICCS 2007, Part III. XXXVII, 1257 pages. 2007.

Vol. 4488: Y. Shi, G.D. van Albada, J.J. Dongarra, P.M.A. Sloot (Eds.), Computational Science – ICCS 2007, Part II. XXXV, 1251 pages. 2007.

Vol. 4487: Y. Shi, G.D. van Albada, J.J. Dongarra, P.M.A. Sloot (Eds.), Computational Science – ICCS 2007, Part I. LXXXI, 1275 pages. 2007.

Vol. 4484: J.-Y. Cai, S.B. Cooper, H. Zhu (Eds.), Theory and Applications of Models of Computation. XIII, 772 pages. 2007.

Vol. 4475: P. Crescenzi, G. Prencipe, G. Pucci (Eds.), Fun with Algorithms. X, 273 pages. 2007.

Vol. 4474: G. Prencipe, S. Zaks (Eds.), Structural Information and Communication Complexity. XI, 342 pages. 2007.

Vol. 4459: C. Cérin, K.-C. Li (Eds.), Advances in Grid and Pervasive Computing. XVI, 759 pages. 2007.

Vol. 4449: Z. Horváth, V. Zsók, A. Butterfield (Eds.), Implementation and Application of Functional Languages. X, 271 pages. 2007.

Vol. 4448: M. Giacobini (Ed.), Applications of Evolutionary Computing. XXIII, 755 pages. 2007.

Vol. 4447: E. Marchiori, J.H. Moore, J.C. Rajapakse (Eds.), Evolutionary Computation, Machine Learning and Data Mining in Bioinformatics. XI, 302 pages. 2007.

Vol. 4446: C. Cotta, J.I. van Hemert (Eds.), Evolutionary Computation in Combinatorial Optimization. XII, 241 pages. 2007.

Vol. 4445: M. Ebner, M. O'Neill, A. Ekárt, L. Vanneschi, A.I. Esparcia-Alcázar (Eds.), Genetic Programming. XI, 382 pages. 2007.

Vol. 4436: C.R. Stephens, M. Toussaint, L.D. Whitley, P.F. Stadler (Eds.), Foundations of Genetic Algorithms. IX, 213 pages. 2007.

Vol. 4433: E. Şahin, W.M. Spears, A.F.T. Winfield (Eds.), Swarm Robotics. XII, 221 pages. 2007.

Vol. 4432: B. Beliczynski, A. Dzielinski, M. Iwanowski, B. Ribeiro (Eds.), Adaptive and Natural Computing Algorithms, Part II. XXVI, 761 pages. 2007.